Sustainable Construction of Future: Opportunities and Challenges for Green and Buildings

Sustainable Construction of Future: Opportunities and Challenges for Green and Buildings

Editors

Wesam Salah Alaloul
Bassam A. Tayeh
Muhammad Ali Musarat

MDPI • Basel • Beijing • Wuhan • Barcelona • Belgrade • Manchester • Tokyo • Cluj • Tianjin

Editors

Wesam Salah Alaloul
Department of Civil and
Environmental Engineering
Universiti Teknologi
PETRONAS
Seri Iskandar
Malaysia

Bassam A. Tayeh
Department of Civil
Engineering
Gaza University
Gaza
Palestine

Muhammad Ali Musarat
Offshore Engineering Centre
Institute of Autonomous System
Universiti Teknologi
PETRONAS
Seri Iskandar
Malaysia

Editorial Office
MDPI
St. Alban-Anlage 66
4052 Basel, Switzerland

This is a reprint of articles from the Special Issue published online in the open access journal *Sustainability* (ISSN 2071-1050) (available at: www.mdpi.com/journal/sustainability/special_issues/Future_Building).

For citation purposes, cite each article independently as indicated on the article page online and as indicated below:

LastName, A.A.; LastName, B.B.; LastName, C.C. Article Title. *Journal Name* **Year**, *Volume Number*, Page Range.

ISBN 978-3-0365-7897-2 (Hbk)
ISBN 978-3-0365-7896-5 (PDF)

© 2023 by the authors. Articles in this book are Open Access and distributed under the Creative Commons Attribution (CC BY) license, which allows users to download, copy and build upon published articles, as long as the author and publisher are properly credited, which ensures maximum dissemination and a wider impact of our publications.

The book as a whole is distributed by MDPI under the terms and conditions of the Creative Commons license CC BY-NC-ND.

Contents

Preface to "Sustainable Construction of Future: Opportunities and Challenges for Green and Buildings" vii

Abdullah O. Baarimah, Wesam Salah Alaloul, M. S. Liew, Widya Kartika, Mohammed A. Al-Sharafi and Muhammad Ali Musarat et al.
A Bibliometric Analysis and Review of Building Information Modelling for Post-Disaster Reconstruction
Reprinted from: *Sustainability* **2021**, *14*, 393, doi:10.3390/su14010393 1

Dhanasingh Sivalinga Vijayan, Parthiban Devarajan, Arvindan Sivasuriyan, Anna Stefańska, Eugeniusz Koda and Aleksandra Jakimiuk et al.
A State of Review on Instigating Resources and Technological Sustainable Approaches in Green Construction
Reprinted from: *Sustainability* **2023**, *15*, 6751, doi:10.3390/su15086751 21

Wesam Salah Alaloul, Muhammad Ali Musarat, Muhammad Babar Ali Rabbani, Muhammad Altaf, Khalid Mhmoud Alzubi and Marsail Al Salaheen
Assessment of Economic Sustainability in the Construction Sector: Evidence from Three Developed Countries (the USA, China, and the UK)
Reprinted from: *Sustainability* **2022**, *14*, 6326, doi:10.3390/su14106326 45

Ahsan Waqar, Abdul Hannan Qureshi and Wesam Salah Alaloul
Barriers to Building Information Modeling (BIM) Deployment in Small Construction Projects: Malaysian Construction Industry
Reprinted from: *Sustainability* **2023**, *15*, 2477, doi:10.3390/su15032477 81

Muhammad Ali Musarat, Wesam Salah Alaloul, Muhammad Irfan, Pravin Sreenivasan and Muhammad Babar Ali Rabbani
Health and Safety Improvement through Industrial Revolution 4.0: Malaysian Construction Industry Case
Reprinted from: *Sustainability* **2022**, *15*, 201, doi:10.3390/su15010201 111

Daniela Durand, Jose Aguilar and Maria D. R-Moreno
An Analysis of the Energy Consumption Forecasting Problem in Smart Buildings Using LSTM
Reprinted from: *Sustainability* **2022**, *14*, 13358, doi:10.3390/su142013358 137

Maria Ghufran, Khurram Iqbal Ahmad Khan, Fahim Ullah, Wesam Salah Alaloul and Muhammad Ali Musarat
Key Enablers of Resilient and Sustainable Construction Supply Chains: A Systems Thinking Approach
Reprinted from: *Sustainability* **2022**, *14*, 11815, doi:10.3390/su141911815 159

Hassan Ashraf, Ahsen Maqsoom, Tayyab Tahir Jajja, Rana Faisal Tufail, Rashid Farooq and Muhammad Atiq Ur Rehman Tariq
Error Management Climate and Job Stress in Project-Based Organizations: An Empirical Evidence from Pakistani Aircraft Manufacturing Industry
Reprinted from: *Sustainability* **2022**, *14*, 17022, doi:10.3390/su142417022 179

Ahsen Maqsoom, Hasnain Mubbasit, Muwaffaq Alqurashi, Iram Shaheen, Wesam Salah Alaloul and Muhammad Ali Musarat et al.
Intrinsic Workforce Diversity and Construction Worker Productivity in Pakistan: Impact of Employee Age and Industry Experience
Reprinted from: *Sustainability* **2021**, *14*, 232, doi:10.3390/su14010232 197

Hifsa Khurshid, Muhammad Raza Ul Mustafa and Mohamed Hasnain Isa
Modified Activated Carbon Synthesized from Oil Palm Leaves Waste as a Novel Green Adsorbent for Chemical Oxygen Demand in Produced Water
Reprinted from: *Sustainability* 2022, 14, 1986, doi:10.3390/su14041986 213

Qaiser Iqbal, Muhammad Ali Musarat, Najeeb Ullah, Wesam Salah Alaloul, Muhammad Babar Ali Rabbani and Wesam Al Madhoun et al.
Marble Dust Effect on the Air Quality: An Environmental Assessment Approach
Reprinted from: *Sustainability* 2022, 14, 3831, doi:10.3390/su14073831 231

Abdus Samad Azad, Rajalingam Sokkalingam, Hanita Daud, Sajal Kumar Adhikary, Hifsa Khurshid and Siti Nur Athirah Mazlan et al.
Water Level Prediction through Hybrid SARIMA and ANN Models Based on Time Series Analysis: Red Hills Reservoir Case Study
Reprinted from: *Sustainability* 2022, 14, 1843, doi:10.3390/su14031843 247

Lahiba Imtiaz, Sardar Kashif-ur-Rehman, Wesam Salah Alaloul, Kashif Nazir, Muhammad Faisal Javed and Fahid Aslam et al.
Life Cycle Impact Assessment of Recycled Aggregate Concrete, Geopolymer Concrete, and Recycled Aggregate-Based Geopolymer Concrete
Reprinted from: *Sustainability* 2021, 13, 13515, doi:10.3390/su132413515 267

Muhammad Hamza, Rai Waqas Azfar, Khwaja Mateen Mazher, Basel Sultan, Ahsen Maqsoom and Shabir Hussain Khahro et al.
Exploring Perceptions of the Adoption of Prefabricated Construction Technology in Pakistan Using the Technology Acceptance Model
Reprinted from: *Sustainability* 2023, 15, 8281, doi:10.3390/su15108281 287

Tahir Ali Akbar, Azka Javed, Siddique Ullah, Waheed Ullah, Arshid Pervez and Raza Ali Akbar et al.
Principal Component Analysis (PCA)–Geographic Information System (GIS) Modeling for Groundwater and Associated Health Risks in Abbottabad, Pakistan
Reprinted from: *Sustainability* 2022, 14, 14572, doi:10.3390/su142114572 313

Waqas Farooq, Muhammad Ali Musarat, Javed Iqbal, Syed Asfandyar Ali Kazmi, Adnan Daud Khan and Wesam Salah Alaloul et al.
Optimized Thin-Film Organic Solar Cell with Enhanced Efficiency
Reprinted from: *Sustainability* 2021, 13, 13087, doi:10.3390/su132313087 335

Preface to "Sustainable Construction of Future: Opportunities and Challenges for Green and Buildings"

The focus of this reprint is to draw attention to the construction industry stakeholders oriented towards green and sustainable construction innovations for the future. Green and sustainable construction has become a necessity in today's society, as well as for the future, wherein there are many possibilities to investigate and encourage reform. However, its implementation and adoption still suffer from various challenges, such as a lack of knowledge, low self-esteem, and lack of resources. Such challenges open gateways for new opportunities to resolve these issues, for which there is huge potential and possibility for development.

Wesam Salah Alaloul, Bassam A. Tayeh, and Muhammad Ali Musarat
Editors

Review

A Bibliometric Analysis and Review of Building Information Modelling for Post-Disaster Reconstruction

Abdullah O. Baarimah [1], Wesam Salah Alaloul [1,*], M. S. Liew [1], Widya Kartika [2], Mohammed A. Al-Sharafi [3], Muhammad Ali Musarat [1], Aawag Mohsen Alawag [1] and Abdul Hannan Qureshi [1]

[1] Department of Civil and Environmental Engineering, Universiti Teknologi PETRONAS, Bandar Seri Iskandar 32610, Perak, Malaysia; abdullah_20000260@utp.edu.my (A.O.B.); shahir_liew@utp.edu.my (M.S.L.); muhammad_19000316@utp.edu.my (M.A.M.); aawag_17006581@utp.edu.my (A.M.A.); abdul_19000967@utp.edu.my (A.H.Q.)
[2] Civil Engineering Department, Universitas Janabadra, Yogyakarta 55231, Indonesia; widya.kartika@janabadra.ac.id
[3] Department of Information Systems, Azman Hashim International Business School, Universiti Teknologi Malaysia, Skudai 81310, Johor, Malaysia; alsharafi@ieee.org
* Correspondence: wesam.alaloul@utp.edu.my

Abstract: Post-disaster reconstruction (PDR) is a dynamic, complex system that is chaotic in nature, and represents many challenges and issues. Recently, building information modelling (BIM) has been commonly utilized in the construction industry to solve complex and dynamic challenges. However, BIM has not been thoroughly considered for managing PDR, and there is a lack of comprehensive scientometric analyses that objectively examine the trends in BIM applications in PDR. A literature search was performed considering studies published from 2010 to March 2021 using the Scopus database. A total of 75 relevant studies were found to meet the inclusion criteria. The collected literature was analyzed using VOSviewer through scientific journals, authors, keywords, citations, and countries. This is the first study in its vital significance and originality that aims to investigate the current states of research on BIM applications in PDR and provide suggestions for potential research directions. The findings showed that "Reconstruction" and "Safety Management" have emerged as mainstream research themes in this field and recently attracted scholars' interest, which could represent the directions of future research. Five major research domains associated with BIM were identified based on the most frequently used keywords, namely "Disasters", "Earthquakes", "HBIM", "Damage Detection", and "Life Cycle". Moreover, a proposed conceptual framework of BIM adoption for PDR is provided. Accordingly, the outcomes of this study will help scholars and practitioners gain clear ideas of the present status and identify the directions of future research.

Keywords: BIM; post-disaster reconstruction; construction industry; scientometric analysis; visualization; PRISMA; review

1. Introduction

Disasters can be defined as "an action that causes a threat to life, well-being, material goods, and the environment from the extremes of natural processes or technology" [1]. Natural and human-made disasters affect the built environment. The large-scale damages caused by infrastructures and houses are accompanied by injuries and fatalities, reversal or stagnation of the local economy, and mislaying of livelihood sources [2]. Post-disaster reconstruction (PDR) has been gaining more attention in the world because of frequent natural environment disasters, such as earthquakes tsunamis, and other activities, caused by human-made factors, such as conflicts and wars, which have raised the importance of PDR [3,4]. Following the increasing occurrence of major disasters, stakeholders are increasingly initiating reconstruction to reduce the effects of those disasters on the built environment; however, reconstruction projects are considered challenging to implement in terms of capacity and resources [5,6]. Nevertheless, PDR is categorized as unpredictable, chaotic, complex, and dynamic; this indicates several difficulties due to its differences when compared to traditional construction [7]. Conventional construction has been used in reconstruction projects, whereas some features, such as a single lifecycle of project and inflexibility in aspects of creating a specified project duration, have proven unsatisfactory for the complications encountered in the aftermath of the disaster [8,9]. Reconstruction projects face immense challenges, such as time and cost overrun, and low quality, due to several factors during the implementation [10,11].

The main target of any PDR project is to attain high levels of beneficiary satisfaction. Nonetheless, PDR projects frequently fail in their pre-planned objectives; for example, only 20% of building requirements are fulfilled, with most buildings being constructed on a temporary instead of permanent basis [12]. Moreover, the efforts of reconstruction projects have lacked any suitable coordination mechanism and monitoring framework [13]. If not properly handled, those challenges can result in ineffective PDR project delivery and often a failure of the project [3,14]. Successful delivery of PDR projects is important to restore essential services and return to normalcy after disasters [15,16]. Controlling the cost, time, and delivery of the projects are the most significant factors in evaluating successful PDR projects, which will be heightened by utilizing one of the emerging technologies or processes that aim to enhance productivity and sustainability in PDR projects.

Building information modelling (BIM) is considered one of these technologies, and has altered the ways of the practices of architecture, engineering, and construction industries during recent years. It brings stakeholders in construction to a single productive platform [17]. Additionally, BIM is not only a technical facility but also an activity concept for efficient project delivery linked to advanced technology [18]. It utilizes information and communication technologies to enhance growth monitoring, increase performance, and boost productivity [19]. Consequently, numerous studies have been conducted on BIM within the construction industry from the perspective of the industry, organizations, and users [20,21]. Other studies have focused on BIM in terms of adoption, implementation, challenges, and benefits, as well as strategies and application [22,23].

In this regard, BIM is an intelligence tool that has a wide range of benefits, such as collaboration among parties, visualization of project execution, enhanced productivity and efficiency, enhanced communications, improved design quality, cost estimation, positive return on investment, enhanced sustainability, faster development while reducing the cost and rework, on-time delivery capabilities, clash detection, better contract documentation, life-cycle cost data management, and competitive edge [24–26]. These and other advantages have prompted governments, institutions, and organizations to implement BIM in their construction industries [27–29].

With the increasing significant challenges in PDR, it is necessary to apply BIM in reconstruction projects after disasters to overcome these challenges [30–33]. Despite some reported evidence on the benefits of BIM within the construction industry, the adoption of BIM for the PDR field has not received adequate attention. Moreover, there is a lack of adoption of modern methods for managing PDR that is a cause of low boost. However, a few studies have focused on utilizing BIM for PDR. For example, Dakhil and Alshawi [31] explored the BIM applications that supported building disaster management, including disaster planning, site planning, and existing condition modelling, and its advantages; they suggested that future research should empirically study all applications of BIM during the project life cycle and identify the advantages of those applications. Nawari and Ravindran [32] listed the potential advantages of BIM in conjunction with blockchain for post-disaster rebuilding. The authors presented a framework for an automated reconstruction permitting process by using Hyperledger Fabric; however, no evidence of actual implementation has been found. Messaoudi and Nawari [33] proposed a virtual framework based on BIM and a generalized adaptive framework for speeding up the permitting process of reconstruction in the aftermath of the disaster in Florida; however, this framework was limited to the time of the permitting process for rebuilding.

On the other hand, scientometric analysis is being used by a growing number of scholars to address subjective problems in literature reviews [34–37]. In contrast, there is a paucity of scientometric analysis in the current studies that explore trends of BIM applications in the PDR projects field. It is worth noting that the PDR in question is still in its early stages for BIM adoption. One of the challenges is determining the need to start setting the foundations for best frameworks and guidelines for promoting the BIM applications in PDR projects. Experience has shown that it is never too early to start preparing for reconstruction in the aftermath of the disaster. Accordingly, it is crucial to map the related literature in order to address this research gap. This study is unique because it explores trends of BIM applications in PDR from an objective standpoint.

Based on the researchers' knowledge, this is the first study that aims to investigate the current state of research on BIM applications in PDR and provide suggestions for potential research directions. To achieve the study aim, the present study analyzed the collected papers through scientific journals, authors, keywords, citations, and countries. In addition, a proposed conceptual framework was developed that underlines the relationship between PDR and BIM adoption, which impacts BIM adoption for PDR. Therefore, the outcomes of this study will assist scholars and practitioners in gaining clear ideas of the present status and identifying the directions of future research. It may promote BIM awareness and offer more opportunities for a positive perception of BIM implementation in reconstruction projects in the aftermath of future disasters.

2. Methodology

The present study used the PRISMA guidelines in order to review the current literature [38]. We followed the PRISMA guidelines without considering meta-analysis approaches. The scoping procedure was employed to extract the most related papers on BIM and PDR. The collected literature was retrieved from the Scopus database. Scopus includes more journals and scientific publications than any other available literature database (such as "Web of Science") [39,40]. A list of keywords relevant to disaster and reconstruction was created. These keywords, along with the keyword "BIM", were utilized for the literature search, with the following query string: TITLE-ABS-KEY ((("BIM" OR "Building Information Model*") AND ("Disaster" OR "post-conflict" OR "post-war" OR "war" OR "earthquake" OR "flood" OR "landslides" OR "Tsunami" OR "storm" OR "cyclone" OR "tornado" OR "hurricane") AND ("construction" OR "reconstruction" OR "recovery" OR "rehabilitation" OR "repair" OR "rebuild*" OR "retrofitting" OR "restoration")). In addition, the time range was set from 2010 to March 2021. Initially, a total of 185 documents were displayed; these included research papers, reviews, book chapters, and other types of documents. The documents were limited to research articles, review papers, conference papers, and book chapters. As result, 150 papers were chosen in this stage as illustrated in Figure 1. Additionally, in the second stage, the data were transferred to an Excel sheet, and after assessing the documents and excluding irrelevant publications as well as non-English publications, a total of 75 papers were finally considered eligible to be included for further analysis. In the end, returned to the Scopus database to select those papers manually in order to export an Excel sheet file that was used for the scientometric analysis. Figure 1 illustrates the framework implementation of the current review.

Moreover, a scientometric analysis was conducted after the literature sample was obtained. With the rapid progress in technology, scientometric analysis can now be performed using a range of existing software. VOSviewer was selected in this study to draw science mappings, since it has remarkable text mining capabilities and is ideal for dealing with larger networks [41]. Currently, VOSviewer is increasingly being used in construction industry research to create science mappings, for example, BIM [42], system dynamics [36], and artificial intelligence [43]. The collected papers were examined through five aspects: scientific journals, authors, keywords, citations, and countries. In the literature review study, according to Ren et al. [44], those five aspects are considered as the main components of the scientometric analysis that can help researchers to grasp the current state of research quickly. The most critical measurements include the citations, documents, average publication year, average citations, and average normalized citations [45], where the last three metrics are inextricably linked to each other. The average publication year mainly refers to the documents that are published in a certain year [42]. The average citations were determined by dividing the total number of citations by documents. Furthermore, average normalized citations indicate the normalized number of citations of a journal, author, keyword, document, and country. This was calculated by dividing the total number of citations by the average number of citations received during the given year; the higher the score, the greater the impact [45].

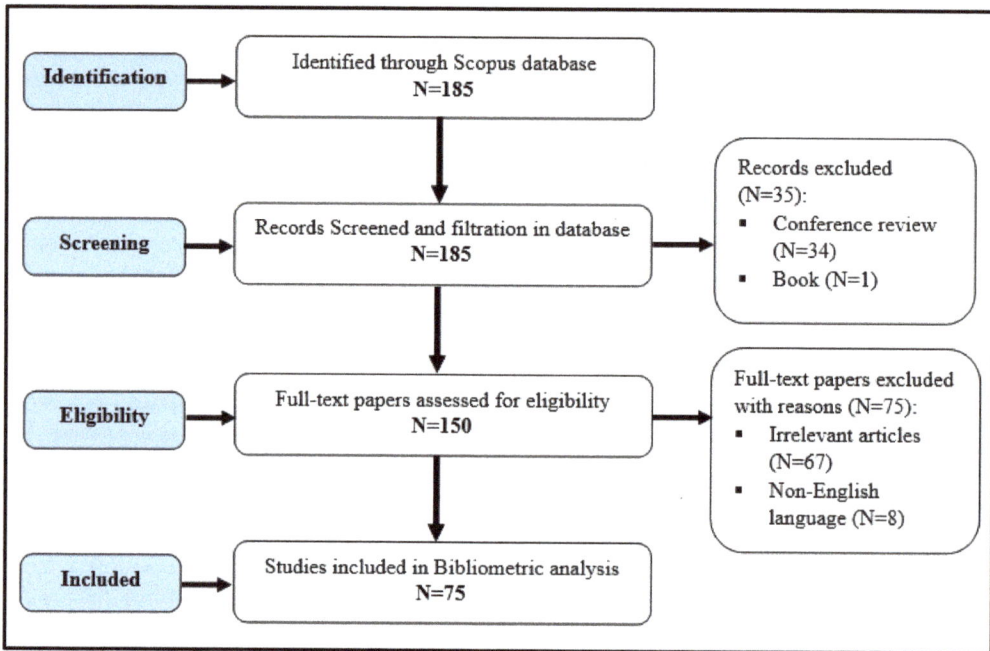

Figure 1. Study flowchart.

3. Results and Interpretation of Articles

This section analyzes the scientific journals, authors, keywords, citations, and countries active in the targeted research. It will provide a clear and succinct explanation of the experimental findings, their analysis, and the experimental conclusions that were reached.

3.1. Analysis of Published Journals

It is common for researchers to share and communicate their research findings through various published journals. In the current study, VOSviewer was utilized to find the source journals of the papers obtained, as shown in Figure 2. The minimum number of published documents and citations of a source was set at two and one, respectively. According to Jin et al. [45], there is no limit to setting the threshold value. Several attempts were made with various threshold values until the most appropriate values for determining the optimal range of sources were discovered. Accordingly, a total of 14 journals out of 58 met the thresholds. There was a total of 11 leading journals linked to each other, as shown in Figure 2. It was found that the nodes of "ISPRS Archives" and "Automation in Construction" were the largest in regard to the category of conferences and journals, respectively, and were linked to most other journals. This indicates that these two journals are leading journals in this research field.

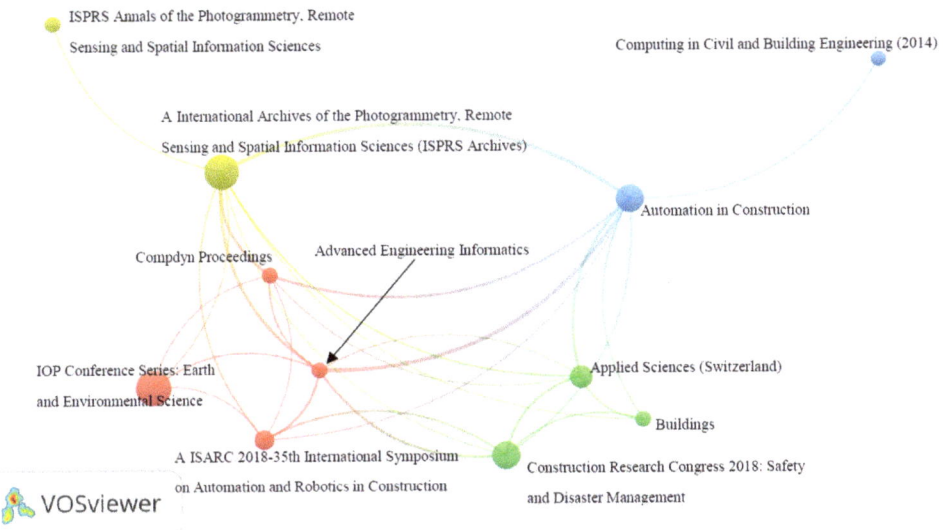

Figure 2. Mapping of published journals.

In Figure 2, the source journals are divided into different clusters in terms of colors. Therefore, these journals were grouped: "Applied Sciences", 'Buildings", and "Construction Research Congress 2018". The journals that appear in the same cluster have a greater level of interconnectedness, which means that articles from these journals cite each other more frequently. A quantitative summary of the journals can be obtained from Table 1.

Table 1. Details of published journals.

No.	Source Journal	Total Link Strength	Documents	Total Citations	Avg. Citations	Avg. Norm. Citations
1	"Advanced Engineering Informatics"	29	2	79	39.5	7.2
2	"Applied Sciences (Switzerland)"	16	4	12	3.0	2.2
3	"Automation in Construction"	36	6	111	18.5	2.3
4	"Buildings"	7	2	25	12.5	2.4
5	"Compdyn Proceedings"	15	2	1	0.5	0.1
6	"Computing in Civil and Building Engineering (2014)"	2	2	9	4.5	0.6
7	"Construction Research Congress 2018: Safety and Disaster Management"	16	6	19	3.2	0.6
8	"International Archives of the Photogrammetry, Remote Sensing and Spatial Information Sciences (ISPRS Archives)"	29	9	63	7.0	1.1
9	"IOP Conference Series: Earth and Environmental Science"	5	10	3	0.3	0.3
10	"ISARC 2018—35th International Symposium on Automation and Robotics in Construction"	12	3	4	1.3	0.3
11	"ISPRS Annals of the Photogrammetry, Remote Sensing and Spatial Information Sciences"	1	2	9	4.5	0.9

The total link strength, number of documents, and total citations, which are all strongly connected can be used to assess the productivity of a given journal's research outputs. Moreover, the importance of a journal's contribution to the research community can be also evaluated through the average citation. The average normalized citation per document is not always linked to the number of documents. As shown in Table 2, it was found that "Advanced Engineering Informatics" is a top-ranked journal, with the greatest average citations and average normalized citations. Moreover, in terms of total citations, "Automation in Construction" showed incredible performance and recorded the maximum number of total citations. Other more significant journals in terms of average normalized citations included "Buildings", "Automation in Construction", and "Applied Sciences", which also have a strong potential influence on BIM applications for the PDR field.

3.2. Analysis of Co-Authorship

Scholars usually collaborate in academic research, which can improve productivity and access to expertise as well as prevent scholars from being isolated [46]. The minimum number of published articles and an author's citations in this study were set at one and ten, respectively. Thus, of the 219 authors, 27 met the thresholds. There was a total of 10 influential authors linked to each other, as visualized in Figure 3.

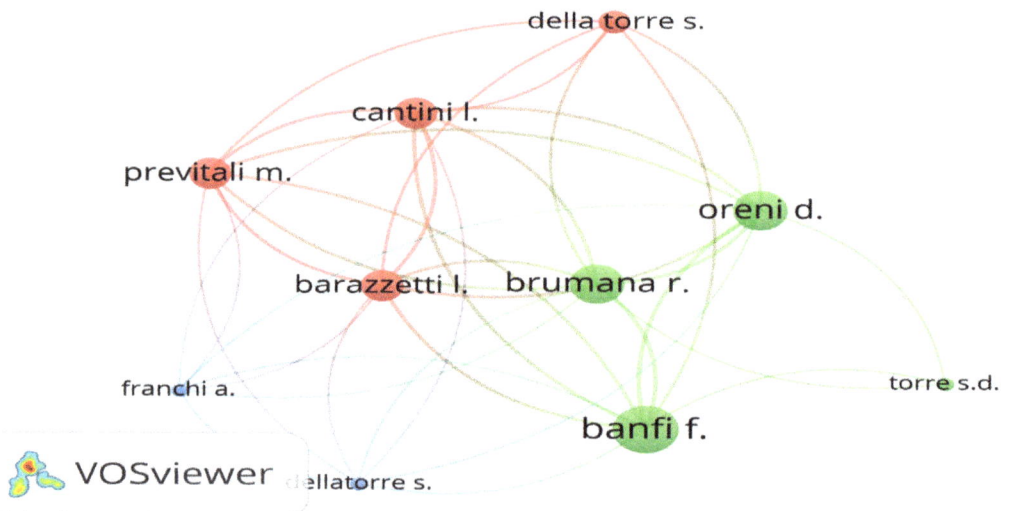

Figure 3. Mapping of co-authorship.

As shown in Figure 3, the authors were explicitly divided into three groups based on colors. Here, Previtali M., Cantini L., Della Torre S., and Barazzetti L. appeared in the same research group. Brumana R. was in the middle of this network and was connected with the other two groups of authors, suggesting that Brumana R. maintains strong academic cooperation with leading researchers in the field. Banfi F. has the largest node compared to other authors, and he collaborates closely with all authors, meaning that Banfi F. is considered one of the top scholars in this domain. The details of these productive authors are presented in Table 2.

Table 2 demonstrates that the most productive author is Banfi F., who published the largest number of documents. Regarding overall research significance, Banfi F., Brumana R., and Oreni D. ranked at the top by achieving 84 citations. In the aspects of collaboration links, again Banfi F., Brumana R., and Oreni D. had the best network of all researchers in this field, with a total link strength of 22. When it comes to average citations, Dellatorre S. and Franchi A. ranked first with 26 average citations, and the authors Barazzetti L., Cantini L., and Previtali M. ranked second with a recorded 24.7 average citations, suggesting their active influence in this field. According to average year publications, the emerging researchers listed in Table 2 have contributed to the research field and have recently linked BIM application to PDR, where the average year publications started from 2017.

Table 2. Details of co-authorship.

Author	Total Link Strength	Documents	Total Citations	Avg. Pub. Year	Avg. Citations	Avg. Norm. Citations
Banfi F.	22	5	84	2018	16.8	2.5
Barazzetti L.	19	3	74	2018	24.7	3.4
Brumana R.	22	4	84	2018	21.0	2.9
Cantini L.	19	3	74	2018	24.7	3.4
Della Torre S.	12	2	48	2018	24.0	3.3
Dellatorre S.	7	1	26	2017	26.0	3.5
Franchi A.	7	1	26	2017	26.0	3.5
Oreni D.	22	4	84	2018	21.0	2.9
Previtali M.	19	3	74	2018	24.7	3.4
Torre S.D.	3	1	10	2017	10.0	1.4

3.3. Analysis of Co-Occurring Keywords

Keywords help scholars understand mainstream topics and emphasize possible research directions [47,48]. The connection of keywords demonstrates the knowledge of the research domain in the aspects of intellectual organization, relationships, and development trends [49]. By setting the minimum frequency of keywords to three in this study, only 73 out of 746 keywords met the thresholds. Those keywords were filtered, and a few keywords that were not relevant to the study were removed. In consequence, a total of 38 keywords were selected and are represented in Figure 4.

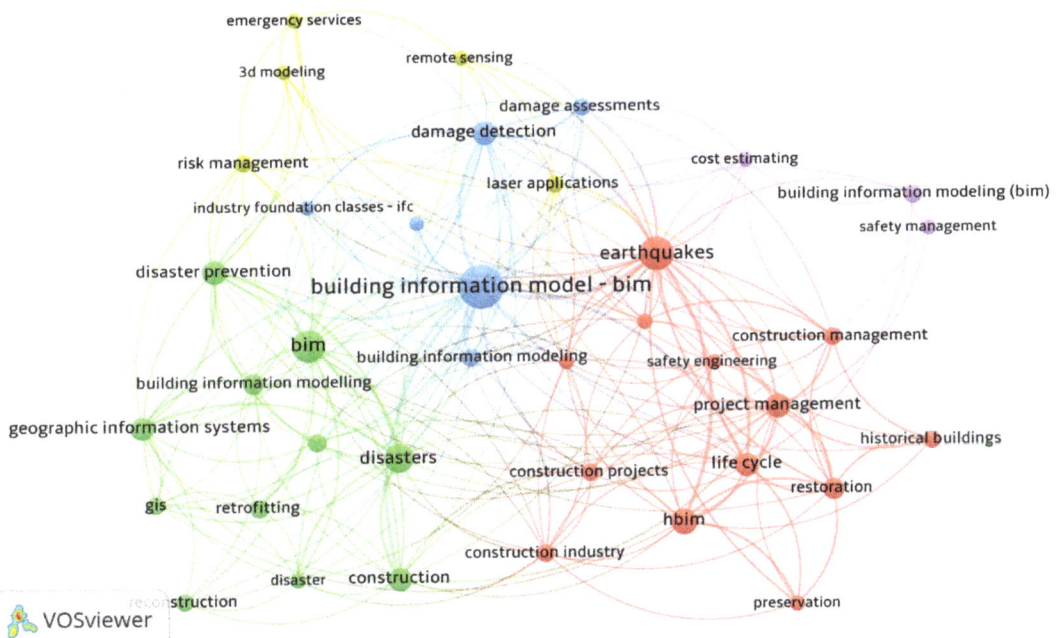

Figure 4. Mapping of co-occurring keywords.

Figure 4 depicts the research directions for the applications of BIM, and it is not surprising to see that "Building Information Model—BIM" was the most commonly listed research keyword. The other keywords associated with BIM, such as "Disasters" and "Earthquakes", indicate that BIM applications were primarily utilized in this field. A quantitative summary of keywords can be obtained from Table 1, and the frequency of keywords conformed to those in Figure 4. It should be noted that "Damage Assessments", "Disaster Prevention", "Historical Buildings", "Reconstruction", "Repair", "Restoration", "Retrofitting", "Risk Management", "Safety Management", "Life Cycle", and "Virtual Reality" were included in the common research areas based on the frequency of keywords. Hence, it can be inferred that the characteristics of BIM applications can evaluate the impacts of different policies.

As shown in Table 3, the average citation and average normalized citation demonstrate the impact of the keyword in the academic community. Following the keywords "Preservation", "Restoration", "Historical Buildings", "Construction Management", and "Repair" attracted more interest according to average citations. Interestingly, the keyword "Preservation" was ranked at the top with 3.4 average normalized citations, suggesting its influence in this field. Besides, the average year publication indicates the novelty of those keywords. For instance, some articles relative to "Reconstruction" and "Safety Management" were published recently in 2020, meaning that they emerged as mainstream research themes in this field and spurred the interest of scholars recently, which could represent future research directions as well.

Table 3. Details of co-occurring keywords.

Keywords	Total Link Strength	Occurrence	Avg. Pub. Year	Avg. Citations	Avg. Norm. Citations
3D Modeling	8	3	2019	2.7	1.3
BIM	39	14	2018	3.6	1.3
Building Information Model—BIM	80	26	2017	3.7	0.9
Building Information Modeling	16	4	2018	8.5	1.2
Building Information Modeling (BIM)	7	5	2019	3.4	2.3
Building Information Modelling	22	6	2018	3.2	0.4
Construction	27	7	2018	1.6	0.4
Construction Industry	16	4	2019	6.5	0.9
Construction Management	15	4	2017	16.8	1.5
Construction Projects	14	4	2018	0.3	0.3
Cost Estimating	13	3	2015	4.3	0.8
Damage Assessments	20	4	2014	9.3	1.4
Damage Detection	34	8	2016	6.3	0.8
Disaster	15	3	2018	1.3	0.9
Disaster Prevention	32	8	2019	1.6	0.3
Disasters	42	11	2018	2.3	0.6
Earthquake	11	3	2018	5.3	0.6
Earthquake Engineering	11	3	2018	1.3	0.6
Earthquakes	55	15	2018	6.3	1.2
Emergency Services	8	3	2019	0.0	0.0
Geographic Information Systems	27	7	2018	2.6	0.5
GIS	14	4	2017	2.8	0.6
HBIM	23	9	2018	10.4	1.7
Historical Buildings	8	4	2017	17.0	1.2
Industry Foundation Classes—IFC	13	3	2017	11.3	2.0
Laser Applications	16	4	2017	7.0	0.9
Life Cycle	24	7	2018	4.6	0.9
Preservation	10	3	2018	24.7	3.4
Reconstruction	7	4	2020	1.5	0.6
Remote Sensing	9	3	2017	2.3	0.2
Repair	12	3	2018	10.7	2.3
Restoration	21	6	2017	17.8	1.7
Retrofitting	12	5	2017	2.6	0.5
Risk Management	12	4	2019	2.3	0.5
Robotics	10	3	2015	1.0	0.2
Safety Engineering	9	3	2019	6.3	1.4
Safety Management	3	3	2020	2.0	2.7
Virtual Reality	15	5	2019	2.0	0.6

3.4. Analysis of Article Citations

Researchers need to find publications that have significant contributions to the academic community. By setting the minimum number of document citations to four, only 26 out of 75 documents met the thresholds. There was a total of 14 crucial articles linked to each other, as presented in Figure 5.

As shown in Figure 5, the node of Biagini C. (2016) [50] was the largest, indicating the most cited papers. However, the node of Newari N.O. (2019b) [32] was small and in the middle of the network, which has strong connections with other most cited papers. The influence of each document was evaluated through the number of links, total citations, and normalized citations as detailed in Table 4.

It is interesting to find that the insightful research results were primarily published in the domain of applications of BIM, especially in the phases of disasters, from pre- to post-disaster. Hence, it can be inferred that the properties of BIM applications are capable of solving challenges related to PDR. It is expected that applications of BIM will be used more frequently in future studies within PDR.

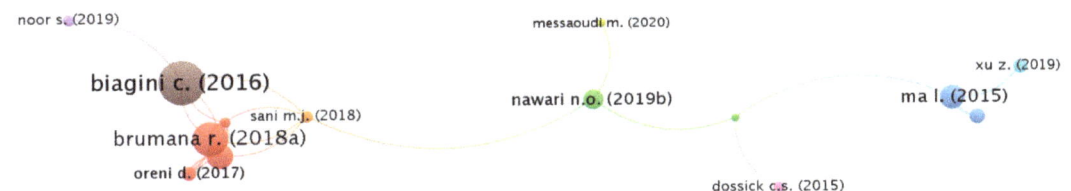

Figure 5. Mapping of article citations.

Table 4. Details of article citations.

No.	Article	Title	Number of Links	Citations	Norm. Citations
1	Messaoudi M. (2020) [33]	"BIM-Based Virtual Permitting Framework (VPF) For Post-Disaster Recovery and Rebuilding in the State of Florida"	1	4	5.4
2	Noor S. (2019) [51]	"Modeling and Representation of Built Cultural Heritage Data Using Semantic Web Technologies and Building Information Model"	1	7	1.4
3	Nawari N.O. (2019a) [52]	"BIM Data Exchange Standard for Hydro-Supported Structures"	3	4	0.8
4	Xu Z. (2019) [53]	"A Prediction Method of Building Seismic Loss Based on BIM and Fema P-58"	1	9	1.9
5	Nawari N.O. [32]	"Blockchain and Building Information Modeling (BIM): Review and Applications in Post-Disaster Recovery"	3	16	3.3
6	Brumana R. (2018a) [54]	"Generative HBIM Modelling to Embody Complexity (Lod, Log, Loa, Loi): Surveying, Preservation, Site Intervention—the Basilica Di Collemaggio (L'aquila)"	5	42	5.8

Table 4. Cont.

No.	Article	Title	Number of Links	Citations	Norm. Citations
7	Sani M.J. (2018) [55]	"GIS And BIM Integration At Data Level: A Review"	5	7	1.0
8	Brumana R. (2018b) [56]	"Scan to HBIM-Post Earthquake Preservation: Informative Model as Sentinel at the Crossroads of Present, Past, And Future"	5	6	0.8
9	Brumana R. (2017) [57]	"HBIM Challenge Among the Paradigm of Complexity, Tools and Preservation: the Basilica Di Collemaggio 8 Years After The Earthquake (L'aquila)"	5	26	3.5
10	Oreni D. (2017) [58]	"Survey, HBIM and Conservation Plan of a Monumental Building Damaged by Earthquake"	3	10	1.4
11	Biagini C. (2016) [50]	"Towards the BIM Implementation for Historical Building Restoration Sites"	5	63	4.2
12	Ma L. (2016) [59]	"Preparation of Synthetic As-Damaged Models for Post-Earthquake BIM Reconstruction Research"	1	11	0.7
13	Ma L. (2015) [60]	"Information Modeling of Earthquake-Damaged Reinforced Concrete Structures"	3	23	4.1
14	Dossick C.S. (2015) [61]	"Learning in Global Teams: BIM Planning and Coordination"	1	6	1.1

Besides research topics, the number of links mentioned in Table 4 demonstrates an article's influence within the academic community. The articles by Brumana R. (2018a) [54], Sani M.J. (2018) [55], Brumana R. (2018b) [56], Brumana R. (2017) [57], and Biagini C. (2016) [50] had the strongest number of links. Moreover, two articles from Biagini C. (2016) [50] and Brumana R. (2018a) [54] earned the highest citations and had the greatest normalized citations, respectively, in highly cited articles. Brumana R. had three out of the fourteen most cited papers regarding the number of citations; hence, this indicates that the author Brumana R. has led an important series of studies on BIM applications for PDR compared to other authors. Additionally, other researchers, including but not limited to Messaoudi M., Xu Z., Ma L., and Dossick C.S., have conducted the most influential research.

3.5. Analysis of Countries

The availability of information on outstanding countries in a research field can help scholars collaborate between them on projects, get grants, and share their findings [46]. In this study, diligent countries were also recognized according to their research contributions. VOSviewer was used to identify, evaluate, and visualize the source countries of researchers [41]. The minimum number of published articles and citations of a country was set at two and ten, respectively. Accordingly, out of 32 countries, 11 met the thresholds. There was a total of nine important countries linked to each other, as presented in Figure 6.

It can be observed in Figure 6 that the countries were classified explicitly into three groups based on colors, where the first group included Canada, South Korea, and the United States. It is interesting to note that Italy had the largest node, implying that Italian academics were the primary contributors to research on the applications of BIM to solve PDR problems. The details of these productive countries are provided in Table 5.

As can be seen in Table 5, Italy, China, and the United States were ranked higher in terms of total documents. Academics from Italy earned the highest total citations compared to the countries active in BIM applications for PDR research, showing that the utilization of BIM applications by Italy's academics was extremely enlightening and aided in the research area of PDR. Based on the average citation, which shows the importance of the study conducted in the country, Canada ranked first with 19.5 average citations, and Israel ranked second with 11.7 average citations recorded. Interestingly, academics from Canada, South Korea, Greece, and Cyprus recorded higher average normalized citations, indicating that they are strongly competitive and provide important contributions to BIM applications for PDR research.

Figure 6. Mapping of countries.

Table 5. Details of countries.

Country	Total Link Strength	Documents	Total Citations	Avg. Pub. Year	Avg. Citations	Avg. Norm. Citations
Canada	196	4	78	2018	19.5	2.9
China	23	14	24	2019	1.7	0.7
Cyprus	40	3	21	2013	7.0	1.6
Germany	5	2	11	2016	5.5	1.1
Greece	37	2	11	2015	5.5	1.9
Israel	11	3	35	2015	11.7	1.7
Italy	114	17	168	2018	9.9	1.3
South Korea	30	4	12	2019	3.0	2.3
United States	126	10	30	2018	3.0	1.3

4. Discussion

Following the bibliometric analysis, this section summarizes the present status of the research and offers directions for future research.

4.1. Post-Disaster Reconstruction

The present study found various expectations for research regarding PDR, with the majority focusing mainly on short-term recovery and ignoring long-term reconstruction. Lyons [62] reported that PDR is primarily unsuccessful in achieving its pre-planned goals, where the failure rate of the reconstruction project is beyond 50% [63]. Delays in the process of reconstruction projects might reduce the effectiveness of the reconstruction and make achieving the goals more challenging [64,65]. Based on the work that needs to be carried out post-disaster, several other challenges can influence the timeframe of the work [8]. PDR mainly relies on economic, cultural, social, environmental, and political elements [66]. Various problems emanate because of inadequate supports from poor governance, local government, poor infrastructure, and insufficient knowledge and preparedness. Moreover, other elements that positively influence the reconstruction processes include addressing technical problems, integrated information, short and long-term approaches, and public participation in dealing with technical problems [67]. PDR is influenced by the locations of the destroyed areas because this influences the assigned funds, technical resources, and labors [5]. Despite the value of community contribution in PDR, it is important to ascertain timely information during the reconstruction period [68]. Moreover, although PDR offers chances to lower vulnerability and enhance sustainability in disaster-affected communities, most reconstruction projects have failed to meet these objectives [69]. Hence, PDR is highly demanding and complex, and needs several well-coordinated and diverse actions.

Moreover, during the past few years, several studies have been conducted in this area to investigate various variables exerting either negative or positive impacts on a PDR project [3–5,9,10,68,70–72]. The unsuccessful outcomes of PDR projects are due to the following: inadequate availability of resources, delays in project implementations, inadequate coordination amidst participation organizations, corruptions, substandard quality of the reconstructed building, inadequate community participation, inadequate road access, inadequate government support, problems with land availability and acquisition, ineffective design, conventional 2D documentation, manual schedule and cost estimation, and inadequate extensive resource database. Meanwhile, available evidence shows that PDR has a discouraging record of performance in recent decades due to ineffective reconstruction strategies that do not consider the concept of collaboration among parties, have coordination procedures, and adopt modern communication technology.

Some examples of good reconstruction practices include the establishment of construction guidelines and permitting processes, as well as the certification of reconstructed housing to ensure safe building construction [33]. However, these practices have taken a long time to obtain approval. Since acting quickly is the priority in a post-disaster situation, the traditional manual cost estimation methods are not feasible. Based on previous studies, there is a need for a tool to calculate the construction cost of the proposed alternatives quickly and automatically [73]. Reconstruction requires avoiding disintegration between stakeholders, such as governments, emergency agencies, builders, relief organizations, designers, and disaster victims [7], and also demands the enhancement of delivery practices in order to provide higher value to stakeholders [6,16]. Moreover, non-participatory reconstruction practices by donors have caused conflicts and resentment among the local people [13]. This study found that the existing practices of reconstruction projects must be compiled and evaluated to determine whether the proposed conceptual designs satisfy construction codes. Accordingly, it is susceptible to personal mistakes, and any updates need to be performed manually, which causes budget overruns and schedule delays in reconstruction projects.

The current status of managing PDR mostly seeks to avert factors causing failures in the PDR projects. However, information about post-disaster management is ongoing, and most desire to learn from past failures. To overcome these challenges, PDR must integrate short and long-term reconstruction to guarantee that housing and infrastructure requirements are fulfilled throughout the short-term recovery phase, while lowering vulnerability and enhancing sustainability and resilience in the long-term reconstruction phase. As a result, the goal of efficient PDR may be attained by integrating the requirements of stakeholders with modern management practices and information technology. Therefore, to accomplish the PDR project on schedule, within the allocated cost and with a high standard of quality, there is a need for more objective studies to boost productivity and sustainability in PDR projects. Besides, it is necessary to adopt new developments in modern management practices and information technologies, such as building information modelling, that can enhance the competitive environment of PDR projects based on the improvement of product quality, on-time project delivery, and cost reduction.

4.2. Building Information Modelling (BIM)

BIM is a powerful tool adopted in construction projects to control project information and building design in digital form throughout the life cycle of buildings. This approach allows for information interchange and interoperability between parties [74]. BIM has been widely promoted as an nD modelling platform to improve collaboration and communication, and its scope has expanded from "geometric models (3D) to include time (4D), costs (5D), sustainability of the environment (6D), and facility management (7D)" [75]. Its significant advantages in terms of cost and time savings, as well as increased performance and boosting of productivity, have compelled construction players to adopt and rapidly implement it within several fields [76]. A 3D building model that identifies the clash detection, improves the schedules of construction, and prepares construction site activities was shown to be useful in everyday operations [77]. Additionally, the utilization of BIM has provided significant benefits to the construction industry over the project's life cycle, from conceptualization to demolition [78]. Significant advantages found in the design phase of the project include improved visualization, efficiency, and productivity, whereas, during the construction phase, BIM can consist of cost analysis and auto-scheduling, allowing for improved project coordination and on-time delivery capabilities [79]. Accordingly, BIM encourages all professionals and stakeholders to contribute and collaborate to produce a high-quality output throughout the project.

Diffusion and implementation are two steps in the BIM adoption process. With effective BIM adoption and implementation, complexities and challenges in project management will be greatly minimized as and partnerships between stakeholders will be enhanced over the project life cycle [28,48,80,81]. The implementation of BIM is expanding rapidly in the international context [80]. The implementation of BIM has reached a significant level in several developed world countries, such as the UK, USA, Australia, and Canada [23,82]. However, there is a low rate of BIM adoption not only in developing nations but also in certain developed world nations. While BIM implementation in developing nations has fulfilled the criteria during the design stage, it is substandard during the construction stage, as highlighted by Memon et al. [83]. Due to the fragmentation in implementation, BIM applications are delayed, and the construction industry remains at a low level of BIM adoption. Numerous factors contribute to this fragmented activity, including a lack of understanding about BIM implementation activity, a lack of coordination and collaboration between different disciplines, a lack of practice standards and guidelines, resistance to changing current working practices, and a lack of skill in preparing BIM plans and the ability to use them with stakeholders effectively [84]. It can be noted that the studies centering on BIM in developing world nations have highlighted the dynamics of BIM adoption and are working to improve BIM maturity in these countries. The barriers to BIM adoption must be addressed, and the benefits must be explained adequately to achieve comprehensive adoption of BIM and avail its benefits. Thus, future studies concentrating on this research field should perform case studies in which real-life adoption of BIM is validated.

Moreover, the utilization of BIM applications has progressively appeared as an essential topic in the PDR field [31,33]. Despite some reported evidence on its benefits within the construction industry, the adoption of BIM for PDR has been limited, and stakeholders and decision-makers in the aftermath of the disasters are not excited to adopt and implement BIM into its reconstruction practices. Therefore, there is a need to transfer from traditional reconstruction practices to BIM-based practices. Using BIM applications in reconstruction projects for planning, design, and construction will create and manage support key data and reports. The application of BIM in reconstruction projects will result in more cost-effective design and enhance communications and collaboration among parties.

It can be concluded that several studies address BIM adoption in general; however, few researchers concentrate on BIM adoption challenges and factors that influence adoption without providing a comprehensive perspective and an in-depth understanding of issues for the adoption of BIM. Nevertheless, there is a lack of studies on factors influencing BIM adoption in various dimensions of BIM research.

4.3. Disaster Management and Building Information Modelling

The role of BIM applications in disaster management is evident through all phases, from pre- to post-disaster. For instance, Drogemuller [85] conducted a study on the benefits of BIM in disaster response. This study looked at a variety of scenarios in which BIM was used during several disaster stages, including prevention, preparation, reaction, and recovery. By employing augmented reality to simulate multiple disaster scenarios and the best method to cope with them, the BIM can aid disaster preparation. Facility managers can utilize BIM to track key building data to undertake maintenance and prevent the failure of the building in the event of a disaster. Additionally, BIM can create 3D visualizations to assist stakeholders and decision-makers in comprehending the larger picture of a disaster's effect and speed up assessments of building damage. This could result in more effective planning and cooperation between stakeholders [85]. Dakhil and Alshawi [31] explored the BIM applications that support building disaster management, including disaster planning, site planning, and existing condition modelling, and their advantages; they suggested that future research should empirically study all applications of BIM during the project life cycle and identify the advantages of those applications. Based on the study by Kim and Hong [86], BIM information is a helpful tool for the response of disaster management. They presented a "BIM-based disaster integration information system" that allowed first responders to locate the event occurrence rapidly. In addition, Wang et al., [87] debated using a BIM-based virtual environment to assist the residents' building management during disasters. This suggested system employed BIM data and the game engine to design a real-time evacuation path for building occupants via a mobile device [87]. Moreover, Boguslawski et al., [88] proposed an algorithm for calculating evacuation routes during disasters. This method used a combination of BIM and geographic information system (GIS) to display the fastest egress route. According to Lyu et al. [89], BIM and GIS integration might be a valuable tool for city managers to identify flooding disasters.

Additionally, Sertyesilisik [90] emphasized the contribution of BIM to the resiliency of disaster from the pre- to the post-disaster stage, particularly by impacting the performance of the rescue operations, rebuild activity, and supply chain. Academic researchers and decision-makers play a vital role in encouraging and educating the public regarding the significance of BIM as a tool for improving the robustness of built environments throughout the disaster management process [90]. Moreover, Kermanshachi and Rouhanizadeh [73] developed a BIM-based automatic cost estimation methodology to aid in reconstruction by providing a precise estimate of the required budget. This tool is limited in that it can only estimate the cost of repairing a damaged structure. The tool has had one upgrade that allows it to reflect changes in estimated costs, although it still has limitations [73]. Furthermore, it was shown that very few studies on BIM integration with blockchain have been undertaken for PDR permits [32]. Nawari and Ravindran [32] listed the potential advantages of BIM in conjunction with blockchain for post-disaster rebuilding. The authors presented a framework for automated reconstruction permitting in any transaction by using Hyperledger Fabric. As a result, paperwork, additional processing fees, and the time it takes to obtain building permits, can all be reduced; however, no evidence of actual implementation has been found. Hence, more time is required to investigate blockchain in PDR. Thus, greater research into BIM and PDR combined with blockchain will provide a major boost in the PDR field.

Furthermore, Biagini et al. [50] recommended using BIM for historic building rehabilitation. The fundamental concern for historical building repair is on-site management. A detailed information plan for executing the restoration plan should be provided in order to provide effective on-site management. While conventional restoration techniques produce 2D maps with inadequate data, BIM allows users to create a 3D model of historic buildings and link it to a range of data. In the case of "Basilica di Collemaggio", which was harshly damaged because of a seismic event, BIM was used for a variety of purposes, including various scenario simulations for making decisions regarding the collapsed dome, structural analyses, and construction site management through various stages of the rebuild [57]. Xu et al. [53] established a BIM-based seismic damage evaluation based on FEMA P-58. Stakeholders can proceed through a virtual walkthrough to see how damage and loss are distributed, given that FEMA P-58 necessitates thorough information in order to forecast building damage. Furthermore, the BIM model for building components may be at various levels of development. Messaoudi and Nawari [33] proposed a virtual framework according to BIM and a generalized adaptive framework for speeding up the permitting process of reconstruction in the aftermath of the disaster in Florida. The fundamental procedure of the framework was to determine the type of construction permission, apply the newly established permitting framework, and decide on the permit result. The framework was able to save around 18 h for each permit, though this framework only considered the time of the permitting process of reconstruction. Finally, Rad et al. [91] provided an integration of BIM and life cycle cost (LCC) to assess building resiliency following an earthquake all through the building service life. The proposed BIM-LCC approach mainly was employed during the conceptual stage. The suggested BIM-LCC approach has limitations in that it only addresses earthquakes as one of the potential events that can occur through a building's life cycle. Nevertheless, in order to completely assess the building resiliency, numerous events of a disaster may need to be examined, and the consequences of each on the structure explored, which might be a topic for future research.

According to the literature, the available evidence shows that utilizing BIM for PDR is still nascent. Even though BIM has a great deal of potential within the construction industry, the challenges of adopting BIM in PDR have not been extensively studied. Moreover, BIM applications in PDR, such as scheduling, communication and collaboration, project delivery, demolition process, and deconstruction have been neglected in past studies. The majority of previous studies on BIM for PDR have been conducted using a qualitative approach. As a result, further primary investigations should be undertaken to adopt BIM for PDR projects fully. We recommend setting the foundations for a better framework and guidelines for BIM adoption through the planning, design, and construction phases of PDR projects. We also suggest that the barriers to adopting BIM for PDR projects within the construction industry be identified. In addition, studies concentrating on this research field in the future should perform quantitative research approaches in which real-life adoption of BIM is validated, which will be helpful for decision-makers. Finally, we recommend the integration of BIM for PDR with blockchain and/or other technology, such as the Internet of Things (IoT), which will provide a significant boost in the PDR field.

5. Conceptual Framework

Based on the above discussions, this study developed the proposed conceptual framework of BIM adoption for PDR, as illustrated in Figure 7. The proposed conceptual framework comprises a variety of key elements, such as the current practices of PDR projects, benefits of BIM adoption, and barriers to BIM adoption. While the previous frameworks focused only on the adoption of BIM for conventional construction [20,21,92–94], and others focused on the management of PDR projects separately [3,12,14,68], our proposed framework is the first attempt to integrate BIM adoption for PDR projects. This framework also incorporates main stakeholders, such as governments, NGOs and donors, disaster victims, and construction players. Those components serve as the foundation for developing the conceptual framework of BIM adoption for PDR.

In contrast, the proposed conceptual framework is not static or complete; instead, it reflects the current knowledge on BIM for PDR. Therefore, it serves as a leading strategy from which future researchers can conduct more comprehensive studies to identify the advantages and disadvantages of the current practices of the PDR, as well as the role and responsibilities of each stakeholder participating in the reconstruction projects. Moreover, we suggest that the most important benefits of BIM adoption and its barriers within the construction industry be identified. Accordingly, the benefits of BIM will mainly be determined to overcome the disadvantages of the current practices of PDR. There is a relationship between PDR and BIM adoption, which impacts BIM adoption for PDR.

Furthermore, the proposed conceptual framework shows how the current practices of PDR and the benefits and barriers of BIM adoption, considering the relationships among the stakeholders, will affect the adoption of BIM for PDR, and how it can be led to success. It also demonstrates that the adoption of BIM in PDR projects occurs through the interactive relationships between stakeholders. As a result, the success of the PDR project is heavily reliant on stakeholder relationships, leadership, decision-makers, and decisions made individually by each stakeholder, who is influenced by the decisions of other stakeholders, in conjunction with the mandatory adoption of BIM. Therefore, for a better understanding of this scenario, the proposed conceptual framework summarizes that the full adoption of BIM through the planning, design, and construction phases of PDR will lead to the success of reconstruction projects based on the success of their components, which is a part of the process of PDR. The proposed conceptual framework can assist to improve collaboration among reconstruction project stakeholders and offer more opportunities for a positive perception of BIM adoption in PDR in the future.

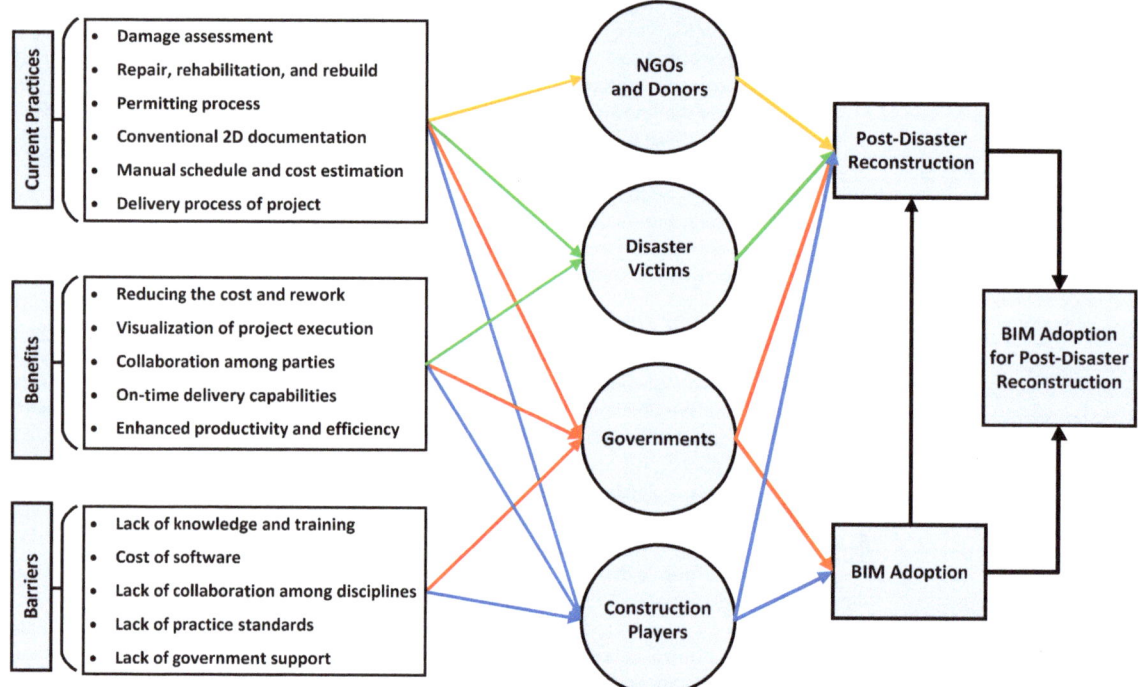

Figure 7. Conceptual framework.

6. Conclusions

Post-disaster reconstruction has been considered as a dynamic, complex system, and represents many challenges and issues. Recently, building information modelling has been extensively utilized in the construction industry due to its efficiency in solving complex and dynamic challenges. To provide a holistic overview of the present status, the current study used VOSviewer to visualize related papers published from 2010 to March 2021. The findings showed that "Automation in Construction" had an outstanding performance in BIM applications for the PDR field, and recorded the highest number of total citations. The analysis of keyword frequency showed that "Reconstruction" and "Safety Management" emerged as mainstream research themes in this field and recently attracted scholars' interest, which could represent the directions of future research. Five major research domains associated with BIM were identified, namely "Disasters", "Earthquakes", "HBIM", "Damage Detection", and "Life Cycle". Finally, a proposed conceptual framework was provided, highlighting the relationship between post-disaster reconstruction and BIM adoption, which impacts BIM adoption for PDR.

The outcomes of this study will assist scholars and practitioners in gaining clear ideas of the present status and identifying the directions of future research in this area. It will develop and open a new research area of BIM in the life cycle phases of PDR projects. Furthermore, while using BIM for reconstruction projects in the aftermath of disasters is still nascent, further primary investigations should be conducted to adopt BIM for post-disaster reconstruction projects fully. We recommend setting foundations for a better framework and guidelines for BIM adoption through the planning, design, and construction phases of PDR projects. We also suggest that the barriers to adopting BIM for PDR within the construction industry be identified. In addition, we recommend the integration of BIM for PDR with blockchain and/or other technology, such as the IoT, which will significantly boost the PDR field. However, there are a few limitations to this research. For instance, this study followed the PRISMA guidelines, without paying attention to meta-analyses. Thus, future studies should be undertaken considering meta-analyses to support the analysis of this study. Moreover, the collected papers were only sourced from the Scopus database, meaning that other related papers may be missing. Thus, prospective studies could include articles from another database. In addition, only English papers were included; however, there could be related papers in other languages that should be considered in future studies to overwhelm these limitations.

Author Contributions: Conceptualization, A.O.B. and W.S.A.; methodology, A.O.B., W.S.A., M.A.A.-S. and M.A.M.; software, A.O.B., A.M.A. and A.H.Q.; validation, A.O.B., W.S.A., M.A.A.-S. and M.S.L.; formal analysis, A.O.B., M.A.M., W.K., A.M.A. and A.H.Q.; investigation, A.O.B., M.A.M., W.K. and W.S.A.; data curation, A.O.B., W.S.A. and A.M.A.; writing—original draft preparation, A.O.B., W.S.A. and M.A.A.-S.; writing—review and editing, A.O.B., W.S.A., M.A.M., M.A.A.-S. and M.S.L.; supervision, W.S.A. and M.S.L.; project administration, A.O.B., W.S.A. and M.A.A.-S.; funding acquisition, W.S.A. All authors have read and agreed to the published version of the manuscript.

Funding: This research received no external funding.

Institutional Review Board Statement: Not applicable.

Informed Consent Statement: Not applicable.

Data Availability Statement: All the data is available within this manuscript.

Acknowledgments: The authors would like to appreciate the Shale Gas Research Group (SGRG) in UTP and Shale PRF project (cost center # 0153AB-A33) awarded to E. Padmanabhan for the support. The authors would also like to thank Universiti Teknologi PETRONAS (UTP) and UNIVERSITAS JANABADRA (UJ), (cost centre #015ME0-274; grant title: Risk Management of Liquefaction Soil Opak Fault Area Patalan Bantul Regency), for the support provided for this research.

Conflicts of Interest: The authors declare no conflict of interest.

References

1. Gunes, A.E.; Kovel, J.P. Using GIS in Emergency Management Operations. *J. Urban Plan. Dev.* **2000**, *126*, 136–149. [CrossRef]
2. Barakat, S. Housing Reconstruction after Conflict and Disaster. *Humanit. Policy Group Netw. Pap.* **2003**, *43*, 1–40.
3. Bilau, A.; Witt, E.; Lill, I. Practice Framework for the Management of Post-Disaster Housing Reconstruction Programmes. *Sustainability* **2018**, *10*, 3929. [CrossRef]
4. Anilkumar, S.; Banerji, H. An Inquiry into Success Factors for Post-Disaster Housing Reconstruction Projects: A Case of Kerala, South India. *Int. J. Disaster Risk Sci.* **2021**, *12*, 24–39. [CrossRef]
5. Ismail, D.; Majid, T.; Roosli, R. Analysis of Variance of the Effects of a Project's Location on Key Issues and Challenges in Post-Disaster Reconstruction Projects. *Economies* **2017**, *5*, 46. [CrossRef]
6. Bilau, A.; Witt, E.; Lill, I. Analysis of Measures for Managing Issues in Post-Disaster Housing Reconstruction. *Buildings* **2017**, *7*, 29. [CrossRef]
7. Ismail, D.; Majid, T.A.; Roosli, R.; Samah, N.A. Project Management Success for Post-Disaster Reconstruction Projects: International NGOs Perspectives. *Procedia Econ. Financ.* **2014**, *18*, 120–127. [CrossRef]
8. Dias, N.T.; Keraminiyage, K.; DeSilva, K.K. Long-Term Satisfaction of Post Disaster Resettled Communities. *Disaster Prev. Manag. Int. J.* **2016**, *25*, 581–594. [CrossRef]
9. Enshassi, A.; Chatat, T.; von Meding, J.; Forino, G. Factors Influencing Post-Disaster Reconstruction Project Management for Housing Provision in the Gaza Strip, Occupied Palestinian Territories. *Int. J. Disaster Risk Sci.* **2017**, *8*, 402–414. [CrossRef]
10. Alaloul, W.S.; Alfaseeh, A.S.; Tayeh, B.A.; Zawawi, N.A.W.A.; Liew, M.S. Reconstruction of Residential Buildings Post-Disaster: A Comparison of Influencing Factors. *AIP Conf. Proc.* **2019**, *2157*, 020035. [CrossRef]
11. Charles, S.H.; Chang-Richards, A.; Yiu, T.W. What Do Post-Disaster Reconstruction Project Success Indicators Look like? End-User's Perspectives. *Int. J. Disaster Resil. Built Environ.* **2021**. [CrossRef]

12. Islam, M.Z.; Kolade, O.; Kibreab, G. Post-Disaster Housing Reconstruction: The Impact of Resourcing in Post-Cyclones Sidr and Aila in Bangladesh. *J. Int. Dev.* **2018**, *30*, 934–960. [CrossRef]
13. Bilau, A.; Witt, E. An Analysis of Issues for the Management of Post-Disaster Housing Reconstruction. *Int. J. Strateg. Prop. Manag.* **2016**, *20*, 265–276. [CrossRef]
14. Capell, T.; Ahmed, I. Improving Post-Disaster Housing Reconstruction Outcomes in the Global South: A Framework for Achieving Greater Beneficiary Satisfaction through Effective Community Consultation. *Buildings* **2021**, *11*, 145. [CrossRef]
15. Sospeter, N.G.; Rwelamila, P.M.D.; Gimbi, J. Project Management Challenges for Post-Disaster Reconstruction Projects in Angola: A Public Sector Perspective. *Int. J. Manag. Proj. Bus.* **2020**, *14*, 767–787. [CrossRef]
16. Aliakbarlou, S.; Wilkinson, S.; Costello, S.B.; Jang, H. Client Values within Post-Disaster Reconstruction Contracting Services. *Disaster Prev. Manag.* **2017**, *26*, 348–360. [CrossRef]
17. Yaakob, M.; Ali, W.N.A.W.; Radzuan, K. Identifying Critical Success Factors (CSFs) of Implementing Building Information Modeling (BIM) in Malaysian Construction Industry. *AIP Conf. Proc.* **2016**, *1761*, 020105.
18. Abdirad, H. Metric-Based BIM Implementation Assessment: A Review of Research and Practice. *Archit. Eng. Des. Manag.* **2017**, *13*, 52–78. [CrossRef]
19. Abd Hamid, A.B.; Mohd Taib, M.Z.; Abdul Razak, A.H.N.; Embi, M.R. Building Information Modelling: Challenges and Barriers in Implement of BIM for Interior Design Industry in Malaysia. *IOP Conf. Ser. Earth Environ. Sci.* **2018**, *140*, 012002. [CrossRef]
20. Howard, R.; Restrepo, L.; Chang, C.-Y. Addressing Individual Perceptions: An Application of the Unified Theory of Acceptance and Use of Technology to Building Information Modelling. *Int. J. Proj. Manag.* **2017**, *35*, 107–120. [CrossRef]
21. Kim, S.; Park, C.H.; Chin, S. Assessment of BIM Acceptance Degree of Korean AEC Participants. *KSCE J. Civ. Eng.* **2016**, *20*, 1163–1177. [CrossRef]
22. Al-Ashmori, Y.Y.; Othman, I.; Rahmawati, Y.; Amran, Y.H.M.; Sabah, S.H.A.; Rafindadi, A.D.; Mikić, M. BIM Benefits and Its Influence on the BIM Implementation in Malaysia. *Ain Shams Eng. J.* **2020**, *11*, 1013–1019. [CrossRef]
23. Georgiadou, M.C. An Overview of Benefits and Challenges of Building Information Modelling (BIM) Adoption in UK Residential Projects. *Constr. Innov.* **2019**, *19*, 298–320. [CrossRef]
24. Wong, J.; Wang, X.; Li, H.; Chan, G. A Review of Cloud-Based BIM Technology in the Construction Sector. *J. Inf. Technol. Constr.* **2014**, *19*, 281–291.
25. Shehzad, H.M.F.; Ibrahim, R.B.; Yusof, A.F.; Khaidzir, K.A.M. Building Information Modeling: Factors Affecting the Adoption in the AEC Industry. In Proceedings of the 2019 6th International Conference on Research and Innovation in Information Systems (ICRIIS), Johor Bahru, Malaysia, 2–3 December 2019; pp. 1–6.
26. Haruna, A.; Shafiq, N.; Montasir, O.A. Building Information Modelling Application for Developing Sustainable Building (Multi Criteria Decision Making Approach). *Ain Shams Eng. J.* **2021**, *12*, 293–302. [CrossRef]
27. Chan, D.W.M.; Olawumi, T.O.; Ho, A.M.L. Perceived Benefits of and Barriers to Building Information Modelling (BIM) Implementation in Construction: The Case of Hong Kong. *J. Build. Eng.* **2019**, *25*, 100764. [CrossRef]
28. Gamil, Y.; Rahman, I.A.R. Awareness and Challenges of Building Information Modelling (BIM) Implementation in the Yemen Construction Industry. *J. Eng. Des. Technol.* **2019**, *17*, 1077–1084. [CrossRef]
29. Saka; Chan A Scientometric Review and Metasynthesis of Building Information Modelling (BIM) Research in Africa. *Buildings* **2019**, *9*, 85. [CrossRef]
30. Gan, L.; Wang, Y.; Lin, Z.; Lev, B. A Loss-Recovery Evaluation Tool for Debris Flow. *Int. J. Disaster Risk Reduct.* **2019**, *37*, 101165. [CrossRef]
31. Dakhil, A.; Alshawi, M. Client's Role in Building Disaster Management through Building Information Modelling. *Procedia Econ. Financ.* **2014**, *18*, 47–54. [CrossRef]
32. Nawari, N.O.; Ravindran, S. Blockchain and Building Information Modeling (BIM): Review and Applications in Post-Disaster Recovery. *Buildings* **2019**, *9*, 149. [CrossRef]
33. Messaoudi, M.; Nawari, N.O. BIM-Based Virtual Permitting Framework (VPF) for Post-Disaster Recovery and Rebuilding in the State of Florida. *Int. J. Disaster Risk Reduct.* **2020**, *42*, 101349. [CrossRef]
34. Sepasgozar, S.M.E.; Hui, F.K.P.; Shirowzhan, S.; Foroozanfar, M.; Yang, L.; Aye, L. Lean Practices Using Building Information Modeling (BIM) and Digital Twinning for Sustainable Construction. *Sustainability* **2020**, *13*, 161. [CrossRef]
35. Adegoriola, M.I.; Lai, J.H.K.; Chan, E.H.; Amos, D. Heritage Building Maintenance Management (HBMM): A Bibliometric-Qualitative Analysis of Literature. *J. Build. Eng.* **2021**, *42*, 102416. [CrossRef]
36. Wu, Z.; Yang, K.; Lai, X.; Antwi-Afari, M.F. A Scientometric Review of System Dynamics Applications in Construction Management Research. *Sustainability* **2020**, *12*, 7474. [CrossRef]
37. Qureshi, M.I.; Khan, N.; Qayyum, S.; Malik, S.; Hishan, S.S.; Ramayah, T. Classifications of Sustainable Manufacturing Practices in ASEAN Region: A Systematic Review and Bibliometric Analysis of the Past Decade of Research. *Sustainability* **2020**, *12*, 8950. [CrossRef]
38. Liberati, A.; Altman, D.G.; Tetzlaff, J.; Mulrow, C.; Gøtzsche, P.C.; Ioannidis, J.P.A.; Clarke, M.; Devereaux, P.J.; Kleijnen, J.; Moher, D. The PRISMA Statement for Reporting Systematic Reviews and Meta-Analyses of Studies That Evaluate Health Care Interventions: Explanation and Elaboration. *J. Clin. Epidemiol.* **2009**, *62*, e1–e34. [CrossRef] [PubMed]
39. Zhao, X.; Zuo, J.; Wu, G.; Huang, C. A Bibliometric Review of Green Building Research 2000–2016. *Archit. Sci. Rev.* **2019**, *62*, 74–88. [CrossRef]

40. Aghaei Chadegani, A.; Salehi, H.; Yunus, M.; Farhadi, H.; Fooladi, M.; Farhadi, M.; Ale Ebrahim, N. A Comparison between Two Main Academic Literature Collections: Web of Science and Scopus Databases. *Asian Soc. Sci.* **2013**, *9*, 18–26. [CrossRef]
41. van Eck, N.J.; Waltman, L. Software Survey: VOSviewer, a Computer Program for Bibliometric Mapping. *Scientometrics* **2010**, *84*, 523–538. [CrossRef] [PubMed]
42. Wu, Z.; Chen, C.; Cai, Y.; Lu, C.; Wang, H.; Yu, T. BIM-Based Visualization Research in the Construction Industry: A Network Analysis. *Int. J. Environ. Res. Public Health* **2019**, *16*, 3473. [CrossRef]
43. Darko, A.; Chan, A.P.C.; Adabre, M.A.; Edwards, D.J.; Hosseini, M.R.; Ameyaw, E.E. Artificial Intelligence in the AEC Industry: Scientometric Analysis and Visualization of Research Activities. *Autom. Constr.* **2020**, *112*, 103081. [CrossRef]
44. Ren, R.; Hu, W.; Dong, J.; Sun, B.; Chen, Y.; Chen, Z. A Systematic Literature Review of Green and Sustainable Logistics: Bibliometric Analysis, Research Trend and Knowledge Taxonomy. *Int. J. Environ. Res. Public Health* **2019**, *17*, 261. [CrossRef]
45. Jin, R.; Gao, S.; Cheshmehzangi, A.; Aboagye-Nimo, E. A Holistic Review of Off-Site Construction Literature Published between 2008 and 2018. *J. Clean. Prod.* **2018**, *202*, 1202–1219. [CrossRef]
46. Hosseini, M.R.; Martek, I.; Zavadskas, E.K.; Aibinu, A.A.; Arashpour, M.; Chileshe, N. Critical Evaluation of Off-Site Construction Research: A Scientometric Analysis. *Autom. Constr.* **2018**, *87*, 235–247. [CrossRef]
47. Su, H.-N.; Lee, P.-C. Mapping Knowledge Structure by Keyword Co-Occurrence: A First Look at Journal Papers in Technology Foresight. *Scientometrics* **2010**, *85*, 65–79. [CrossRef]
48. He, Q.; Wang, G.; Luo, L.; Shi, Q.; Xie, J.; Meng, X. Mapping the Managerial Areas of Building Information Modeling (BIM) Using Scientometric Analysis. *Int. J. Proj. Manag.* **2017**, *35*, 670–685. [CrossRef]
49. van Eck, N.J.; Waltman, L. Visualizing Bibliometric Networks. In *Measuring Scholarly Impact*; Springer International Publishing: Cham, Germany, 2014; pp. 285–320.
50. Biagini, C.; Capone, P.; Donato, V.; Facchini, N. Towards the BIM Implementation for Historical Building Restoration Sites. *Autom. Constr.* **2016**, *71*, 74–86. [CrossRef]
51. Noor, S.; Shah, L.; Adil, M.; Gohar, N.; Saman, G.E.; Jamil, S.; Qayum, F. Modeling and Representation of Built Cultural Heritage Data Using Semantic Web Technologies and Building Information Model. *Comput. Math. Organ. Theory* **2019**, *25*, 247–270. [CrossRef]
52. Nawari, N.O. BIM Data Exchange Standard for Hydro-Supported Structures. *J. Archit. Eng.* **2019**, *25*, 04019015. [CrossRef]
53. Xu, Z.; Zhang, H.; Lu, X.; Xu, Y.; Zhang, Z.; Li, Y. A Prediction Method of Building Seismic Loss Based on BIM and FEMA P-58. *Autom. Constr.* **2019**, *102*, 245–257. [CrossRef]
54. Brumana, R.; Della Torre, S.; Previtali, M.; Barazzetti, L.; Cantini, L.; Oreni, D.; Banfi, F. Generative HBIM Modelling to Embody Complexity (LOD, LOG, LOA, LOI): Surveying, Preservation, Site Intervention—The Basilica Di Collemaggio (L'Aquila). *Appl. Geomatics* **2018**, *10*, 545–567. [CrossRef]
55. Sani, M.J.; Abdul Rahman, A. GIS and BIM Integration at Data Level: A Review. *ISPRS Int. Arch. Photogramm. Remote Sens. Spat. Inf. Sci.* **2018**, *42*, 299–306. [CrossRef]
56. Brumana, R.; Della Torre, S.; Oreni, D.; Cantini, L.; Previtali, M.; Barazzetti, L.; Banfi, F. SCAN to HBIM-Post Earthquake Preservation: Informative Model as Sentinel at the Crossroads of Present, Past, and Future. In *Lecture Notes in Computer Science (Including Subseries Lecture Notes in Artificial Intelligence and Lecture Notes in Bioinformatics)*; Springer International Publishing: Cham, Germany, 2018; Volume 11196 LNCS, pp. 39–51. ISBN 9783030017613.
57. Brumana, R.; Della Torre, S.; Oreni, D.; Previtali, M.; Cantini, L.; Barazzetti, L.; Franchi, A.; Banfi, F. HBIM Challenge among the Paradigm of Complexity, Tools and Preservation: The Basilica Di Collemaggio 8 Years after the Earthquake (L'Aquila). *ISPRS Int. Arch. Photogramm. Remote Sens. Spat. Inf. Sci.* **2017**, *42*, 97–104. [CrossRef]
58. Oreni, D.; Brumana, R.; Della Torre, S.; Banfi, F. Survey, HBIM and Conservation Plan of a Monumental Building Damaged by Earthquake. *ISPRS Int. Arch. Photogramm. Remote Sens. Spat. Inf. Sci.* **2017**, *XLII-5/W1*, 337–342. [CrossRef]
59. Ma, L.; Sacks, R.; Zeibak-Shini, R.; Aryal, A.; Filin, S. Preparation of Synthetic As-Damaged Models for Post-Earthquake BIM Reconstruction Research. *J. Comput. Civ. Eng.* **2016**, *30*, 04015032. [CrossRef]
60. Ma, L.; Sacks, R.; Zeibak-Shini, R. Information Modeling of Earthquake-Damaged Reinforced Concrete Structures. *Adv. Eng. Inform.* **2015**, *29*, 396–407. [CrossRef]
61. Carrie Sturts, D. Learning in Global Teams: BIM Planning and Coordination. *Int. J. Autom. Smart Technol.* **2015**, *5*, 119–135. [CrossRef]
62. Lyons, M. Building Back Better: The Large-Scale Impact of Small-Scale Approaches to Reconstruction. *World Dev.* **2009**, *37*, 385–398. [CrossRef]
63. Ika, L.A.; Diallo, A.; Thuillier, D. Critical Success Factors for World Bank Projects: An Empirical Investigation. *Int. J. Proj. Manag.* **2012**, *30*, 105–116. [CrossRef]
64. Tagliacozzo, S. Government Agency Communication during Postdisaster Reconstruction: Insights from the Christchurch Earthquakes Recovery. *Nat. Hazards Rev.* **2018**, *19*, 04018001. [CrossRef]
65. Rouhanizadeh, B.; Kermanshachi, S.; Dhamangaonkar, V.S. Identification and Categorization of Policy and Legal Barriers to Long-Term Timely Postdisaster Reconstruction. *J. Leg. Aff. Disput. Resolut. Eng. Constr.* **2019**, *11*, 1–10. [CrossRef]
66. Chang, Y.; Wilkinson, S.; Potangaroa, R.; Seville, E. Identifying Factors Affecting Resource Availability for Post-disaster Reconstruction: A Case Study in China. *Constr. Manag. Econ.* **2011**, *29*, 37–48. [CrossRef]

67. Sharma, K.; KC, A.; Subedi, M.; Pokharel, B. Post Disaster Reconstruction after 2015 Gorkha Earthquake: Challenges and Influencing Factors. *J. Inst. Eng.* **2018**, *14*, 52–63. [CrossRef]
68. Sadiqi, Z.; Trigunarsyah, B.; Coffey, V. A Framework for Community Participation in Post-Disaster Housing Reconstruction Projects: A Case of Afghanistan. *Int. J. Proj. Manag.* **2017**, *35*, 900–912. [CrossRef]
69. Shi, M.; Cao, Q.; Ran, B.; Wei, L. A Conceptual Framework Integrating "Building Back Better" and Post-Earthquake Needs for Recovery and Reconstruction. *Sustainability* **2021**, *13*, 5608. [CrossRef]
70. Rouhanizadeh, B.; Kermanshachi, S.; Nipa, T.J. Exploratory Analysis of Barriers to Effective Post-Disaster Recovery. *Int. J. Disaster Risk Reduct.* **2020**, *50*, 101735. [CrossRef]
71. Rouhanizadeh, B.; Kermanshachi, S. Barriers to an Effective Post-Recovery Process: A Comparative Analysis of the Public's and Experts' Perspectives. *Int. J. Disaster Risk Reduct.* **2021**, *57*, 102181. [CrossRef]
72. Baarimah, A.O.; Alaloul, W.S.; Liew, M.S. Post-Disaster Reconstruction Projects in Developing Countries: An Overview. *Lect. Notes Electr. Eng.* **2022**, *758*. [CrossRef]
73. Kermanshachi, S.; Rouhanizadeh, B. Feasibility Analysis of Post Disaster Reconstruction Alternatives Using Automated BIM-Based Construction Cost Estimation Tool. In Proceedings of the CSCE 6th International Disaster Mitigation Specialty Conference, Montreal: Canadian Society of Civil Engineering, Fredericton, NB, Canada, 13–16 June 2018; pp. 13–16.
74. Ilhan, B.; Yaman, H. Meta-Analysis of Building Information Modeling Literature in Construction. *Int. J. Eng. Innov. Technol.* **2013**, *3*, 373–379.
75. Montiel-Santiago, F.J.; Hermoso-Orzáez, M.J.; Terrados-Cepeda, J. Sustainability and Energy Efficiency: BIM 6D. Study of the BIM Methodology Applied to Hospital Buildings. Value of Interior Lighting and Daylight in Energy Simulation. *Sustainability* **2020**, *12*, 5731. [CrossRef]
76. Bui, N.; Merschbrock, C.; Munkvold, B.E. A Review of Building Information Modelling for Construction in Developing Countries. *Procedia Eng.* **2016**, *164*, 487–494. [CrossRef]
77. Kim, K.P.; Freda, R.; Nguyen, T.H.D. Building Information Modelling Feasibility Study for Building Surveying. *Sustainability* **2020**, *12*, 4791. [CrossRef]
78. Obi, L.; Awuzie, B.; Obi, C.; Omotayo, T.S.; Oke, A.; Osobajo, O. BIM for Deconstruction: An Interpretive Structural Model of Factors Influencing Implementation. *Buildings* **2021**, *11*, 227. [CrossRef]
79. Ghaffarianhoseini, A.; Tookey, J.; Ghaffarianhoseini, A.; Naismith, N.; Azhar, S.; Efimova, O.; Raahemifar, K. Building Information Modelling (BIM) Uptake: Clear Benefits, Understanding Its Implementation, Risks and Challenges. *Renew. Sustain. Energy Rev.* **2017**, *75*, 1046–1053. [CrossRef]
80. Adekunle, S.A.; Ejohwomu, O.; Aigbavboa, C.O. Building Information Modelling Diffusion Research in Developing Countries: A User Meta-Model Approach. *Buildings* **2021**, *11*, 264. [CrossRef]
81. *The Business Value of BIM for Construction in Major Global Markets*; SmartMarket Report; McGraw Hill Construction: New York, NY, USA, 2014; pp. 1–60.
82. Doan, D.T.; GhaffarianHoseini, A.; Naismith, N.; Ghaffarianhoseini, A.; Zhang, T.; Tookey, J. Examining Critical Perspectives on Building Information Modelling (BIM) Adoption in New Zealand. *Smart Sustain. Built Environ.* **2020**. [CrossRef]
83. Hameed Memon, A.; Abdul Rahman, I.; Memon, I.; Iffah Aqilah Azman, N. BIM in Malaysian Construction Industry: Status, Advantages, Barriers and Strategies to Enhance the Implementation Level. *Res. J. Appl. Sci. Eng. Technol.* **2014**, *8*, 606–614. [CrossRef]
84. Construction Industry Transformation Programme 2016–2020. *CIDB Kuala Lumpur Malays.* **2016**, *184*, 1–71.
85. Drogemuller, R. BIM Support for Disaster Response. In *Risk-Informed Disaster Management: Planning for Response, Recovery and Resilience*; 2015; pp. 391–405. Available online: https://eprints.qut.edu.au/76145/1/E3.2.pdf (accessed on 5 December 2021).
86. Kim, J.-E.; Hong, C.-H. A Study on the Application Service of 3D BIM-Based Disaster Integrated Information System Management for Effective Disaster Response. *J. Korea Acad. Coop. Soc.* **2018**, *19*, 143–150.
87. Wang, B.; Li, H.; Rezgui, Y.; Bradley, A.; Ong, H.N. BIM Based Virtual Environment for Fire Emergency Evacuation. *Sci. World J.* **2014**, *2014*, 589016. [CrossRef] [PubMed]
88. Boguslawski, P.; Mahdjoubi, L.; Zverovich, V.; Fadli, F.; Barki, H. BIM-GIS Modelling in Support of Emergency Response Applications. *Build. Inf. Model. Des. Constr. Oper.* **2015**, *149*, 381. [CrossRef]
89. Lyu, H.-M.; Wang, G.-F.; Shen, J.; Lu, L.-H.; Wang, G.-Q. Analysis and GIS Mapping of Flooding Hazards on 10 May 2016, Guangzhou, China. *Water* **2016**, *8*, 447. [CrossRef]
90. Sertyesilisik, B. Building Information Modeling as a Tool for Enhancing Disaster Resilience of the Construction Industry. *Trans. VŠB Tech. Univ. Ostrava Saf. Eng. Ser.* **2017**, *12*, 9–18. [CrossRef]
91. Rad, M.A.H.; Jalaei, F.; Golpour, A.; Varzande, S.S.H.; Guest, G. BIM-Based Approach to Conduct Life Cycle Cost Analysis of Resilient Buildings at the Conceptual Stage. *Autom. Constr.* **2021**, *123*, 103480. [CrossRef]
92. Ahuja, R.; Jain, M.; Sawhney, A.; Arif, M. Adoption of BIM by Architectural Firms in India: Technology–Organization–Environment Perspective. *Archit. Eng. Des. Manag.* **2016**, *12*, 311–330. [CrossRef]
93. Chen, Y.; Yin, Y.; Browne, G.J.; Li, D. Adoption of Building Information Modeling in Chinese Construction Industry. *Eng. Constr. Archit. Manag.* **2019**, *26*, 1878–1898. [CrossRef]
94. Ahmed, S.H.A.; Suliman, S.M.A. A Structure Equation Model of Indicators Driving BIM Adoption in the Bahraini Construction Industry. *Constr. Innov.* **2020**, *20*, 61–78. [CrossRef]

Review

A State of Review on Instigating Resources and Technological Sustainable Approaches in Green Construction

Dhanasingh Sivalinga Vijayan [1], Parthiban Devarajan [1], Arvindan Sivasuriyan [1], Anna Stefańska [2,3,*], Eugeniusz Koda [2], Aleksandra Jakimiuk [2], Magdalena Daria Vaverková [2,4], Jan Winkler [5], Carlos C. Duarte [3,6] and Nuno D. Corticos [3,6]

1. Department of Civil Engineering, Aarupadai Veedu Institute of Technology-Vinayaka Mission Research Foundation, Paiyanoor 603104, India; vijayan@avit.ac.in (D.S.V.); parthi92bhde@gmail.com (P.D.); sivarvind@gmail.com (A.S.)
2. Institute of Civil Engineering, Warsaw University of Life Sciences, 02-787 Warsaw, Poland; eugeniusz_koda@sggw.edu.pl (E.K.); aleksandra_jakimiuk@sggw.edu.pl (A.J.); magdalena_vaverkova@sggw.edu.pl (M.D.V.)
3. BSTS Lab, Building Science, Technology and Sustainability Laboratory, Rua Sá Nogueira, Polo Universitário do Alto da Ajuda, 1349-063 Lisboa, Portugal; carlosfcduarte@campus.ul.pt (C.C.D.); ncorticos@campus.ul.pt (N.D.C.)
4. Department of Applied and Landscape Ecology, Faculty of AgriSciences, Mendel University in Brno, Zemědělská 1, 613 00 Brno, Czech Republic
5. Department of Plan Biology, Faculty of AgriSciences, Mendel University in Brno, Zemědělská 1, 613 00 Brno, Czech Republic; jan.winkler@mendelu.cz
6. CIAUD, Research Centre for Architecture, Urbanism and Design, Department of Technology in Architecture, Urbanism and Design, Lisbon School of Architecture, Universidade de Lisboa, Rua Sá Nogueira, Polo Universitário do Alto da Ajuda, 1349-063 Lisboa, Portugal
* Correspondence: anna_stefanska@sggw.edu.pl

Abstract: Green building is a way to reduce the impact of the building stock on the environment, society, and economy. Despite the significance of a systematic review for the upcoming project, few studies have been conducted. Studies within the eco-friendly construction scope have been boosted in the past few decades. The present review study intends to critically analyse the available literature on green buildings by identifying the prevalent research approaches and themes. Among these recurring issues are the definition and scope of green buildings, the quantification of green buildings' advantages over conventional ones, and several green building production strategies. The study concludes that the available research focuses mainly on the environmental side of green buildings. In contrast, other crucial points of green building sustainability, such as social impacts, are often neglected. Future research objectives include the effects of climate on the effectiveness of green building assessment methods; verification of the actual performance of green buildings; specific demographic requirements; and future-proofing.

Keywords: economy; eco-materials; energy; environmental impact; lean construction; pollution; sustainability

1. Introduction

In this modern era, it is necessary to protect the environment. The demand for housing and its development plan has increased daily since the population growth rate is higher than ever. It becomes a challenge for engineers and architects to find a way to preserve resources while also incorporating environmentally friendly technologies into the building [1]. The Earth's resources are classified into renewable and non-renewable materials. With the advancement of modern technology, the construction industry has surpassed all other sectors in exploiting non-renewable natural resources, consuming approximately 3000 metric tonnes per year [2]. Continuing to extract raw materials for construction activities ultimately deprives them in the long run, resulting in a significant environmental burden. Among the several environmental impact agents, the construction sector was identified as one with the greatest failures and shortcomings related to the sustainability issue.

The extraction of raw materials for construction depletes the environment and has emerged as a significant source of pollution. It was long believed that construction activity harmed our environment through adverse impacts such as biodiversity loss. Such a result is mainly due to raw material extraction, landfill problems, worker inefficiency, resource depletion, acid rain, global warming, poor air quality, and smog production during the manufacture and transportation of building products [3]. Due to rapid availability and low cost, concrete and mortar are modern construction's most employed building resources. As the daily demand for cement increased, it led to a corresponding boost in production [4]. Each year, it is estimated that cement manufacturing alone causes 7% of global CO_2 emissions and other harmful greenhouse gases (GHG) to the atmosphere.

The current quantity of CO_2 in the atmosphere is 550 ppm (Parts Per million), with an annual increase of 2.5 ppm. As a result, the mean air temperature is rising alarmingly. Construction impacts are categorised by their effects on the ecology, natural resources, and human health. Activities like quarrying and river sand mining can harm the ecosystem and human health. Fine contaminants are created in a dust form during large-scale development and exploration. Workers and tenants still confront health hazards despite building bylaws and environmental controls. Toxic compounds and particles in the materials may alter indoor air quality, increasing the risk of early death and long-term respiratory disorders such as asthma and silicosis. Inhaling "respirable crystalline silica" in limestone and aggregates might cause health problems. To address this issue, sustainable and green construction display a set of practices that focus on raising environmental awareness and promoting eco-friendly labelling to battle pollution [5].

To fulfil sustainable guidelines, the Architecture, Engineering, and Construction sector (AEC) must operate within the planet's capacity to absorb the waste and pollutants generated by its activities and continuously develop stricter requirements for raw materials extraction and transformation. The created environment presents us with a tremendous obstacle [6]. The construction, operation, and eventual demolition of buildings have significant environmental impacts through material and energy consumption. These processes result in pollution and waste, which often strain inadequate infrastructure. Additionally, the built environment substantially impacts individuals, communities, and organisations' both physically and financially. A high standard and an aesthetic structure enhance the surroundings and teach designers how to manufacture more sustainable facilities. On the contrary, a substandard building will have the opposite effect. Buildings and physical environments are undesirable and unsustainable when they degrade the community, contributing to poor health and social isolation and generating excessive financial obligation.

Numerous corporations, industries, and local, national, and international governmental entities have chosen sustainable development as their official policy. After more than three decades since the establishment of World Environment Day by the United Nations General Assembly in 1972, the environmental movement appears to be progressing in reversing unsustainable development tendencies. To meet the challenge, designers must enhance the quality of life for everybody by developing wholesome structures and environments that fit individuals and communities of the present and future. To fulfil the responsibility to protect other species and ecosystems, we must minimise resource use, waste, and pollution. Consequently, buildings and the built environment will be subject to increasing criteria, such as energy efficiency. The need for multi-criteria analysis for buildings and housing is necessary to assess the quality of housing, which requires consideration of several factors. Identifying and categorising quality factors is one of the most challenging tasks [7].

A wealth of materials is available to all professions to create sustainable structures. Still, the current practice barely applies to the basic sustainability concepts in most existing buildings. This results in less efficient, more expensive, and less environmentally friendly structures. Improving the sectors that supply building designers with materials, services, and knowledge might significantly influence the environment and quality of life [8]. The present review study intends to present the most accessible and accurate source of information on how to design and develop sustainable buildings and built environments. It provides a comprehensive review concerning the recent changes in sustainable and green construction, particularly in reducing the use of non-renewable materials and mitigating the environmental impact of construction and building operations. Additionally, it aims to answer how to make well-informed, universal judgments on a building design that benefits people's health and the environment in a sustainable way.

The goal of the green building movement is to lessen the destruction of natural habitats during the building process. Waste decreases, and eco-friendly materials and methods are utilized [9]. A more sustainable future can be achieved by implementing resources and sustainability principles in green construction. One of the most crucial parts of an eco-friendly building is using appropriate materials. Since bamboo and recycled plastic are used in construction, they require less power and display less carbon impact than conventional alternatives like cement and steel. The success of green construction approaches in mitigating environmental damage depends on careful monitoring of resource consumption at every stage of the project. The time it takes for materials to disintegrate after being removed from a site, the amount of energy needed during manufacture, and the volume of trash created during installation and maintenance are all factors to be considered.

Building priorities show a clear diversity of goals, which comes from putting emphasis on different environmental issues. One looks at things from an environmental point of view, and the other from a humanistic point of view. The evolving ecological goals in architecture bring with them potential dangers if there is a limited understanding of associated issues. Defining ecological goals consciously and correctly lays the groundwork for designing sustainable architecture. Modern examples of ecologically sound architecture should be based on a balance between human and environmental concerns [10].

The implementation of sustainable practices in buildings can contribute to a variety of goals and causes. Firstly, it seeks to lessen the negative effect that buildings have on the surrounding environment by reducing the number of non-renewable resources and materials that are utilized. Secondly, it has the potential to reduce energy consumption both during the construction and operation phases throughout the structure's lifetime. Thirdly, it can improve the health and well-being of building inhabitants by producing a healthier interior atmosphere. This, in turn, has the potential to increase productivity and decrease absenteeism. Last but not least, it can help to create a constructed environment that is more sustainable and resilient, capable to endure the effects of climate change. In general, the use of sustainable building practices can help to lessen the negative effects that buildings have on the surrounding environment, improve the health and well-being of building occupants, and contribute to a more sustainable and resilient built environment.

The paper is divided as follows:

Section 1 describes and provides an overview of the current state of research in the field. It can be useful for both researchers and students who are looking to gain a better understanding of the latest developments in a specific area.

Section 2 describes different methods for writing review articles, such as literature search, selection criteria, data extraction and analysis. We will also discuss how these methods can be used effectively to produce meaningful results.

Section 3 describes the idea of green building, which emphasizes the use of energy-efficient materials and technologies to reduce energy consumption and waste production. Additionally, modern construction projects often consider factors such as water conservation, air quality, and noise pollution when designing buildings.

Section 4 describes the components used in sustainable construction that play an important role in reducing the environmental footprint of a building. These components include green building materials such as recycled steel, bamboo, and timber; energy efficient insulation; renewable energy sources; water management systems; and air quality systems. Each component contributes to the overall sustainability of a structure by reducing its environmental impact.

Section 5 discusses the various phases of sustainability practice in modern construction, including resource management, energy efficiency strategies, green building materials, and waste management. It will also explore how these practices can create a more sustainable built environment.

Section 6 discusses how conventional construction techniques can significantly impact the environment. The effects of traditional construction can be far-reaching, from the use of energy-intensive materials to the release of pollutants.

Section 7 examines the current economic growth status in sustainable construction and potential scenarios where it can be applied. It also provides insights into how this type of construction can benefit both businesses and consumers alike.

Section 8 explores the life cycle analysis (LCA) of sustainable construction, looking at how it can be used to assess the sustainability of different materials and processes used in construction projects. It will also discuss potential scenarios for LCA and how it can help enhance sustainable practices in the industry.

2. Methodology

The construction of environmentally friendly buildings is gaining significance as people become more conscious of their impact on the surrounding environment. As a result, it is of the utmost importance to devise a stringent and comprehensive technique for evaluating green building construction projects. During this review process, consideration should be given to several issues, including energy efficiency, water conservation, choice of materials, indoor air quality, and noise control. Plus, the review procedure should consider the use of sustainable design techniques in addition to renewable energy sources. If a targeted and thorough evaluation plan is developed, we can assume that green building construction projects are designed and built with environmental stewardship in mind, addressing all requirements for health and safety.

The present study aims to cover peer-reviewed articles published primarily within the last ten years. 42.1% of the papers were published within the previous five years, 38.6% within the last ten years, and 19.3% more than ten years ago. The latter were thoroughly examined and discussed among all authors concerning their relevance to the topic and content quality. We resorted to the Web of Science, SCOPUS, MDPI and ProQuest search engines, complimented with unformatted searches to fill specific issues. Following this approach, we conducted searches using isolated keywords or keyword combinations related to each chapter. The keyword terms were: Brick, Building Energy Efficiency, Cement and Concrete, Demolition and Construction Waste, Eco-friendly Materials, Environmental Impact, Green Construction, Hempcrete, Lean Construction, Life Cycle Analysis, Natural Materials, Reuse, Recycle, Reduce, Sustainable Construction, Water Efficiency Strategies. Any uncertainties regarding the content and publications' relevancy were discussed between the authors until reaching a consensus.

3. Concept of Sustainability in Modern Construction

All living beings are connected and reliant on one another following the law of nature. According to the ecological principle, supervision of a healthy environment is based on effectively and efficiently handling its resources [11]. The term "sustainability" denotes the ability of an ecosystem, a society, or any framed community to function while limiting the use of available resources without harming the environment. The concept of green sustainable development is realized by honing the ability to live in harmony with the environment; thus, the green-sustainable idea in the AEC field has aided in achieving a stable, economically viable environment. While an engineer must design a building with the concept of green sustainable construction in mind, it is critical to understand the goal of sustainability, its history, and its economic implication [12,13]. The theory of "Ecological

Development" was followed by many countries from 1987 onwards due to the growth of industrialization, economics, and environmental sustainability. It is a concept that emphasizes the conservation of biodiversity and ecosystems, the responsible use of natural resources, and the advancement of environmental sustainability. All activities must be carried out in a way that protects the environment and provides enough resources to meet human needs. Given this, it demands meticulous planning and execution to accomplish this. Developing beneficial strategies requires incorporating stakeholders from various sectors, including governmental and non-governmental organizations, corporations, and communities. To ensure the long-term sustainability of our environment also entails monitoring efforts for any alterations or advancements made towards achieving ecological balance.

Green buildings are typically defined as structures that protect the environment and resources efficiently throughout a building's life cycle by carefully considering aspects such as design, construction, maintenance, operation, repair, and rehabilitation. The concept of "*green building*" has gained popularity and spread worldwide in recent decades, focusing on issues such as global warming, unpredictably changing monsoons, and controlling emissions from the construction industry [14]. The traditional construction method assumed that GHG were only emitted when fossil fuels were used directly for heating and electricity. However, construction materials emit GHG in most cases during manufacturing, transportation, construction, operation, and demolition [15]. In sustainable buildings, indoor air quality was naturally better, and occupants showed higher levels of comfort and satisfaction, positively affecting their happiness and health.

Generally, a green and sustainable building positively impacts human well-being over its lifetime. This building provides increased durability, reduced maintenance, and a pleasant indoor environment for the owner and users. Green sustainable construction is not the same in all countries because of regional features that differentiate based on their culture, climate, tradition, building types, economic status, social condition, and environmental priorities [16]. Green sustainable construction has several design, operation, assembly, and maintenance strategies. Choosing and utilizing appropriate materials is crucial during the design of such buildings. According to a recent report, implementing the Green Sustainable Concept in developing countries like India can save up to 8500 MW of power annually. Additionally, new green sustainable buildings enhance the local ecosystem by using locally sourced materials [17].

When creating green buildings, it is necessary to remember that we are generating a new ecosystem, the basis of which is plants. Green roofs are a typical anthropogenic ecosystem where human civilization determines the composition of the soil profile (substrate) and vegetation as well as the water regime. These factors interact and are confronted with the surrounding ecosystems. The results of the interactions and confrontations shape the further development of the green roof ecosystem [18]. In particular, the composition of the substrate layers and the water regime must comply with the requirements of the planned vegetation. The environmental requirements of individual plant species and the possible effects of the urban environment on vegetation have been studied by numerous authors [19,20].

Most engineers prioritize aesthetic, social, economic, and ecological features over designing green sustainable construction to achieve the best possible building standard. It was believed that through this construction method, chance would displace human dominance over nature and establish a fulfilled connection with the natural world. To ensure the successful implementation of green sustainable construction, several guidelines must be followed to eliminate the issues connected to impeding performance [21]. A separate chapter for green futures and green infrastructure is urban agriculture which aims to promote ecological sustainability, social justice, and local food security [22]. Implementing urban agriculture has architectural and urban planning potential for using green zones and vacant parts of the land, revitalizing post-industrial areas, and developing aesthetic and social possibilities. Nonetheless, it requires access to specific needs (soil, water, nutrients, human energy) and avoiding contact with environmentally hazardous substances [23,24].

While implementing this concept, it is crucial to incorporate renewable energy sources into buildings to maximize energy conservation and promote environmental protection. Throughout implementation, it is necessary to deliberate on green sustainable technologies, products, systems, and equipment utilization that preserve natural resources and the environment. The cost of a green sustainable building may appear higher during the design stage. However, during the construction phase, the output demonstrated the possibilities of savings through energy conservation, reduced maintenance, and effective utilization of waste materials [25]. It has been shown that incorporating green sustainable technologies into construction results in high building performance and productivity.

In 2022, the global market for green construction materials was valued at an estimated USD 200 billion, and this figure is projected to rise at a CAGR of 11.2% over the next few years. Increasing construction activity and government efforts to enact green and energy-efficient building rules are expected to drive market growth over the forecast period. As a stimulus against the recent global crisis, governments have emphasized energy efficiency and green construction, which may impact the demand for sustainable construction products. The availability of high-performance green building products and the rising energy cost are the primary factors propelling the market. Yet, expansion could be stunted by consumers who are too focused on price and erratic energy Policymaking.

The following statistics reveal the global green building materials market revenue until 2022. Due to their low carbon footprint and long service life at a reasonable price, structural products are predicted to expand at 11.9% throughout the projection period. Developing the construction market, especially in emerging economies, is anticipated to boost the market overall. The second largest market is expected to be interior materials, thanks to increasing consumer awareness of the environmental benefits of these products, such as increased aesthetics, lighting, and air quality. The ability to regulate humidity and improve air quality using these materials is also anticipated to contribute to the industry's expansion during the next few years. Due to its ability to conserve energy, insulation is predicted to be the largest application segment, with a value of USD 92 billion by 2025. This market is expected to expand throughout the projected period thanks to rising demand from the residential and commercial construction industries. Roofing finishing was the subsequent most common use. Rubber, slag, sludge, stone granules, and corrugated mixed paper are some of the recyclables that go into making these items. Non-toxic recycled rubber roofing is predicted to see rising demand from the roofing industry because of its resistance to weather and long lifespan.

North America is anticipated to represent more than 40% of the global market. Building rules and supportive laws governing the use of green building materials in the construction industry and rising rehabilitation efforts are projected to keep this pattern going for the foreseeable future. As a result of the expanding residential construction industry in the region, Asia Pacific is projected to grow at the quickest rate during the forecast period, with a CAGR of 15.0%. Product demand in the area is anticipated to be stimulated by the Paris climate accord signed by India and China to combat climate change and the expanding infrastructure development in both nations. The business increase in the application of green building is expected to rise by various governments across the area. Green building materials are expected to see increased demand in Europe over the projected period as efforts to cut maintenance and operational costs of structures gain prominence.

Raw material suppliers increasingly integrate forward in response to the market's rising demand and limitless expansion potential. The rising demand for imports will likely lead to a decline in the regional dominance of industry participants. During the following years, this pattern should persist, which will drive market expansion. Raw material suppliers increasingly integrate forward to address the rising demand and limitless expansion potential. The increasing demand for imports leads to a decline in the regional dominance of industry participants. During the next few years, this pattern should persist, which should help drive market expansion. The following Figure 1 shows the global green building material application area's annual revenue for 2022.

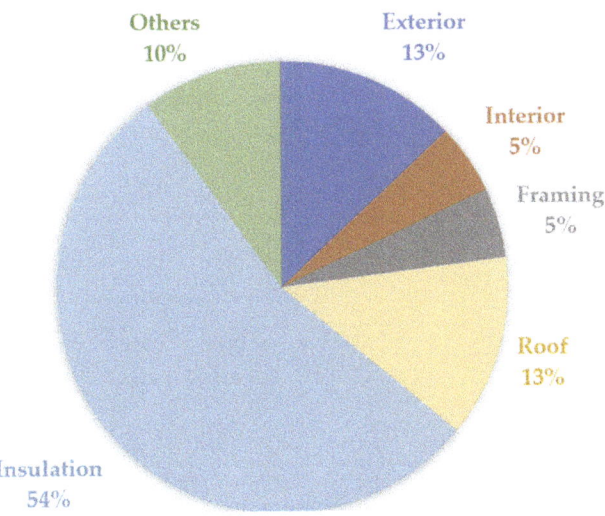

Figure 1. Different phases of sustainable construction (authors compilation).

4. Major Components in a Sustainable Construction

The planet Earth serves as a shelter for human beings, continuously protecting us from natural disasters while providing life-sustaining resources. However, life on earth is on the verge of extinction, and our actions as a species have been identified as the primary cause. Daily activities and common objects we use in our daily lives are among the primary contributors to the rate of carbon footprint and are considered a substantial contributor to climate change [26]. Our buildings' unmetered consumption of resources, energy, and water contributes significantly to environmental problems.

The green building concept can be a solution since it has the fewest negative consequences at every level, from the built and natural ecosystems of its immediate surroundings to broader regional and global contexts [27]. Using natural resources judiciously and effectively managing the building stock will save scarce resources, reduce energy consumption, and improve the overall environmental quality. Although this definition is straightforward, it is still too widely used in this context [28]. Therefore, it is essential to have quantifiable criteria for determining whether or not a building is "green". Understanding how a sustainable building is classified will allow us to recognize its qualities and the requirements that designate it as *"green construction"*. Sustainable construction is a method of reducing the environmental impact of construction projects. It entails the use of environmentally friendly, energy-efficient, and cost-effective materials [29]. As illustrated in Figure 2, specific components must be considered to ensure sustainable construction. These include using renewable energy, water conservation techniques, and waste management strategies. It is also necessary to understand how to design buildings in such a way that they reduce their carbon footprint and use natural resources wisely. By incorporating these essential components into construction projects, we can create more sustainable buildings that benefit both people and the environment. Throughout the building's life cycle, including the predesign, design, construction, and operation phases, this should be executed with diligence, rigor, and thorough evaluation and judgement [30].

The performance indicators for buildings under design and operation/retrofit must be employed in green buildings. These relate the efficient use of energy and water, waste reduction, sustainable materials, structure durability, and indoor air quality improvement under the green construction design as a path to decrease the building's CO_2 emissions and environmental footprint shown in Figure 2.

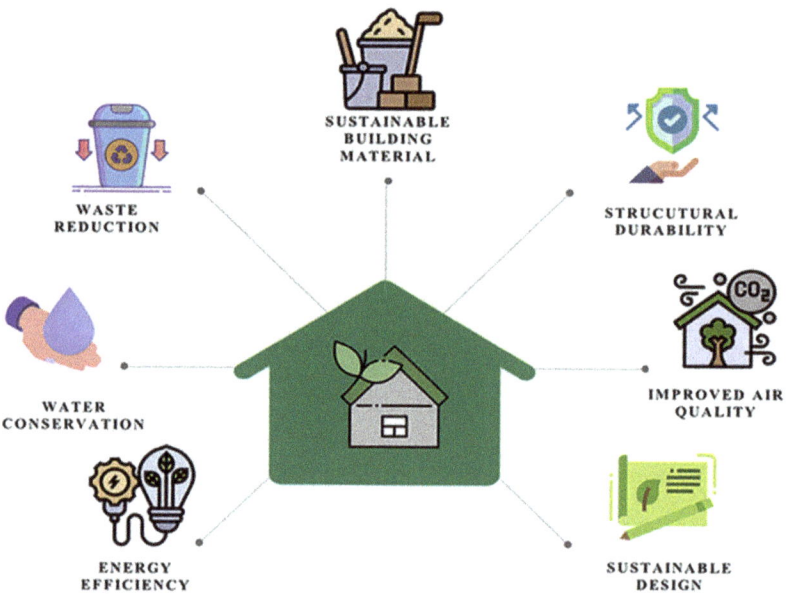

Figure 2. Essential components of sustainable construction (authors compilation).

5. Varied Phases of Sustainability Practice in Modern Construction

There are several approaches to modify and regulate the current nature of construction to make it less damaging to the environment without diminishing the positive outcomes of building activities. Potocnik's study suggests adopting a multidisciplinary strategy spanning a variety of components to construct a sustainable future in the building sector [31]. Other authors refer that to acquire an advantage from environmentally friendly construction techniques, they should be conducted in the context of the whole life cycle of a building [32]. Sustainable construction is becoming increasingly important as we strive to reduce our environmental impact. It entails using more environmentally friendly materials, processes, and methods than traditional building techniques. A construction project must go through several stages to attain sustainability, as shown in Figure 3. Planning, design, construction, and maintenance are examples of these. Each phase presents challenges that must be carefully considered to ensure the project meets all goals. The natural environment must be considered in planning, and energy efficiency and resource conservation must be considered in the design. During construction, materials should be chosen with an eye towards durability and recyclability while maintenance should focus on reducing waste and pollution. Understanding the stages of sustainable construction is the first step to designing eco-friendly projects.

Wise material and construction process decisions might help lower a building's energy consumption through reduced solar heat gain or loss, which reduces air-conditioning loads. The energy needed for extraction, processing, manufacturing, and transportation can be reduced by selecting materials with low embodied energy [33]. Due to the fragmentation of natural areas and ecosystems brought on by construction activities, the extraction and consumption of natural resources such as building materials, raw materials for producing building materials, and building materials themselves directly impact natural biodiversity. The built environment consumes a significant quantity of mineral resources, primarily non-renewable. Therefore, it is crucial to use fewer non-renewable resources. This approach should be considered during the design phase when selecting and prescribing materials and should account for and assess their environmental footprint, mainly their embodied carbon and energy.

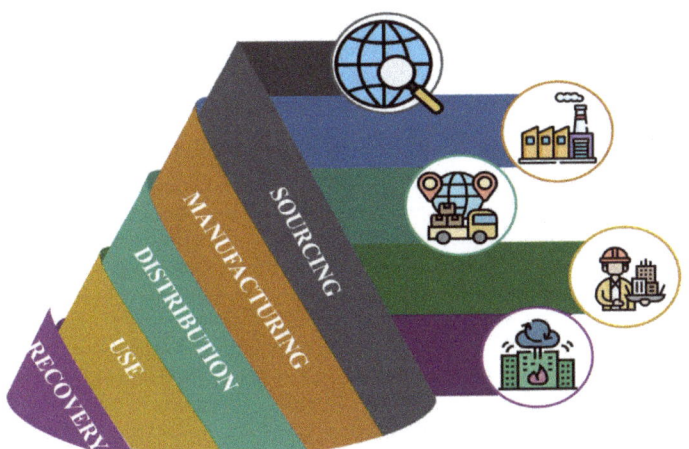

Figure 3. Phases of sustainable construction (authors compilation).

On the other hand, recycling positively impacts the environment by reducing waste output and resource usage. Reusing and recycling construction materials and components as alternatives to reintroducing them into the production chain has previously been established in the literature [34]. In this way, reusing materials such as bricks, glass, and tiles while repairing and dismantling buildings is an alternative for reducing Construction and Demolition Waste (CDW).

5.1. Sustainable Practice during Procurement of Construction Materials

Material procurement has caused problems for domestic economies and the environment affecting the construction sector. Hence it is necessary to highlight material procurement following sustainability requirements, and as a result, the construction industry's performance in material management, which right now is far from ideal [35]. When implementing a sustainable management strategy, the environmental, economic, and social impacts during the supply and flow of materials at various stages of a structure's life cycle should be considered [36]. Sustainable materials request a paradigm shift beginning with procuring raw materials and continuing throughout the journey of a constructed facility to reduce their environmental impact [37]. By affecting the amount of embodied energy and carbon released into the environment and diminishing natural resources upstream, the facility's positive effects can last well beyond its life cycle if the process starts with suitable materials [38].

In general, purchasing goods and services in a way that considers the economic, environmental and social impact of the purchase on people and communities is known as sustainable procurement [39]. It is also necessary for these purchases to increase the efficiency and value of the resources that are under use. Traditionally, procurement decisions were made exclusively based on price, quality, and time [40]. These factors still matter; however, in today's construction markets, emerging and established sustainable initiatives are combined with these well-established elements [41]. Environmentally friendly procurement should reduce material waste, CO_2 emissions and energy and water consumption, promoting biodiversity and equitable and sustainable economic development. Lastly, the sustainable way of acquiring construction materials must result in social benefits.

The issue of sustainability raises the question of energy and CO_2 consumption as well as landfill waste disposal. In the European Union, 38 million tonnes of glass waste are produced annually, and new goals for more sustainable waste handling were set for 2020. In particular, soda-lime, borosilicate, and lead-crystal glass are the glass families that may be combined in this experimental effort to be recycled as cast glass components. The kiln-cast technique prepared several mixes for melting at 970 °C, 1120 °C, and 1200 °C. An experimental splitting test was used for each sample to determine the force trend and fracture behaviour. Soda-lime-silica glass was discovered to be the most compliant glass recipe with the necessary physical and mechanical qualities [42].

5.2. Sustainable Water Conservation in Construction

Groundwater is one of the most common water sources for construction, as it is pumped from aquifers and used for various purposes. In remote areas, due to some natural and artificial obstacles, such as mountainous terrain, and groundwater for mining and construction activities, water supply is not always accessible and available [43]. Pumping groundwater is required for its use, requiring energy that contributes to GHG emissions [44]. It is essential to continuously supply water to ensure the community's needs for drinking, washing, waste disposal and sewages, swimming, and acclimatization in buildings. Also, stormwater that runoff from a driveway that connects roadways and roofs pollutes the waterways and causes flooding [45]. Direct and indirect water consumption is linked to water usage in a typical building. Generally, its immediate use can be calculated by the water consumed by construction workers during several activities, such as washing aggregates, preparing and curing concrete, suppressing dust, washing equipment, and cleaning hard surfaces [46].

The "indirect use" refers to using embodied water to make construction materials. The water conservation index assesses the average consumption of a building and compares it to overall water use, resulting in a "water savings rate". This rate indicates how much water is being conserved [47]. It is essential to evaluate the water-saving efficiency of kitchens and bathroom taps, including and considering the recycling of rain and second-hand intermediate water. At the design stage, it is essential to implement control systems for water resources and select more sustainable planning options [48]. As a result, it is necessary to align the building design and construction method to optimize water usage. During the project planning stage, we should consider water preservation measures such as using water-saving plumbing fixtures, rainwater harvesting systems, and a greywater system instalment [43]. Also, reducing water consumption during material production should be prioritized. Water reuse and recycling can help reduce embodied water usage in building materials manufacturing. Water pollution control on construction sites is another feature that should be evaluated.

Environmental assessment tools for construction must consider water pollution to minimize ecological impacts. The water consumed in buildings is determined mainly by the usage method and the sector type. Introducing environmentally friendly devices can improve water efficiency and decrease consumption through recycling or harvesting systems, one of the most important features to consider in new construction [49] under the "green" label. In the AEC sector, water sustainability differs from water conservation in that it reduces "waste" water while not interfering with daily needs or customer satisfaction. Rainwater and greywater harvesting, water-wise landscaping, and high-efficiency flow and flush fittings enhance building water sustainability [50]. Installing water-efficient fittings is now recommended to ensure that potable water is conserved in buildings rather than the traditional high-flow ones. These are measures that need some additional up-front capital investment. Nonetheless, studies have shown that this investment paid off in most cases, mainly in building with frequently used fixtures [51].

Water is required for both direct construction and embodied water production. Due to this, recycling water usage in construction is critical. To save water in the long run, it is imperative to calculate water usage throughout the infrastructure's life during the planning phase [52]. In many cases, such as collecting rainwater and heating it for domestic use, water usage is linked with GHG emissions. As a result, zero-energy and green buildings have gained popularity among the population [53,54]. It is also advisable to optimize the system by preventing water loss due to leaking pipes and faucets [55]. With increased awareness from an environmental view, favourable government policies, and ongoing lectures on the subject, efficiency related to water consumption in construction is expected to improve in the future.

5.3. Sustainability during the Usage of Alternative Construction Materials

The concrete industry has begun incorporating alternative industrial waste materials and CDW into structural concrete applications (Table 1). Its application has been more frequent in recent years due to the availability of waste from demolitions and the reduction in aggregate acquisition costs [56]. Those new applications will help the concrete industry reduce its carbon footprint and pursue sustainability.

Table 1. List of Environmental reutilized waste in construction.

Category of Waste	Type of Waste	Countries Acquire A Large Quantity of Waste	Citation
Industrial waste	Fly Ash produced from the coal power plant as a by-product	China, Canada, India and Vietnam	[11]
	Salvage steel fibre as a by-product of steel manufacturing	USA and Cameron	[25]
	Granulated blast furnace slag By-product of iron and steel-making	India and Great Britain	[28]
	Polyethylene terephthalate derived from Shredded waste plastic bottles	Nigeria and USA	[35]
	Calcium carbide residues from industrial gas	Burkina Faso	[36]
	Magnesium oxide derivative as By-product of mining & industrial company	USA & Spain	[37]
	Crumb rubber from recycled industry of transportation waste	India, China, Australia, and Spain	[38]
	Glass fibre reinforced polymer waste from water boxes manufacturing company	India and Brazil	[40]
	Waterworks sludge waste from water treatment plants	India and China	[41]
	Alumina filler and coal ash from an Aluminium foundry plant	USA and Spain	[45]
	Eucalyptus pulp microfibre from paper manufacturing	China, India, and Brazil	[46]
Agricultural waste	Straw bales from wheat, rice and barley	China, India, Egypt, Italy, Germany, Japan, France, Peru, Spain, Turkey, and Morocco	[49]
	Oil palm fruit bunch fibre after the extraction of palm oil	Indonesia and Malaysia	[50]
	Sugarcane bagasse left extracted after the juice in the sugar industry	India, Brazil, Ghana, Portugal, and Sri Lanka	[51]
	Cassava peels	Kenya and Colombia	[52]
	Henequen fibre (leaf)	Great Britain	[53]
	Sisal fibre	Brazil and Kenya	[46]
	Spent coffee ground and processed tea waste	China, India, and Turkey	[57]
	Bio briquettes	India and China	[58]
	Tobacco residue from tobacco industry by-product	Cuba and Turkey	[59]
	Olive waste fibre	Italy and Morocco	[60]
	Seaweeds fibre extraction from alginate	Italy and Great Britain	[61]
	Corn husk ash	USA & Nigeria	[62]
	Saw dust from wood-based industries	India, Brazil, and Mexico	[63]

The construction industry is ranked as the most environmentally impactful as it consumes vast energy and natural resources and generates massive waste [64]. In addition, conventional construction also incorporates a set of materials and substances that are harmful to living beings in the short and long term, leading to a series of health and environmental consequences (Table 2).

Table 2. Summary of non-eco-friendly conventional construction materials and their effects.

Conventional Constructional Material	Environmental and Health Impacts	Reference
Acrylonitrile	Irritates mucous membranes of the lung walls Affect habitants of aquatic organisms	[65]
Ammonia	Make water more acidic Increasing corrosive nature	[66]
Arsenic and its compounds	May damage the growth of the foetus Lead to cancer	[67]
Bitumen	Lung cancer Block percolation of groundwater	[68]
Borax and its substances	Poison to all kinds of living organisms	[69]
Cadmium	Damage to the function of the kidney, liver, and lungs	[70]
Copper and its substances	Affect habitants of aquatic organisms A Bio accumulative material	[71]
Epoxy	May trigger a strong allergy reaction	[72]
Fluorides	Decay the growth of the plant and aquatic organisms May reduce the strength of bones	[73]
Lead and its compounds	Related to kidney and brain damages	[74]
Nonyl phenol	React as environment oestrogen Increase the water acidity declining the growth of aquatic organisms	[75]
Styrene	Affect the lungs' function Cause damage to the reproductive system	[76]
Vinyl acetate	Increasing corrosive nature Affect habitants of aquatic organisms	[77]
Wood dust	Lung cancer May trigger robust allergy	[78]
Demolition Dust	Create eye irritation and breathing problem Reduce plant photosynthesis ability	[79]

Concrete is a commonly consumed product in the entire construction industry mainly due to its versatility and ability to be easily altered. Decreasing the concrete environmental burden can pave the way towards a more sustainable construction industry [80]. Most of this resource consumption occurs in developing countries such as China, India, and Brazil. From an environmental perspective, China and India appear to be struggling regarding this matter as the requirement for concrete rises, and natural resource depletion becomes an alarming problem. As a result, in the early twenty-first century, the work on green and sustainable technologies has taken center stage.

Due to the rapid urbanization of industrial areas, old buildings are being demolished and replaced with new ones with higher standards. In traditional construction, the CDW would have been disposed of in landfills or repurposed for the construction of pavements. The mining, processing, and transportation operations required to acquire and haul large amounts of aggregate consume significant amounts of energy, emit substantial amounts of carbon dioxide and harm the ecology of forested areas and riverbeds. Thus, finding a substitute for virgin aggregate has been a long-standing concern [81]. Recently, extensive research has been done on recycling demolition waste to determine whether it can be used instead of natural aggregates. In general, the recycled aggregates are extracted from discarded waste, generated by the demolition of concrete buildings, the use of unfinished concrete, the failure of precast concrete members, the expiration of concrete pavements, and the testing of samples in several laboratories [82]. Recycled aggregates include tiles, brick aggregates, concrete, marbles, bitumen, and asphalt. The term "recycled concrete aggregate" (RCA) replacement stands for typically processed aggregates made by crushing old or parent concrete like demolished concrete wastes [83]. Still, there has only been limited research on the partial usage of concrete wastes as a replacement for cement in concrete, which requires further investigation. Repurposing and reincorporating this kind of debris in construction will be a step forward to improve environmental safety [84].

Natural fibers have a major role in developing environmentally friendly composites. Their main advantages are low density, renewability, cost-effectiveness, flexibility, and recyclability. Several kinds of natural fibers are used to construct building materials; these include bamboo, palmyra, crushed coconut shell, peel of banana skin, sisal and jute fibers, bagasse, and fabric [85]. Using natural fibers as construction materials added the benefits of being eco-friendly and improving their properties. The rice husk is a highly efficient and widely used fuel in many countries in energy generation units [46]. Rice husk ash is a pozzolanic material formed due to this burning process. It has over 75% silica and retains rice husk at about 20%. Typically, this process dumps ash into nearby waterways, contaminating the water and causing environmental pollution [57]. Due to incomplete ignition and unburned carbon, ash made from rice husk has a lower pozzolanic effect at temperatures below 500 °C. Due to the transformation of silica to a non-crystalline or amorphous form, the temperature of 550–700 °C results in ash having improved pozzolanic characters. Numerous studies have been conducted on using rice husk ash in partially replacing cement with or without replacing fine aggregates in cementitious composites.

Another example is in Sakhare and Ralegaonkar's study, which investigated the possibility of combining cotton waste and lime powder to create an innovative lightweight and low-cost composite material for construction. The mechanical and physical characteristics of concrete composites containing a large concentration of cotton waste and Lime powder were analyzed. Results indicate that replacing the lime with the cotton waste does not result in immediate brittle fracture, even moving beyond the failure loads. Moreover, it demonstrates an energy absorption capacity of excellent levels, significantly reducing unit weights and introducing a smoother surface compared to currently available bricks of concrete [58]. Demir, aiming to promote environmental stewardship and sustainability in the long-term use of ferrocement, discussed the implications of previous research on using industrial waste materials in ferrocement works. The authors examined how different industrial waste materials affected mortar and mesh reinforcement behaviour [59]. Lamrani et al. have studied the viability of a novel method for making lightweight composite elements in a sandwich arrangement by encasing an aerated lightweight concrete core in a high-performance ferrocement box. The results are analyzed using control mixtures made entirely of aerated-type concrete. Results marked a significant increase in flexural and compressive strength and reduced water absorption to fractions of the specimens used as controls for the control specimens [60]. Salleh et al. study present agricultural, industrial and food wastes as agents forming pores in producing porous ceramics. Identifying waste material and clay confirms that processing conditions like sintering temperature, compaction pressure and material composition affect pore formation [86]. Zero waste in food, agriculture, and industry can alleviate environmental concerns and ease production in a closed loop. Porous ceramics of waste origin can help create more sustainable environments while expanding the economy, particularly in alternative building materials.

5.4. Sustainability in Masonry Materials

Brick masonry is a construction and building material widely used not only in the past in Egyptian, Mesopotamian, and Roman constructions but still today at a worldwide scale. Bricks are traditionally made by combining earth-based raw materials, then moulded, dried and fired until they reach a specified strength parameter or by using ordinary Portland cement (OPC) for concrete bricks [87]. Dove developed a variety of bricks from several waste and by-product materials to reduce pollution, waste generation, and raw materials depletion, thus contributing to a more sustainable and environmental practice. As a result of the limitations of the traditional brick-making method, over the last two decades, the brick-making process has shifted toward the use of waste materials [61]. To exemplify, Goel and Kalamdhad showed the feasibility of using municipal solid wastes in degraded form as input material in manufacturing bricks burnt at a concentration of 5 to 20 weight % and then burning the product at temperatures between 850 and 900 °C [88]. Another study by SP Raut et al. examined that different waste materials of varying compositions were combined with the raw material at several levels to create waste-origin bricks. Other waste materials have been used in brick production, including paper processing residues, cigarette butts, textile effluent sludges from the treatment plant, polystyrene foam, plastic fiber, fly ash, polystyrene and straw fabric [89]. Batagarawa et al. modified lightweight and porous bricks with adequate compressive strength and low thermal conductivity by adding paper processing residues to an earthen brick. Chemical analysis was done for raw material and paper waste. The mixtures of paper waste and brick raw materials were made in several proportions, up to 30% by weight [62]. Ayodele et al. probed the possibility of reusing sludge from textile effluent treatment plants in building materials. The engineering and physicochemical properties of this composite sludge specimen from South India were investigated to determine the sludge's suitability for both non-structural and structural applications with cement replacement [63]. Juel et al. investigated the possibility of using tannery sludge for manufacturing clay bricks. Various quantities of tannery sludge in 10%, 20%, 30%, and 40% were used to replace the clay. It was determined that 10% of tannery sludge by weight is the optimal constituent for it -amended tannery sludge. Raising the tannery sludge proportions and firing temperature caused a decrease in shrinkage, weight, and bulk density during the firing process. Additionally, it was demonstrated that incorporating a 10–40% tannery sludge content can save up to 15–47% of the firing energy. The findings show that combining tannery sludge allows to the production of bricks that meet the quality standards of the American Society for Testing and Materials (ASTM) [90]. Adazabra et al. inspected the use of spent shea waste replaced at a rate of 5–20% of weight while manufacturing clay bricks that are moulded, compacted, and ablaze for more than an hour at 900–1200 °C. Increasing the amount of used shea waste in clay material increased water absorption values in every tested scenario. As the produced brick showed lower strength by incorporating shea waste it was classified as a non-load-bearing part of structural construction [91]. Sutcu et al. analyzed the physiomechanical and thermal performance of porous clay bricks tested after adding an olive mill waste concentration of 0–10% of the weight to the mixture. With a 10% waste from olive mills, the samples' bulk density was reduced by up to 1450 kg/m^3. The porosity of the modified samples raised from 30.8 to 47.0 as the olive mill waste was taken from 0% to 10%. At 950 °C, the compressive strength fell from 36.9 MPa to just 10.26 MPa. The study demonstrated that olive mill waste effectively creates pores in bricks [92]. Ornam et al. studied the impact of waste sago husk on manufacturing bricks made of fly ash. Samples were moulded and dried in the sun before being burned at 550 °C for two hours with a zinc stove plate and aluminium foil. As the amount of sago husk in the bricks was higher, the bricks' strength gradually decreased. On the other hand, the specimen's density may decrease as the sago husk content increases, down to 1810 kg/m^3. The developed fly ash brick had the lowest initial absorption rate while adhering to ASTM C67 specifications and requirements. While waste-derived bricks have a few commercial applications and fabrication, a literature review reveals their potential and versatility as a partial or complete substitute for traditional raw materials as produced bricks meet a wide range of quality standards. Researchers can scale up promising findings by including all necessary data and methodologies and planning and designing experiments according to industrial manufacturing procedures [93].

6. Strategies for Reducing the Impact of Conventional Construction

Based on the previous scenario, building materials and construction processes should be targeted to reduce environmental impact. Several strategies were developed to assist sustainable forms of construction, aiming at building material manufacture and construction methods [65] (Figure 4). The advanced material manufacturing process has resulted in waste reduction. Resource conservation efforts are visible through reduced primary level production and increased recycling, development and use of partial and complete substitutes in case of using novel construction methods, materials of high impact, high-performance materials development coupled with eco-friendly materials [94]. Conventional construction has contributed significantly to global carbon emissions and climate change. To conserve the environment, we must limit the impact of the traditional construction industry. As depicted in Figure 4, many tactics can be applied to accomplish this objective. They include using environmentally friendly building materials, enhanced energy efficiency, and waste management strategies [95]. In addition, breakthrough technologies, such as AI-assisted design tools, can lessen the environmental effect of construction projects by increasing their efficiency and precision. By implementing these strategies, we are ensuring that the conventional construction industry will decrease its environmental footprint and no longer be a global warming and climate change protagonist.

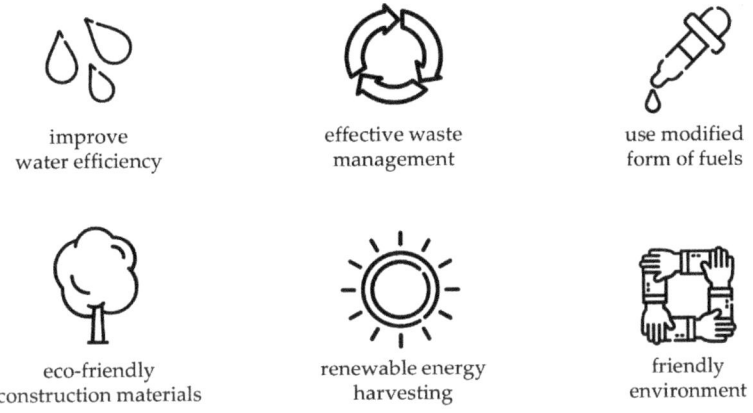

Figure 4. Strategies to control the environmental impact of conventional construction (authors compilation).

6.1. Impact Control through Improved Materials Production Processes

Implementing improved manufacturing processes reduces the amount of carbon emitted and energy consumed during the production of the materials. According to LCA studies, ferrous, cement and nonferrous metals are the materials with the highest environmental impact. Energy and carbon emission savings can be achieved by updating or optimizing existing manufacturing technology, including reutilized residual or waste heat in the furnace to generate electricity. The cement industry is an excellent example of this, as it uses waste heat recovery in manufacturing [66,96].

Only six of the eleven emerging economies with a significant cement production capacity have heat recovery systems, with India leading in total installed capacity and designed systems. An alternative to mitigate the harms of cement manufacturing on the environment is to reduce the clinker quantity in the final product [97]. This can be accomplished by replacing a certain amount of the clinker with materials that display a low impact and significant pozzolanic characteristics, as described above when the LCA of Portland cement production was conducted. To exemplify, cement made from pulverized fly ash and clinker has less environmental impact than traditional cement [98]. Furthermore, according to Garcia-Segura et al. blended cement not only produces fewer GHG emissions than conventional cement manufacturing, but it [97] is proven to have higher durability.

Edmundson and Horsfall and Rajdev et al. have described how blended combinations can significantly decrease the impact of cement production on the environment by substituting low-impact supplementary cementitious materials for a portion of the clinker [67,99]. Upgrading and improving cement manufacturing machinery continuously is a way to mitigate environmental damage progressively.

6.2. Impact Control by the Materials Recycling

The process by which used materials are transformed into new products that would otherwise be discarded is known as recycling. It's a powerful tool for increasing the efficiency of energy usage and lowering CO_2 emissions caused by the material industry [68,99]. The demand for raw materials is reduced by recycling, which saves energy and reduces carbon emissions. According to the World Steel Association recycling is crucial to produce metals like steel and iron and nonferrous ones like copper and aluminium because it reduces the reliance on natural resources [100].

Additionally, energy costs in steel production are almost 20% to 40% of the total price, according to the source of energy [100]. On the other hand, steel is 100% recyclable, and recycling this material can save up to 25% of energy. Steel, as one of the carbon-intensive building materials, emits 1.9 tonnes of CO_2 as a by-product of one tonne of raw steel manufactured [101].

Given this, recycling contributes to reducing the carbon footprint of steel production. Recycled steel has a carbon intensity of approximately 16% of virgin steel, according to the Inventory of Carbon and Energy (ICE), a database for building materials' carbon coefficients and embodied energy [102]. Another industry that employs recycling for reducing carbon footprint and energy is the aluminium industry. In their LCA study, Grimaud et al. discovered significant environmental advantages in recycling aluminium shredder cables [103]. Furthermore, Rajadesingu and Arunachalam described that the recycling of aluminium has benefited both the environment and the economy. Energy savings and a reduction in bauxite mining are two of the advantages. It also helps countries where secondary aluminium production is the primary metal source [69].

6.3. Impact Control through Material Substitution

Cement production is an energy-demanding and carbon-emitting process. Carbon emissions are also produced by the chemical processes involved. Most construction industry stakeholders may not have control over minimizing the impact of cement production on the environment through the option of renewable energy coupled with waste heat recovery processes. The construction industry's contribution to reducing the environmental impact of cement production by selecting suitable substitutes minimises the need for OPC. To reach low-impact alternative materials, the following additions can be used: rice husk ash, calcinated shale, volcanic ash, and calcinated clay. Incorporating supplementary cementitious material (SCM) into OPC has improved the material's durability, long-term strength, and workability [70]. Additionally, SCM improves cement-based structures' corrosion resistance and decreases their permeability and absorption [71]. As a result, incorporating SCMs as an OPC blend in structures reduces the use of OPC, thereby avoiding carbon emissions and saving energy associated with its production.

6.4. Innovative Construction Techniques for Impact Mitigation

The construction industry now has a predominance of materials harming the environment using energy consumption and emissions. Additionally, inefficient construction processes generate significant waste, accounting for 10% and 30% of total landfill waste [72] urging innovation in this field. Along with the search in finding durable and low-impact materials, novel building techniques that promote sustainability by resource efficiency are being evaluated. Both carbon- and energy-intensive construction materials, like steel reinforcement bars and cement, will always play a significant role in the AEC sector. The strategy is to discover new ways to use these materials to mitigate their total impact on the environment. Sustainable construction methods like offsite manufacturing, prefabrication and lean construction are recommended to minimize the waste produced due to in situ construction.

The concept of lean or lean thinking was created and implemented in the automotive industry to help maximize value by reducing waste [104]. Lean manufacturing and assembly processes have been dubbed revolutionary due to their application of lean thinking. The use of these lean manufacturing principles in the construction industry is termed lean construction, aiming to increase productivity while decreasing waste. Ismail and AbdelKareem have demonstrated that construction was the first industry to adopt the lean philosophy, which perceives construction as a process of transformation, flow, and value creation. Lean construction's primary goal is to boost productivity while lowering waste. Integrated project delivery is frequently related to lean construction. It is a process for delivering projects that jointly leverage all stakeholders' knowledge, insights, and abilities to increase product value, maximize efficiency and reduce waste. As a result, the term "Lean" has come to mean "Sustainability" and "any innovative way to improve the efficiency of building design and construction" [73].

6.5. Impact Reduction Using Eco-Friendly Renewable Materials

Sustainable materials may not necessarily satisfy the demand of a less technologically advanced society. When available, renewable products present a great opportunity. Studies show that using locally sourced materials efficiently reduces carbon emissions and energy during the embodied phase of construction [105]. Myers et al. conducted a study which discovered that by substituting renewable components for some traditional building components and materials, embodied energy could be reduced by almost 28%, i.e., 7.5 to 5.4 GJ/m^2 [106].

Hemp is being accepted as a renewable building component, particularly in the United States and Europe, as hemp cultivation is now legally encouraged [29]. For non-load-bearing walls, hemp concrete is more environmentally friendly than conventional concrete panels [107]. Pretot et al. conducted an LCA of a wall made of hemp concrete. They concluded that natural fibers of plant origin, like kenaf, showed promising results to be considered an eco-friendly construction product. According to this author, hemp house buildings in South Africa appear to be Africa's most sustainable structure [107].

Batouli and Zhu found that insulation materials made from kenaf fibers have less environmental impact than synthetic insulation components [108]. Bahranifard et al. study compared earthen and conventional plaster and discovered that earthen plaster made of clay has a significantly lower environmental impact than conventional plaster made of hydraulic lime or Portland cement [76]. Melià et al. concluded that although the production phase was found to have the most significant environmental impact, hemp concrete has a lesser effect on the environment than conventional construction materials. Additionally, the hemp plant's capacity for carbon sequestration via photosynthesis aids in mitigating climate change [109].

7. Status of Economy Growth in Sustainable Construction

Numerous studies indicate that green construction leads to significant economic savings through high employee productivity, enhancing health and safety benefits, and cutting down maintenance, operational and energy costs [77]. Mounting proofs show sustainable buildings economically benefit occupants, operators and building owners. Such buildings mark lower yearly operating costs due to insufficient water, energy, and repair/maintenance churn with operating expenses. Cost savings in this form do not have to be offset by higher initial costs. The initial cost for such a building would be similar to or lower than a typical traditional building through integrated design and innovative use of sustainable materials and equipment [78]. The economy of most nations relies heavily on sustainable building practices (Figure 5) as they have the potential to create jobs, lower energy prices, and enhance the environment. Nations adopt sustainable building methods, knowing how this affects their economies. This chapter will examine the economic impact of sustainable construction in several countries worldwide. Future economic growth could be aided by adopting sustainable building materials and techniques, which we will also consider.

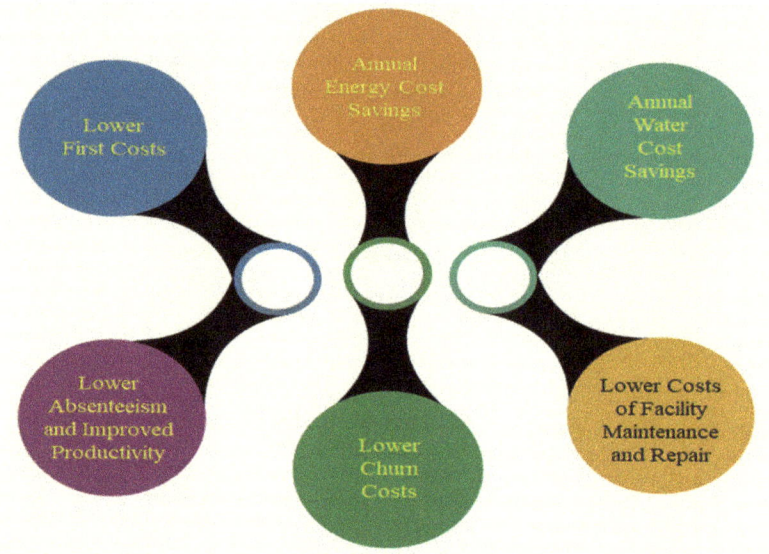

Figure 5. Economy status of sustainable construction (authors compilation).

Along with direct cost savings, sustainable buildings can provide dwelling owners with indirect economic perks just as features of a sustainable building can improve occupants' comfort, health and well-being, lower absenteeism, and increase productivity [79]. Building owners can benefit financially from various advantages, such as lower risks and longer-lasting structures. New chances to attract new employees, decreased costs related to complaint handling, reduced project permitting time and expenses, and community acceptance and support, will generate a consistent valuation of assets. Sustainable buildings also benefit society economically through smaller fees associated with damages related to air pollution and reduced infrastructure costs. Generally, investment is minimal, and the cost of life-cycle time is typically less than traditional buildings [110].

8. Life Cycle Assessment of Sustainable Construction

Life Cycle Assessment (LCA) is a technique for assessing the environmental implications of a product or service over its entire life cycle. It is used to detect and quantify the environmental implications of sustainable construction projects, such as energy usage, water consumption, and waste production. With LCA, we can evaluate the viability of construction projects before their construction. This ensures materials are responsibly obtained, and buildings are constructed to be as energy efficient as feasible. In addition, it enables us to discover areas with the potential to reduce the environmental impact of construction projects. There is a wide range of tools to assess the built environment, energy labelling, material selection, and indoor air quality [111]. In this section, we will examine how LCA may be used to evaluate the sustainability of building projects, as seen in Figure 6, and highlight some of the most important factors to consider when conducting an LCA analysis.

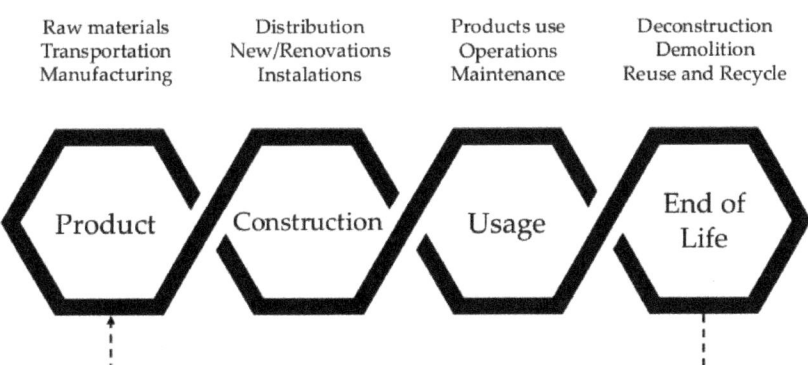

Figure 6. Life Cycle Assessment of Sustainable Construction.

LCA was first used for product comparison in the United States and Europe, but it is now widely used in product design, strategic planning, and government policy. Apart from the environmental assessment provided by LCA, it also delivers a method for reducing environmental impacts using trade-off analysis [112]. LCA is a complex concept focusing on energy, pollutants, and material flow at the inner and outer levels from a life cycle perspective to support better decision-making [113]. The effects in the construction industry occur at various life cycle stages, including manufacturing, processing, functional, and disposal of building materials. Recently, LCA was introduced to assess GHG emissions and embodied energy consumption during the initial stage of construction when using steel, concrete, and wood structural members in any form of building [114,115].

9. Future Study of Sustainable Practice in Construction

Sustainable and environmentally friendly building practices are becoming increasingly important as the global population rises. The challenge is meeting user needs while minimizing construction's negative environmental effects. The present study discusses how future buildings could be made more environmentally friendly, mainly through green construction practices. We also examined some real-world applications of green buildings and discussed how they might be implemented to improve the planet's long-term viability. "Green construction" refers to any technique used to construct a building that minimizes its negative effects on the natural environment. Despite the definition, there is no single approach to a green building; instead, it combines sustainable practices that consider local cultural norms to create a more sustainable future for our planet. These kinds of practices are rising in today's world. Green buildings can be made even more productive and economical with the help of digital transformation techniques like 3D printing and optimization strategies for design. This method of building has the potential to alter the construction industry pushing for higher qualitative and environmental standards. Architects can save time, energy and cut costs using 3D printing technology to create buildings bioclimatic suited to each location. These digital transformation methods pave the way toward environmentally responsible building practices as they become widely available worldwide.

10. Conclusions

After examining several studies, the following findings were developed. Nowadays, there is a massive disparity in sustainable design and construction research. A more comprehensive strategy must be devised to appreciate the interaction between urban design, buildings, building systems, and materials. Along with being spread throughout the building delivery process, from planning and design, to construction, operation and maintenance, this understanding is critical. The main objectives of this study were to comprehend sustainable building and its benefits. According to the study's findings, utilizing environmentally friendly materials and technologies can reduce the environmental impact that traditional buildings have on the environment, the economy, and people. New buildings should use resources like energy, water, and locally sourced materials more efficiently than in the past. Sustainable structures can acquire larger amounts of natural light and improve ventilation resulting in healthier indoor spaces. Plus, they incorporate high-performance systems, efficient rainwater collection equipment and harness renewable energy sources. The holistic approach creates a building with a reduced carbon footprint and lower energy consumption. Tools and rating systems like life cycle assessment, must be used during the process as they are key to understanding and implementing a sustainable approach to the construction industry.

Author Contributions: Conceptualization, D.S.V., P.D. and A.S. (Arvindan Sivasuriyan); methodology, E.K., A.S. (Anna Stefańska) and C.C.D.; validation, A.S. (Arvindan Sivasuriyan), E.K., M.D.V. and N.D.C.; investigation, D.S.V., P.D., M.D.V., A.J. and J.W.; resources, D.S.V., P.D., A.J., M.D.V., J.W. and C.C.D.; data curation, A.S. (Arvindan Sivasuriyan), A.S. (Anna Stefańska) and C.C.D.; writing—original draft preparation, D.S.V., P.D. and A.S. (Arvindan Sivasuriyan), A.S. (Anna Stefańska), E.K., A.J., M.D.V., J.W., N.D.C. and C.C.D.; writing—review and editing, A.S. (Anna Stefańska) and C.C.D.; visualisation, D.S.V., P.D. and A.S. (Arvindan Sivasuriyan); supervision, A.S. (Arvindan Sivasuriyan), E.K. and N.D.C.; project administration, A.S. (Arvindan Sivasuriyan), A.S. (Anna Stefańska), E.K. and N.D.C. All authors have read and agreed to the published version of the manuscript.

Funding: This research received no external funding.

Institutional Review Board Statement: Not applicable.

Informed Consent Statement: Not applicable.

Data Availability Statement: The data presented in this study are available on request from the corresponding author and shared after his consideration.

Conflicts of Interest: The authors declare no conflict of interest.

References

1. Pacheco-Torgal, F. *Eco-Efficient Construction and Building Materials*; Woodhead Publishing: Cambridge, UK, 2011. [CrossRef]
2. Windapo, A.; Ogunsanmi, O. Construction sector views of sustainable building materials. *Proc. Inst. Civ. Eng. Eng. Sustain.* **2014**, *167*, 64–75. [CrossRef]
3. Song, Y.; Zhang, H. Research on sustainability of building materials. *IOP Conf. Ser. Mater. Sci. Eng.* **2018**, *452*, 022169. [CrossRef]
4. Kumar Sharma, N. Sustainable Building Material for Green Building Construction, Conservation and Refurbishing. *Int. J. Adv. Sci. Technol.* **2020**, *29*, 5343–5350. Available online: https://www.researchgate.net/publication/342946652 (accessed on 15 January 2023).
5. Muthusamy, M.; Subburaj, R.G.; Shanmughan, S.; Arunachalam, S.; Raja, S. Go green by 'cement less technology in construction industry': A review. *AIP Conf. Proc.* **2020**, *2240*, 060003. [CrossRef]
6. Li, L.; Liu, W.; You, Q.; Chen, M.; Zeng, Q. Waste ceramic powder as a pozzolanic supplementary filler of cement for developing sustainable building materials. *J. Clean. Prod.* **2020**, *259*, 120853. [CrossRef]
7. Lachauer, L.; Kotnik, T. Geometry of Structural form. In *Advacements in Architectural Geometry 2010*; Cecato, C., Hesselgren, L., Pauly, M., Pottmann, H., Wallner, J., Eds.; Springer: Vienna, Austria, 2010; pp. 193–203. [CrossRef]
8. Pawlik, E.; Ijomah, W.; Corney, J.; Powell, D. Exploring the Application of Lean Best Practices in Remanufacturing: Empirical Insights into the Benefits and Barriers. *Sustainability* **2022**, *14*, 149. [CrossRef]
9. Vaverková, M.D.; Radziemska, M.; Bartoň, S.; Cerdà, A.; Koda, E. The use of vegetation as a natural strategy for landfill restoration. *L. Degrad. Dev.* **2018**, *29*, 3674–3680. [CrossRef]
10. Kobylarczyk, J.; Marchwiński, J. Pluralism of goals of proecological architecture. *Bud. I Archit.* **2020**, *19*, 5–14. [CrossRef]
11. Huynh, T.P.; Nguyen, T.C.; Do, N.D.; Hwang, C.L.; Bui, L.A.T. Strength and thermal properties of unfired four-hole hollow bricks manufactured from a mixture of cement, low-calcium fly ash and blended fine aggregates. *IOP Conf. Ser. Mater. Sci. Eng.* **2019**, *625*, 012010. [CrossRef]
12. Klemm, A.; Wiggins, D. *Sustainability of Natural Stone as a Construction Material*, 2nd ed.; Elsevier Ltd.: Amsterdam, The Netherlands, 2016. [CrossRef]
13. Vale, B. *Materials and Buildings*; Elsevier Ltd.: Amsterdam, The Netherlands, 2017. [CrossRef]
14. Yildirim, N.; Erdonmez, F.S.; Ozen, E.; Avci, E.; Yeniocak, M.; Acar, M.; Dalkilic, B.; Ergun, M.E. *Fire-Retardant Bioproducts for Green Buildings*; Woodhead Publishing: Cambridge, UK, 2020. [CrossRef]

15. Pacheco-Torgal, F.; Ivanov, V.; Tsang, D. *Bio-Based Materials and Biotechnologies for Eco-Efficient Construction*; Woodhead Publishing: Cambridge, UK, 2020.
16. Dipasquale, L.; Rovero, L.; Fratini, F. *Ancient Stone Masonry Constructions*; Elsevier: Amsterdam, The Netherlands, 2016. [CrossRef]
17. Correia, V.C.; Santos, S.F.; Tonoli, G.H.D.; Savastano, H. *Characterization of Vegetable Fibers and Their Application in Cementitious Composites*; Elsevier: Amsterdam, The Netherlands, 2016. [CrossRef]
18. Boas Berg, A.; Černý, M.; Hurajová, E.; Winkler, J. Anthropogenic ecosystem of green roofs from the perspective of rainwater management. *ACTA Sci. Pol.—Archit. Bud.* **2022**, *21*, 9–19. [CrossRef]
19. Ellenberg, H.; Weber, H.E.; Düll, R.; Wirth, V.; Werner, W.; Paulißen, D. *Zeigwerte von Pflanzen in MittelEuropa (Pointer Values of Plants in Central Europe)*; Scripta Geobotanica; Verlag Wrich Goltze: Goöttingen, Germany, 1992; pp. 1–248.
20. Koda, E.; Winkler, J.; Wołkonowicz, P.; Černý, M.; Kiersnowska, A.; Pasternak, G.; Vaverková, M.D. Vegetation changes as indicators of landfill leachate seepage locations: Case study. *Ecol. Eng.* **2022**, *174*, 106448. [CrossRef]
21. Zuo, J.; Zhao, Z.Y. Green building research-current status and future agenda: A review. *Renew. Sustain. Energy Rev.* **2014**, *30*, 271–281. [CrossRef]
22. Born, B.; Purcell, M. Avoiding the Local Trap: Scale and Food Systems in Planning Research. *J. Plan. Educ. Res.* **2016**, *26*, 195–207. [CrossRef]
23. Sieczka, A.; Bujakowski, F.; Falkowski, T.; Koda, E. Morphogenesis of a floodplain as a criterion for assessing the susceptibility to water pollution in an agriculturally rich valley of a lowland river. *Water* **2018**, *10*, 399. [CrossRef]
24. Nogeire-McRae, T.; Ryan, E.P.; Jablonski, B.; Carolan, M.; Arathi, H.S.; Brown, C.S.; Saki, H.H.; McKeen, S.; Lapansky, E.; Schipanski, M.E. The role of urban agriculture in a secure, healthy, and sustainable food system. *Bioscience* **2018**, *68*, 748–759. [CrossRef]
25. Medjo Eko, R.; Offa, E.D.; Yatchoupou Ngatcha, T.; Seba Minsili, L. Potential of salvaged steel fibers for reinforcement of unfired earth blocks. *Constr. Build. Mater.* **2012**, *35*, 340–346. [CrossRef]
26. Fini, A.A.F.; Akbarnezhad, A. *Sustainable Procurement and Transport of Construction Materials*; Elsevier Inc.: Amsterdam, The Netherlands, 2019. [CrossRef]
27. Koda, E.; Podlasek, A. Sustainable Use of Construction and Demolition Wastes in a Circular Economy Perspective. In *Sustainable and Circular Management of Resources and Waste towards a Green Deal*; Prasad, M.N.V., Smol, M., Eds.; Elsevier: Amsterdam, The Netherlands, 2023.
28. Oti, J.E.; Kinuthia, J.M.; Bai, J. Design thermal values for unfired clay bricks. *Mater. Des.* **2010**, *31*, 104–112. [CrossRef]
29. Gołębiewski, M.; Pietruszka, B. Risk of interstitial condensation in outer walls made of hemp-lime composite in Polish climatic conditions. *Arch. Civ. Eng.* **2021**, *67*, 107–121. [CrossRef]
30. Ogunsanya, O.A.; Aigbavboa, C.O.; Thwala, D.W.; Edwards, D.J. Barriers to sustainable procurement in the Nigerian construction industry: An exploratory factor analysis. *Int. J. Constr. Manag.* **2022**, *22*, 861–872. [CrossRef]
31. Potocnik, J. Towards a Circular Economy: Business Rationale for an Accelerated Transition. 2015. Available online: https://kidv.nl/media/rapportages/towards_a_circular_economy.pdf?1.2.1 (accessed on 17 January 2023).
32. Tam, V.W.Y.; Le, K.N. *Sustainable Constructio Technologies Life-Cycle Assessment*; Elsevier: Oxford, UK; Cambridge, MA, USA, 2019. Available online: https://ebin.pub/sustainable-construction-technologies-life-cycle-assessment-0128117494-9780128117491-0128117491.html (accessed on 15 February 2023).
33. Tabrizikahou, A.; Nowotarski, P. Mitigating the energy consumption and the carbon emission in the building structures by optimization of the construction processes. *Energies* **2021**, *14*, 3287. [CrossRef]
34. Rybak-Niedziółka, K.; Starzyk, A.; Łacek, P.; Mazur, Ł.; Myszka, I.; Stefańska, A.; Kurcjusz, M.; Nowysz, A.; Langie, K. Use of Waste Building Materials in Architecture and Urban Planning—A Review of Selected Examples. *Sustainability* **2023**, *15*, 5047. [CrossRef]
35. Akinwumi, I.I.; Domo-Spiff, A.H.; Salami, A. Marine plastic pollution and affordable housing challenge: Shredded waste plastic stabilized soil for producing compressed earth bricks. *Case Stud. Constr. Mater.* **2019**, *11*, e00241. [CrossRef]
36. Moussa, H.S.; Nshimiyimana, P.; Hema, C.; Zoungrana, O.; Messan, A.; Courard, L. Comparative Study of Thermal Comfort Induced from Masonry Made of Stabilized Compressed Earth Block vs Conventional Cementitious Material. *J. Miner. Mater. Charact. Eng.* **2019**, *7*, 385–403. [CrossRef]
37. Espuelas, S.; Omer, J.; Marcelino, S.; Echeverría, A.M.; Seco, A. Magnesium oxide as alternative binder for unfired clay bricks manufacturing. *Appl. Clay Sci.* **2017**, *146*, 23–26. [CrossRef]
38. Werkheiser, I.; Piso, Z. People Work to Sustain Systems: A Framework for Understanding Sustainability. *J. Water Resour. Plan. Manag.* **2015**, *141*, A4015002. [CrossRef]
39. Porter, H.; Blake, J.; Dhami, N.K.; Mukherjee, A. Rammed earth blocks with improved multifunctional performance. *Cem. Concr. Compos.* **2018**, *92*, 36–46. [CrossRef]
40. Gandia, R.M.; Gomes, F.C.; Corrêa, A.A.R.; Rodrigues, M.C.; Mendes, R.F. Physical, mechanical and thermal behavior of adobe stabilized with glass fiber reinforced polymer waste. *Constr. Build. Mater.* **2019**, *222*, 168–182. [CrossRef]
41. Xie, M.; Gao, D.; Liu, X.; Li, F.; Huang, C. Utilization of waterworks sludge in the production of fired/unfired water permeable bricks. *Adv. Mater. Res.* **2012**, *531*, 316–319. [CrossRef]
42. Maria, G.; Bristogianni, T.; Oikonomopoulou, F.; Rigone, P.; Enrico Sergio, M. Recycled Glass Mixtures as Cast Glass Components for Structural Applications, towards Sustainability. *Chall. Glass* **2020**, *7*. [CrossRef]

43. Bardhan, S. Assessment of water resource consumption in building construction in India. *WIT Trans. Ecol. Environ.* **2011**, *144*, 93–101. [CrossRef]
44. Achal, V.; Pan, X.; Lee, D.J.; Kumari, D.; Zhang, D. Remediation of Cr(VI) from chromium slag by biocementation. *Chemosphere* **2013**, *93*, 1352–1358. [CrossRef] [PubMed]
45. Miqueleiz, L.; Ramirez, F.; Oti, J.E.; Seco, A.; Kinuthia, J.M.; Oreja, I.; Urmeneta, P. Alumina filler waste as clay replacement material for unfired brick production. *Eng. Geol.* **2013**, *163*, 68–74. [CrossRef]
46. Ojo, E.B.; Bello, K.O.; Mustapha, K.; Teixeira, R.S.; Santos, S.F.; Savastano, H. Effects of fibre reinforcements on properties of extruded alkali activated earthen building materials. *Constr. Build. Mater.* **2019**, *227*, 116778. [CrossRef]
47. Silva, M.; Naik, T.R. Sustainable use of resources—Recycling of sewage treatment plant water in concrete. In Proceedings of the Second International Conference on Sustainable Construction Materials and Technologies, Ancona, Italy, 28–30 June 2010; pp. 1731–1740.
48. Thornback, J.; Snowden, C.; Anderson, J.; Foster, C. *Water Efficiency—The Contribution of Construction Products*; Construction Products Association: London, UK, 2015.
49. Türkmen, İ.; Ekinci, E.; Kantarcı, F.; Sarıcı, T. The mechanical and physical properties of unfired earth bricks stabilized with gypsum and Elazığ Ferrochrome slag. *Int. J. Sustain. Built Environ.* **2017**, *6*, 565–573. [CrossRef]
50. Chan, C.M. Effect of natural fibres inclusion in clay bricks: Physico-mechanical properties. *World Acad. Sci. Eng. Technol.* **2011**, *73*, 51–57.
51. Lima, S.A.; Varum, H.; Sales, A.; Neto, V.F. Analysis of the mechanical properties of compressed earth block masonry using the sugarcane bagasse ash. *Constr. Build. Mater.* **2012**, *35*, 829–837. [CrossRef]
52. Namango, S. Development of Cost-Effective Earthen Building Material for Housing Wall Construction: Investigations into the Properties of Compressed Earth Blocks Stabilized with Sis. 2016. Available online: https://www.researchgate.net/publication/283068265_Development_of_cost-effective_earthen_building_material_for_housing_wall_construction (accessed on 19 January 2023).
53. NSW Government. Soils and Construction. 2004. Available online: www.environment.nsw.gov.au/resources/water/BlueBookV1Chapters.pdf (accessed on 15 January 2023).
54. Burkhardt, M.; Zuleeg, S.; Vonbank, R.; Schmid, P.; Hean, S.; Lamani, X.; Bester, K.; Boller, M. Leaching of additives from construction materials to urban storm water runoff. *Water Sci. Technol.* **2011**, *63*, 1974–1982. [CrossRef]
55. Spigarelli, M. *10 Ways to Save Water in Commercial Buildings*; CCJM Engineers Ltd.: Chicago, IL, USA, 2012.
56. Zamri, M.F.M.A.; Bahru, R.; Amin, R.; Khan, M.U.A.; Razak, S.I.A.; Abu Hassan, S.; Kadir, M.R.A.; Nayan, N.H.M. Waste to health: A review of waste derived materials for tissue engineering. *J. Clean. Prod.* **2021**, *290*, 125792. [CrossRef]
57. Demir, I. An investigation on the production of construction brick with processed waste tea. *Build. Environ.* **2006**, *41*, 1274–1278. [CrossRef]
58. Sakhare, V.V.; Ralegaonkar, R.V. Use of bio-briquette ash for the development of bricks. *J. Clean. Prod.* **2016**, *112*, 684–689. [CrossRef]
59. Demir, I. Effect of organic residues addition on the technological properties of clay bricks. *Waste Manag.* **2008**, *28*, 622–627. [CrossRef] [PubMed]
60. Lamrani, M.; Mansour, M.; Laaroussi, N.; Khalfaoui, M. Thermal study of clay bricks reinforced by three ecological materials in south of Morocco. *Energy Procedia* **2019**, *156*, 273–277. [CrossRef]
61. Dove, C. The development of unfired earth bricks using seaweed biopolymers. *WIT Trans. Built Environ.* **2014**, *142*, 219–230. [CrossRef]
62. Batagarawa, A.L.; Abodunrin, J.A.; Sagada, M.L. *Enhancing the Thermophysical Properties of Rammed Earth by Stabilizing with Corn Husk Ash*; Springer: Cham, Switzerland, 2019; pp. 405–413. [CrossRef]
63. Ayodele, A.L.; Oketope, O.M.; Olatunde, O.S. Effect of sawdust ash and eggshell ash on selected engineering properties of lateralized bricks for low cost housing. *Niger. J. Technol.* **2019**, *38*, 278. [CrossRef]
64. Khan, M.S.; Sohail, M.; Khattak, N.S.; Sayed, M. Industrial ceramic waste in Pakistan, valuable material for possible applications. *J. Clean. Prod.* **2016**, *139*, 1520–1528. [CrossRef]
65. Achanzar, W.; Mangipudy, R. Acrylonitrile. In *Encyclopedia of Toxicology*, 3rd ed.; Academic Press: Cambridge, MA, USA, 2014; pp. 76–78. [CrossRef]
66. Krystsina, B.; Natallia, Y. Ammonia migration from building materials to indoor air; Technical, economical and environmental aspects. *E3S Web Conf.* **2019**, *136*, 8–11. [CrossRef]
67. Edmundson, M.C.; Horsfall, L. Construction of a modular arsenic-resistance operon in E. coli and the production of arsenic nanoparticles. *Front. Bioeng. Biotechnol.* **2015**, *3*, 160. [CrossRef]
68. Widyatmoko, D. Sustainability of Bituminous Materials. In *Sustainability of Construction Materials*; Khatib, J.M., Ed.; Woodhead Publishing: Cambridge, UK, 2016. [CrossRef]
69. Rajadesingu, S.; Arunachalam, K.D. Pulverization and characterizationof nano borax decahydrateand shielding efficiency of gamma and neutron radiation in bio-caulk enriched high-performance concrete. *Mater. Lett.* **2021**, *302*, 130400. [CrossRef]
70. Lin, H.; Shi, J.; Dong, Y.; Li, B.; Yin, T. Construction of bifunctional bacterial community for co-contamination remediation: Pyrene biodegradation and cadmium biomineralization. *Chemosphere* **2022**, *304*, 135319. [CrossRef]

71. Ghazi, A.B.; Jamshidi-Zanjani, A.; Nejati, H. Clinkerisation of copper tailings to replace Portland cement in concrete construction. *J. Build. Eng.* **2022**, *51*, 104275. [CrossRef]
72. Daneshvar, D.; Deix, K.; Robisson, A. Effect of casting and curing temperature on the interfacial bond strength of epoxy bonded concretes. *Constr. Build. Mater.* **2021**, *307*, 124328. [CrossRef]
73. Ismail, Z.Z.; AbdelKareem, H.N. Sustainable approach for recycling waste lamb and chicken bones for fluoride removal from water followed by reusing fluoride-bearing waste in concrete. *Waste Manag.* **2015**, *45*, 66–75. [CrossRef] [PubMed]
74. Saedi, A.; Jamshidi-Zanjani, A.; Darban, A.K.; Mohseni, M.; Nejati, H. Utilization of lead–zinc mine tailings as cement substitutes in concrete construction: Effect of sulfide content. *J. Build. Eng.* **2022**, *57*, 104865. [CrossRef]
75. Sahu, S.S.; Gandhi, I.S.R. Studies on influence of characteristics of surfactant and foam on foam concrete behaviour. *J. Build. Eng.* **2021**, *40*, 102333. [CrossRef]
76. Bahranifard, Z.; Vosoughi, A.R.; Farshchi Tabrizi, F.; Shariati, K. Effects of water-cement ratio and superplasticizer dosage on mechanical and microstructure formation of styrene-butyl acrylate copolymer concrete. *Constr. Build. Mater.* **2022**, *318*, 125889. [CrossRef]
77. Al-Kheetan, M.J.; Rahman, M.M.; Ghaffar, S.H.; Al-Tarawneh, M.; Jweihan, Y.S. Comprehensive investigation of the long-term performance of internally integrated concrete pavement with sodium acetate. *Results Eng.* **2020**, *6*, 100110. [CrossRef]
78. Siddique, R.; Singh, M.; Mehta, S.; Belarbi, R. Utilization of treated saw dust in concrete as partial replacement of natural sand. *J. Clean. Prod.* **2020**, *261*, 121226. [CrossRef]
79. Abera, Y.S.A. Performance of concrete materials containing recycled aggregate from construction and demolition waste. *Results Mater.* **2022**, *14*, 100278. [CrossRef]
80. Barbieri, L.; Andreola, F.; Lancellotti, I.; Taurino, R. Management of agricultural biomass wastes: Preliminary study on characterization and valorisation in clay matrix bricks. *Waste Manag.* **2013**, *33*, 2307–2315. [CrossRef]
81. Eliche-Quesada, D.; Azevedo-Da Cunha, R.; Corpas-Iglesias, F.A. Effect of sludge from oil refining industry or sludge from pomace oil extraction industry addition to clay ceramics. *Appl. Clay Sci.* **2015**, *114*, 202–211. [CrossRef]
82. Shibib, K.S. Effects of waste paper usage on thermal and mechanical properties of fired brick. *Heat Mass Transf. und Stoffuebertragung* **2015**, *51*, 685–690. [CrossRef]
83. Njeumen Nkayem, D.E.; Mbey, J.A.; Kenne Diffo, B.B.; Njopwouo, D. Preliminary study on the use of corn cob as pore forming agent in lightweight clay bricks: Physical and mechanical features. *J. Build. Eng.* **2016**, *5*, 254–259. [CrossRef]
84. Subashi De Silva, G.H.M.J.; Hansamali, E. Eco-friendly fired clay bricks incorporated with porcelain ceramic sludge. *Constr. Build. Mater.* **2019**, *228*, 116754. [CrossRef]
85. He, H.; Yue, Q.; Su, Y.; Gao, B.; Gao, Y.; Wang, J.; Yu, H. Preparation and mechanism of the sintered bricks produced from Yellow River silt and red mud. *J. Hazard. Mater.* **2012**, *203–204*, 53–61. [CrossRef]
86. Salleh, S.Z.; Kechik, A.A.; Yusoff, A.H.; Taib, M.A.A.; Nor, M.M.; Mohamad, M.; Tan, T.G.; Ali, A.; Masri, M.N.; Mohamed, J.J.; et al. Recycling food, agricultural, and industrial wastes as pore-forming agents for sustainable porous ceramic production: A review. *J. Clean. Prod.* **2021**, *306*, 127264. [CrossRef]
87. Fernandes, F.M.; Lourenço, P.B.; Castro, F. Ancient Clay Bricks: Manufacture and Properties. In *Materials, Technologies and Practice in Historic Heritage Structures*; Springer: Berlin/Heidelberg, Germany, 2010. [CrossRef]
88. Goel, G.; Kalamdhad, A.S. Degraded municipal solid waste as partial substitute for manufacturing fired bricks. *Constr. Build. Mater.* **2017**, *155*, 259–266. [CrossRef]
89. Raut, S.P.; Ralegaonkar, R.V.; Mandavgane, S.A. Development of sustainable construction material using industrial and agricultural solid waste: A review of waste-create bricks. *Constr. Build. Mater.* **2011**, *25*, 4037–4042. [CrossRef]
90. Juel, M.A.I.; Mizan, A.; Ahmed, T. Sustainable use of tannery sludge in brick manufacturing in Bangladesh. *Waste Manag.* **2017**, *60*, 259–269. [CrossRef]
91. Adazabra, A.N.; Viruthagiri, G.; Kannan, P. Influence of spent shea waste addition on the technological properties of fired clay bricks. *J. Build. Eng.* **2017**, *11*, 166–177. [CrossRef]
92. Sutcu, M.; Ozturk, S.; Yalamac, E.; Gencel, O. Effect of olive mill waste addition on the properties of porous fired clay bricks using Taguchi method. *J. Environ. Manag.* **2016**, *181*, 185–192. [CrossRef]
93. Ornam, K.; Kimsan, M.; Ngkoimani, L.O.; Santi. Study on Physical and Mechanical Properties with Its Environmental Impact in Konawe—Indonesia upon Utilization of Sago Husk as Filler in Modified Structural Fly Ash—Bricks. *Procedia Comput. Sci.* **2017**, *111*, 420–426. [CrossRef]
94. Madlool, N.A.; Saidur, R.; Rahim, N.A. Investigation of Waste Heat Recovery in Cement Industry: A Case Study. *Int. J. Eng. Technol.* **2012**, *4*, 665–667. [CrossRef]
95. Sizirici, B.; Fseha, Y.; Cho, C.S.; Yildiz, I.; Byon, Y.J. A review of carbon footprint reduction in construction industry, from design to operation. *Materials* **2021**, *14*, 6094. [CrossRef]
96. International Finance Corporation (IFC). *Improving Thermal and Electric Energy Efficiency at Cement Plants: International Best Practice*; International Finance Corporation (IFC): Washington, DC, USA, 2017.
97. García-Segura, T.; Yepes, V.; Alcalá, J. Life cycle greenhouse gas emissions of blended cement concrete including carbonation and durability. *Int. J. Life Cycle Assess.* **2014**, *19*, 3–12. [CrossRef]
98. Rutkowska, G.; Chalecki, M.; Żółtowski, M. Fly ash from thermal conversion of sludge as a cement substitute in concrete manufacturing. *Sustainability* **2021**, *13*, 4182. [CrossRef]

99. Rajdev, R.; Yadav, S.; Sakale, R. Comparison between Portland Pozzolana Cement & Processed Fly Ash blended Ordinary Portland Cement. *Int. Conf. Recent Trends Appl. Sci. Engg. Appl.* **2013**, *3*, 24–30.
100. World Steel Association (WSA). Fact Sheet: Energy Use in the Steel Industry. 2013, pp. 1–3. Available online: https://www.steel.org.au/getattachment/bfc722b0-1320-4d04-84df-30bdd500aa5b/fact_energy_2016.pdf (accessed on 17 January 2023).
101. World Steel Association. Steel's Contribution to a Low Carbon Future and Climate Resilient Societies. Worldsteel Position Paper. 2020, pp. 1–6. Available online: https://www.acero.org.ar/wp-content/uploads/2020/02/Position_paper_climate_2020_vfinal.pdf (accessed on 16 January 2023).
102. Hammond, G.; Jones, C. The Inventory of Carbon and Energy (ICE). In *A BRIA Guide Embodied Carbon*; Lowrie, F., Tse, P., Eds.; Sustainalbe Energy Research Team, University of Bath: Bath, UK, 2014; pp. 22, 25, 66. Available online: https://greenbuildingencyclopaedia.uk/wp-content/uploads/2014/07/Full-BSRIA-ICE-guide.pdf (accessed on 17 January 2023).
103. Grimaud, G.; Perry, N.; Laratte, B. Life Cycle Assessment of Aluminium Recycling Process: Case of Shredder Cables. *Procedia CIRP* **2016**, *48*, 212–218. [CrossRef]
104. Aziz, R.F.; Hafez, S.M. Applying lean thinking in construction and performance improvement. *Alexandria Eng. J.* **2013**, *52*, 679–695. [CrossRef]
105. Venkatarama Reddy, B.V.; Prasanna Kumar, P. Embodied energy in cement stabilised rammed earth walls. *Energy Build.* **2010**, *42*, 380–385. [CrossRef]
106. Myers, F.; Fuller, R.; Crawford, R.H. The potential to reduce the embodied energy in construction through the use of renewable materials. In Proceedings of the ASA 2012: The 46th Annual Conference of the Architectural Science Association (Formerly ANZAScA)—Building on Knowledge: Theory and Practice, Gold Coast, Australia, 14–16 November 2012; pp. 1–8.
107. Pretot, S.; Collet, F.; Garnier, C. Life cycle assessment of a hemp concrete wall: Impact of thickness and coating. *Build. Environ.* **2014**, *72*, 223–231. [CrossRef]
108. Batouli, S.M.; Zhu, Y. Comparative life-cycle assessment study of kenaf fiber-based and glass fiber-based structural insulation panels. ICCREM 2013: Construction and Operation in the Context of Sustainability. In Proceedings of the International Conference on Construction and Real Estate Management, Karlsruhe, Germany, 10–11 October 2013; pp. 377–388. [CrossRef]
109. Melià, P.; Ruggieri, G.; Sabbadini, S.; Dotelli, G. Environmental impacts of natural and conventional building materials: A case study on earth plasters. *J. Clean. Prod.* **2014**, *80*, 179–186. [CrossRef]
110. Goh, B.H.; Sun, Y. The development of life-cycle costing for buildings. *Build. Res. Inf.* **2016**, *44*, 319–333. [CrossRef]
111. Zabalza Bribián, I.; Valero Capilla, A.; Aranda Usón, A. Life cycle assessment of building materials: Comparative analysis of energy and environmental impacts and evaluation of the eco-efficiency improvement potential. *Build. Environ.* **2011**, *46*, 1133–1140. [CrossRef]
112. Samad, M.H.A.; Yahya, H.Y. Life Cycle Analysis of Building Materials. In *Renewable Energy and Sustainable Technologies for Building and Environmental Applications: Options for a Greener Future*; Springer: Cham, Switzerland, 2016; pp. 1–252. [CrossRef]
113. Vijayan, D.S.; Parthiban, D. Effect of Solid waste based stabilizing material for strengthening of Expansive soil—A review. *Environ. Technol. Innov.* **2020**, *20*, 101108. [CrossRef]
114. Parthiban, D.; Vijayan, D.S.; Koda, E.; Vaverkova, M.D.; Piechowicz, K.; Osinski, P.; Van Duc, B. Role of industrial based precursors in the stabilization of weak soils with geopolymer—A review. *Case Stud. Constr. Mater.* **2022**, *16*, e00886. [CrossRef]
115. Smol, M.; Marcinek, P.; Koda, E. Drivers and Barriers for a Circular Economy (CE) Implementation in Poland—A Case Study of Raw Materials Recovery Sector. *Energies* **2021**, *14*, 2219. [CrossRef]

Disclaimer/Publisher's Note: The statements, opinions and data contained in all publications are solely those of the individual author(s) and contributor(s) and not of MDPI and/or the editor(s). MDPI and/or the editor(s) disclaim responsibility for any injury to people or property resulting from any ideas, methods, instructions or products referred to in the content.

Article

Assessment of Economic Sustainability in the Construction Sector: Evidence from Three Developed Countries (the USA, China, and the UK)

Wesam Salah Alaloul [1], Muhammad Ali Musarat [1,*], Muhammad Babar Ali Rabbani [2], Muhammad Altaf [1], Khalid Mhmoud Alzubi [1] and Marsail Al Salaheen [1]

1 Department of Civil and Environmental Engineering, Universiti Teknologi PETRONAS, Bandar Seri Iskandar 32610, Malaysia; wesam.alaloul@utp.edu.my (W.S.A.); muhammad_20000250@utp.edu.my (M.A.); khalid_20001254@utp.edu.my (K.M.A.); marsail_20001253@utp.edu.my (M.A.S.)
2 College of Science & Engineering, Flinders University, Adelaide 5042, Australia; babaralirabbani@yahoo.com
* Correspondence: muhammad_19000316@utp.edu.my

Abstract: The construction sector plays a significant role in contributing to uplifts in economic stability by generating employment and providing standardized social development. Economic sustainability in the construction sector has been less addressed despite its wide applicability in the economy. This study aimed to perform a comparative analysis to determine the application of a circular economy in the construction sector toward economic sustainability, along with its long-term forecasting. A time series analysis was used on the construction sector of the United States of America (USA), China, and the United Kingdom (UK) from 1970 to 2020, by taking into account individual effects to propose a framework with global validity. Statistical analysis was performed to analyze the dependence of the construction sector and determine its short- and long-term contributions. The results revealed that the construction sectors in these countries tend to bounce back to equilibrium in the case of short-term effects; however, the construction sector behaves differently with respect to each sector after experiencing long-term effects. The results show that the explanatory power of the forecasting model (R^2) was found to be 0.997, 0.992, and 0.996 for the USA, China, and the UK. Based on the concept of the circular economy, it was concluded that the USA will become a leader in attaining sustainability in construction owing to its ability to recover quickly from shocks, and that the USA will become the largest construction sector in terms of GDP, with a USD 0.3 trillion higher GDP than that of the Chinese sector. Meanwhile, there will be no significant change in the construction GDP of the UK up to the end of 2050. Moreover, the speeds of the construction sector toward equilibrium in the long run in the USA, China, and the UK, and regaining of their original positions, is 0.267%, 1.04%, and 0.41% of their original positions, respectively. This study has a significance in acting as a guideline for introducing economic and environmental sustainability in construction policies, because of the potential of the construction sectors to recover from possible recessions in their respective countries.

Keywords: econometric analysis; construction sector; sustainable construction; circular economy; forecasting

1. Introduction

The activities of each economic sector in a country contribute in many ways to the national gross domestic product (GDP). One of these sectors is construction, which has a deciding role in socioeconomic development by providing infrastructures, transport, employment opportunities, energy demand, telecommunications, and investments [1]. The construction sector is one of the most complex and dynamic sectors of the economy. It contributes to the sustainable objectives of the country, including revenue generation, employment opportunities, and social needs; therefore, an analysis of the construction sector's effect on the GPD is necessary [2]. The role of the construction sector in cumulative GDP was analyzed, and it was found that the sector is greatly affected by a lack of privatization, skilled labor, inaccessible immigration rules, and the influences of bureaucrats [3]. In the early support of the economy, construction has an influential role in its capability to uplift the GDP of the country and is responsible for its modernization because of its role in improving infrastructure [4]. This sector is regarded as the backbone of the country because it influences every level of the economy [5]. It also has the potential to uplift the economy because it does not only include construction projects but also technological and social change, client demands, and the increasing use of every sector in its execution [6]. With the construction sector making use of resources from all other sectors of the economy, it is therefore considered a driving factor toward prosperity in every country [7]. Owing to the importance of the construction sector, its influences have forward and backward linkages with other sectors. Any negative change in its performance will produce a recession in the economy [8].

For example, in 2018, construction activities in Turkey decreased by 4.8% due to high-interest rates on construction activities and, as a result, the country suffered major losses in the construction sector [9]. Table 1 presents the values of construction and percentage change in various developed countries in 2018–2019 and the corresponding employment levels.

Table 1. Construction output of developed countries for 2018–2019.

Country	Value Added in Construction (Current Price, Trillion USD)		Percent Change in Construction (Current Price, Trillion USD)		People Employed in Construction (Millions)	Reference
	2018	2019	2018	2019		
Japan	0.281	0.284	1.33%	1.31%	2.93	[10]
Canada	0.126	0.126	3.78%	0.34%	1.20	[11]
Germany	0.175	0.186	12.27%	6.59%	2.13	[12]
France	0.138	0.139	8.55%	0.46%	1.51	[13]
Italy	0.079	0.076	6.15%	−3.38%	1.33	[14]
Australia	0.144	0.142	1.62%	−1.41%	1.10	[15]
Finland	0.017	0.017	11.87%	−1.75%	1.83	[16]
UK	0.123	0.129	3.05%	5.28%	2.30	[17]
USA	0.848	0.892	6.32%	5.25%	9.08	[18]

1.1. Construction Sector in the USA, China, and the UK

The construction sector of the United States of America (USA) is regarded as one of the largest marketplaces all over the world [19]. Through its linkages, the US construction sector supports investments, transportation, manufacturing, growth, output, and other related building material industries; moreover, it provides jobs to plenty of workers, which generate income opportunities and reduce poverty [20]. The job losses in workers of all classes in the sector increased toward the end of the first quarter of 2020, and then decreased toward the end of all the remaining quarters of 2020 [21].

In the United Kingdom (UK), the construction sector likewise contributes considerably to the national economy. The UK construction sector has a diverse range of sectors, such as manufacturing, mining, and services. Considered one of the largest sectors of the UK, construction employed over 9% of the workforce in 2019 [22]. The construction sector contributes an estimated 15.3% to the national GDP of the UK and has an economic generation of 6% [23]. Furthermore, it employs 2.3 million people, which is 7.1% of the entire UK workforce, because the average wages of construction sector workers are 5% higher than those of workers in other sectors [24]. With the UK construction sector immensely affected by the COVID-19 crisis, its construction output contracted 19.5% in the first half of 2020 [25].

Figure 1 illustrates the effects of the unseen event on construction sector output in the UK. The growth of construction output was highest in pre-COVID-19 times. However, after COVID-19, the level of construction output decreased up to the end of March 2020 [26].

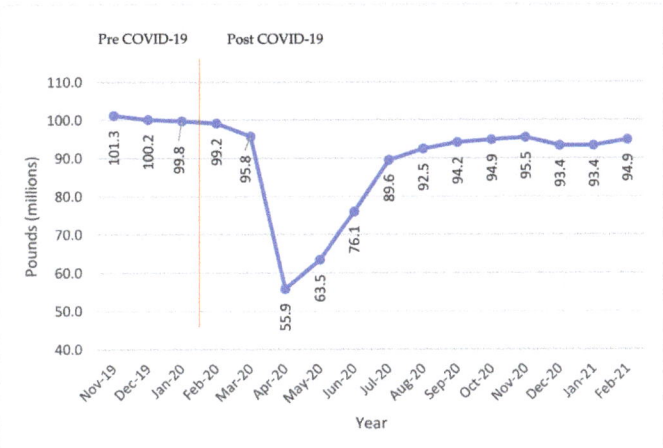

Figure 1. Construction output of the UK.

China is regarded as one of the largest construction output sectors globally with a construction growth of USD 1.04 trillion in 2020 [27]. The sector accounts for a 6.9% contribution to the national GDP of China [28]. In China, the construction industry is fully dependent on materials and services from other sectors that make up the construction workflow in its 30 provinces [29]. Construction activities stopped due to the COVID-19 crisis, which greatly affected the output in the first quarter of 2020 [30]. According to the National Bureau of Statistics, in the Chinese construction industry, there was also a 3.5% growth in the sector in Q4 of 2020, with a growth of 6.6% year on year in the same quarter [31].

Based on construction development trends, it is evident that the construction industry causes the country's economy to thrive but, on the other hand, it is also the main source of adverse effects to the environment and human existence because of carbon emissions. For example, the current use of resources in the construction sector is problematic, as energy-related CO_2 emissions soared by 9.95 gigatons in 2019, a figure that comprises approximately 38% of global greenhouse gas (GHG) emissions [32]. Therefore, there is a need for a concept that could incorporate the fundamentals of sustainability by minimizing waste products and increasing the efficiency of materials. One such concept is the circular economy.

1.2. Circular Economy

The concept of the circular economy is not new. It was first introduced by Kenneth. E. Boulding in the book The Economics of Natural Resources [33]. The concept is a type of economic development that takes into account the scarcity of materials and its effects on the environment and social aspects [34]. The circular economy addresses three main points, namely, reducing the use of raw materials, recycling demolished materials, and generating less waste debris, thereby reducing environmental pollution and achieving cleaner production [35]. It has been shown that sustainability and the circular economy are related to at least eight relationships [36]. Hence, the innovative aspects of sustainable development need to be introduced into the circular economy in every sector [37]. Circular construction framework techniques will lead to the minimal use of locally available resources, water and energy, less waste generation, and guarantee the reuse and recycling of demolished materials. By reducing the negative impacts on the environment, such a circular framework will pave the way for cleaner construction practices [38]. This study introduces the cleaner circular model for construction practices that can drive countries toward a sustainable construction sector.

This study shows how circularity could be embedded in the construction sector across the globe. The construction sector is the most daunting as it the most unresolved sector with regard to circularity, with a potential for sustainability in terms of the economy and the environment. Meanwhile, investing in the construction sector will ensure its adoption and the invention of sustainable techniques that will ensure sustainability in the economy and construction practices [39]. First, the sector has many kinds of problematic effects on health, the resilience of our communities, and equality. Second, the current rate of change—or even the direction of change—in the construction sector is inadequate, owing to several issues, which have increasingly negative effects on social barriers and social resistance [40]. As an example, carbon emissions were tracked using time series data and carbon indicators were identified to set targets to reach the carbon reduction targets [41], which shows the potential of using time series in achieving sustainability. Hence, the increasing popularity of the circular economy concept means it is starting to be accepted as a coherent strategy to respond to the resource-related and environmental challenges in front of us.

Based on the importance of the influence of the construction (CONST) sector on the national GDP and its role in the road to sustainability, the following research questions, shown in Table 2, were established.

Table 2. Research questions and their hypothesis.

Research Objective	Research Question	Research Hypothesis
1	How would the cumulative economy of developed countries, such as the USA, China, and the UK, react when an external shock is experienced in the CONST sector?	A shock to the CONST sector does not affect the cumulative economy.
2	In which direction must the CONST sector move to ensure economic sustainability?	Sustainability in the CONST sector is impossible after recovering from the shock.
3	What steps should investors and policy makers take to impose sustainability in the shock-absorbed sector and ensure the sustainable progress of the sector after recovering from the shock?	It is the responsibility of the government rather than the investors and policy makers to drive the sector toward a sustainable economy.
4	What concrete actions should be followed for the application of statistical circularity in the CONST sector?	Time series cannot be used for data analysis in sustainability.

The objective of this study is to assess the direction of the construction sector after a shock (e.g., recession or pandemic) has been received in three different countries, namely, the USA, the UK, and China, and how much time the sector will require to move toward economic sustainability. The criteria to determine economic sustainability include environmentally friendly processes, profitability, and social inclusion [42]. The significance of this study is that it will enable policy makers to comprehend the underlying concepts behind the short- and long-term effects of the construction sector and the need for collaboration among investors, to work in closer partnership with innovators to create sustainable economies and generate employment opportunities.

2. Literature Review

The Turkish construction industry has a positive influence on GDP growth. The net GDP of the country increased to 7.3% in 1987 and 11.1% from 2002–2012; this increase was twice the national GDP of the country, which also increased due to the exponential growth in the construction sector [43]. A very important question was analyzed: "Does construction output contribute to economic growth?" It was found that bidirectional linkages exist between the construction sector and the economy of the country, and a positive relationship exists between the short- and long-run effects [44]. The Malaysian construction industry experienced a considerable rise in the construction sector due to the use of highly mechanized modern equipment, which increased construction growth. Residential and non-residential growth increased by 30% and 17.8%, respectively. The productivity indicators showed a rise in the GDP of the economy [45]. Time-series data from 1990 to 2009 were collected in Nigeria. The results revealed that GDP and the construction sector have bidirectional Granger causality, meaning that any change in one sector will affect the performance of other sectors. Hence, the construction sector plays a vital role in contributing to the national economy [46]. Based on the time series of Hong Kong data from 1983 to 2013, a bidirectional correlation in the long-run effect was found between GDP and the construction sector. The long-term linkages suggest that policies, industrial development, and innovation must be introduced in the construction sector to ensure consistent growth [47]. A similar study was conducted in Ghana using data collected from 1968 to 2004. The results revealed that construction growth was linked to overall GDP performance with a three-year lag. Moreover, Ghana's GDP showed high performance after two years of growth in the construction sector, confirming the causality between construction and GDP [48]. In another study, 50-year period data from 1968 to 2017 were selected to examine an economic shock and its effects on the construction industry. The construction industry was found to have thrived when there was political stability, optimum weather conditions, and less energy shortage. During the military dictatorship, there was a decline of 21.6% in the construction industry [49]. According to a survey conducted in Afghanistan, 25% of construction projects employed 0.5% of the labor workforce and contributed 0.5% to the national economy [50]. Turkey's construction industry was analyzed based on a data sample from 1998 to 2014. The industry was found to have short-term effects that lasted for just five years on real GDP [51].

2.1. Use of Granger and VECM

In another study, the results showed that construction is a Granger cause of GDP growth and mortgages. It was concluded that these two factors can be used as early indicators for construction performance [52]. A similar study was conducted on statistical data from Malaysia from 1970 to 2019 to study construction sector effects. An impulse response function (IRF) and a vector error correction model (VECM) were used to study shock behavior for short- and long-term effects on the Malaysian economy. A sustainability framework with a global application was also proposed [53]. The need for a post-epidemic prevention system must be integrated to make the construction sector resilient. A standard procedure must be developed that can balance the cost of halted activities in the sector and the probability of a disaster occurring. For this purpose, digital innovation must be employed in the sector to satisfy the need for sustainability [54]. To define the circular economy, a mathematical approach was used that takes into account the recycling of demolition waste for the construction sector in the UK. It was found that government policies could be the only solution to achieving circularity in the construction built environment [55]. An investigative study was performed in Hong Kong to answer the question of which factors are hurdles in adopting cleaner and sustainable processes in the construction sector. From over 140 construction site interviews, the adoption of sustainable processes was found to be greatly related to financial profitability, managerial decisions, and regulatory bodies [56].

2.2. Sustainable Process

The factors contributing to the application of sustainable processes in construction management were reviewed. Based on the interviews, a total of 82 indicators were selected that could relate to the application of cleaner practices in the construction sector. These indicators belong to the areas of social interaction, economic strategies, and environmental processes [57]. To reduce CO_2 emissions, two approaches, namely, production-based and consumption-based, were used to assess global construction carbon emissions. Based on the data from 1995 to 2009, the forward linkage of global construction with CO_2 emissions was found to be between 16% and 20% while the backward linkage was between 37% and 46%. Based on the findings, structural optimization, low-carbon emission processes, and mechanism for transportation were proposed to reduce CO_2 production in the high-pollution sector [58]. In evaluating the number of carbon emissions from the conventional method of construction used in Pakistan, modular CO_2 emissions were found to be 3449.73 kg CO_2-equivalent GHG while conventional building practices generated 6501.91 kg CO_2 GHG emissions. It was concluded that modular construction practices result in 46.9% of CO_2 emissions and must be adopted to achieve sustainability in the system [59]. The waste reduction behavior of construction workers was assessed using a system dynamics approach. Reactive actions and prioritization at the construction site were found to be effective in reducing waste generation. A model based on the policies and measures on the construction site was also proposed to reduce construction waste [60]. An input-output method was adopted to assess the reuse of construction materials. Data from Ontario indicated that reusing construction materials will pose fewer environmental effects and prove beneficial to the economy. It will also increase the GDP and employment opportunities in Canada [61]. To assess the relationship between carbon emissions and the economic prosperity of China's construction sector, the standard deviational ellipse method was used on data from 2005 to 2015. The carbon emissions of 30 provinces of China were studied, and it was found that the economic development in most provinces has a forward linkage direction with carbon emissions, meaning that low carbon emissions indicate slow economic development in these particular provinces. Therefore, the need for policy making for sustainable development was proposed [62]. Statistical analysis was performed based on a questionnaire survey in the Indian construction sector. It was found that resource policies, eco-friendly practices, industry green technologies, and an institutional framework for the application of sustainability in the construction sector are the driving factors toward attaining sustainability and could prove helpful for policy makers and project managers in the construction sector [63]. The construction sector of China and its neighboring countries were investigated for the linkages between the economy, environment, and resources. Compared to other countries, the construction sector of China generates more carbon emissions. It uses more resources and generates more emissions compared to economic profit for the countries in which Chinese firms utilize energy. Based on such findings, protective measures such as energy structure, practices for sustainable development, and efficient allocation of resources were proposed [64]. A low-carbon emission construction sector can only be achieved by incorporating sustainable technologies, input from social sciences on sustainable construction processes, and data exchange, as well as evaluating techniques and decisions based on leadership, project managers, and researchers [65].

2.3. Use of Impulse Response Function (IRF)

An empirical analysis using multivariate models was conducted between the GDP of the USA and unemployment rate. It was revealed that the IRF captures shock behavior and the skewness of plot shows the density of the shock received [66]. The IRF analysis suggests that structural observation is essential for revealing the results, along with proper selection of lags. The IRF was performed to study the effect of interest rate with respect to price hike. By selecting lag = 1, it was found that after a shock, liquidity increases but as the money stabilizes, the interest rate touches the highest level, signifying the application of IRF in macro-econometric dynamics [67].

This study investigates the behavior and influence of the construction sector on other sectors using two theories. As evident from the previous work, the effects of other econometric relationships can be used in research by measuring their strength of linkages using the Granger methodology proposed by Granger and Engle. This method is used to model relationships involving economic issues. Since this study also deals with the output of the construction sector, the Granger concept fully satisfies the methodology to be followed. Another concept this study uses is the Impulse Response Function, which works on the principle of input-output behavior of a system by keeping constraints on input and studying the future output as a result of impulse response.

3. Methodology

This study utilizes a quantitative research approach because it estimates linkage direction and short- and long-run relationships to estimate the vector error correction model (VECM) between the construction sector and other key economic sectors of the USA, China, and the UK. This study also uses a quantitative statistical method to determine the contribution and effects of the construction sector on the aggregate economy. Finally, forecasting from 2020 to 2050 is performed to estimate the output growth of the construction sector of the USA, China, and the UK. The research steps that are followed for this study are shown in Figure 2.

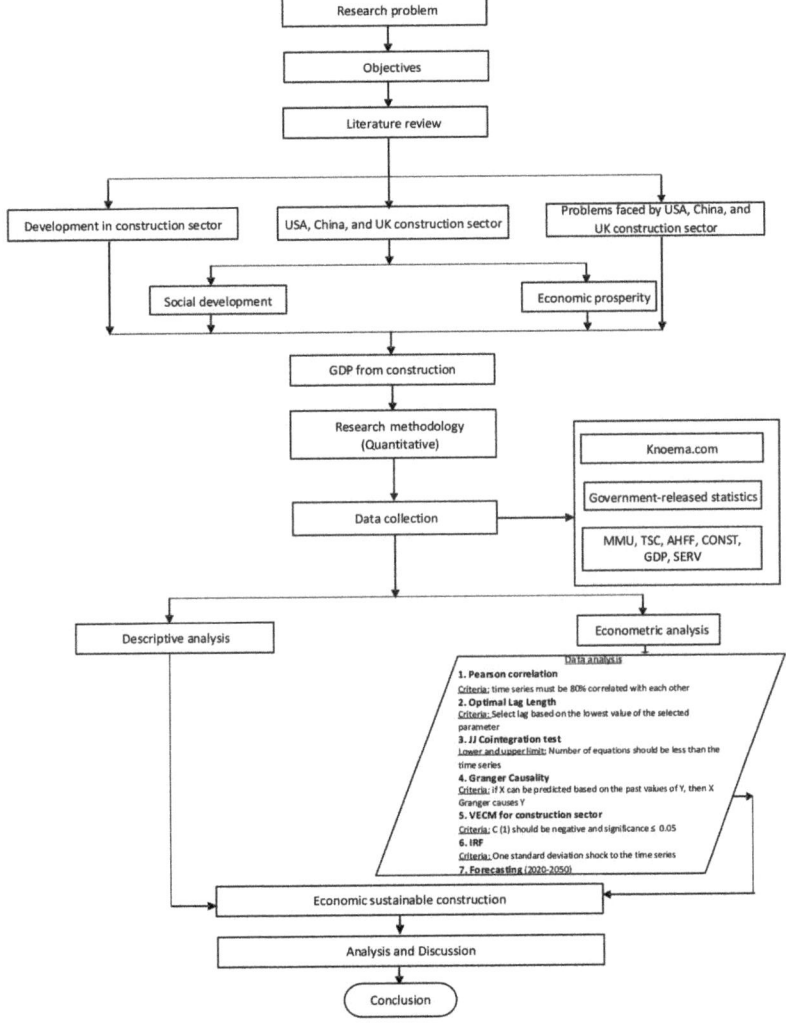

Figure 2. A research framework.

The theoretical framework of this study can be explained as: the use of the cointegration technique to assess how the multivariate data is dependent on each other. The Granger technique is used to assess whether the effect of one sector affects another sector or just the primary sector. To measure the behavior of a sector towards a shock, the impulse response is measured to assess the behavior of construction and other sectors. Based on these, the results from these former techniques are used in the error correction model to create an equation for long-term forecasting of the univariate series. In the light of previous studies, it is evident that there exists a link between CONST and other sectors, which must be analyzed to study the behavior of the CONST sector.

3.1. Collection of Econometric Data

The data for this study were collected from the government statistical department and Knoema from the years 1970 to 2020 [68]. The cut-off for data selection was 2020 instead of 2021 because of the unavailability of officially released statistics and the constant change in numbers due to COVID-19. The descriptive data collected were used to understand the general dynamics of the data. The collected data consisted of: construction (CONST); agriculture, hunting, forestry, and fishing (AHFF); mining, manufacturing, and utilities (MMU); services (SERV); transport, storage, and communication (TSC); and GDP. The data collected for the USA, China, and the UK are shown in Appendices A and B, respectively.

3.2. Data Analysis

After the data collection, the Granger causality test, VECM, and IRF were performed. The structural integrity of the time series was tested using cumulative sum control (CUSUM) tests. The explanatory power was checked using R^2. Validation of the time series was performed using residual correlograms and heteroskedasticity and serial correlation tests.

3.2.1. Johansen Juselius (JJ) Cointegration Technique

Cointegration involves the stationary time series being tested for linear relations among the variables. The null hypothesis for the Johansen Juselius (JJ) cointegration test is that there exists no cointegrating equation. If the value of significance is greater than 0.05, then we fail to reject the null hypothesis [69]. The advantage of using the JJ cointegration test instead of other tests is that it does not need a dependent variable and it diminishes the effects of errors that could be carried over to other steps. If there are no cointegrating equations, then the series does not exhibit long-run relations and VECM cannot be applied.

The mathematical expression can be given in Equation (1) by [70]:

$$J_{Trace} = -T \sum_{i=r+1}^{n} \ln(1 - \hat{\lambda}_i), \qquad (1)$$

where T is the time series size, and λ_i is the largest eigenvalue.

3.2.2. Granger Causality Using the Pairwise Function

This test was developed by Granger [71] to test for causation between two variables. The underlying principle behind the test is that if any X can be predicted based on the past values of Y, then X Granger causes Y. In other words, the past values of X have the power to predict growth in the Y variable [72]. The Granger test can be expressed mathematically by Equations (2) and (3):

$$Y_t = \beta_0 + \sum_{j=1}^{J} \beta_j Y_{t-j} + \sum_{k=1}^{K} \gamma_k X_{t-k} + \mu_t \qquad (2)$$

$$X_t = \beta_0 + \sum_{j=1}^{J} \beta_j X_{t-j} + \sum_{k=1}^{K} \gamma_k Y_{t-k} + \mu_t, \qquad (3)$$

where μ_t is uncorrelated white noise, and γ_k is a measure of the influence of X_{t-k} on Y_t. If γ_k is statistically significant for both equations, then causality is bidirectional. If X does not cause Y while causes X, then it is regarded as unidirectional causality. However, if both X and Y are non-significant, then they have no causal relationships. The null hypothesis is that no causality exists among the time series. However, rejecting the null hypothesis indicates the presence of causality.

3.2.3. Error Correction Model

The error correction model (ECM) is used when the variables have unit roots and are cointegrated. When there is no equilibrium, ECM is used to introduce adjustments for short- and long-term equilibriums. The ECM along with the direction of adjustments is called VECM, which is also called a restricted vector autoregressive (VAR) system. Its mathematical expression was given by Gujarati [73] and Granger [74] as follows in Equation (4):

$$\Delta Y_t = \prod Y_{t-1} + \sum_{i=1}^{m-1} \Phi_i \Delta Y_{t-i} D_i + \mu_t \tag{4}$$

where ΔY_t is the independent variable, $\prod$ is the matrix of cointegrating vectors, and Φ represents a matrix of independent variables. The procedure for conducting VECM is the selection of appropriate lags using selected parameters, selection of many equations, and finally, estimation of VECM using (p-1) lags.

VECM was used in this study for two reasons: first, if the equations are cointegrated in the system, then there will be an accurate representation of short- and long-term interdependencies of the variables; second, its wide applicability in multivariate time series [75].

3.2.4. Structural Stability Analysis

This test, which was first used by Brown et al. [76], was conducted to test for structural stability in the developed VECM model. This analysis is an important step of this study as it shows the presence of structural breaks in the system due to the unit root, which will produce misleading results [77]. The mathematical form is given in Equation (5):

$$w_m = \frac{1}{\hat{\sigma}} \sum_{t=k+1}^{T} w_t, \ m = k+1, \ldots, t, \tag{5}$$

where w_t is the recursive residual, and m is the sample number. The analysis is rejected if the plot deviates from the suggested boundary by the test confidence level of 95%.

3.2.5. Shock Responses of the Construction Sector

IRF was used to introduce a shock to the sector (variable), and the behavior of the sector was evaluated after receiving a shock of one standard deviation, as well as the behaviour of other sectors after receiving the shock [78]. This function also stated the amount of time required for the variables to return to their original position.

This study used the Cholesky dof (degree of freedom) as an IRF function, which is used for intersectoral linkages [79]. As the study focused on the construction sectors of the USA, China, and the UK, the CONST variable was thus used as an exogenous and endogenous variable.

3.2.6. Forecasting Using the VECM Equation

Forecasting was performed using a VECM equation that took into account the short- and long-term effects. This forecasting was preferred due to its structural integrity, absence of autoregressive conditional heteroskedasticity (ARCH), autocorrelation, and serial correlation in the series. It was performed from the years 2020 to 2050.

The forecast predictive power can be checked using Theil statistics, which was first developed by Theil [80]. If the forecasted values and actual values are 0, then the model has reliable predictive power. The value of 1 suggests that both entities will move in the opposite direction and, hence, that the model is unreliable. The model is shown in Equations (6)–(8):

$$U = \sqrt{\frac{\sum_{i=1}^{n-1}(FPE_{t+1} - APE_{t+1})^2}{\sum_{i=1}^{n-1}(APE_{t+1})^2}} \tag{6}$$

$$FPE_{t+1} = \frac{F_{t+1} - Y_t}{Y_t}, \quad (7)$$

$$APE_{t+1} = \frac{Y_{t+1} - Y_t}{Y_t}, \quad (8)$$

where FPE_{t+1} is the forecast percentage and APE_{t+1} is the actual percentage error.

3.2.7. Validation of the Estimated Model

Serial Correlation Analysis

This test is considered an alternative to Q-statistics in serial correlation and is used for large multiplier (LM) tests. Therefore, it is regarded as the Breusch-Godfrey serial correlation LM Test. This test is preferred when there is a possibility of autocorrelation in errors. Hence, it is effective in determining the autocorrelation for lagged dependent variables [81]. It is given in Equations (9) and (10):

$$y_t = X_t \beta + \epsilon_t \quad (9)$$

$$\epsilon_t = X_t \gamma + \left(\sum_{s=1}^{p} \alpha_s \epsilon_{t-s} \right) + v_t, \quad (10)$$

where X_t is the lagged residuals, p is the order of lags, α, β, and γ are the coefficients, v_t is the white noise, and ϵ_t is the error term [82].

Heteroskedasticity Test: Breusch-Pagan-Godfrey

Among many heteroskedasticity tests, this study used the Breusch-Pagan-Godfrey (BPG) test. The term "heteroskedasticity" means differently scattered. It is commonly used for checking errors in regressors. The null hypothesis for this test is that error variances are equal. Based on the value of probability chi-square value, if the value is more than 0.05, then the data have homoskedasticity and are fit for regression [83]. The BPG test is expressed in Equation (11):

$$BPG = nR^2_{\hat{u}^2_i}, \quad (11)$$

where n is the number of observations, and $R^2_{\hat{u}^2_i}$ is the coefficient of determination of the regressors [84].

4. Results

4.1. Correlation among the Sectors

The dependence and relationship of the construction sector with other sectors can be judged from the Pearson correlation test. The reason this test was performed was to check how much influence one sector will have on the other sectors.

The comparison of correlation values, as shown in Table 3, shows that all the sectors of the USA, China, and the UK are highly correlated with the other sectors within each respective country, i.e., above 80%. Hence, the behavior of other sectors (AHFF, MMU, SERV, TSC, and GDP) can adequately be modeled based on the behavior of one sector (CONST).

Table 3. Pearson correlation test.

	USA					
	CONST	AHFF	MMU	SERV	TSC	GDP
CONST	1	-	-	-	-	-
AHFF	0.824162	1	-	-	-	-
MMU	0.984415	0.833575	1	-	-	-
SERV	0.992240	0.832636	0.993558	1	-	-
TSC	0.992444	0.839118	0.992722	0.998396	1	-
GDP	0.992894	0.835452	0.995092	0.999686	0.999136	1

Table 3. Cont.

	China					
	CONST	**AHFF**	**GDP**	**MMU**	**SERV**	**TSC**
CONST	1	-	-	-	-	-
AHFF	0.9787043	1	-	-	-	-
GDP	0.9983059	0.9873097	1	-	-	-
MMU	0.9901612	0.9939498	0.99531938	1	-	-
SERV	0.9960696	0.9641160	0.99294770	0.9773136	1	-
TSC	0.9915226	0.9923408	0.99721276	0.9957675	0.985313	1

	UK					
	CONST	**MMU**	**SERV**	**TSC**	**AHFF**	**GDP**
CONST	1	-	-	-	-	-
MENU	0.947675196	1	-	-	-	-
SERV	0.992610869	0.950618572	1	-	-	-
TSC	0.994261823	0.952857278	0.998081	1	-	-
AHFF	0.857882606	0.923748829	0.864765904	0.865759	1	-
GDP	0.992974491	0.96256965	0.998552932	0.998712557	0.874481454	1

4.2. Granger Causality Using the Pairwise Function

The Granger causality test checks the data for the presence of the null hypothesis: that CONST does not Granger cause AHFF. A probability level of less than 0.05 shows that the null hypothesis is rejected, which means that CONST does cause Granger AHFF.

Table 4 indicates that the null hypothesis was rejected for CONST-AHFF, SERV-CONST, TSC-CONST, and GDP-CONST, meaning that any change in these sectors will show a change in the corresponding sectors because the value of probability is less than 0.05. Meanwhile, the other sector values are greater than the significance level of 0.05, indicating that any change in the sector will not affect the behavior of other sectors.

Table 4. Empirical results of Granger causality.

USA						
Null Hypothesis	Lag	Alternate Hypothesis	F-Stat	Prob.	Null Hypo Result
AHFF does not Granger cause CONST	2	-	0.9347	0.400	Accept
CONST does not Granger cause AHFF	2	CONST Granger causes AHFF	6.5459	0.003	Reject
MMU does not Granger cause CONST	2	-	1.8112	0.175	Accept
CONST does not Granger cause MMU	2	-	0.4914	0.615	Accept
SERV does not Granger cause CONS	2	SERV Granger causes CONST	7.7330	0.001	Reject
CONST does not Granger cause SERV	2	-	2.0743	0.138	Accept
TSC does not Granger cause CONS	2	TSC Granger causes CONST	7.1754	0.002	Reject
CONST does not Granger cause TSC	2	-	3.0496	0.057	Accept
GDP does not Granger cause CONS	2	GDP Granger causes CONST	6.5634	0.003	Reject
CONST does not Granger cause GDP	2	-	2.0258	0.144	Accept

Table 4. Cont.

China					
Null Hypothesis	Lag	Alternate Hypothesis	F-Stat	Prob.	Null Hypo Result
AHFF does not Granger cause CONST	2	-	1.535	0.227	Accept
CONST does not Granger cause AHFF	2	CONST Granger causes AHFF	8.990	0.000	Reject
MMU does not Granger cause CONST	2	MMU Granger causes CONST	6.386	0.003	Reject
CONST does not Granger cause MMU	2	-	0.548	0.581	Accept
SERV does not Granger cause CONS	2	SERV Granger causes CONST	9.548	0.000	Reject
CONST does not Granger cause SERV	2	CONST Granger causes SERV	6.766	0.002	Reject
TSC does not Granger cause CONS	2	TSC Granger causes CONST	4.511	0.016	Reject
CONST does not Granger cause TSC	2	-	1.230	0.302	Accept
GDP does not Granger cause CONS	2	GDP Granger causes CONST	6.628	0.003	Reject
CONST does not Granger cause GDP	2	-	1.038	0.362	Accept
UK					
Null Hypothesis	Lag	Alternate Hypothesis	F-Stat	Prob.	Null Hypo Result
AHFF does not Granger cause CONST	2	AHFF Granger causes CONST	3.0429	0.028	Reject
CONST does not Granger cause AHFF	2	-	0.8779	0.486	Accept
MMU does not Granger cause CONST	2	MMU Granger causes CONST	4.0740	0.007	Reject
CONST does not Granger cause MMU	2	-	0.9368	0.453	Accept
SERV does not Granger cause CONS	2	-	2.3194	0.075	Accept
CONST does not Granger cause SERV	2	-	1.1108	0.366	Accept
TSC does not Granger cause CONS	2	TSC Granger causes CONST	7.396	0.000	Reject
CONST does not Granger cause TSC	2	CONST Granger causes TSC	5.1160	0.002	Reject
GDP does not Granger cause CONS	2	-	2.2285	0.084	Accept
CONST does not Granger cause GDP	2	-	0.6251	0.647	Accept

The comparison of results indicates that the CONST sector has considerable influence on other sectors in China compared to that in the USA and the UK. The CONST sector of the UK is not considerably affected by the performance of other sectors. Meanwhile, the USA sectors have more influence on other sectors, which means that China and USA CONST sectors are more volatile than the UK CONST sector.

4.3. JJ Cointegration Examination

The JJ test was performed to test for the null hypothesis that there exist no cointegrating equations if the significance level is less than 0.05. The null hypothesis was rejected for four cointegrating equations based on probability values of less than 0.05.

The trace test and rank test results were generated. In Table 5, the trace test results illustrate the number of selected integrating equations for the USA, China, and the UK. The number of cointegrating equations for VECM using USA, China, and UK construction data was selected as four, based on p-values of less than 0.05 using the rank trace test.

Table 5. Unrestricted Cointegration Rank Test (Trace).

	USA			
Hypothesized No. of CE(s)	Eigenvalue	Trace Statistics	0.05 Critical Value	Prob (p-Value)
None *	0.619333	138.9555	95.75366	0.0000
At most 1 *	0.558057	92.59568	69.81889	0.0003
At most 2 *	0.37982	53.40008	47.85613	0.0138
At most 3 *	0.308523	30.46832	29.79707	0.0418
At most 4	0.225141	12.75991	15.49471	0.1239
	China			
Hypothesized No. of CE(s)	Eigenvalue	Trace Statistics	0.05 Critical Value	Prob (p-Value)
None *	0.759897	197.5713	95.75366	0.0000
At most 1 *	0.686021	130.5171	69.81889	0.0000
At most 2 *	0.568508	76.07092	47.85613	0.0000
At most 3 *	0.369329	36.56708	29.79707	0.0071
At most 4	0.265984	14.90145	15.49471	0.0613
	UK			
Hypothesized No. of CE(s)	Eigenvalue	Trace Statistics	0.05 Critical Value	Prob (p-Value)
None *	0.902859	247.6914	95.75366	0.0000
At most 1 *	0.724218	142.7696	69.81889	0.0000
At most 2 *	0.635345	84.80308	47.85613	0.0000
At most 3 *	0.415794	39.40696	29.79707	0.0029
At most 4	0.214404	15.21936	15.49471	0.0550

* Is the rejection of the hypothesis at the 0.05 level.

The absence of cointegration was performed using unrestricted VAR. However, the presence of cointegrated equations can only be modeled using VECM (restricted VAR). This study used VECM for analysis based on the presence of four cointegrating equations for the USA, China, and the UK.

4.4. Identification and Analysis of Short- and Long-Run Coefficients

Validation of the VECM model equation is necessary to check for the presence of errors in the model. This can be performed by making a system of coefficients of the produced model. The C(1) coefficient value should always be negative and the probability level should be less than 0.05. The negative value shows the ability to bounce back to its initial position and the absence of any error within the VECM system.

The coefficient system and its estimation for the USA are presented in Table 6. Based on the VECM equation, the coefficients for China were generated, as shown in Table 7. Similarly, the system of coefficients for the UK is shown in Table 8.

Table 6. Coefficient Values and Probabilities of CONST-USA.

	Coefficient	Std. Error	t-Statistic	Prob.
C(1)	−0.266972	0.083635	−3.192121	0.0017
C(2)	−0.009694	0.032714	−0.296308	0.7673
C(3)	−0.811232	0.248583	−3.26343	0.0013

Table 6. *Cont.*

	Coefficient	Std. Error	t-Statistic	Prob.
C(4)	1.316046	0.215376	6.110451	0.000
C(5)	−0.032232	0.191056	−0.168702	0.8662
C(6)	0.007221	0.026844	0.268985	0.7882
C(7)	0.024828	0.01807	1.374041	0.1711
C(8)	0.745002	0.397053	1.87633	0.0622
C(9)	0.790626	0.324114	2.439347	0.0157
C(10)	3.022295	0.833919	3.624206	0.0004
C(11)	1.837517	0.984832	1.865817	0.0636
C(12)	0.178122	0.351363	0.506945	0.6128
C(13)	0.498085	0.32868	1.515412	0.1314
C(14)	−4.494491	1.635197	−2.748592	0.0066
C(15)	−4.444107	1.343703	−3.307359	0.0011
C(16)	0.078439	0.042387	1.850532	0.0658

Table 7. Coefficient Values and Probabilities of CONST-China.

	Coefficient	Std. Error	t-Statistic	Prob.
C(1)	−1.014322	0.289812	−3.499928	0.0006
C(2)	−0.092923	0.183677	−0.505903	0.6138
C(3)	0.188934	0.126361	1.495191	0.1373
C(4)	−0.218396	0.170481	−1.281056	0.2024
C(5)	0.219318	0.397163	0.552211	0.5817
C(6)	0.349016	0.368886	0.946135	0.3458
C(7)	0.532093	0.278034	1.913773	0.0578
C(8)	−0.120549	0.156527	−0.770144	0.4426
C(9)	−0.148638	0.110337	−1.347126	0.1802
C(10)	−0.383467	0.104807	−3.658794	0.0004
C(11)	−0.041635	0.107735	−0.386455	0.6998
C(12)	0.006208	0.082337	0.075394	0.9400
C(13)	0.084532	0.067312	1.255824	0.2114
C(14)	0.210847	0.160852	1.310807	0.1922
C(15)	0.079435	0.123087	0.645357	0.5198
C(16)	−0.032542	0.097744	−0.332933	0.7397

Table 8. Coefficient Values and Probabilities of CONST-UK.

	Coefficient	Std. Error	t-Statistic	Prob.
C(1)	−0.412881	0.208946	−1.976014	0.0501
C(2)	0.039678	0.127414	0.311412	0.7560
C(3)	−0.020529	0.080231	−0.255872	0.7984
C(4)	−0.47974	0.490898	−0.97727	0.3301
C(5)	−0.227722	0.400833	−0.568122	0.5709
C(6)	−0.099155	0.316624	−0.313163	0.7546
C(7)	−0.466073	0.406674	−1.146059	0.2538
C(8)	−0.122356	0.20325	−0.602	0.5482
C(9)	0.108997	0.191311	0.569736	0.5698
C(10)	0.027216	0.176419	0.154268	0.8776
C(11)	0.138414	0.14796	0.93548	0.3512
C(12)	0.217524	0.168988	1.287215	0.2002
C(13)	−0.121834	0.107699	−1.131244	0.2599
C(14)	0.673255	0.550924	1.222047	0.2238
C(15)	0.663833	0.388275	1.709696	0.0896
C(16)	0.505628	0.436445	1.158514	0.2487

The comparison of these tables shows that the CONST sector of China will bounce back 1.01% quicker than those of the USA and the UK because it has less volatility and is capable of supporting itself when it is hit by a recession. The CONST sectors of the USA and the UK are more volatile and will not recover as quickly, at 0.26% and 0.41%, respectively.

4.5. Tests for Assessing the Explanatory Power and Efficiency of the VECM Equation

The explanatory power and efficiency of the VECM equation were tested using the coefficient of R^2 and the F-statistic. In the case of the USA, the value of R^2 indicated explanatory power with the value of 0.997, which was sufficient to extract useful information from the statistical data. The F-statistic value was less than 0.05, so the efficiency of the model was also acceptable. The rule of thumb for autocorrelation suggests an absence of autocorrelation in the model based on the value of the Durbin-Watson (DW) test of 1.606.

Similarly, the value of R^2 for China was also satisfactory, at 0.992. The significance was tested using the *p*-value of the F-statistic, which was recorded as lower than 0.05. The presence of autocorrelation was tested using the DW test and the value was 1.88, which signified the absence of autocorrelation in the system.

Similarly, the value of R^2 for the UK model was obtained as 0.996, with an F-statistic value less than 0.05, hence confirming its significance. The DW test with a value of 1.87 indicated that the model does not suffer from autocorrelation. The results are shown in Table 9.

Table 9. Result of CONST equation: USA, China, and the UK.

Parameters	USA	China	UK
Coefficient of determination (R^2)	0.997	0.992	0.996
Adjusted R^2	0.996	0.984	0.995
Probability of F-statistic	0.000	0.009	0.000
Durbin-Watson statistics	1.606	1.88	1.872

4.6. Validation of the Estimated Equation for the CONST Model

The VECM equation should be non-spurious and non-biased. Various checks can be performed to validate the VECM equation. This study selected the following three tests to check for consistency in the VECM system by performing residual diagnosis checks: Breusch-Godfrey serial correlation LM Test; and heteroskedasticity test (Breusch-Pagan-Godfrey).

4.6.1. Serial Correlation Test

The presence of serial correlation in residuals was tested using the Breusch-Godfrey test. The results indicated that the series was free from serial correlation based on the chi-square value of probability being greater than 0.05. Table 10 depicts the statistical evidence for the USA.

Table 10. Breusch-Godfrey Serial Correlation LM Test—USA.

F-Statistic	0.423895	Prob. F(2,27)	0.6588
Obs*R-squared	1.400412	Prob. Chi-Square(2)	0.4965

The same test was applied to the China series, which was found free from serial correlation based on its chi-square probability value being greater than 0.05, as shown in Table 11.

Table 11. Breusch-Godfrey Serial Correlation LM Test—China.

F-Statistic	1.740809	Prob. F(2,41)	0.1881
Obs*R-squared	3.835276	Prob. Chi-Square(2)	0.1470

Table 12 shows the chi-square value of probability is greater than 0.5, hence showing no sign of serial correlation. The data were also fit for forecasting.

Table 12. Breusch-Godfrey Serial Correlation LM Test—UK.

F-Statistic	0.050215	Prob. F(1,16)	0.8255
Obs*R-squared	0.143918	Prob. Chi-Square(1)	0.7044

4.6.2. Heteroskedasticity Test

The presence of heteroskedasticity was tested using ARCH. The null hypothesis, that the series was homoscedastic, was tested. The chi-square probability test value is greater than 0.05, meaning that the series is not heteroskedastic. The results for the USA, China, and the UK are shown in Tables 13–15, respectively.

Table 13. Heteroskedasticity Test: Breusch-Pagan-Godfrey—USA.

F-statistic	1.364560	Prob. F(16,29)	0.2268
Obs*R-squared	19.75718	Prob. Chi-Square(16)	0.2314
Scaled explained SS	5.167185	Prob. Chi-Square(16)	0.9949

Table 14. Heteroskedasticity Test: Breusch-Pagan-Godfrey—China.

F-statistic	0.109815	Prob. F(19,25)	1.0000
Obs*R-squared	3.466360	Prob. Chi-Square(19)	1.0000
Scaled explained SS	12.97102	Prob. Chi-Square(19)	0.8401

Table 15. Heteroskedasticity Test: Breusch-Pagan-Godfrey—UK.

F-statistic	0.543013	Prob. F(28,17)	0.9263
Obs*R-squared	21.71758	Prob. Chi-Square(28)	0.7942
Scaled explained SS	3.437171	Prob. Chi-Square(28)	1.0000

4.7. Structural Stability Analysis

The structural stability of the VECM model was tested using CUSUM tests, which were performed with a 5% significance level. The null hypothesis is that "there are no structural breaks in the system", with a significance level of 5%. The results show that there are no structural breaks and the presence of stability is fit for IRF. Figure 3a depicts the results of the CUSUM test.

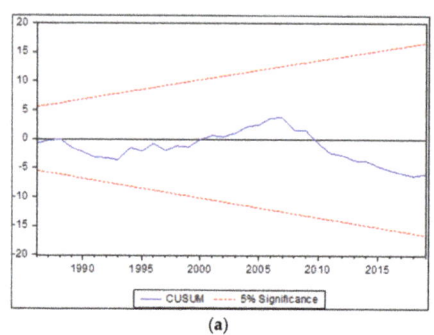
(a)

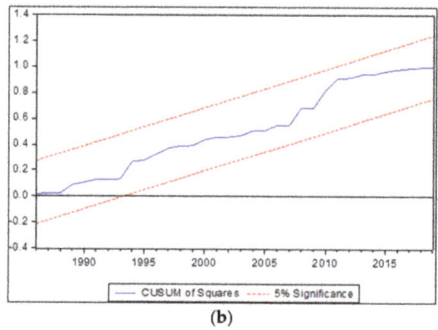
(b)

Figure 3. Structural stability test—USA. (**a**): CUSUM test; (**b**): CUSUM square test.

The CUSUM square test indicates the lower and upper bounds of the 5% level of significance for residuals. The results show there is no structural break in the system as residuals are inside the percentage level of significance, making it fit for IRF and forecasting, as shown in Figure 3b.

Figure 4a,b The structural integrity of the China series. No structural break exists in the system based on the CUSUM and CUSUM square lines that lie within the 5% significance level.

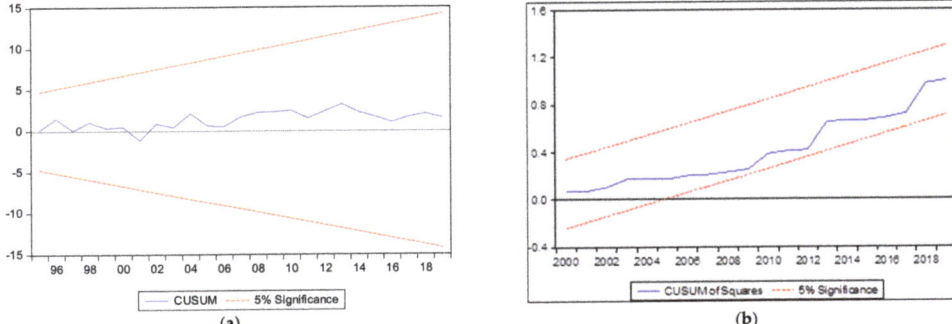

Figure 4. Structural stability test—China. (**a**): CUSUM test; (**b**): CUSUM square test.

The structural integrity for the UK series was also tested. The results revealed that the CUSUM and CUSUM square lines lie inside the 5% significance level and the series is fit for forecasting, as shown in Figure 5a,b, respectively.

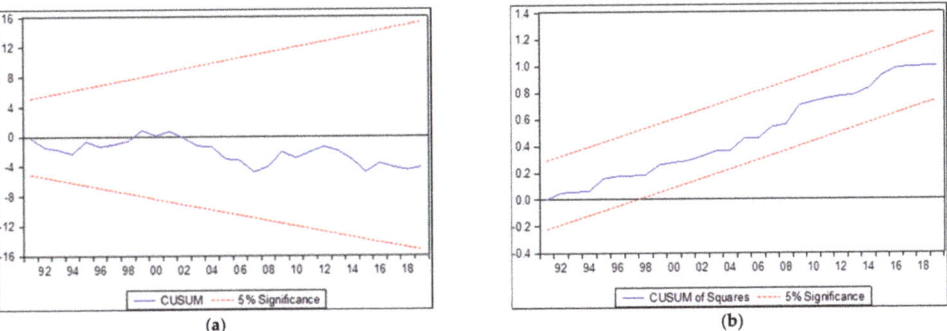

Figure 5. Structural stability test—UK. (**a**): CUSUM test; (**b**): CUSUM square test.

4.8. Shock Responses of the Construction Sector

The IRF produces a shock of one time period. In this case, CONST is the endogenous variable. A one-time period shock is given to CONST, and the behavior of other sectors is recorded. The IRF also shows how much time is required for any sector to absorb this shock. In this study, one positive standard deviation shock is produced in CONST, and its behavior is measured in AHFF, MMU, SERV, CONST, and GDP.

In Figure 6a, the response of CONST is shown after a shock in AHFF in the USA. After the second period (second year), there is a positive trend in the response of CONST. The outcome shows that expansion in AHFF will negatively affect the output of CONST owing to the presence of backward linkages. It will take almost 10 years for the CONST industry to regain its original position from before the shock. Figure 6b reveals that it will take 10 years for the AHFF sector in the USA to recover from a shock produced by the CONST sector. As the linkage is unidirectional, there is no forward linkage between AHFF and CONST; therefore, CONST will not affect the activity of AHFF. Additionally, this sector is not sensitive to activity in the CONST sector.

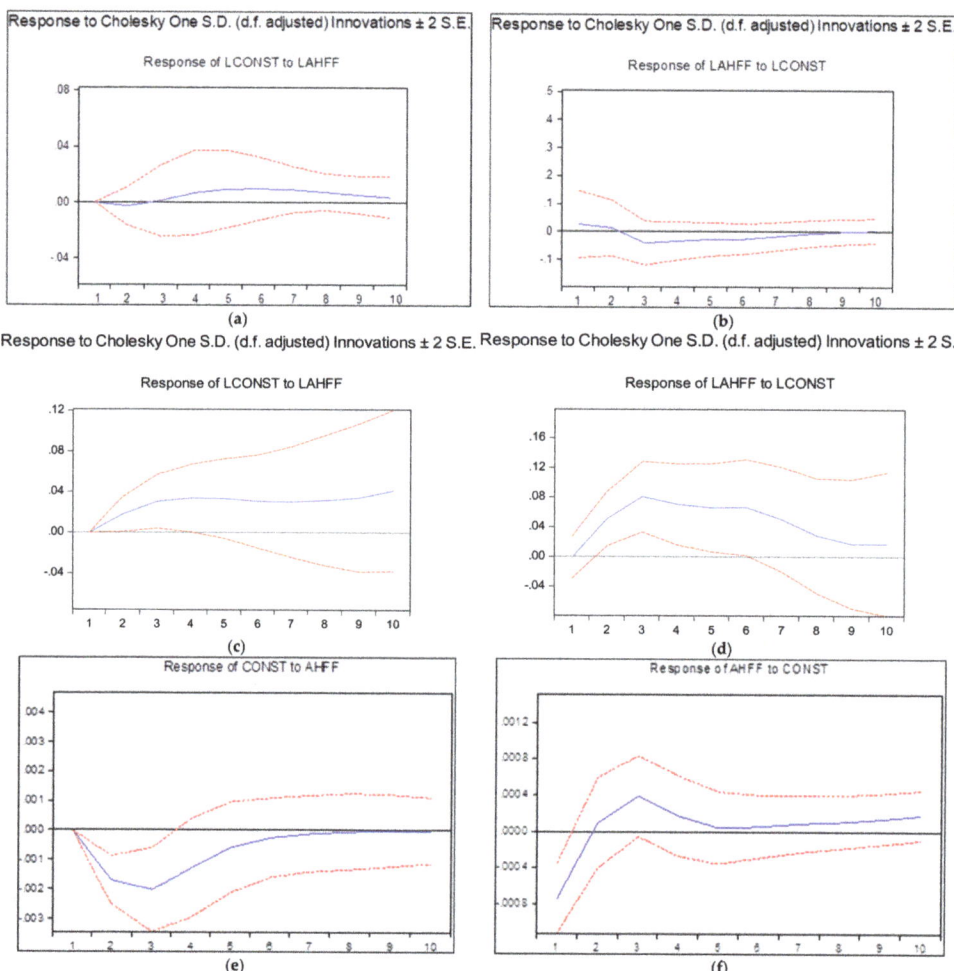

Figure 6. Impact on CONST-AHFF and AHFF-CONST from a shock. (**a**) Response of CONST to AHFF—USA; (**b**) Response of AHFF to CONST—USA; (**c**) Response of CONST to AHFF—China; (**d**) Response of AHFF to CONST—China; (**e**) Response of CONST to AHFF—UK; (**f**) Response of AHFF to CONST—UK.

However, the Chinese construction sector will react differently to the USA construction sector. There will be a positive behavioral shock in AHFF when CONST experiences a shock and vice versa. This result shows that the construction industry of China is more flexible than that of the USA and can drive the construction sector to sustainability. This is shown in Figure 6c,d.

In the case of the UK construction sector, any shock in AHFF will first produce negative effects in the CONST sector and will underperform until the fifth period, and will become stable after the sixth period, as shown in Figure 6e. Meanwhile, the AHFF response will first start with negative effects when the CONST sector experiences any change in its output. After that, there will be a positive effect in the AHFF sector for at least 10 years, which can be seen in Figure 6f.

Figure 7a shows the behavior of CONST when a shock is experienced by MMU in the USA. There is no significant positive trend in the behavior of CONST, and it does not deviate from zero lines, indicating that CONST is less likely to be affected by any change in the MMU sector. The CONST will have less effect on MMU and will not deviate much from the zero lines. Figure 7b indicates the response of MMU when a shock is produced in CONST in the USA. There will be a positive effect on MMU and it will return to its original state without much difference. There will be a significant shoot-up in the MMU sector, which will decrease with time and return to its original position due to the shock in CONST.

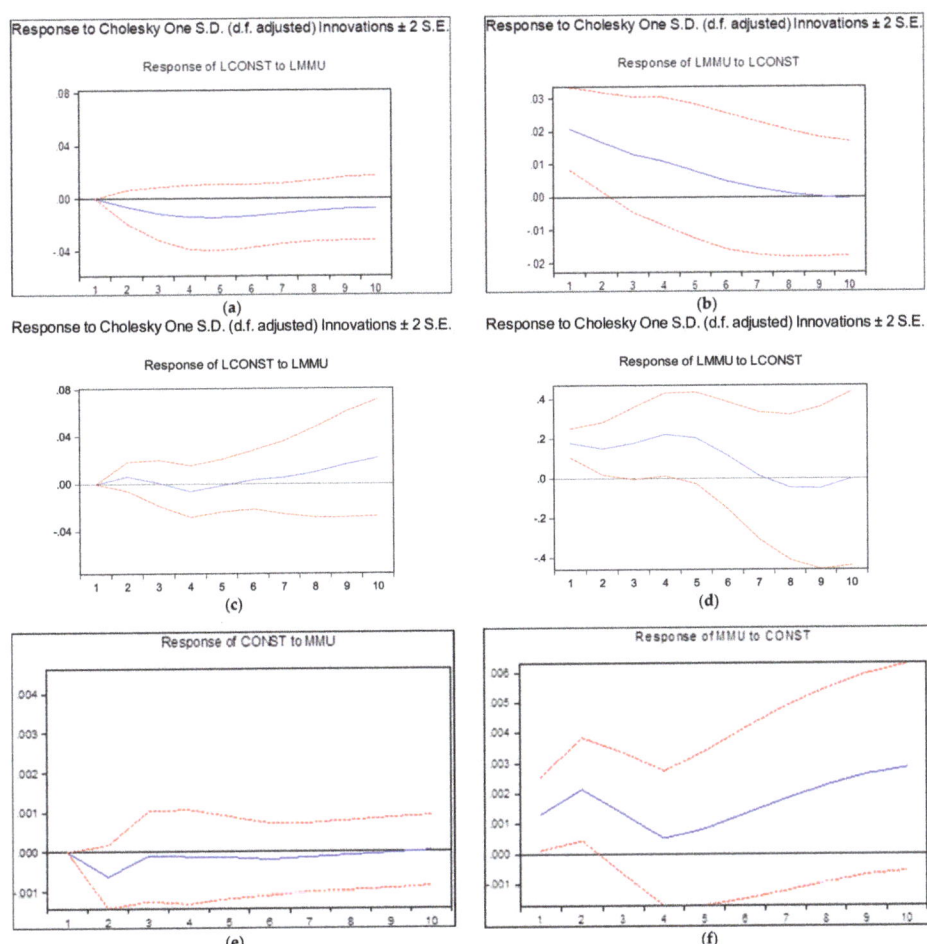

Figure 7. Impact on CONST-MMU and MMU-CONST from a shock. (**a**) Response of CONST to MMU—USA; (**b**) Response of MMU to CONST—USA; (**c**) Response of CONST to MMU—China; (**d**) Response of MMU to CONST—China; (**e**) Response of CONST to MMU—UK; (**f**) Response of MMU to CONST—UK.

Figure 7c indicates the behavior of CONST after any shock is experienced by the MMU sector in China. As evidence, the response of the CONST sector will be stable for the next five years, and then the subsequent five years will show a positive output in CONST. Figure 7d shows the behavior of MMU toward the shock in the CONST sector, where there will be a positive change in the production and activities of MMU, which will stabilize as the period approaches its 10th year.

Figure 7e shows that the first two periods will negatively impact the CONST sector when the MMU sector receives a shock. However, this change will dissipate with time and there will be no major effects on the CONST sector in the UK. Figure 7f illustrates the positive response in MMU when the CONST sector receives a shock; there will be an increase in production for at least 10 years in the UK.

Figure 8a shows the behavior of CONST after a shock is produced in the SERV sector in the USA. There will be a positive trend in CONST when SERV experiences a shock of one time period. As SERV-CONST has a forward linkage, any positive shock will result in a positive output. The shock will stabilize after the eighth period. Figure 8b reveals that the shock produced in SERV will positively affect CONST, which will decrease with time due to the stored services used in the construction industry, such as petrol, transportation, and material supply.

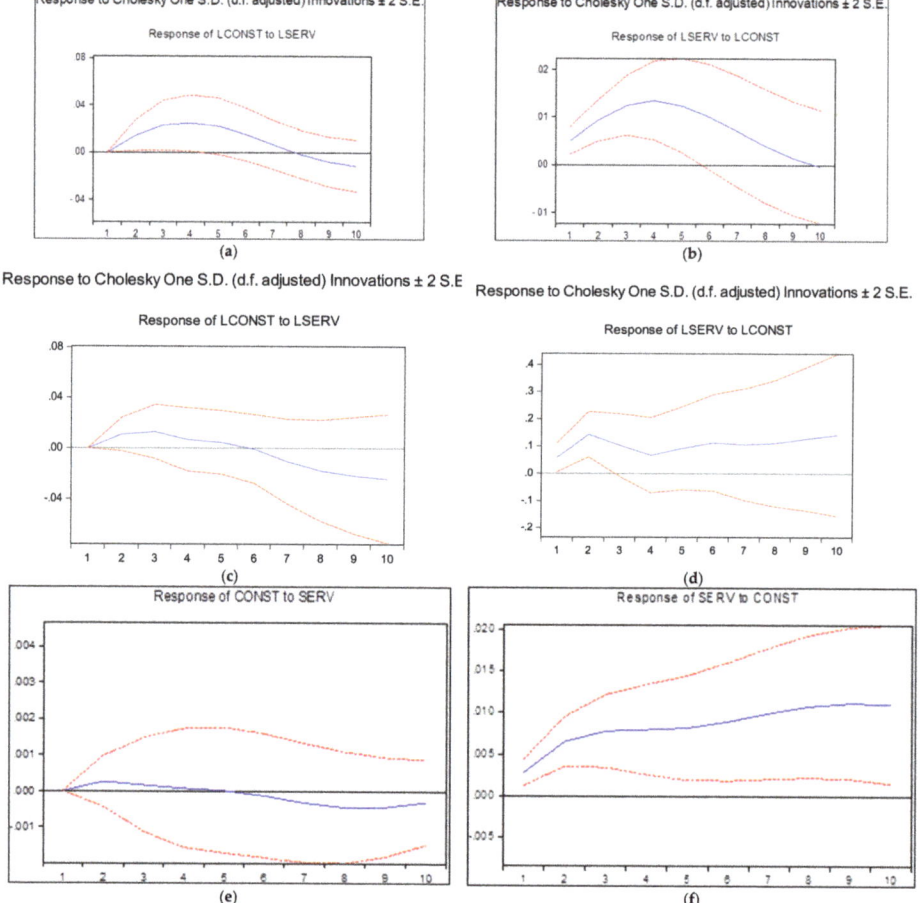

Figure 8. Impact on CONST-SERV and SERV-CONST from a shock. (**a**) Response of CONST to SERV—USA; (**b**) Response of SERV to CONST—USA; (**c**) Response of CONST to SERV—China; (**d**) Response of SERV to CONST—China; (**e**) Response of CONST to SERV—UK; (**f**) Response of SERV to CONST—UK.

Figure 8c signifies the positive behavior in the CONST sector in China when there is a unit shock in the SERV sector during the first five years. After that, the CONST sector will lose its productivity because it is greatly dependent on services for the timely execution of projects. Figure 8d shows the positive behavior in the SERV sector when there is a lack of funding or recession in the CONST sector, positively affecting the performance of the SERV sector.

Figure 8e shows that the UK SERV sector will produce marginal positive effects in the CONST sector until the fifth period, after which it will decrease as it approaches the tenth period, and will stabilize after this period. Figure 8f indicates the positive behavior in the SERV sector of the UK after the CONST sector experiences a shock. The positive effect will continue beyond the tenth period.

Figure 9a illustrates that no significant changes will occur in the CONST sector when a shock is received in the TSC sector of the USA. There will be a slight positive trend, but this trend will not affect productivity and production in the CONST sector. This behavior validates the findings of the Granger causality, in which there is no linkage of CONST-TSC. Figure 9b shows the behavior of the TSC sector to a shock in the CONST sector, which will positively affect the performance of the TSC sector until the first half of the fifth period (year). After that, there will be a negative effect on the TSC sector due to the shortage of transportation of materials and vehicles, expensive storage, and expensive communication.

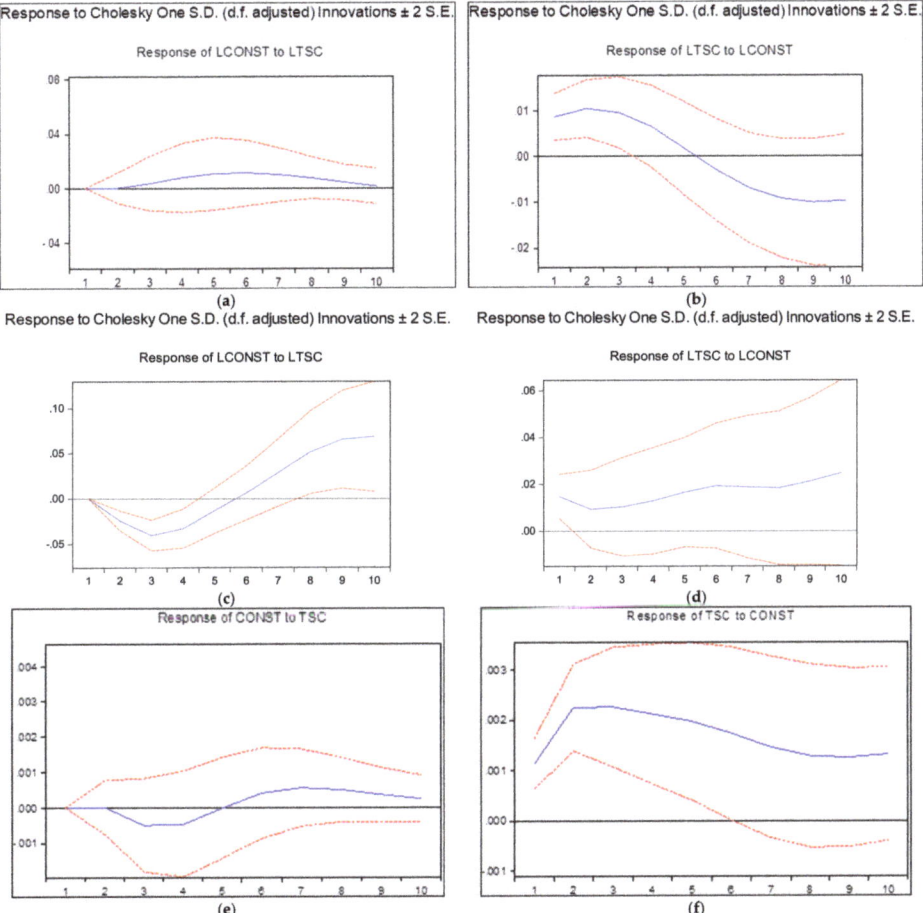

Figure 9. Impact on CONST-TSC and TSC-CONST from a shock. (**a**) Response of CONST to TSC—USA; (**b**) Response of TSC to CONST—USA; (**c**) Response of CONST to TSC—China; (**d**) Response of TSC to CONST—China; (**e**) Response of CONST to TSC—UK; (**f**) Response of TSC to CONST—UK.

Figure 9c shows the negative behavior of the CONST sector toward the end of the fifth period after a shock is received in the TSC of the Chinese economy. However, a recovery will be made in the sixth period and positive effects will start to manifest themselves. Figure 9d shows a positive response for 10 consecutive time periods in the TSC sector when a shock is received in the CONST sector of China.

Figure 9e indicates the negative and positive behavior of the CONST sector after the TSC sector of the UK receives a shock. The negative effects in the CONST sector will continue for five time periods and, after that, there will be a positive output in the CONST sector owing to any change in the TSC sector. Figure 9f reveals that a shock in the CONST industry will positively impact the growth of the TSC and the sector will grow efficiently.

Figure 10a shows the non-significance of CONST-GDP in the USA. As there were no Granger cause linkages, there is only a minimum effect on the performance of CONST by any changes in the GDP. The shock in the GDP of the country will move to negative, which returns to its original position in the eighth year. A positive trend will also be shown toward the end of the tenth year. Figure 10b indicates the Granger causes' forward linkage direction effects. Any shock to the CONST industry will greatly affect the performance of GDP.

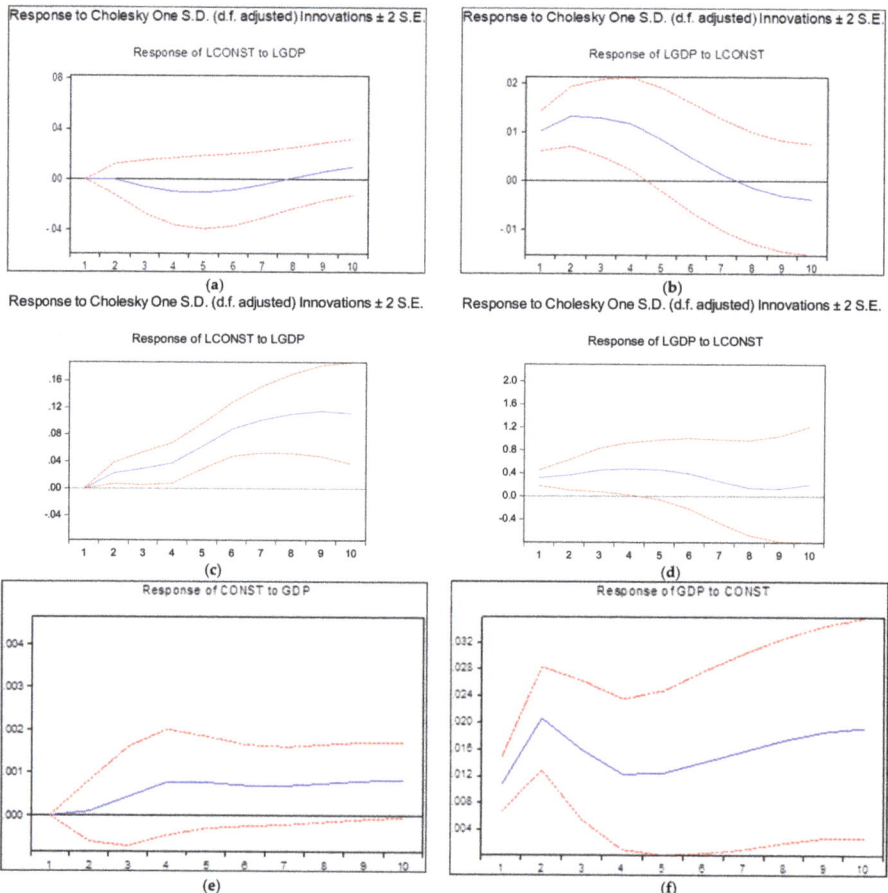

Figure 10. Impact on CONST-GDP and CONST-GDP from a shock. (**a**) Response of CONST to GDP—USA; (**b**) Response of GDP to CONST—USA; (**c**) Response of CONST to GDP—China; (**d**) Response of GDP to CONST—China; (**e**) Response of CONST to GDP—USA—UK; (**f**) Response of GDP to CONST—USA—UK.

Similarly, the Chinese CONST sector will grow considerably after the overall GDP is affected due to any shock. This effect will be longer because the CONST sector of China is the largest in the country, and China will invest all the capital in CONST to boost its overall economy, as shown in Figure 10c. Figure 10d shows the positive change in GDP after a shock is experienced by the CONST sector, although it will be short-lived, and will stabilize as it approaches the end of the tenth period.

Figure 10e shows the positive impact of the CONST sector after the overall GDP of the UK is affected by any shock. There will be a slight increase in the output of the CONST sector, which will continue beyond 10 time periods. Figure 10f illustrates that any shock in the CONST sector will positively affect the performance of the GDP of the UK. Like China, the UK will also support the CONST sector in the case of any shock, which will increase the overall performance of the UK's GDP.

4.9. VECM Forecasting

The forecasting was performed for the USA, China, and UK construction sectors from 2021 to 2050. Based on the findings, the construction sector of the USA is predicted to grow more than twice as much in 2050 as it did in 2020.

Similarly, with China and the USA being at the same point in construction output in 2019, this gap will grow significantly over three decades. The contribution of China's construction sector to the national GDP will increase a little over USD 2 trillion in 2050. However, there will be no significant change in construction output in the UK in 2050. This forecast was made by considering no effects to the laws, policies, and unseen events such as COVID-19. The forecasted values will change significantly if there are pandemics in the future, which will seriously affect the national and global trade-off. The forecasting estimation is shown in Figure 11.

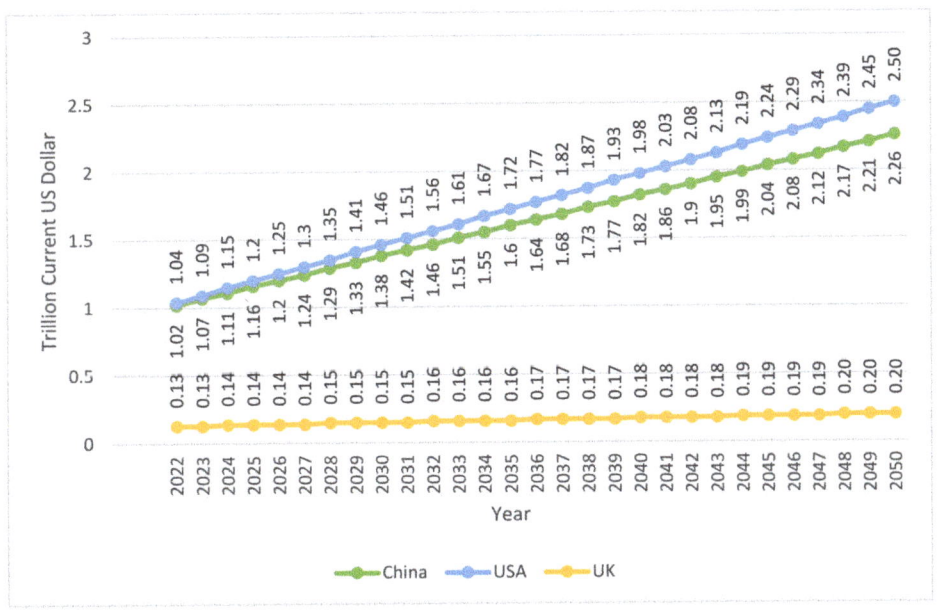

Figure 11. Forecast of the construction sectors of the USA, China, and the UK from 2020 to 2050.

5. Conceptual Framework

By studying the effects on these countries, a sustainable framework was suggested, as shown in Figure 12 that could combine the effects on these countries and be applied on a large scale following the practices of sustainability and economic development in the

CONST sector. First, the directions of linkages among the various sectors are identified [85]. As this study focuses on the CONST sector, this sector is taken as a reference. After the linkages are identified using the Granger test, a pressure point is selected in the CONST sector (e.g., plastic). The short- and long-term effects of the pressure point (such as plastic) are evaluated using VECM. The effects of pressure points are taken into account and their behavior is assessed using IRF when a shock is experienced due to the change in pressure point in the sector. This step constitutes the statistical approach of the framework. The statistical approach and sustainability approach of the construction sector are linked through the enablers of a new ecosystem. The novel economical ecosystem can be created by the conglomeration of circular construction techniques that are financially viable for the pressure points. This includes determining the short- and long-term effects of the pressure points through industrial ecology, eco-economical techniques, outreach and collaboration of new business models, and the eco-designs of the pressure points. Hence, the combined support of government officials and construction bodies can make the necessary shift required for the implementation of sustainability in the economy and the processes of the pressure points. This shift can be regarded as a pocket of change, which can be seen in recyclable materials, optimization of energy efficiency, and circular use of waste materials from construction activities [86]. This will result in the economic resilience of the pressure points, which will directly produce resilience in the design, cost, business models, processes leadership, and government policies of the sector.

The statistical results obtained from this study could be merged with the sustainable procedures. One similar study showing the sustainability outcomes using time series analysis by calculating the carbon footprints was performed. It was found that the time series analysis could provide solutions for sustainable product designs, and sustainable procurement of raw materials [87]. The novelty of this study is that it shows how statistical results could be used to pave a way for sustainability. The use of Granger causality to assess the direction, and the usage of VECM to determine the long- and short-term effects of the behavior of the sector make this framework applicable. The results obtained from the statistical portion would show the fluctuation in the sector after receiving a shock, which could be studied; thus, sustainability concepts could be applied to counter the shocks in the sector. Moreover, the sustainability part of the framework is mostly theoretical, which could be applied to drive the sector towards sustainability.

Comparison of the construction sectors of three developed countries was performed to assess the behavior of the construction sector towards a shock. By setting a standard of behavior for these countries, other countries could benefit by observing how much their respective countries could sustain a shock and how the aftershocks would affect the national GDP; in this way, a great deal of economic recession could be prevented. The comparison was also performed to study the effects of the construction sector of each country on the IRF with subsequent proposal of a global circular framework with suitable application globally.

The value changes over time also entail environmental development. For example, the use of asbestos in construction was once lauded for its fireproofing properties but is now considered dangerous to environmental and social sustainability, as is the case with conventional construction procedures. Hence, the concept of cleaner production, such as social sustainability, environmental impact, and the built environment, is a continuous process. The application of circular strategy implications intensifies the use of "a balance of minimizing the carbon production with economic energy methods" and enables the intensification of redesigning, reusing, and recycling resources. Selective strategies for repairing and maintaining should likewise be applied to ensure efficient upgrading in the construction sector. Finally, the combination of recycling and reusing renewable materials must be made a part of circular strategies to achieve cleaner construction practices.

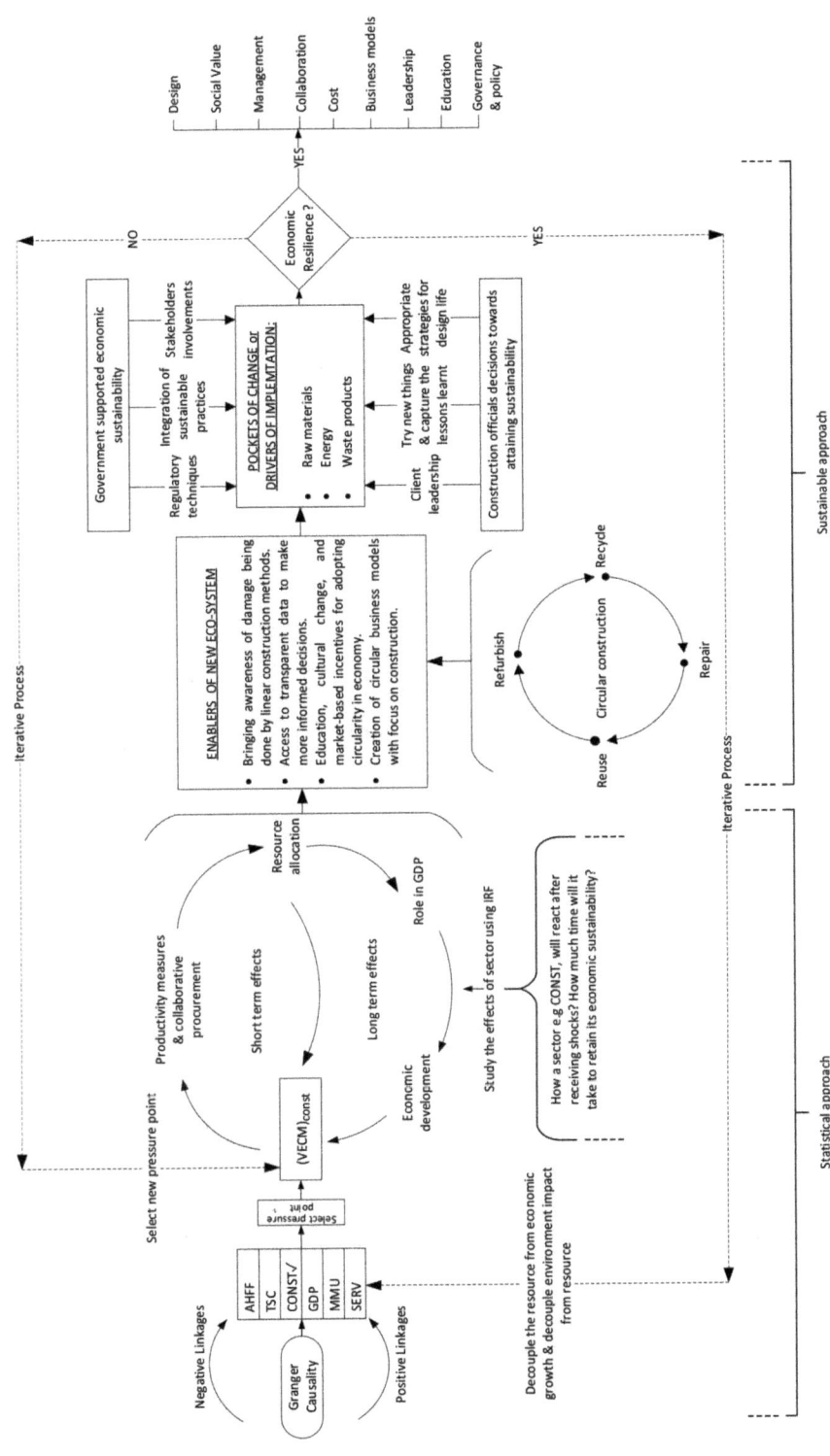

Figure 12. Economic sustainability framework.

6. Discussion

The quantitative analysis of this statistical analysis revealed the contribution, output per period, and average growth of all sectors in relation to the overall GDP of the USA. The average contribution of the CONST sector to the AHFF sector is the smallest with regard to the overall GDP, compared to other sectors of the USA. Similarly, the CONST and AHFF sectors of the UK had a correlation of 85.7%, which were the least correlated, similar to the USA. All the sectors of the USA and UK had first-order integration while China's order of integration was two, suggesting the long-term effects on the variables in these countries. The VECM was selected for modeling long-term relationships. A VECM system indicated the presence of an error correction system in the variables. The long-run coefficient C(1) shows an ability to bounce back to equilibrium. The first coefficient C(1) in the system of variables indicates the speed of adjustment toward the long-run equilibrium of the CONST industry as 0.267%. The negative value of coefficient C(1) implies a negative long-run association with the CONST industry while SERV has a positive association with the CONST sector. This finding means a positive growth or expansion in AHFF, MMU, and TSC will have negative impacts on the growth of the CONST sector. The coefficient C(2) is the short-term speed of adjustment, which means that a percent increase in AHFF will result in a decline of 0.09%. Similarly, a percent increase in C(3) (MMU) will result in a decrease of 0.8% in the CONST sector.

The analysis period selected in this study was from 1970 to 2021. After COVID-19 was declared a pandemic by the World Health Organization (WHO) in 2020, most countries adopted lockdowns nationwide as a preventive measure, which negatively impacted the performance of all sectors. The inclusion of COVID-19-impacted data in this analysis would have produced different results, but it was not made a part of this study due to the unavailability of the officially released data for 2022. It was likewise impossible to analyze the hidden factors that are constantly changing, which would lead to uncertainty in the results. This study performed statistical analysis on the construction sector and other sectors associated with it, and identified the obstacles that should be addressed to make the sectors self-sufficient. It also identified the challenges that are hurdles in achieving sustainability in the construction sector. Any percent change in the construction sector will affect the overall output of the country, depending on how each country takes effective measures to prevent the sector from slipping into recession. The analysis of the construction sector of three countries shows that a sector without any sustainable vision will produce satisfying growth in the short term. However, in the long term, if the rapid implementation of legislation for reducing hydrocarbon emissions is needed, it will prove detrimental to the sector.

6.1. USA Economy

It is expected that the construction industry environment in the USA will become favorable in terms of decreased tariffs under Joe Biden's presidency, given that Trump had increased the cost of construction during his administration, though he did announce a USD 2 trillion relief package for infrastructure development [88]. Accordingly, the USA pledges to cut the use of hydrocarbon fuel and its emissions up to 52% by 2030 [89]. The residential construction sector plays an influential role in the development of construction. Owing to the shock in the economy from the COVID-19 outbreak, millions of people were laid off from construction, which forced first-time buyers to look for low-cost, large spaces for living [90]. However, a question remains about the uncertainty of expansion of residential construction because of COVID-19 lockdowns, which have paralyzed business in this field. It is estimated that the global construction output will increase to USD 8 trillion by 2030, and the top three contributors to construction will include the USA and China, which have a combined construction output of 57% of global growth [91]. Construction in the southern states of the USA is expected to increase, reflecting the higher population growth and catch-up potential of the region. In the next 15 years (up to 2035) [92], the construction industry of the USA will grow faster than in China and will become more dynamic, which will influence the evolution of the prosperity of the society as it will create a vast number of jobs and will ensure wealth and a healthy living standard of the people [93]. Hence, output growth in the construction sector will increase the overall GDP and bring socio-economic prosperity to the country.

6.2. Chinese Economy

Similarly, it is anticipated that the Chinese construction sector will rise to become the largest construction market in the world and will generate USD 13 trillion in revenue by 2030. On the flip side of this marvel development in the construction sector, there will be a draconian production of 28% of the global energy emissions in the absence of sustainable processes [94]. The five-year plan for the sustainability of China limits the use of fossil fuels to 20% with each subsequent passing year [95]. China plans to rely on renewable sources for its energy demands, on-site renewable zero carbon emission energy practices and making public and private real estate sustainable, all of which will drive the country toward a 70% reduction in carbon emission by 2060 [96]. China plans to achieve cleaner production in the construction sector through sustainable urbanization and human settlement, reducing the use of fossil fuels, managing household wastes for rural and urban areas, formulating urban air quality standards, constructing ecological corridors, restoring wetlands, and improving energy savings for existing buildings [97]. To embark on the cleaner environment strategy, as the primary carbon polluters of the world, China and the USA have agreed to cooperate to tackle the climate crisis based on its urgency and reduce fossil fuel emissions by half [98].

6.3. United Kingdom Economy

It is expected that the UK construction sector will surpass the German sector and become the sixth-largest construction sector and the largest one in Europe by 2030 [99]. The UK has introduced the vision Construction 2025, which includes benefits such as reducing GHG emissions by 50%, reducing construction costs by 33%, building cheaper homes and ensuring their fast delivery by 50%, designing smart and safer buildings, sustainable practices, and cleaner production [22]. The World Economic Forum introduced the Infrastructure and Urban Development Industry Vision 2050, which envisions minimum carbon and resilient construction solutions. The application of low-carbon emissions and innovation of cleaner practices can be implemented through the collaboration of the stakeholders, performance-based delivery of construction practices, a skilled workforce, use of digital systems to optimize social, economic, and social benefits, reducing risks by responding quickly to the losses, and application of a long-term lifecycle of optimized solutions [100].

The VECM analysis of Australian construction markets shows that other sectors' price hikes increase construction prices, which is also evident in this study, with other sectors such as SERV, TSC, MMU, and AHFF increasing *in price* [101]. In another study to forecast construction demand in Australia, it was found that the VECM model of forecasting construction demand was affected by GDP, population, and exports [102]. The results of this study also indicate that the construction sector has forward linkages to GDP and other sectors. An empirical investigation of the construction sector and economic growth was performed in Saudi Arabia. It was found that there is a unidirectional causality and long-run effects in the construction sector [103]. This study also revealed that there would be short-run effects of a shock in the construction sector and long-run effects on other sectors' behavior along with economic growth.

Although many studies were conducted to study the behavior of the construction sector and its contribution to the cumulative economy, there is an absence of studies that use statistical processes such as IRF to study the behavior of the construction sector toward the road to sustainability and cleaner production. Therefore, this study closes the literature gap by proposing the use of statistical inferences that can be employed in a circular economy to bring the construction sector one step closer to sustainability.

The methodology followed by the current study suggests that IRF can adequately be used to forecast shock behavior by creating an impulse in the multivariate vector autoregressive function. This statement validates the methodology followed by previous studies, indicating that a single variable can have instantaneous effects on the second to last variables; therefore, the concept of the shock is the correct representation within the economic system.

How, then, can the construction sector be made a sustainable sector? First, we have to invest in work density, which is the only way to be energy-productive, resource-productive, flow optimally, and use the stock of resources optimally. Second, we have to make sure it is diverse in a way that can be used within the work density. Third, work density must be reintegrated back into the city. Finally, the novel availability of renewable resources and the accessibility of micro-mobility must be returned to construction sites.

7. Conclusions

This study used a dynamic statistical approach to determine the role of each sector in the cumulative economy. Three countries (the USA, China, and the UK) were selected as the scope of analysis to study the behavior of the CONST sector. Based on the results, it was concluded that the CONST sector is volatile, can trigger a recession in the economy, and behaves differently to other sectors due to its nature. The CONST sector has short- and long-term effects on itself and other sectors. To achieve sustainability, the CONST sector would have to make use of carbon processes to return to its original position after experiencing a shock. After regaining its former position, only then could the CONST sector move towards sustainability; hence, for this purpose, a combination of statistical processes and the pillars of sustainability must be followed. Adopting new pressure points and analyzing their effects in the sector, along with their quantitative role in the CONST sector, could lead to sustainability in the sector. However, the absence of a sustainable framework to support these sectors will lead to environmental degradation, which can be prevented by implementing the proposed framework.

The USA is facing challenges in the CONST sector due to the hike in prices of major materials for construction and lack of employment. Aside from other reasons, the COVID-19 pandemic has also disrupted the overall sector. Currently, the hurdle in the UK construction sector is to enable the local skilled workforce to adopt digital mechanistic approaches while retaining the current labor force and its traditional knowledge of construction techniques. This scenario leads to poor productivity and insufficient innovation, which draw attention to and emphasize sustainable novel procedures in designing and planning. The challenges faced by the Chinese construction sector include environmental problems, such as excessive pollution, the difference in income between Eastern and Western construction workforces, insufficient capital funding, and strict banking loans for supporting the construction sector.

The theoretical framework selected for this study best suited and was proved adequate to answer the research problems. The use of Granger and impulse response concepts established that the construction sector is linked to other sectors, with dynamic output effects on future series; hence, the variable satisfies the hypothesis of linkages and the effects of the study of economic sectors.

In light of the above conclusion, the following research hypotheses were concluded. Firstly, the CONST sector is greatly dependent on other sectors for its material supply; any major change in this sector would also affect the performance of other sectors; hence, the related hypothesis is rejected. Also, recovering from a major shock would require an increased demand in energy, which would increase the threshold to attain sustainability in the CONST sector. Secondly, the construction sector must move beyond sustainable processes to achieve its former position and construction should move in the same direction as the national economy to regain its former GDP growth. Thirdly, investors and policy makers should pay attention to emphasizing a cleaner construction sector in terms of policies, stakeholder trust, human resources, and capacity-building programs; hence, the related hypothesis is rejected. Finally, the results of the IRF and VECM could be incorporated to assess the behavior of the construction sector if sustainability is introduced in this sector; hence, the related hypothesis is rejected. Hence, the short- and long-term effects of the construction sector on employment, output, and circularity indicate that to regain its original position, the construction sector would have to make excessive use of CO_2 to run its engine; therefore, in the long term, the construction sector would become unsustainable.

8. Recommendation and Future Prospects

This study recommends an in-depth examination into the identification of problems facing the construction sector, which has the potential to generate a considerable output and prove fruitful in helping the country recover from times of economic recession. Joint venture programs between multinational firms and local authorities must be increased to grow human capital, increase skilled labor for cleaner production, execute required work in decreased time, achieve resource efficiency, and develop machine-operating skilled workers. Knowledge sharing should be encouraged to increase the trust of local and overseas investors, which will raise foreign direct investments in the construction sector that could pave the way for a circular strategy. Finally, the process of tendering, sustainable construction methods, integrated solutions for the development of methods, loans and funding for research, contract agreement, and transparent payment procedures must be applied to the construction sector, which will attract many stakeholders to invest in innovative procedures in the sector.

Future research should focus on construction multipliers. This study used the GDP of the CONST sector but data were limited. In the future, other contributing factors, such as health, revenues, employment rates, biodiversity loss, and material consumption, must be analyzed to reflect the accomplishment of suitable practices in the CONST sector. This will help determine the factors for lagging in sustainable development, hence enabling the government and private firms to make informed policies for sustainable development in the CONST sector. Future work should focus on replicating the statistical procedure in individual geological regions (states or neighboring countries). This individual focus will not only help identify differences at the small scale in the drivers of change and their enablers, but also develop deeper insight into understanding corporate and stakeholder roles.

Author Contributions: Data curation, M.A.M., M.B.A.R., K.M.A. and M.A.S.; Formal analysis, M.A.M., M.B.A.R. and M.A.; Funding acquisition, W.S.A.; Investigation, M.A.M. and M.B.A.R.; Methodology, W.S.A., M.A.M., M.B.A.R., K.M.A. and M.A.S.; Project administration, W.S.A. and K.M.A.; Resources, W.S.A.; Software, M.A.M. and M.A.; Validation, M.A.; Visualization, M.A. and M.A.S.; Writing—original draft, M.B.A.R.; Writing—review & editing, W.S.A., M.A.M. and K.M.A. All authors have read and agreed to the published version of the manuscript.

Funding: This research received no external funding.

Institutional Review Board Statement: Not applicable.

Informed Consent Statement: Not applicable.

Data Availability Statement: Not applicable.

Acknowledgments: The authors would like to appreciate the University Internal Research Funding (URIF) project (cost center # 015LB0-051) in Universiti Teknologi PETRONAS (UTP) awarded to Wesam Alaloul for the support.

Conflicts of Interest: The authors declare no conflict of interest.

Appendix A. USA Data

Table A1. Data description of the USA & China sectors (in USD trillion).

YEAR	CONST	MMU	AHFF	SERV	TSC	GDP	CONST	MMU	AHFF	SERV	TSC	GDP
1970	0.051	0.288	0.025	0.416	0.104	1.07	0.0084	0.010	0.08	0.083	0.026	0.22
1971	0.056	0.305	0.026	0.457	0.113	1.16	0.009	0.0108	0.083	0.093	0.039	0.245
1972	0.063	0.335	0.031	0.5004	0.127	1.27	0.009	0.012	0.083	0.100	0.0292	0.255
1973	0.07	0.374	0.045	0.552	0.141	1.42	0.01	0.013	0.091	0.108	0.0301	0.275
1974	0.074	0.401	0.044	0.604	0.155	1.54	0.01	0.0132	0.095	0.109	0.031	0.282
1975	0.076	0.432	0.0458	0.666	0.162	1.68	0.012	0.0128	0.097	0.125	0.033	0.303
1976	0.085	0.491	0.044	0.736	0.184	1.87	0.103	0.0147	0.0975	0.121	0.0351	0.298
1977	0.094	0.558	0.045	0.824	0.205	2.08	0.013	0.0165	0.095	0.138	0.37	0.325
1978	0.11	0.625	0.052	0.936	0.233	2.35	0.013	0.018	0.102	0.162	0.0421	0.367
1979	0.126	0.69	0.062	1.04	0.257	2.63	0.014	0.019	0.127	0.178	0.046	0.41
1980	0.13	0.746	0.056	1.18	0.279	2.85	0.019	0.021	0.137	0.201	0.0553	0.458
1981	0.131	0.861	0.068	1.32	0.31	3.2	0.02	0.022	0.155	0.206	0.059	0.493
1982	0.132	0.866	0.065	1.44	0.322	3.34	0.022	0.024	0.177	0.218	0.071	0.537
1983	0.14	0.903	0.052	1.58	0.353	3.63	0.027	0.027	0.197	0.239	0.082	0.602
1984	0.167	1.09	0.07	1.75	0.386	4.03	0.031	0.033	0.231	0.281	0.102	0.727
1985	0.188	1.03	0.07	1.92	0.409	4.33	0.041	0.0421	0.256	0.347	0.127	0.909
1986	0.21	1.03	0.068	2.09	0.428	4.57	0.052	0.049	0.278	0.4	0.154	1.037
1987	0.22	1.09	0.074	2.28	0.461	4.855	0.066	0.056	0.323	0.462	0.183	1.21
1988	0.238	1.19	0.074	2.49	0.491	5.23	0.081	0.068	0.386	0.581	0.227	1.51
1989	0.246	1.24	0.085	2.69	0.505	5.64	0.079	0.081	0.426	0.652	0.296	1.71
1990	0.249	1.28	0.09	2.89	0.529	5.96	0.086	0.116	0.506	0.69	0.33	1.88
1991	0.233	1.28	0.085	3.06	0.562	6.15	0.100	0.142	0.534	0.813	0.381	2.2
1992	0.235	1.32	0.093	3.27	0.591	6.52	0.14	0.168	0.586	1.03	0.489	2.71
1993	0.249	1.38	0.09	3.43	0.627	6.85	0.22	0.217	0.696	1.42	0.648	3.56
1994	0.276	1.48	0.098	3.59	0.674	7.28	0.29	0.278	0.957	1.95	0.897	4.86
1995	0.291	1.56	0.091	3.81	0.711	7.63	0.37	0.324	1.21	2.5	1.12	6.13
1996	0.317	1.6	0.108	4.03	0.742	8.07	0.43	0.378	1.4	2.95	1.31	7.18
1997	0.339	1.67	0.108	4.34	0.784	8.57	0.46	0.414	1.44	3.3	1.55	7.97
1998	0.379	1.69	0.998	4.59	0.863	9.06	0.49	0.466	1.48	3.41	1.788	8.51
1999	0.417	1.77	0.092	4.9	0.939	9.63	0.51	0.517	1.47	3.6	1.99	9.05
2000	0.461	1.86	0.098	5.27	0.956	10.25	0.55	0.616	1.49	4.02	2.307	10.02
2001	0.486	1.8	0.099	5.58	0.988	10.58	0.59	0.687	1.57	4.38	2.689	11.086
2002	0.493	1.78	0.095	5.87	1.03	10.93	0.64	0.749	1.65	4.77	3.07	12.171
2003	0.525	1.87	0.114	6.14	1.06	11.45	0.75	0.791	1.73	5.53	3.495	13.742
2004	0.594	2.005	0.142	6.48	1.16	12.21	0.87	0.93	2.14	6.57	4.05	16.184
2005	0.651	2.15	0.128	6.93	1.21	13.03	1.04	1.066	2.241	7.79	4.05	18.731
2006	0.697	2.32	0.125	7.32	1.27	13.81	1.24	1.21	2.4	9.22	5.72	21.94
2007	0.715	2.42	0.144	7.68	1.35	14.45	1.53	1.46	2.84	11.16	7.34	27.009
2008	0.648	2.47	0.147	7.89	1.41	14.71	1.88	1.63	3.34	12.172	8.611	31.92

Table A1. Cont.

YEAR	CONST	MMU	AHFF	SERV	TSC	GDP	CONST	MMU	AHFF	SERV	TSC	GDP
2009	0.565	2.27	0.13	8.05	1.37	14.44	2.26	1.65	3.46	13.8	10.06	34.85
2010	0.525	2.42	0.146	8.3	1.44	14.99	2.72	1.87	3.96	16.51	11.771	41.21
2011	0.524	2.55	0.18	8.56	1.48	15.54	3.29	2.18	4.61	19.51	13.96	48.79
2012	0.553	2.608	0.179	8.96	1.53	16.19	3.68	2.37	5.05	20.89	15.9	53.85
2013	0.587	2.707	0.215	9.19	1.62	16.78	4.08	2.6	5.46	22.23	18.24	59.29
2014	0.636	2.81	0.201	9.63	1.67	17.52	4.54	2.85	5.47	23.31	20.45	64.35
2015	0.694	2.738	0.182	10.09	1.81	18.22	4.77	3.05	5.98	23.49	23.57	68.88
2016	0.746	2.66	0.166	10.48	1.91	18.71	5.14	3.3	6.24	24.54	26.66	74.63
2017	0.79	2.82	0.176	10.89	1.99	19.51	5.79	3.71	6.46	27.51	30.101	83.2
2018	0.84	3.02	0.718	11.47	2.12	20.58	6.54	4.03	6.75	30.108	33.93	91.92
2019	0.892	3.05	0.175	11.95	2.26	21.43	7.09	4.28	7.35	31.71	37.00	99.08
2020	0.895	2.85	0.174	11.84	2.17	20.89	6.10	4.53	7.11	30.02	40.22	101.59

Appendix B. UK Data

Table A2. Data description of the UK sectors (in USD trillion).

YEAR	CONST	MMU	AHFF	SERV	TSC	GDP
1970	0.0036	0.0173	0.0012	0.019	0.0053	0.054
1971	0.0042	0.019	0.0013	0.022	0.00609	0.06
1972	0.0049	0.021	0.0015	0.025	0.0069	0.068
1973	0.0059	0.0239	0.0017	0.03	0.00813	0.078
1974	0.007	0.026	0.0019	0.035	0.00939	0.088
1975	0.0089	0.0328	0.0024	0.044	0.0118	0.109
1976	0.01	0.039	0.0027	0.053	0.0134	0.129
1977	0.0115	0.0462	0.003	0.062	0.0149	0.15
1978	0.012	0.054	0.0033	0.072	0.016	0.175
1979	0.014	0.064	0.0037	0.085	0.018	0.207
1980	0.016	0.075	0.004	0.1003	0.0205	0.243
1981	0.017	0.082	0.0043	0.112	0.022	0.269
1982	0.019	0.088	0.0045	0.124	0.024	0.294
1983	0.0204	0.095	0.0047	0.138	0.026	0.323
1984	0.021	0.1002	0.0048	0.149	0.027	0.346
1985	0.023	0.107	0.005	0.165	0.029	0.381
1986	0.026	0.111	0.0055	0.177	0.033	0.41
1987	0.03	0.117	0.0062	0.194	0.038	0.455
1988	0.035	0.126	0.007	0.216	0.045	0.511
1989	0.041	0.134	0.008	0.24	0.052	0.566
1990	0.042	0.14	0.0083	0.269	0.056	0.615
1991	0.039	0.141	0.0083	0.289	0.059	0.647
1992	0.036	0.144	0.0087	0.313	0.061	0.672
1993	0.035	0.151	0.0092	0.334	0.0632	0.707

Table A2. Cont.

YEAR	CONST	MMU	AHFF	SERV	TSC	GDP
2015	0.109	0.2437	0.012	0.943	0.175	1.91
2016	0.113	0.243	0.0114	0.987	0.187	1.99
2017	0.119	0.256	0.0118	1.01	0.194	2.06
2018	0.123	0.264	0.012	1.05	0.202	2.14
2019	0.129	0.265	0.013	1.08	0.213	2.21
1994	0.038	0.161	0.0093	0.355	0.067	0.745
1995	0.0407	0.169	0.0107	0.374	0.069	0.85
1996	0.043	0.182	0.0099	0.398	0.074	0.907
1994	0.038	0.161	0.0093	0.355	0.067	0.745
1995	0.0407	0.169	0.0107	0.374	0.069	0.85
1996	0.043	0.182	0.0099	0.398	0.074	0.907
1997	0.0434	0.1849	0.0092	0.415	0.081	0.951
1998	0.0491	0.184	0.0096	0.433	0.089	0.997
1999	0.055	0.1846	0.0091	0.45	0.095	1.03
2000	0.059	0.193	0.0093	0.474	0.103	1.09
2001	0.062	0.188	0.0089	0.504	0.107	1.13
2002	0.068	0.189	0.0109	0.532	0.111	1.18
2003	0.071	0.193	0.011	0.571	0.119	1.25
2004	0.072	0.194	0.0101	0.612	0.124	1.31
2005	0.0803	0.202	0.0076	0.659	0.128	1.39
2006	0.085	0.2135	0.0088	0.699	0.132	1.47
2007	0.0921	0.2133	0.0087	0.743	0.142	1.54
2008	0.094	0.223	0.018	0.766	0.147	1.58
2009	0.081	0.211	0.016	0.772	0.144	1.54
2010	0.082	0.222	0.0097	0.785	0.148	1.6
2011	0.085	0.224	0.0115	0.807	0.154	1.66
2012	0.088	0.231	0.011	0.837	0.158	1.71
2013	0.095	0.241	0.0114	0.868	0.163	1.78
2014	0.101	0.2437	0.014	0.91	0.17	1.86

References

1. Huang, L.; Krigsvoll, G.; Johansen, F.; Liu, Y.; Zhang, X. Carbon emission of global construction sector. *Renew. Sustain. Energy Rev.* **2018**, *81*, 1906–1916. [CrossRef]
2. Durdyev, S.; Ismail, S. Role of the construction industry in economic development of Turkmenistan. *J. Chang.* **2012**, *64*, 883–890.
3. Giang, D.T.; Pheng, L.S. Role of construction in economic development: Review of key concepts in the past 40 years. *Habitat Int.* **2011**, *35*, 118–125. [CrossRef]
4. Lewis, T.M. Quantifying the GDP-construction relationship. *Econ. Mod. Built Environ.* **2009**, 34–59. [CrossRef]
5. Lean, C.S. Empirical tests to discern linkages between construction and other economic sectors in Singapore. *Constr. Manag. Econ.* **2001**, *19*, 355–363. [CrossRef]
6. Rangelova, F. *Fundamentals of Economics in Sustainable Construction*; Bultest Standard Ltd.: Sofia, Bulgaria, 2015.
7. Lopes, J. Construction in the economy and its role in socio-economic development. In *New Perspectives on Construction in Develop Countries*, 1st ed.; Routledge: London, UK, 2012; pp. 41–71.
8. Ofori, G. Construction industry and economic growth in Singapore. *J. Constr. Manag. Econ.* **1988**, *6*, 57–70. [CrossRef]
9. Stewardson, L. Turkey's Construction Industry Plunges into Recession. 2019. Available online: https://www.worldcement.com/africa-middle-east/05042019/turkeys-construction-industry-plunges-into-recession/ (accessed on 19 April 2021).

10. Knoema. Japan-Value Added in Construction in Current Prices. 2020. Available online: https://knoema.com/atlas/Japan/topics/Economy/National-Accounts-Value-Added-by-Economic-Activity-at-current-prices-US-Dollars/Value-added-in-construction (accessed on 20 February 2020).
11. Knoema. Canada-Value Added in Construction in Current Prices. 2020. Available online: https://knoema.com/atlas/Canada/topics/Economy/National-Accounts-Value-Added-by-Economic-Activity-at-current-prices-US-Dollars/Value-added-in-construction (accessed on 24 February 2020).
12. Knoema. Turkey-Value Added in Construction in Current Prices. 2020. Available online: https://knoema.com/atlas/Turkey/topics/Economy/National-Accounts-Value-Added-by-Economic-Activity-at-current-prices-US-Dollars/Value-added-in-construction (accessed on 20 February 2020).
13. Knoema. India-Value Added in Construction. 2020. Available online: https://knoema.com/atlas/India/topics/Economy/National-Accounts-Value-Added-by-Economic-Activity-at-current-prices-US-Dollars/Value-added-in-construction (accessed on 20 February 2020).
14. Knoema. Pakistan-Value Added in Construction. 2020. Available online: https://knoema.com/atlas/Pakistan/topics/Economy/National-Accounts-Value-Added-by-Economic-Activity-at-current-prices-US-Dollars/Value-added-in-construction (accessed on 20 February 2020).
15. Knoema. Australia-Value Added in Construction. 2020. Available online: https://knoema.com/atlas/Australia/topics/Economy/National-Accounts-Value-Added-by-Economic-Activity-at-current-prices-US-Dollars/Value-added-in-construction (accessed on 20 February 2020).
16. Knoema. Finland-Value Added in Construction in Current Prices. 2020. Available online: https://knoema.com/atlas/Finland/topics/Economy/National-Accounts-Value-Added-by-Economic-Activity-at-current-prices-US-Dollars/Value-added-in-construction (accessed on 20 February 2020).
17. Knoema. United Kingdom-Value Added in Construction in Current Prices. 2021. Available online: https://knoema.com/atlas/United-Kingdom/topics/Economy/National-Accounts-Value-Added-by-Economic-Activity-at-current-prices-US-Dollars/Value-added-in-construction (accessed on 20 February 2020).
18. Knoema. United States of America-Value Added in Construction in Current Prices. 2021. Available online: https://knoema.com/atlas/United-States-of-America/topics/Economy/National-Accounts-Value-Added-by-Economic-Activity-at-current-prices-US-Dollars/Value-added-in-construction (accessed on 20 February 2020).
19. Statista Research Department. U.S. Construction Industry-Statistics & Facts. 2020. Available online: https://www.statista.com/topics/974/construction/ (accessed on 20 February 2020).
20. Khan, R.A. Role of construction sector in economic growth: Empirical evidence from Pakistan economy. In Proceedings of the First International Conference on Construction in Developing Countries (ICCIDC), Karachi, Pakistan, 4–5 August 2008.
21. Alsharef, A.; Banerjee, S.; Uddin, S.; Albert, A.; Jaselskis, E. Early Impacts of the COVID-19 Pandemic on the United States Construction Industry. *Int. J. Environ. Res. Public Health* **2021**, *18*, 1559. [CrossRef]
22. Department for Business, Energy & Industrial Strategy. *Construction Sector Deal*; Department for Business: London, UK, 2019.
23. Green, B. *The Real Face of Construction 2020*; The Chartered Institute of Building: Berkshire, UK, 2020.
24. CIOB. *The Real Face of Construction 2020, Socio-Economic Analysis of the True Value of the Built Environment*; CIOB: Hong Kong, China, 2020.
25. Wood, L. *United Kingdom Construction Market Outlook 2020–2024: Disruptions Caused by COVID-19 Pandemic and Brexit Implications*; Research and Markets: Dublin, Ireland, 2020.
26. Allcoat, J. *Construction Output in Great Britain: February 2021*; Office for National Statistics: Newport, UK, 2021.
27. Department, S.R. Market Value of the Construction Industry in China from 2018 to 2021. 2021. Available online: https://www.statista.com/statistics/1068161/china-construction-industry-value/#:~{}:text=China%27s%20construction%20industry%20value%202018%20to%202021&text=China%20has%20become%20the%20largest,to%201%2C049%20billion%20U.S.%20dollars (accessed on 2 December 2021).
28. Department, S.R. GDP Share of the Construction Industry Value in China from 2018 to 2021. 2019. Available online: https://www.statista.com/statistics/1068213/china-construction-industry-gdp-contribution-share/#:~{}:text=China%27s%20construction%20industry%20contribution%20share%20to%20GDP%202018%20to%202021&text=In%202018%2C%20the%20construction%20industry,a%20five%20percent%20annual%20growth (accessed on 2 December 2021).
29. Gao, J.; Tang, X.; Ren, H.; Cai, W. Evolution of the Construction Industry in China from the Perspectives of the Driving and Driven Ability. *Sustainability* **2019**, *11*, 1772. [CrossRef]
30. IBISWord. Building Construction Industry in China-Market Research Report. 2020. Available online: https://www.ibisworld.com/china/market-research-reports/building-construction-industry/ (accessed on 2 December 2021).
31. GlobalData. *China Construction Market Report 2021-Key Trends and Opportunities to 2025 N.B.o.S*; GlobalData: London, UK, 2021.
32. Neill, P. Construction Industry Accounts for 38% of CO_2 Emissions. 2019. Available online: https://webcache.googleusercontent.com/search?q=cache:oa0aspGTU_kJ:https://environmentjournal.online/articles/emissions-from-the-construction-industry-reach-highest-levels/+&cd=3&hl=zh-CN&ct=clnk&gl=kr (accessed on 2 December 2021).
33. Kneese, A.V. The economics of natural resources. *Popul. Dev. Rev.* **1988**, *14*, 281–309. [CrossRef]
34. ICLEI. Circular Development Pathway. 2020. Available online: https://africa.iclei.org/pathways_cat/circular-development-pathway/ (accessed on 2 December 2021).

35. L'économie Circulaire. Minstere De La Transition Eologique. 2020. Available online: https://www.ecologie.gouv.fr/leconomie-circulaire (accessed on 5 May 2021).
36. Geissdoerfer, M.; Savaget, P.; Bocken, N.M.P.; Jan Hultink, E. The Circular Economy—A new sustainability paradigm? *J. Clean. Prod.* **2017**, *143*, 757–768. [CrossRef]
37. Hysa, E.; Kruja, A.; Rehman, N.U.; Laurenti, R. Circular Economy Innovation and Environmental Sustainability Impact on Economic Growth: An Integrated Model for Sustainable Development. *Sustainability* **2020**, *12*, 4831. [CrossRef]
38. Corona, B.; Shen, L.; Reike, D.; Carreón, J.R.; Worrell, E. Towards sustainable development through the circular economy—A review and critical assessment on current circularity metrics. *Resour. Conserv. Recycl.* **2019**, *151*, 104498. [CrossRef]
39. CIOB. Socio-Economic Impact of Construction Industry. 2020. Available online: https://www.ciob.org/industry/policy-research/policy-positions/socio-economic-impact-construction (accessed on 2 December 2021).
40. Mok, K.Y.; Shen, G.Q.; Yang, J. Stakeholder management studies in mega construction projects: A review and future directions. *Int. J. Proj. Manag.* **2015**, *33*, 446–457. [CrossRef]
41. Schipper, L.; Unander, F.; Murtishaw, S.; Ting, M. Indicators of energy use and carbon emissions: Explaining the Energy Economy Link. *Annu. Rev. Energy Environ.* **2001**, *26*, 49–81. [CrossRef]
42. RMIT University. The Four Pillars of Sustainability. 2021. Available online: https://www.futurelearn.com/info/courses/sustainable-business/0/steps/78337 (accessed on 22 March 2022).
43. Gül, Z.; Çağatay, S.; Taşdoğan, C. Input-Output Analysis of Turkish Construction Industry by using World Input-Output Database for 2002–2011 Period. In Proceedings of the 22nd International Input-Output Conference, Lisbon, Portugal, 14–18 July 2020.
44. Osei, D.B.; Aglobitse, P.B.; Bentum-Ennin, I. Relationship between construction expenditure and economic growth in sub-Saharan Africa. *Ghana. J. Econ.* **2017**, *5*, 28–55.
45. Chia, F.C. Construction and economic development: The case of Malaysia. *Int. J. Constr. Manag.* **2012**, *12*, 23–35. [CrossRef]
46. Oladinrin, T.; Ogunsemi, D.; Aje, I.J.F. Role of construction sector in economic growth: Empirical evidence from Nigeria. *FUTY J. Environ.* **2012**, *7*, 50–60. [CrossRef]
47. Chiang, Y.-H.; Tao, L.; Wong, F.K. Causal relationship between construction activities, employment and GDP: The case of Hong Kong. *Habitat Int.* **2015**, *46*, 1–12. [CrossRef]
48. Anaman, K.A.; Osei-Amponsah, C. How The Growth of The Construction Industry Can Help Accelerate Economic Development. *Ghanaian Times Newspaper*, 2017.
49. Anaman, K.A.; Egyir, I.S. Economic Shocks and the Growth of the Construction Industry in Ghana Over the 50-Year Period From 1968 to 2017. *Res. World Econ.* **2019**, *10*, 1. [CrossRef]
50. Nasir, M.H.a.H. Effect of construction industry on country economy and its economic growth analysis-evidence from Afghanistan. *Int. J. Creat. Res. Thoughts* **2020**, *8*, 2786–2799.
51. Erol, I.; Unal, U. *Role of Construction Sector in Economic Growth: New Evidence from Turkey*; MPRA Pap. 68263; University Library of Munich: Munich, Germany, 2015. Available online: https://mpra.ub.uni-muenchen.de/68263/ (accessed on 8 September 2021).
52. Göksu, S.; Şen, M.A.; Gücek, S. İnşaat Sektörü, faiz orani ve ekonomik büyüme ilişkisinin analizi: Türkiye örneği (2002–2019). *Ekev Akademi Dergisi* **2019**, *80*, 465–482. [CrossRef]
53. Alaloul, W.; Musarat, M.; Rabbani, M.; Iqbal, Q.; Maqsoom, A.; Farooq, W. Construction Sector Contribution to Economic Stability: Malaysian GDP Distribution. *Sustainability* **2021**, *13*, 5012. [CrossRef]
54. Wang, J. Vision of China's future urban construction reform: In the perspective of comprehensive prevention and control for multi disasters. *Sustain. Cities Soc.* **2020**, *64*, 102511. [CrossRef] [PubMed]
55. Ghaffar, S.H.; Burman, M.; Braimah, N. Pathways to circular construction: An integrated management of construction and demolition waste for resource recovery. *J. Clean. Prod.* **2020**, *244*, 118710. [CrossRef]
56. Hazarika, N.; Zhang, X. Factors that drive and sustain eco-innovation in the construction industry: The case of Hong Kong. *J. Clean. Prod.* **2019**, *238*, 117816. [CrossRef]
57. Stanitsas, M.; Kirytopoulos, K.; Leopoulos, V. Integrating sustainability indicators into project management: The case of construction industry. *J. Clean. Prod.* **2020**, *279*, 123774. [CrossRef]
58. Zhang, L.; Liu, B.; Du, J.; Liu, C.; Li, H.; Wang, S. Internationalization trends of carbon emission linkages: A case study on the construction sector. *J. Clean. Prod.* **2020**, *270*, 122433. [CrossRef]
59. Pervez, H.; Ali, Y.; Petrillo, A. A quantitative assessment of greenhouse gas (GHG) emissions from conventional and modular construction: A case of developing country. *J. Clean. Prod.* **2021**, *294*, 126210. [CrossRef]
60. Yang, B.; Song, X.; Yuan, H.; Zuo, J. A model for investigating construction workers' waste reduction behaviors. *J. Clean. Prod.* **2020**, *265*, 121841. [CrossRef]
61. Chan, J.; Bachmann, C.; Haas, C. Potential economic and energy impacts of substituting adaptive reuse for new building construction: A case study of Ontario. *J. Clean. Prod.* **2020**, *259*, 120939. [CrossRef]
62. Du, Q.; Zhou, J.; Pan, T.; Sun, Q.; Wu, M. Relationship of carbon emissions and economic growth in China's construction industry. *J. Clean. Prod.* **2019**, *220*, 99–109. [CrossRef]
63. Mojumder, A.; Singh, A. An exploratory study of the adaptation of green supply chain management in construction industry: The case of Indian Construction Companies. *J. Clean. Prod.* **2021**, *295*, 126400. [CrossRef]
64. Chuai, X.; Lu, Q.; Huang, X.; Gao, R.; Zhao, R. China's construction industry-linked economy-resources-environment flow in international trade. *J. Clean. Prod.* **2020**, *278*, 123990. [CrossRef]

65. Murtagh, N.; Scott, L.; Fan, J. VSI editorial-Sustainable and resilient construction: Current status and future challenges. *J. Clean. Prod.* **2020**, *268*, 122264. [CrossRef]
66. Koop, G.; Pesaran, M.H.; Potter, S.M. Impulse response analysis in nonlinear multivariate models. *J. Econom.* **1996**, *74*, 119–147. [CrossRef]
67. Ronayne, D. *Which Impulse Response Function?* University of Warwick: Coventry, UK, 2011.
68. Knoema. USA-Value Added in Construction. 2020. Available online: https://knoema.com/atlas/United-States-of-America/topics/Economy/National-Accounts-Value-Added-by-Economic-Activity-at-current-prices-US-Dollars/Value-added-in-agriculture (accessed on 20 February 2020).
69. Johansen, S.; Juselius, K. Testing structural hypotheses in a multivariate cointegration analysis of the PPP and the UIP for UK. *J. Econ.* **1992**, *53*, 211–244. [CrossRef]
70. Johansen, S.; Juselius, K. Maximum likelihood estimation and inference on cointegration—with appucations to the demand for money. *Oxf. Bull. Econ. Stat.* **1990**, *52*, 169–210. [CrossRef]
71. Granger, C.W.J. Investigating Causal Relations by Econometric Models and Cross-spectral Methods. *Econometrica* **1969**, *37*, 424–438. [CrossRef]
72. Koop, G.; Quinlivan, R. *Analysis of Economic Data*; John Wiley & Sons: Hoboken, NJ, USA, 2005.
73. Dawson, P.J. The demand for calories in developing countries. *Oxf. Dev. Stud.* **1997**, *25*, 361–369. [CrossRef]
74. Granger, C. Some properties of time series data and their use in econometric model specification. *J. Econ.* **1981**, *16*, 121–130. [CrossRef]
75. Suharsono, A.; Aziza, A.; Pramesti, W. Comparison of vector autoregressive (VAR) and vector error correction models (VECM) for index of ASEAN stock price. In *AIP Conference Proceedings*; AIP Publishing: Woodbury, NY, USA, 2017.
76. Brown, R.L.; Durbin, J.; Evans, J.M. Techniques for Testing the Constancy of Regression Relationships Over Time. *J. R. Stat. Soc. Ser. B Stat. Methodol.* **1975**, *37*, 149–163. [CrossRef]
77. Perron, P. The great crash, the oil price shock, and the unit root hypothesis. *Econom. J. Econom. Soc.* **1989**, *57*, 1361–1401. [CrossRef]
78. James, H.A. *Time Series Analysis*; Oxford Unviersity: Princeton, NJ, USA, 1944.
79. Lu, C.; Xin, Z. *Impulse-Response Function Analysis: An Application to Macroeconomic Data of China*; Dalarna University: Dalarna, Sweden, 2010.
80. Theil, H. *Economic Policy and Forecasts*; North-Holland: Amsterdam, The Netherlands, 1958.
81. Godfrey, L.G. Testing for higher order serial correlation in regression equations when the regressors include lagged dependent variables. *Econom. J. Econom. Soc.* **1978**, *46*, 1303–1310. [CrossRef]
82. Eviews. Eviews Help and Documnetation. 2020. Available online: http://www.eviews.com/help/helpintro.html#page/content%2Ftesting-Residual_Diagnostics.html%23ww182888 (accessed on 2 December 2021).
83. Breusch, T.S.; Pagan, A.R. A simple test for heteroscedasticity and random coefficient variation. *Econom. J. Econom. Soc.* **1979**, *47*, 1287–1294. [CrossRef]
84. Glen, S. Elementary Statistics for the Rest of Us! 2016. Available online: https://www.statisticshowto.com/breusch-pagan-godfrey-test/ (accessed on 2 December 2021).
85. Munaro, M.R.; Tavares, S.F.; Bragança, L. Towards circular and more sustainable buildings: A systematic literature review on the circular economy in the built environment. *J. Clean. Prod.* **2020**, *260*, 121134. [CrossRef]
86. Osobajo, O.; Omotayo, T.; Oke, A.; Obi, L. Transition towards circular economy implementation in the construction industry: A systematic review. *Smart Sustain. Built Environ.* **2020**, *4*, 34.
87. Hale, J.; Long, S. A Time Series Sustainability Assessment of a Partial Energy Portfolio Transition. *Energies* **2020**, *14*, 141. [CrossRef]
88. Pramuk, J. *Trump Calls for $2 Trillion Infrastructure Package as Part of Coronavirus Response*; CNBC: Englewood Cliffs, NJ, USA, 2020.
89. Sydney Kalich, A.H. President Biden Pledges to Cut US Fossil Fuel Emissions up to 52% by 2030. Newstation. 22 April 2021. Available online: https://www.newsnationnow.com/climate/biden-pushes-for-momentum-as-us-returns-to-climate-fight/ (accessed on 2 December 2021).
90. Lescohier, J. Survey: Over 25% of US Construction Firms Laying Off Workers. Information that Builds and Powers the World. 6 April 2020. Available online: https://www.khl.com/1143291.article (accessed on 2 December 2021).
91. Robinson, G. *Global Construction Market to Grow $8 Trillion by 2030: Driven by China, US and India*; Global Construction Perspectives Ltd.: London, UK, 2021.
92. Markets, R.A. United States Construction Market Outlook to 2024: The Impact of COVID-19 and Projected Recovery. 2020. Available online: https://www.globenewswire.com/news-release/2020/08/25/2083079/0/en/United-States-Construction-Market-Outlook-to-2024-The-Impact-of-COVID-19-and-Projected-Recovery.html (accessed on 2 December 2021).
93. García, M.Á. Challenges of the construction sector in the global economy and the knowledge society. *Int. J. Strateg. Prop. Manag.* **2005**, *9*, 65–77. [CrossRef]
94. Gonçalves, T.; Weyl, D. China Is Investing $13 Trillion in Construction. Will It Pursue Zero Carbon Buildings? 2019. Available online: https://www.wri.org/insights/china-investing-13-trillion-construction-will-it-pursue-zero-carbon-buildings (accessed on 2 December 2021).
95. China Takes Fossil Fuel Projects Off Green Bond List. 2021. Available online: https://www.argusmedia.com/en/news/2208896-china-takes-fossil-fuel-projects-off-green-bond-list (accessed on 2 December 2021).

96. Luxner, L. What China's March to Net-Zero Emissions Means for the World. 2021. Available online: https://www.atlanticcouncil.org/blogs/new-atlanticist/what-chinas-march-to-net-zero-emissions-means-for-the-world/#:~{}:text=Chinese%20carbon-dioxide%20emissions%20could%20peak%20around%202025&text=By%202035%2C%20he%20predicted%2C%20China,to%20carbon%20neutrality%20by%202060 (accessed on 2 December 2021).
97. Chinadaily. *China's National Plan on Implementation of the 2030 Agenda for Sustainable Development*; Chinadaily: Beijing, China, 2016.
98. Kim, H.-J. US, China Agree to Cooperate on Climate Crisis with Urgency. Available online: https://www.npr.org/2021/04/18/988493971/u-s-china-agree-to-cooperate-on-climate-crisis-with-urgency (accessed on 2 December 2021).
99. Genç, E.B.A.H. *Global Construction 2030*; PwC: London, UK, 2020.
100. Davos. Infrastructure and Urban Development Industry Vision 2050. In Proceedings of the World Economic Forum, Klosters, Switzerland, 23–26 January 2018.
101. Liu, J.; London, K.A. Convergence among the sub-markets in Australian regional building construction sector. *Australas. J. Constr. Econ. Build.* **2010**, *10*, 11–33.
102. Jiang, H.; Liu, C. Forecasting construction demand: A vector error correction model with dummy variables. *Constr. Manag. Econ.* **2011**, *29*, 969–979. [CrossRef]
103. Alhowaish, A.K. Causality between the construction sector and economic growth: The case of Saudi Arabia. *Int. Real Estate Rev.* **2015**, *18*, 131–147. [CrossRef]

Article

Barriers to Building Information Modeling (BIM) Deployment in Small Construction Projects: Malaysian Construction Industry

Ahsan Waqar, Abdul Hannan Qureshi and Wesam Salah Alaloul *

Department of Civil and Environmental Engineering, Universiti Teknologi PETRONAS, Seri Iskandar 32610, Perak Darul Ridzuan, Malaysia
* Correspondence: wesam.alaloul@utp.edu.my

Abstract: Building information modeling (BIM) application in construction projects is considered beneficial for effective decision making throughout the project lifecycle, as it maximizes benefits without compromising practicality. The Malaysian construction industry is also keen on the adoption of BIM culture. However, various identified and unidentified barriers are hindering its practical implementation. In light of this, this study identified and analyzed critical obstacles to using BIM in Malaysian small construction projects. Through the use of semi-structured interviews and a pilot study using the exploratory factor analysis (EFA) method, the critical BIM barriers (CBBs) have been identified. Based on the findings of the EFA, CBBs were classified into five categories, i.e., technical adoption barrier, behavioral barrier, implementation barrier, management barrier, and digital education barrier. Following the questionnaire survey, feedback of 235 professionals was collected with vested interests in the Malaysian construction business, and the CBBs model was created using analysis of moment structures (AMOS). The findings revealed that although Malaysian experts with little experience in practice were fairly educated about BIM, technical adoption barriers, behavioral barriers, management barriers, and implementation hurdles were critical for adopting BIM. The study's findings will help policymakers eliminate CBBs and use BIM in Malaysia's modest construction projects to save costs, save time, boost productivity, and improve quality and sustainability.

Keywords: construction sector; barriers; BIM; small contractors; SEM; Malaysia

1. Introduction

Building information modeling (BIM) allows construction and design teams to maximize their existing technological infrastructure. By consolidating all relevant multidisciplinary construction and design documents into a single repository, the BIM process facilitates the development and administration of data across the entire architecture engineering and construction (AEC) project lifecycle [1]. Oyuga et al. [2] described the application of BIM as reviewing and checking the daily on-site performance of work activities in comparison to the created plans and confirming the expected performance before or throughout the project. Moreover, Durdyev et al. [3] emphasized the BIM application in small construction projects to be essential, as it allows construction managers to make choices quickly and accurately based on critical inputs. Successful building projects correlate to how well BIM metrics are used. BIM has the capability to integrate with imaging (videogrammetry, laser scanning, and photogrammetry), geospatial (geographic information system (GIS) and global positioning system (GPS), ultra-wideband (UWB), radio frequency identification (RFID), and barcode), and virtual and augmented reality (VR/AR) technologies [4,5]. Hyarat et al. [6] identified that four-dimensional BIM models are necessary to monitor and analyze the building processes. In addition, BIM has been regarded as the first step toward digital construction and has been integrated with a wide range of construction operations, including facility elevations, prefabricated construction projects, and project management activities [7]. According to Olanrewaju et al. [8], one of BIM's core tasks is effective progress management of construction operations, which was not possible due to the many challenges encountered throughout its adoption. According to Berges-Alvarez et al. [9], there is less room for error and less time to think things through when making sustainability-related decisions. A connection is made between the environment and the economy using BIM. A solid proof of concept may be used to push BIM software into the conceptual design phase. The method is only partially automated but nevertheless allows for well-considered choices to be made during the preliminary stages of a building's design. In Olanrewaju et al. [10], the areas of uncertainty, omission and misuse in BIM-based projects have been identified.

In light of the above discussion, it is clear that further research into the dimensions and technical qualities determining the effective deployment of BIM is necessary to improve the knowledge and trust of stakeholders in the construction sector. According to Abu-Hamdeh et al. [11], the contemporary building business recognizes the need to enhance building energy efficiency and use cutting-edge technology. Lin et al. [12] clarified that the use of three-dimensional modeling has also been proven to be helpful in lowering buildings' harmful effects on the environment. According to Chen et al. [13], the construction sector stakeholders' resistance to embracing technology stems mainly from a need for knowledge of management system standards, requirements, and reference frameworks. The fourth industrial revolution (IR4.0) has accelerated the building industry's transition to digital methods [14]. To realize the vision of a fully digitalized construction environment and to advance the IR4.0 environment, it is necessary to encourage the construction industry and other relevant stakeholders to adopt BIM systems for construction processes by addressing the uncertainties they may have about doing so [15].

Charef et al. [16], Shirowzhan et al. [17], Ahmed and Hosque [18] and Hamid and Embi [19] have identified the barriers to BIM implementation in construction projects without categorization on the basis of the scale of the project. It is indicated by the implications of Hamid and Embi [19] and Alwee et al. [20] that the challenges of BIM implementation are not always the same for small-scale and large-scale construction projects, even in the international context. The aforementioned facts provide a rationale for this study, and the scope has been narrowed down to small construction projects. No specific study, such as Taat et al. [21], Manzoor et al. [22], Belayutham et al. [23] and Chen et al. [13], has targeted the small construction projects from Malaysia indicating BIM barriers. For effective identification of barriers relative to any subject variable, there is always a need for non-parametric statistical evaluation and structural equation modeling (SEM), as indicated by Ringle et al. [24] and Wang and Rhemtulla [25]. According to Arif et al. [26] and Yaakob et al. [27], the Malaysian construction industry contributes majorly to the economy by which small construction projects must adopt modern technologies in which BIM is on top. Small construction projects cannot contribute generously to the Malaysian economy without considering the restructuring of small construction projects with modern construction technologies, including BIM. Following the gap indicated by the abovementioned studies, this research involves exploratory factor analysis (EFA) and SEM, which make it unique in specific Malaysian small construction projects.

To effectively deploy BIM technologies, this research intends to establish the research framework by addressing the theoretical-based technical constraints to adoption. The SEM methodology has been adopted to develop a conceptual framework for BIM barriers for each component and to highlight relevant BIM criteria. This research has focused on the localized obstacles to using BIM technology on modest building projects and how those obstacles might be overcome using different approaches. To help construction industry experts and stakeholders develop confidence in using innovative BIM technologies, this study has built the model taking BIM barriers for effective implementation on small construction projects before or after evaluation activities. The power of this research lies in its capacity to push this area of knowledge toward creating a foundational technical model that would facilitate the more effective use of BIM tools in Malaysia's smaller building projects.

2. Methodology

The technique followed a step-by-step procedure that required thorough identification and evaluation of BIM hurdles in order to resolve the concerns discovered in the research. Figure 1 shows the overall study workflow pipeline adopted to achieve a conceptual model reflecting BIM barriers.

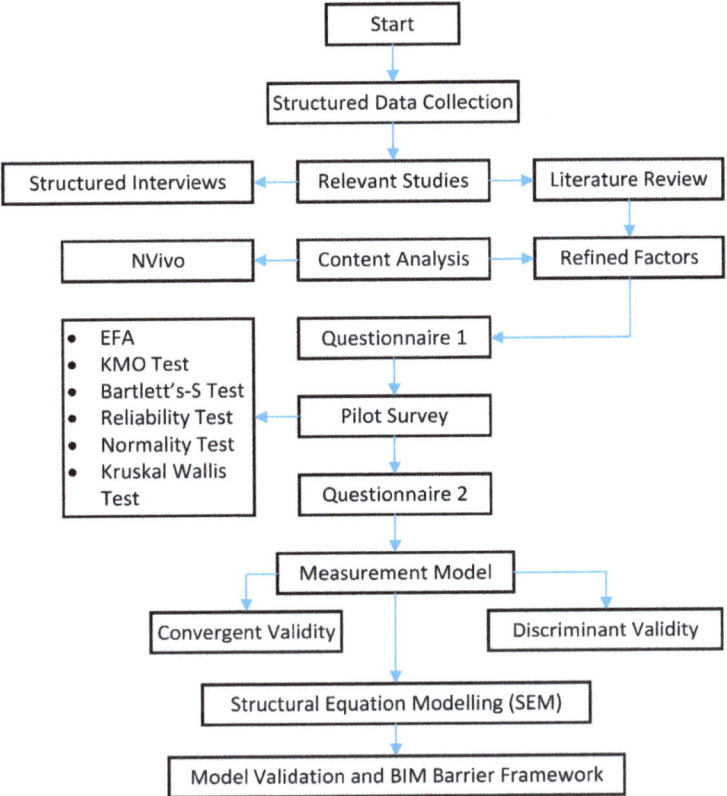

Figure 1. Study flow chart and the successive stages.

First, critical evaluation of the scholarly literature was performed to identify the obstacles toward BIM adoption in small construction projects considering the Malaysian construction industry. Challenges with BIM were identified, and then, those barriers were fine-tuned through semi-structured interviews with five BIM experts and seven experts in small construction projects in Malaysia.

For this study on the challenges of using BIM for small construction projects, it was required to develop a framework. Methods under consideration include multiple linear regression (MLR), structural equation modeling (SEM), system dynamics (SD), exploratory factor analysis (EFA), and artificial neural networks (ANN). According to Julian et el. [28], there is a relationship between unobserved variables; hence, the MLR was not selected. In accordance with Kiraly et al. [29], due to the nature of the research's presented data, SD could not be used. According to Abiodun et al. [30], ANN is a prediction tool, and the purpose of this study is to analyze the difficulties associated with employing BIM on small construction projects. Using the SEM method, several observable and unobservable variables may be defined. SEM has proven to be a beneficial technique in the face of variable inaccuracy. In this study, the SEM approach was utilized to develop a model to identify and find the relationship between BIM barriers and small-scale construction projects. In the social sciences, SEM data are frequently used and acknowledged. Exploratory factor analysis (EFA) and reliability analyses are then used on pilot survey data to see whether any more barriers can be eliminated. At the end of the process, the primary survey with questionnaires was conducted. The confirmatory factor analysis (CFA) technique was used to develop a measurement model by performing convergent and discriminant validity for determining the most significant barriers to implementing BIM in Malaysia's small construction projects.

2.1. Structured Literature Collection

The data-gathering process began with a thorough assessment of BIM barriers in the available literature. The primary goal was to identify the key BIM barriers affecting small building projects. Furthermore, the total strategy was built to find the most agreed-upon BIM hurdles by previous researchers since the critical assessment of current studies is essential for achieving any result. Information was gathered from six different sources, including Springer, Web of Science (WoS), American Society of Civil Engineers (ASCE), Science Direct, Multidisciplinary Digital Publishing Institute (MDPI), and Scopus, while keeping the research timeline between 2011 and 2022. Articles were searched using a number of different keyword combinations, with the flexibility of those keywords being adjusted to fit the overarching subject of the study. We looked for BIM barriers mentioned in previous research publications but restricted our scope to projects of small scale. With all possible keyword combinations, existing papers were searched from the perspective of BIM barriers. The number of studies gathered and relevant studies discovered in the aforementioned databases are summarized in Table 1.

Table 1. Data collection summary.

Database	Keywords Combination	Total Collected Studies	Relevant Studies
Springer	"Small Construction Projects AND BIM OR Challenges or Barriers in BIM Adoption in Small Construction Projects"	827	44
WoS	"BIM Challenges AND Small Construction Projects OR BIM Adoption Barriers in Small Construction Projects"	230	35
ASCE	"Building Information Projects OR BIM in Small Construction Projects OR BIM Barriers OR BIM Challenges"	456	21
Science Direct	"BIM AND Small Construction Projects OR BIM Problems AND Small Construction Projects"	153	87
Scopus	"BIM Challenges OR BIM Hurdles in Small Construction OR Building Information Modeling"	24	6
Google Scholar	"Building Information Modeling OR BIM OR Barriers AND Challenges OR Small Construction Projects"	790	55

With several keyword iterations, we were able to pull in a total of 2480 items from all databases. There were 248 papers that met the criteria for this study after examining their titles, abstracts, and potential barriers. With the goal of identifying BIM hurdles in small building projects, we performed a comprehensive literature study of 248 papers. The majority of barriers identified in the articles had common ground with those that were omitted. There were only 34 barriers that were found to be significant while dealing with small construction projects in Malaysia after the literature research was completed. Table 2 provides a summary of the data collected, including the various types of obstacles found and the categories into which they fall. Barriers to BIM were classified into the following categories, as determined by a review of the relevant literature: human resource barriers, technology barriers, safety barriers, regulatory barriers and financial barriers.

Table 2. Identified BIM barriers and related details.

Categories	Coding	Parameters	Sources
Human Resource Barriers	B1	Lack of digital education and training	[31–35]
	B3	Aside from the construction team leader, no other team members need BIM	[36–40]
	B5	Insufficient teamwork from upper management	[41–44]
	B8	Lack of awareness about the benefits of BIM	[45–51]
	B9	No facilitation and training center for BIM	[18,20,52–58]
	B16	High diversity of workforce in projects	[59–64]
Technology Barriers	B22	Not enough expertise in safety management	[65–68]
	B34	High risk of conflicts in construction contracts	[69–72]
	B21	Poor BIM ability to integrate with project operations	[73–78]
	B23	The construction industry's lackluster adoption of technology	[79,80]
	B31	Lack of flexible modeling capability in BIM tools	[81–85]
	B32	Existing computer-aided design (CAD) tools are appropriate for work	[69–72]
Safety Barriers	B27	Inability to foresee digital technology's positive effects on the safety management process	[8,22,86–88]
	B28	The need for affordable digital tools hinders the safety management process	[16,89–91]
Construction Environment Barriers	B4	Inadequate working processes and quality control standards	[26,92–95]
	B6	Aversion to adopting BIM	[96–102]
	B7	Impractical theoretical evidence from research	[23,103–107]
	B10	Possibility of delays in construction	[17,108,109]
	B11	Absence of a structured methodology that is supportive	[110–114]
	B12	Neither a simple nor universal strategy for BIM Usage exists	[115–117]
	B13	Lack of legal regulations	[118–121]
	B14	In the workplace, resistance to BIM adoption remains strong	[122–129]
	B15	The lack of demand for or insistence on BIM from customers	[21,130–133]
	B17	Too many complexities in design produced by BIM	[27,134–138]
	B20	The integration of BIM will change present levels of efficiency	[139–144]
	B24	Inadequate access to decision-making resources	[145–151]
	B25	The decision to utilize depends on the specifics of each case	[152–156]
Financial Barriers	B2	Competition is high, and profit margins are low	[157–161]
	B18	Impact of COVID-19 on small construction projects	[3,162,163]
	B19	High cost of BIM implementation	[19,164–171]
	B29	Inappropriate rate of return (ROR) and rate of investment (ROI) data	[172–175]
	B26	High ongoing investment in digital infrastructure	[2,176–184]
	B30	Productivity loss when adopting BIM in place of traditional construction	[13,185–188]
	B33	Financial uncertainty related with BIM adoption	[6,189–194]

2.2. Qualitative Analysis (Interview)

For semi-structured interviews, a qualitative questionnaire was prepared involving barriers related to BIM implementation in Malaysian small construction projects. The qualitative questionnaire included all five categories of BIM barriers (human resource barriers, technology barriers, safety barriers, construction environment barriers and financial barriers) based on the BIM categorization from the literature. The required sample size for a semi-structured interview was reported differently in previous studies. Because of the descriptive nature of interviews, the purpose is always to collect as much information as possible. Time is also a factor that limits the number of people involved in interviews. According to literature, the minimum sample size for qualitative interviews must lie between 10 and 20 subjects. In contrast, Dworkin [195] suggested that the minimum number of experts involved in interviews should be between 5 and 50. Furthermore, Hesse-Biber [196] recommended that the minimum number of experts should be 10 with respect to sample size. As a result, 12 experts from Malaysia's small construction industry stakeholders were invited for semi-structured interviews. Higher-level roles in projects, such as executives and project managers, were decidedly interviewed because the implementation of BIM falls under their responsibilities in any construction project. Based on the unavailability of three interviewees, they were interviewed online via conference call, and the remaining were interviewed via a face-to-face meeting.

From the interview, there was total disagreement among the interviewees on the BIM barriers such as inadequate access to decision-making resources (B24), the decision to utilize depends on the specifics of each case (B25), high ongoing investment in digital infrastructure (B26), inability to foresee digital technology's positive effects on the safety management process (B27), the need for affordable digital tools hinders the safety management process (B28), inappropriate rate of return (ROR) and rate of investment (ROI) data (B29), productivity loss when adopting BIM in place of traditional construction (B30), lack of flexible modeling capability in BIM tools (B31), existing computer-aided design (CAD) tools are appropriate for work (B32), financial uncertainty related with BIM adoption (B33) and high risk of conflicts in construction contracts (B34). Out of 34 BIM barriers investigated in the interview, only 23 were identified by experts to be suitable for further investigation.

NVivo 12, a qualitative analysis software, was used to perform detailed content analysis and to categorize the words said by interviewees. The analysis found ten primary categories: complexity, cost, culture, digital adoption, expertise, interest, legislation, safety, safety management resources, and technology, as shown in Figure 2. A total of 23 parameters were extracted, divided into ten prime categories, from the content analysis, which were used to develop the final colligated framework involving all outcomes of the literature review and interview analysis, as shown in Table 3.

Table 3. Final colligated framework for BIM barriers.

Categories	Coding	Parameters	Sources
Complexity	B25	The decision to utilize depends on the specifics of each case	Deleted
	B31	Lack of flexible modeling capability in BIM tools	Deleted
	B32	Existing computer-aided design (CAD) tools are appropriate for work	Deleted
	B34	High risk of conflicts in construction contracts	Deleted
	B10	Possibility of delays in construction	Maintained
	B4	Inadequate working processes and quality control standards	Maintained
Cost	B2	Competition is high, and profit margins are low	Maintained
	B19	High cost of BIM implementation	Maintained
	B7	Impractical theoretical evidence from research	Maintained
	B13	Lack of legal regulations	Maintained
	B26	High ongoing investment in digital infrastructure	Deleted
	B29	Inappropriate rate of return (ROR) and rate of investment (ROI) data	Deleted
	B30	Productivity loss when adopting BIM in place of traditional construction	Deleted
	B33	Financial uncertainty related with BIM adoption	Deleted

Table 3. *Cont.*

Categories	Coding	Parameters	Sources
Culture	B23	The construction industry's lackluster adoption of technology	Maintained
Digital Adoption	B18	Impact of COVID-19 on small construction projects	Maintained
	B24	Inadequate access to decision-making resources	Deleted
Expertise	B15	The lack of demand for or insistence on BIM from customers	Maintained
	B9	No facilitation and training center for BIM	Maintained
	B22	Not enough expertise in safety management	Maintained
Interest	B17	Too many complexities in design produced by BIM	Maintained
	B21	Poor BIM ability to integrate with project operations	Maintained
Legislation	B16	High diversity of workforce in projects	Maintained
	B20	The integration of BIM will change present levels of efficiency	Maintained
Safety	B8	Lack of awareness about the benefits of BIM	Maintained
	B12	Neither a simple nor universal strategy for BIM Usage exists	Maintained
	B27	Inability to foresee digital technology's positive effects on the safety management process	Deleted
Safety Management Resources	B1	Lack of digital education and training	Maintained
	B5	Insufficient teamwork from upper management	Maintained
	B28	The need for affordable digital tools hinders the safety management process	Deleted
Technology	B3	Aside from the construction team leader, no other team members need BIM	Maintained
	B6	Aversion to adopting BIM	Maintained
	B11	Absence of a structured methodology that is supportive	Maintained
	B14	In the workplace, resistance to BIM adoption remains strong	Maintained

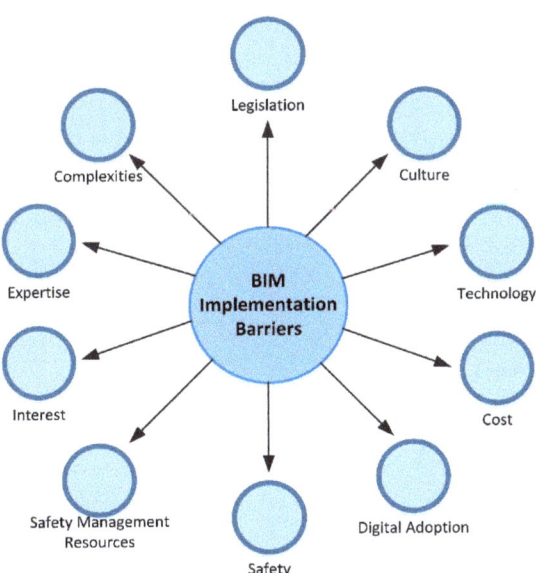

Figure 2. Qualitative analysis outcomes via NVivo.

2.3. Quantitative Analysis (Pilot Survey and Main Survey Questionnaire)

According to the Construction Industry Development Board (CIDB), there were a total of 39,158 registered small construction companies in Malaysia in 2021, and around 80% of them are actively operating on small construction projects. Perak was selected as the research area in which construction companies from grades G1 to G4 were selected. A complete random sampling method was adopted to determine the sample size. A pilot survey was conducted on the 23 BIM barriers identified during interviews. A pilot questionnaire was constructed involving closed-ended questions based on 23 BIM barriers. The sample size was decided to be a minimum of 100 respondents, while the distributed pilot survey questionnaires were 200. Respondents were from small construction companies only operating in Malaysia. Out of 200 distributed pilot questionnaires, 166 were obtained, meeting the validity criteria of more than 50%. Exploratory factor analysis (EFA) was conducted on the obtained dataset of the pilot survey questionnaire. Rather than putting a predetermined structure on the data, EFA looks into whether or not the recommended combination of variables or characteristics is acceptable, and it also looks into the probable underlying factor structure of a collection of observed variables. EFA was a suitable test in this case because the sample size was between the 150 and 300 range, and the BIM barriers were found to be 23, which was in the acceptable range of 20 to 50. Furthermore, the sample size (166) should be greater than the product of the number of responses (5) and the number of survey questions (23). For this pilot survey questionnaire, 166 was greater than $23 \times 5 = 115$, which qualified the data for EFA analysis. Data were also subjected to the Kaiser–Mayer–Olkin (KMO) and Bartlett's Tests to assess the representativeness and homogeneity of the sample. The KMO test has a range of 0–1 for its index, with results above 0.6 being considered satisfactory for revealing the character of correlations between variables. A p value of less than 0.05 for Bartlett's Test, which evaluates the sphericity of data through factor analysis, is considered to be acceptable. SPSS 24.0 was used to conduct both EFA and KMO and Bartlett's Test.

For the main questionnaire analysis by quantitative survey, the determined sample size is 240, while 100 is the minimum. A total of 20 BIM barriers were involved in the main questionnaire survey resulting from EFA. Demographics data were also collected to efficiently analyze the frequency of respondents. The questionnaire was distributed to 500 contractor companies in Malaysia working on small construction projects. SEM was performed for analytical purposes. In order to evaluate hypotheses about the connections between latent variables and the observed data, SEM was created in the 1980s. The first model in SEM is the measurement model, and it employs confirmatory factor analysis (CFA) to enrich the model by confirming the validity and reliability of the measuring variables against pre-set criteria, consequently linking the constructs with the latent components. The second model, a structural model, evaluates the relationships between the latent components by computing variances, testing hypotheses, and changing the model as necessary. By swapping out the correlation between the components for the hypothesized causal links, the conceptual model may be fine-tuned until it can be used to test the hypothesis. This study developed a conceptual framework for SEM evaluation using the findings of an EFA analysis on the previously identified barriers to BIM collected from the literature.

3. Analysis and Discussion

3.1. Background Information of Respondents

The background information of respondents is presented in Table 4. From a professional perspective, 7.23% were architects, 8.43% were quantity surveyors, 59.64% were civil engineers, 4.22% were M&E engineers, 18.07% were project managers, and the remaining 2.41% were from other profession types. A high percentage of civil engineers were involved in the study, which corresponds to the better judgement of BIM implementation barriers in small construction projects. From an organization perspective, 48.19% of the respondents worked in contractor organizations, 45.78% worked in consultant organizations, and the remaining 6.02% were directly working with the client. From experience in the Malaysian construction industry perspective, 23.49% had 0–5 years, 31.33% had 10–15 years, 6.63% had 15–20 years, and 4.82% had over 20 years of experience. More young workers involved in the study were people dealing with the implementation of BIM. It was also found that 99.40% of the respondents were working on small construction projects, and the remaining 0.6% were not relevant to small construction projects in Malaysia.

Table 4. Background information of respondents showing category, classification, frequency and percentage.

Category	Classification	Frequency	%
Profession	Architect	12	7.23%
	Quantity Surveyor	14	8.43%
	Civil Engineer	99	59.64%
	M&E Engineer	7	4.22%
	Project Manager	30	18.07%
	Other	4	2.41%
Organization	Contractor	80	48.19%
	Consultant	76	45.78%
	Client	10	6.02%
Experience in the Malaysian Construction Industry	0–5 Years	39	23.49%
	5–10 Years	56	33.73%
	1–15 Years	52	31.33%
	15–20 Years	11	6.63%
	Over 20 Years	8	4.82%
Experience in Small Construction Projects	Yes	165	99.40%
	No	1	0.60%

3.2. Level of Frequency of BIM in Small Construction Projects

The frequency of BIM implementation in small construction projects is presented in Figure 3. By following Halim et al. [197] and Hyarat et al. [6], the five levels were used to measure the response. According to the results, 43% of the respondents indicated very low implementation of BIM in small construction projects in Malaysia. Moving further, 28% of the respondents indicated low, 6% indicated average, 10% indicated high, and 13% indicated very high implementation of BIM in small construction projects in Malaysia. If seen from the collective perspective, the disagreement regarding the implementation of BIM in small construction projects is 71%, while the agreement is only 23%. It can be interpreted from the results that Malaysia does not have a significant implementation of BIM in small construction projects. Oslanrewaju et al. [8] also indicated the lack of the latest construction technologies in developing countries such as Malaysia. Similar findings were obtained from a primary research perspective showing that the Malaysian small construction industry still lacks BIM. It verifies the research gap and provides adequate comparative insights with existing research where the BIM implementation in small construction projects was found to be very low [20,198].

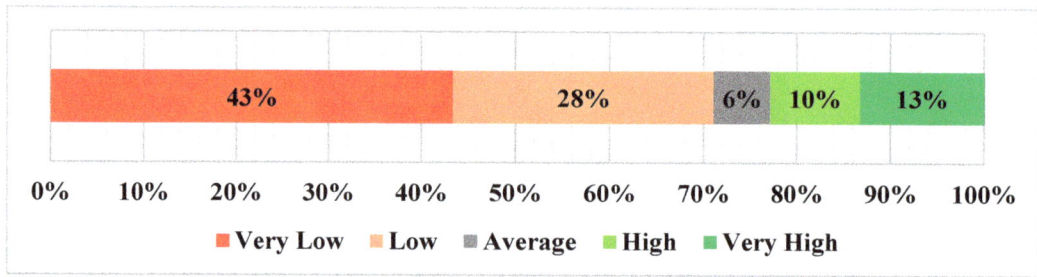

Figure 3. Level of BIM in small construction projects.

3.3. Barriers to BIM in Small Construction Projects

3.3.1. Reliability and Normality of Data

Whole data from the questionnaire were tested for reliability by measuring Cronbach's Alpha value. The initial test on the reliability indicated a Cronbach's Alpha above 0.8, which indicated high reliability of data. Further interpretation of the reliability constant for each of the BIM barriers indicated high reliability. The Shapiro–Wilk test was carried out to measure the significance of each barrier from a reliability perspective, as shown in Table 5. All the values were less than 0.05, indicating a high significance of data. With high reliability and significance, the test statistics confirmed further use of nonparametric tests.

Table 5. Reliability (Cronbach Alpha Test) and normality (Shapiro–Wilk Test) results.

Code	Cronbach's Alpha	Shapiro–Wilk Test		
		Statistic	df	p Value
B01	0.841	0.872	166	0.000
B02	0.815	0.864	166	0.000
B03	0.815	0.838	166	0.000
B04	0.815	0.867	166	0.000
B05	0.818	0.861	166	0.000
B06	0.817	0.849	166	0.000
B07	0.821	0.849	166	0.000
B08	0.818	0.846	166	0.000
B09	0.819	0.835	166	0.000
B10	0.825	0.847	166	0.000
B11	0.820	0.863	166	0.000
B12	0.816	0.861	166	0.000
B13	0.821	0.847	166	0.000
B14	0.822	0.850	166	0.000
B15	0.820	0.847	166	0.000
B16	0.817	0.831	166	0.000
B17	0.821	0.857	166	0.000
B18	0.824	0.840	166	0.000
B19	0.823	0.836	166	0.000
B20	0.817	0.842	166	0.000
B21	0.820	0.850	166	0.000

Note: p value of Shapiro–Wilk test is significant at the level of 0.05.

3.3.2. Mean Score Ranking of BIM Barriers

By following the descriptive statistics, the mean score for each of the BIM barriers was calculated. The purpose was to determine if any BIM barriers have a mean score of less than 3. The lowest mean observed was 3.05, and the maximum was 3.49. Both are greater than 3, which confirms that the mean of data is completely fine, and there is no irregularity in the data. Table 6 shows the mean score ranking of barriers to BIM in small construction projects showing mean, standard deviation (SD), rank, median and Kruskal–Wallis test results for intergroup comparisons. Moreover, the Wilcoxon signed-rank test was conducted to determine the significance of responses with respect to the sample mean values. The findings were satisfactory and indicated that all the BIM barriers are appropriately considered concerning the judgement of respondents and the research objectives.

Table 6. Mean score ranking of barriers to BIM in small construction projects.

Code	Mean	SD	Rank	Median	p Value [1]	p Value [2]	p Value [3]	p Value [4]
B01	3.23	1.452	14	3.00	0.052	0.504	0.139	0.196
B02	3.20	1.470	18	3.00	0.166	0.014 *	0.025 *	0.856
B03	3.31	1.545	5	4.00	0.004 *	0.011 *	0.000 *	0.414
B04	3.28	1.434	9	4.00	0.103 *	0.459	0.036	0.143
B05	3.19	1.493	19	3.00	0.330	0.124 *	0.086	0.189
B06	3.42	1.462	2	4.00	0.044 *	0.021 *	0.023 *	0.838
B07	3.28	1.512	9	4.00	0.376	0.453	0.020 *	0.772
B08	3.30	1.511	7	4.00	0.076	0.121	0.513	0.403
B09	3.39	1.512	4	4.00	0.040 *	0.000 *	0.460	0.143
B10	3.28	1.512	9	4.00	0.876	0.033 *	0.006 *	0.427
B11	3.21	1.476	17	3.00	0.065	0.497	0.122	0.856
B12	3.05	1.509	21	3.00	0.054	0.074	0.465	0.172
B13	3.25	1.531	13	3.00	0.101	0.192	0.093	0.148
B14	3.25	1.508	12	4.00	0.305	0.116	0.299	0.732
B15	3.16	1.545	20	3.00	0.028 *	0.064	0.002 *	0.195
B16	3.40	1.529	3	4.00	0.024 *	0.026 *	0.498	0.872
B17	3.22	1.507	15	3.50	0.006 *	0.999	0.039 *	0.806
B18	3.22	1.543	15	4.00	0.409	0.474	0.175	0.472
B19	3.49	1.464	1	4.00	0.020 *	0.551	0.065	0.131
B20	3.30	1.527	7	4.00	0.128	0.091	0.040 *	0.723
B21	3.31	1.504	6	4.00	0.002 *	0.436	0.012 *	0.397

[1] p value of Kruskal–Wallis test for intergroup comparison of respondents of different professions. [2] p value of Kruskal–Wallis test for intergroup comparison of respondents of different organizations. [3] p value of Kruskal–Wallis test for intergroup comparison of respondents of different experiences in the construction industry. [4] p value of Kruskal–Wallis test for intergroup comparison of respondents of experience in small projects. * p value of the corresponding test is significant at the level of 0.05.

The Kruskal–Wallis test was conducted for intergroup comparison based on profession, organization, experience and level of frequency of BIM in small construction projects. The classification of respondents was different in each group corresponding to the data collected in the demographics section of the questionnaire [8,21]. The interdependent groups are present in the data, which paved the way for choosing this test and determining the significance value to validate that the data are not normally distributed [26]. Significant results were produced for 14 barriers in different intergroup comparisons. Values larger than 0.05 indicated that the respondents from different professions, organizations, experience and perceived frequency of BIM in small construction were working under similar circumstances. Values lower than 0.05 indicated variation in the perception of BIM barriers with respect to their distribution in groups. The perceived values from the perspective of experience indicated more significant results. This confirms that experience influences people to understand more about the BIM barriers in small construction projects in Malaysia. This further strengthens the concept that the BIM implementation is not only dependent on some legislative measure, but the experienced professionals in small construction projects of Malaysia widely accept the barriers. Similar circumstances in the work environment pave the concept of facing similar kinds of barriers in small construction projects, as demonstrated by Kruskal–Willis test statistics. The values were found to be less deviated from the mean score obtained, confirming the agreement on BIM barriers considered in analysis.

According to calculated rank from descriptive mean analysis, the five most crucial barriers were found that are significantly affecting the implementation of BIM in small construction in Malaysia. These were B19 "High cost of BIM implementation" (mean = 3.49, rank = 1), B06 "Reluctance to transition to BIM" (mean = 3.402 rank = 2), B16 "High diversity of the workforce in projects" (mean = 3.40, rank = 3), B09 "Possibility of delays in construction" (mean = 3.39, rank = 4) and B03 "Lack of BIM experts" (mean = 3.31, rank = 5). In reality, BIM applications involve significant work and operations that do not always relate to the requirements of small construction projects. Durdyev et al. [3] stated that small construction projects mostly have operations in which the integration of BIM applications is inefficient because BIM tools are made commercially for heavy construction projects. Construction workers face problems when integrating BIM tools for small construction projects, where they cannot even find the BIM modules that could solve the problem effectively in small construction projects. B06 "Reluctance to transition to BIM" validated the ongoing trend in the construction sector of Malaysia where construction professionals are resultant to change their construction methods. They always want to stick with conventional methods because implementing new technologies such as BIM requires more resources and input from construction practitioners, which may not always be feasible [20]. This reluctant behavior contributed to putting another barrier in implementing BIM in Malaysia's small construction projects. B16 "High diversity of the workforce in projects" indicated that construction professionals are not implementing BIM, as it is a time-consuming process for small construction projects. In small construction projects, procurement needs to be performed on a timely basis because time is short, and construction professionals always want to start the work as soon as possible [127]. This behavior creates a barrier to the implementation of BIM because construction professionals do not spend time getting into difficulties associated with the time-consuming aspect of BIM. B09 "Possibility of delays in construction" indicates that BIM implementation in small construction may increase the possibility of delays in projects. These delays are not acceptable in any case for small construction professionals because it places profits at stake. Any possible difficulty while working with BIM can easily create problems in the schedule of projects. B03 "Lack of BIM experts" indicates that experts are always needed to implement BIM in small construction projects in Malaysia. This is because many construction workers in small projects do not have experience working with BIM [197]. It makes it difficult and uncertain for construction workers to adopt BIM in all construction operations. B "Neither a simple nor universal strategy for BIM usage exists." (mean = 3.05, rank = 21) was found to be the barrier with the lowest impact on implementation of BIM in small construction projects of Malaysia. It is understandable that BIM does not always require a universal method to be implemented in small construction projects because the requirements are totally different. A universal method can complicate the processes but can also help, depending on the situation where BIM is implemented. It also indicates a positive attitude present among the construction professionals because they are not demanding universal BIM methodology, and therefore, B12 cannot be taken as a significant barrier.

3.3.3. Factor Analysis of BIM Barriers

From the existing literature, 21 BIM implementation barriers were found, and statistically obtained data after survey analysis were significant. However, the possibility of having a similar impact on each barrier cannot be ignored. EFA can solve this problem and perform the grouping to some subgroups of barriers that can be practically feasible to explain concerning small construction projects in Malaysia. Suitability for EFA analysis was determined before conducting the analysis and obtaining subgroups of barriers. According to Al-Aidrous et al. [198] and Alwee et al. [20], the EFA should be conducted when the sample size is greater than 150 but less than 300. Further, it is also necessary to have a greater sample size than the number of questions multiplied by the number of responses each question has in the quantitative survey. For this study, that number is 105, less than 166, which is the sample size. The number of variables being employed in factor analysis must be at least 20 and greater than 50. The criteria for factor analysis are met; therefore, the factor analysis was conducted on 21 variables corresponding to BIM barriers, excluding the risk of inaccurate factor analysis results. The subject-to-variable ratio was found to be 7.90:1.00. Greater than 5:1 is required, and the validity of results from factor analysis is further confirmed [26].

Kaiser–Mayer–Olkin (KMO) and Bartlett's test were applied to the data of 21 variables to measure sampling adequacy and sphericity. The index range of the KMO test is from 0 to 1, in which the acceptable results lie above 0.6, telling the nature of correlations among the variables [199]. For Bartlett's test, the required significance value should be less than 0.05 for good factor analysis results, measuring the sphericity of data. SPSS 24.0 was used, and the findings are presented in Table 7 for both tests. KMO index was found to be 0.853, which is greater than 0.6 and is therefore acceptable. The significance of Bartlett's test was found to be 0.000, which is less than 0.05, indicating that EFA can be adopted for making a subgroup of variables considered in this study.

Table 7. KMO and Bartlett's test results.

Kaiser–Meyer–Olkin Measure of Sampling Adequacy		0.853
Bartlett's Test of Sphericity	Approx. Chi-Square	853.740
	df	210
	Sig.	0.000

Principal component analysis (PCA) was used to conduct EFA analysis, and factor structure was obtained for the 21 variables. Varimax rotation was applied to obtain the rotated component structure. EFA results are presented in Table 8, from which five components have an Eigen value greater than 1. The scree plot presented in Figure 4 indicates the same behavior of variables involved in the analysis. The first five components on the x-axis have an Eigen value greater than one, indicating the possible division of BIM barriers in four groups. The cumulative variance obtained for the five groups is 50.878%, which is greater than 50% and indicates acceptable components. The minimum factor loading cutoff limit of 0.4 was applied to obtain the results corresponding to the rotated component structure.

After examining the component structure obtained from EFA, the five subgroups were devised based on the number of components. They were named behavioral barriers, technical adoption barriers, management barriers, implementation barriers and digital education barriers. The corresponding mean of each barrier in the subgroup was used to calculate the mean for each subgroup. Table 9 shows the mean score ranking of the BIM barriers subgroup, indicating barriers, subgroup mean and subgroup rank of all subgroups. The final ranking of BIM barriers was performed based on the mean subgroup score, and it is discussed as follows.

Table 8. Factor loadings indicating 5 components based on PCA with varimax rotation.

Barriers	1	2	3	4	5
B15	0.704				
B14	0.652				
B6	0.640				
B3	0.571				
B12	0.506				
B11		0.674			
B7		0.627			
B2		0.588			
B21		0.542			
B13		0.513			
B19		0.466			
B4		0.406			
B8		0.375			
B18			0.754		
B20			0.562		
B16			0.471		
B10				0.714	
B9				0.493	
B5				0.362	
B17				0.336	
B1					0.897
Eigen Values	5.883	1.269	1.228	1.190	1.115
% of Variance	28.015	6.041	5.847	5.665	5.311
Cumulative Variance %	28.015	34.055	39.903	45.567	50.878

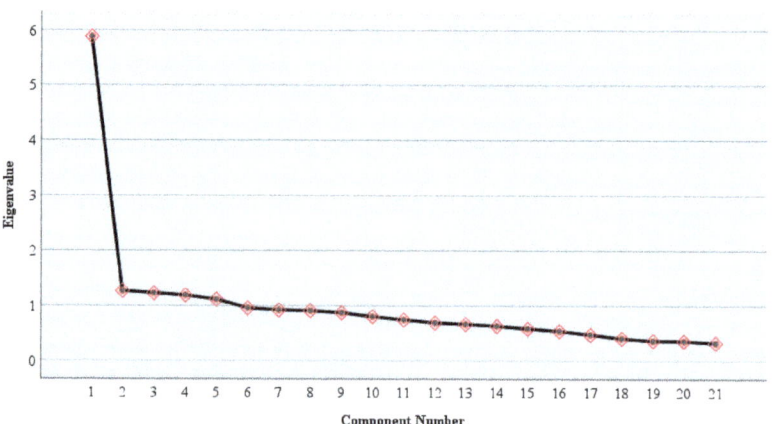

Figure 4. Scree plot results.

Table 9. Mean score ranking of BIM barriers and distribution.

Subgroup	Code	Barriers	Mean	Subgroup Mean	Subgroup Rank
Behavioral Barriers	B15	The lack of demand for or insistence on BIM from customers	3.16	3.24	4
	B14	In the workplace, resistance to BIM adoption remains strong	3.25		
	B6	Aversion to adopting BIM	3.42		
	B3	Aside from the construction team leader, no other team members need BIM	3.31		
	B12	Neither a simple nor universal strategy for BIM Usage exists	3.05		
Technical Adoption Barriers	B11	Absence of a structured methodology that is supportive	3.21	3.29	2
	B7	Impractical theoretical evidence from research	3.28		
	B2	No facilitation and training center for BIM	3.20		
	B21	Poor BIM ability to integrate with project operations	3.31		
	B13	Lack of legal regulations	3.25		
	B19	High cost of BIM	3.49		
	B4	Inadequate working processes and quality control standards	3.28		
	B8	Lack of awareness about the benefits of BIM	3.30		
Management Barriers	B18	Impact of COVID-19 on small construction projects	3.22	3.31	1
	B20	The integration of BIM will change present levels of efficiency.	3.30		
	B16	High diversity of workforce in projects	3.40		
Implementation Barriers	B10	Possibility of delays in construction	3.28	3.27	3
	B9	No financial support for small construction projects	3.39		
	B5	Insufficient teamwork from upper management	3.19		
	B17	Too many complexities in design produced by BIM	3.22		
Digital Education Barrier	B1	Lack of digital education and training	3.23	3.23	5

Management Barriers (mean = 3.31, rank = 1,): The first-ranked subgroup consists of barriers related to management issues that construction workers face when implementing BIM in small construction projects in Malaysia. In total, 28.015% of the variance is explained by this subgroup. The specific items in this subgroup are B18 "Impact of COVID-19 on small construction projects", B20 "The integration of BIM will change present levels of efficiency, and B16 "High diversity of the workforce in projects". The overall impact of management barriers is strong from the perspective of affecting the implementation of BIM in small construction projects in Malaysia. It is a reality that after the COVID-19 pandemic, the situation of small construction companies was not favorable for adopting new technology, which acted as one of the management barriers to implementing BIM [128,199]. Further, Malaysia's work environment is diverse, as most of the workers employed by the small construction companies are from other nations such as Bangladesh and India. The diverse workforce makes it very difficult for construction companies to manage the new technology's implementation, which further creates a major barrier to implementing BIM [72,76]. Implementing BIM is also found to be very time-consuming by the latest research. When small companies try to implement a BIM, it creates unexpected delays in the project schedule. Similarly, leadership issues are always present from the senior management perspective, which contributes to decreasing the adoption rate of BIM by small construction companies.

Technical Adoption Barriers (mean = 3.29, rank = 2): The second-ranked subgroup consists of barriers related to technical issues that construction workers face implementing BIM in small construction projects. In total, 6.041% of the variance is explained by this subgroup. The specific items in this subgroup are: B11 "Absence of a structured methodology that is supportive", B7 "Impractical theoretical evidence from research", B2 "No facilitation and training center for BIM", B21 "Poor BIM ability to integrate with project operations", B13 "Lack of legal regulations", B19 "High cost of BIM implementation", B4 "Inadequate working processes and quality control standards" and B8 "Lack of awareness about the benefits of BIM". The technical barriers are ranked second because most difficulties with implementing the BIM are related to ineffective management controls. Existing methodologies for implementing BIM in construction projects must fully support small construction companies. Further, the literature must provide evidence of practically improving BIM implementation in Malaysia's small construction industry [122,200]. This is because the current environment is changing rapidly after 2020, and the existing protocols in construction may only sometimes work. Similarly, the lack of awareness contributes to increasing the technical difficulties while there is no existing mechanism for training to bridge the gap between large and small construction projects [72,76]. The nature of the client is also relevant to maintaining the project's efficiency on a low budget, due to which they only sometimes demand the implementation of BIM. The integration difficulties are also present because of an inappropriate way of integrating BIM with project operations [200]. This will raise the cost of implementing BIM and impose obligations on small construction firms that need to be more technically prepared to implement it.

Implementation Barriers (mean = 3.27, rank = 3): The third-ranked subgroup consists of barriers related to practical adoption difficulties existing on the construction sites of small projects in Malaysia. In total, 5.847% of the variance is explained by this subgroup, and the specific items are: B9 "No financial support for small construction projects", B10 "Possibility of delays in construction", B5 "Insufficient teamwork from upper management" and B17 "Too much complexities in design produced by BIM". In terms of implementation, a corporation is always required by the leadership, which unfortunately only happens in small construction companies, contributing to the implementation barrier. The risk of delay always exists, due to which the implementation can become uncertain and can even create ambiguity among the responsible workers in decision making [129]. Implementation can also be difficult because there needs to be more financial support available for small construction companies, which is relevant to government policy. The complexities in design also act as implementation barriers because they combine with other factors, such as a lack of awareness among the project members, which ultimately increases the problems in implementing the BIM.

Behavioral Barriers (mean = 3.24, rank = 4): The fourth-ranked subgroup consists of barriers related to behavioral difficulties that construction workers face when implementing BIM in small construction projects of Malaysia. In total, 5.665% of the variance is explained by this subgroup, and the specific items are: B15 "The lack of demand for or insistence on BIM from customers", B14 "In the workplace, resistance to BIM adoption remains strong", B6 "Aversion to adopting BIM", B3 "Aside from the construction team leader, no other team members need BIM." and B12 "Neither a simple nor universal strategy for BIM usage exists". The subcontractor support is greatly affected when the implementation is not set according to plan for small construction projects, and ultimately, it sets a very inappropriate tone between the project stakeholders [122]. Workers in construction projects do not want to change the existing environment, which has direct consequences in increasing the behavior barrier. Most workers need to be more skilled in understanding the requirements of implementing BIMs in construction projects, which causes problems if a universal method is available.

Digital Education Barriers (mean = 3.23, rank = 5): The fifth-ranked subgroup consists of barriers relevant to poor digital education. In total, 5.311% of the variance is explained by this subgroup, and it only has one item: B1 "Lack of digital education and training". Because many tools are available that have reduced the complexities of adopting BIM, digital education may only sometimes be required to understand the requirements of implementing BIM in small construction companies. As a result, the "digital education barriers" subgroup harms Malaysia's environment.

Five subgroups were found from the mean square analysis impacting the implementation of BIM in small construction projects in Malaysia. A clear understanding was developed from the analysis regarding the rank of each subgroup in affecting the implementation of BIM.

3.4. Quantitative Survey

EFA variables having cross-loadings or loadings less than 0.4 were not included; afterward, the main questionnaire was developed, 235 individuals completed the main questionnaire, and new data were gathered. By using AMOS 22, CFA is used to assess the conceptual framework's validity and dependability (CV-DV). In the CFA, the observed variables with loadings below 0.6 were eliminated. The measurement model's final fit for the BIM barrier and parameters for the effective application are shown in Figure 5. Four constructs: "Technical Adoption Barriers (TAB)," "Behavioral Barriers (BB)," "Implementation Barriers (IB) and "Management Barriers (MB)", were used to group the final refined parameters/variables. Variables B1 and B19 were removed from the finished framework since they had low factor loadings between the observed variable and the construct on CFA. Figure 5 illustrates the measurement model involving four groups, BB, TAB, IB and MB and 16 BIM implementation barriers. The value of significance for all barriers was significant, as they are above 0.6. The intergroup correlation is moderately significant but still above 0.3, indicating a high acceptability of outcomes.

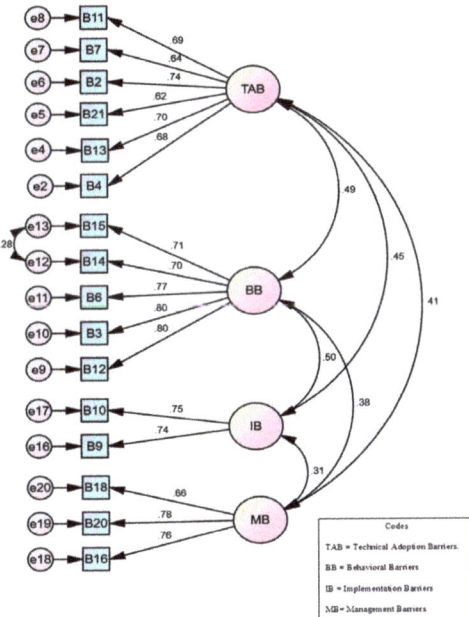

Figure 5. Measurement model for BIM implementation barriers.

In the model, for the improvement, error correlations were established for the variables B14–B15; however, correlated variables are unique parameters and have no similarity. Table 10 shows the reliability and validity tests for the measurement model. The goodness of fit (GOF) is shown in Tables 11 and 12 for the measurement model and structural model, respectively. Based on the model fit of the measurement model, the structural model (SM) was developed, as shown in Figure 6. All values of correlations were significant between the barriers and their associated group. Further, the values were significant even between the latent variable and all categories involved in the structural model. The most significant group of BIM implementation barriers was BB, involving the maximum number of variables with high significance.

Table 10. Validity and reliability of CBBs showing acceptable statistics for all constructs.

Constructs	CR	AVE	MSV	MaxR(H)	BB	TAB	IMB	MB
BB	0.877	0.589	0.230	0.879	0.767			
TAB	0.838	0.500	0.228	0.841	0.478	0.681		
IMB	0.715	0.557	0.230	0.715	0.480	0.448	0.746	
MB	0.777	0.539	0.171	0.786	0.382	0.413	0.306	0.734

Table 11. Goodness of fit (GOF) for the measurement model.

Index	Acceptance	Attained
RMSEA	<0.08	0.47
GFI	>0.90	0.925
CFI	>0.90	0.966
TLI	>0.90	0.958
Cmin/df	<2, 3	1.447
ChiSq	$p > 0.05, p > 0.01$	151.12

Table 12. Goodness of fit (GOF) for the structural model.

Index	Acceptance	Attained
RMSEA	<0.08	0.47
GFI	>0.90	0.925
CFI	>0.90	0.966
TLI	>0.90	0.958
Cmin/df	<2, 3	1.447
ChiSq	$p > 0.05, p > 0.01$	140.350

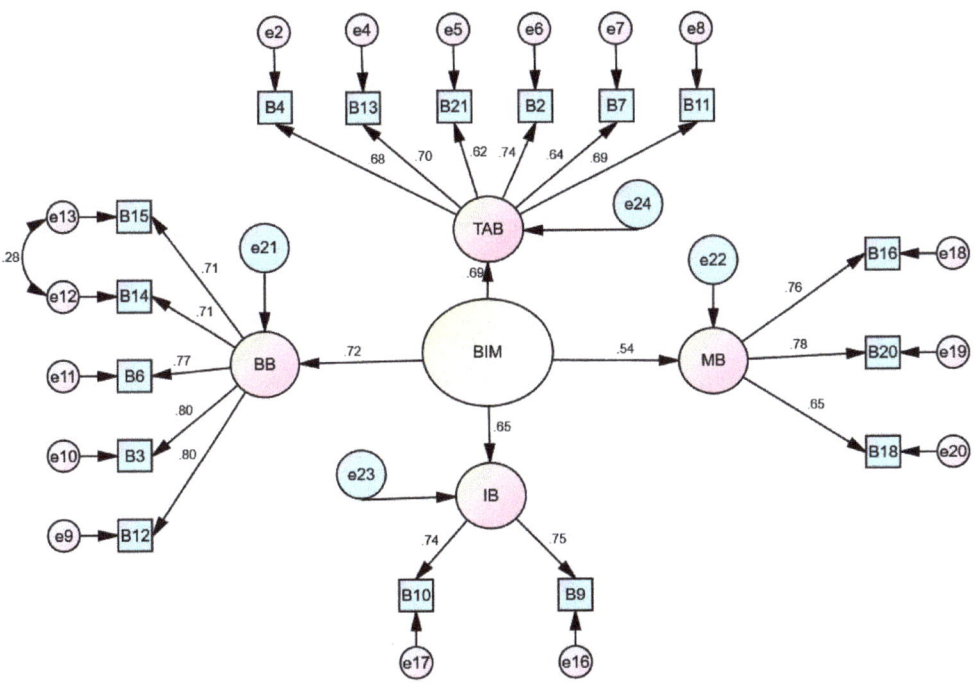

Figure 6. Structural model for BIM implementation barriers.

4. Discussion

Using a combination of a literature review, semi-structured interviews, and questionnaire surveys, the authors of this study developed knowledge-based standards and a reference model that highlights barriers to workers' adoption of building information technology in Malaysia's small construction projects. In addition, SEM was used to focus on and eliminate the most significant obstacles for a smooth BIM rollout. This research aimed to provide a theoretical framework that would emphasize existing challenges to implementing BIM processes. A workable conceptual framework was attained by analyzing and refining 23 SM parameters across five constructs to 16 general characteristics across four constructs. This is especially true given that some parameters appear conceptually similar but actually differ based on technical considerations. Figure 7 illustrates a framework emphasizing the overarching challenges of using BIM successfully in building projects. The framework includes all identified significant BIM barriers divided into groups by which the small construction industry cannot adopt BIM. The framework's parameters and stumbling blocks are now technology agnostic. They were simplified by consolidating similar parameters under a single construct or by eliminating them entirely. According to the systematic literature assessment results, few researchers have only aimed to identify even the most fundamental technological hurdles in the BIM-based construction industry. However, performed research has focused on the primary hurdles in BIM technologies on building processes using key performance indicators or BIM following SEM.

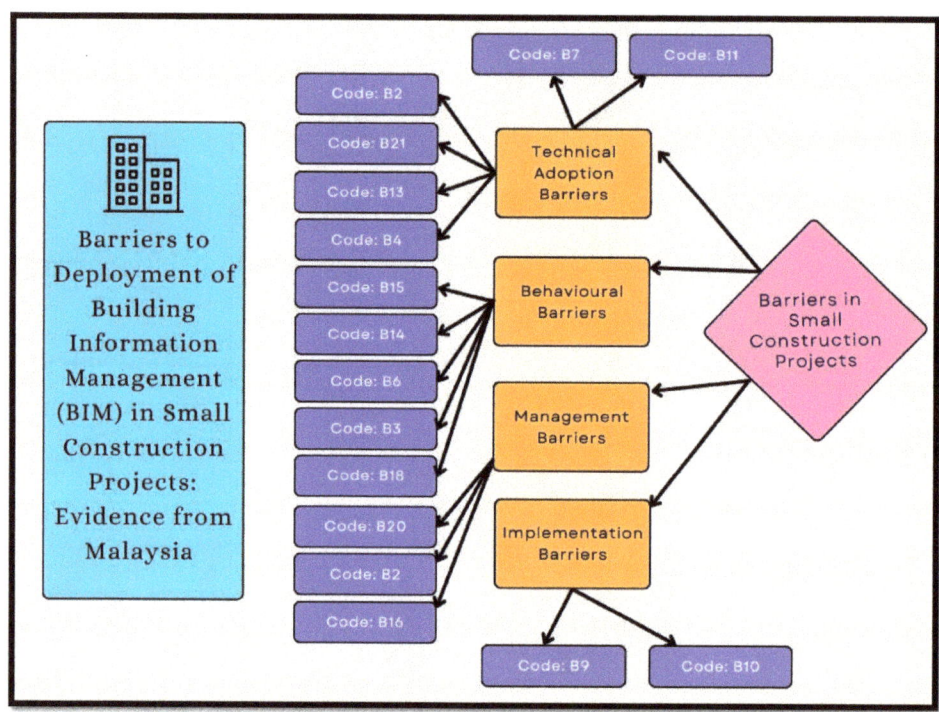

Figure 7. BIM barrier-based framework for small construction projects in Malaysia.

The overarching goal of this model is to demonstrate familiarity with the obstacles that prevent BIM technology from being used effectively. Stakeholders in BIM technology and its application can use this model as a general reference. Studies by Arif et al. [26] and Sriyolja et al. [76]. are just two examples of the many that have evaluated the effects of digital technologies on the efficiency of construction operations by comparing performance- or ranking-related factors using the relative importance index (RII) technique. Therefore, in comparison with the aforementioned studies, this conceptual framework has been devised via performing a mathematical modeling technique, i.e., SEM, which underlines the precise variables and barriers related to the BIM process, to gain confidence in its application, basic operational guidelines, and to educate construction industry stakeholders. The model is novel since it is simple to grasp by practitioners, yet it addresses the broad factors that are genuine roadblocks to digital construction operation (BIM) efficiency. While King et al. [121], Hedayati et al. [78] and Taat et al. [21] have identified BIM roadblocks throughout Malaysia's construction sector as a whole, their implications suggest that future studies conducted on a more granular scale might provide quite different results. Because of this, the results of this research vary when applied to the setting of solely Malaysian small building projects. Additionally, the findings from Taat et al. [21] and Belayutham et al. [23] do not employ the identical methods as in this study to identify the BIM barriers. Given that this study was limited in scope to very small building projects, it stands to reason that the results are highly distinctive.

4.1. Managerial Implications

Identifying significant BIM barriers may facilitate the development of a method that stakeholders such as project owners and contractors may use to integrate BIM into their small construction projects better. In addition, these small construction projects in Malaysia may make great progress by tackling the identified BIM barriers. This will replace the usual method of construction in Malaysian small construction projects. Small construction enterprises in Malaysia must apply BIM in order to have a lasting impact on the economy since the economy is often related to the success of the small construction sector. If the construction sector continues to expand, Malaysia may be able to enter the top 20 economies in the world. These research findings may potentially be utilized to encourage the use of BIM in other developing nations with comparable adoption rates for construction projects. This is particularly relevant in adjacent countries and the global context, where smaller construction projects will be better suited to focus on solving particular barriers. Therefore, countries with difficulty adopting BIM for small construction projects may benefit from employing BIM. Nonetheless, this study provides an enormous contribution that has significant implications for small construction project businesses in the following ways:

- It offers a collection of knowledge on the BIM barriers that small construction industries are currently facing.
- It assists small construction project owners, consultants, and contractors in analyzing and selecting the most effective BIM implementation to enhance project planning, efficiency, and consistency.
- Presented are factual data that might benefit Malaysia and other countries in effectively using BIM for small construction projects.
- In Malaysia, no research has been undertaken on the usage of BIM. This research is important because it reveals a connection between BIM hurdles and Malaysia's small construction industry. This establishes a good platform for a discussion on how BIM might be used to enhance the safety of low-cost construction projects and overcome the knowledge gap.
- The findings given here are relevant only to BIM implementation in small construction projects. Consequently, the project's stakeholders may collaborate to overcome the BIM-related cost, time, and efficiency concerns. Achieving a high degree of sustainability in a project has positive long-term implications.
- This study also establishes a benchmark for measuring the effectiveness of BIM in the administration of a small building project.
- Local communities will be positively affected by the outcomes of this study, as BIM will help in increasing project efficiency and ultimately move small construction projects toward sustainability for Malaysian society.

4.2. Theoretical Implications

Although BIM has been available for a while, its significance is growing even for relatively small construction projects. Small construction projects, in particular, are highlighted by the proposed BIM barriers framework as requiring BIM adoption. This study uses the proposed model to shed insight on the challenges that prevent the use of BIM. These challenges actually work in favor of bringing BIM to Malaysia's relatively small construction industry. Thus, the findings of this study will assist in closing the gap between theoretical and practical BIM implementation. We are not aware of any research that has looked at the barriers to using BIM in the Malaysian construction industry. This finding provides a starting point for researchers, particularly those in the field of construction management, to examine the difficulties of BIM in the context of the small construction industry. Because of this, the theoretical outcomes of this study give a mathematical foundation for precisely recognizing the barriers of BIM, which might be effectively implemented in Malaysia and abroad. The results will be under fair principle, as the study is aimed to improve the implementation of BIM in small construction projects. The results can be used by future researchers in any possible way to improve the implementation of BIM in small construction projects.

5. Conclusions

This research aims to identify and highlight the most fundamental factors or barriers preventing the widespread adoption of building information technology in Malaysia's smaller-scale building projects. In this study, a structured literature review analysis was used with a systematic approach to review the literature and to choose the papers that would be included. After reviewing the data, 34 barriers were singled out as particularly troubling for using BIM in Malaysian small building projects. After conducting semi-structured interviews and evaluating the data using NVIVO, we narrowed our list down to 23 barriers. Afterward, a survey was conducted, and an EFA was performed on the collected data. Following this, the main questionnaire was developed to capture more relevant data from the industry experts and academia. Based on the replies, a structural equation modeling (SEM) technique was selected for statistical analysis, with a specific emphasis on the characteristics of BIM implementation that are hindered by barriers. After conducting SEM tests, the model was updated to reflect the factors that have led many to conclude that BIM is not useful for less-scaled building projects. The conceptual framework established by EFA represents the 16 aspects affecting the deployment of the BIM and is based on a statistical study of 23 parameters. Later, applying convergent and discriminant reliability (CFA), 16 variables were left that accurately reflected the most pressing issues preventing the adoption of BIM in Malaysia's small construction industry. These variables were organized into four categories: technical adoption barriers; behavioral barriers; management barriers; and implementation barriers.

This research uniquely contributes to the current literature by identifying obstacles to BIM's use in Malaysia's smaller building projects. The results are useful for closing the knowledge gap between the existing theoretical literature and the actual use of BIM by the Malaysian small construction sector. Because of the specificity of the study's methodology, sample size, industry size, and possible stakeholders, its findings can only be applied to the small construction sector. The stakeholders better grasp the overall barriers to the deployment of BIM with this model, which depicts the factors that enable the proper implementation of BIM. This study's systematic literature evaluation revealed that few prior investigations into the challenges of implementing BIM in Malaysia used appropriate SEM analytic methods and procedures. Therefore, this research aimed to provide a knowledge framework to close the information gap that contributes to stakeholders' skepticism in the construction sector toward technology. The completed model will persuade construction business professionals to use BIM tools, aiding the IR 4.0 ecosystem and saving money in the long term. This research contributes to the theory and practice for adopting BIM in small construction projects; however, limitations and future research opportunities exist for this study. The study included a sample population only from Perak, Malaysia; the scope of the study can be widened by considering other states or countries, as more factors can be identified. In addition, key factors may vary for other countries, as such factors are dependent on the construction environment, practices, and technological culture. Future studies can be conducted by adopting a more advanced quantitative research method, and effective mitigation techniques can be devised for individual BIM barriers presented in the final framework of this research. Moreover, SM can be modified in terms of project performance control or key performance indicators (cost, time, and quality), primary or secondary processes related to the project (safety management, project planning, supply chain management, etc.), and external implications (CO_2 emissions), considering the barriers to BIM implementation in aforementioned processes.

Author Contributions: Conceptualization, W.S.A. and A.W.; methodology, A.W.; software, A.H.Q. and A.W.; validation, W.S.A., A.H.Q. and A.W.; resources, W.S.A.; data curation, A.W. and A.H.Q.; writing—original draft preparation, A.W.; writing—review and editing, W.S.A. and A.H.Q.; supervision, W.S.A.; funding acquisition, W.S.A. All authors have read and agreed to the published version of the manuscript.

Funding: The authors would like to appreciate the YUTP-FRG 1/2021 (015LC0-369) in Universiti Teknologi PETRONAS (UTP) awarded to Wesam Alaloul for the support.

Data Availability Statement: All data, models, and code generated or used during the study appear in the submitted article.

Conflicts of Interest: The authors declare no conflict of interest.

References

1. Alaloul, W.S.; Qureshi, A.H.; Musarat, M.A.; Saad, S. Evolution of close-range detection and data acquisition technologies towards automation in construction progress monitoring. *J. Build. Eng.* **2021**, *43*, 102877. [CrossRef]
2. Oyuga, J.O.; Gwaya, A.; Njuguna, M.B. Investigation of the current usage of BIM capabilities by large-sized building contractors in Kenya based on theory of innovation diffusion. *Constr. Innov.* **2021**, *23*, 155–177. [CrossRef]
3. Durdyev, S.; Ashour, M.; Connelly, S.; Mahdiyar, A. Barriers to the implementation of Building Information Modelling (BIM) for facility management. *J. Build. Eng.* **2022**, *46*, 103736. [CrossRef]
4. Qureshi, A.H.; Alaloul, W.S.; Wing, W.K.; Saad, S.; Ammad, S.; Musarat, M.A. Factors impacting the implementation process of automated construction progress monitoring. *Ain Shams Eng. J.* **2022**, *13*, 101808. [CrossRef]
5. Qureshi, A.H.; Alaloul, W.S.; Wing, W.K.; Saad, S.; Ammad, S.; Altaf, M. Characteristics-Based Framework of Effective Automated Monitoring Parameters in Construction Projects. *Arab. J. Sci. Eng.* **2022**, 1–19. [CrossRef] [PubMed]
6. Hyarat, E.; Hyarat, T.; Al Kuisi, M. Barriers to the Implementation of Building Information Modeling among Jordanian AEC Companies. *Buildings* **2022**, *12*, 150. [CrossRef]
7. Qureshi, A.H.; Alaloul, W.S.; Wing, W.K.; Saad, S.; Alzubi, K.M.; Musarat, M.A. Factors affecting the implementation of automated progress monitoring of rebar using vision-based technologies. *Constr. Innov.* **2022**. ahead-of-print. [CrossRef]
8. Olanrewaju, O.I.; Kineber, A.F.; Chileshe, N.; Edwards, D.J. Modelling the relationship between Building Information Modelling (BIM) implementation barriers, usage and awareness on building project lifecycle. *Build. Environ.* **2022**, *207*, 108556. [CrossRef]
9. Berges-Alvarez, I.; Muñoz Sanguinetti, C.; Giraldi, S.; Marín-Restrepo, L. Environmental and economic criteria in early phases of building design through Building Information Modeling: A workflow exploration in developing countries. *Build. Environ.* **2022**, *226*, 109718. [CrossRef]
10. Olanrewaju, O.I.; Enegbuma, W.I.; Donn, M.; Chileshe, N. Building information modelling and green building certification systems: A systematic literature review and gap spotting. *Sustain. Cities Soc.* **2022**, *81*, 103865. [CrossRef]
11. Abu-Hamdeh, N.H.; Alsulami, R.A.; Hatamleh, R.I. A case study in the field of building sustainability energy: Performance enhancement of solar air heater equipped with PCM: A trade-off between energy consumption and absorbed energy. *J. Build. Eng.* **2022**, *48*, 103903. [CrossRef]
12. Lin, J.; Lu, S.; He, X.; Wang, F. Analyzing the impact of three-dimensional building structure on CO_2 emissions based on random forest regression. *Energy* **2021**, *236*, 121502. [CrossRef]
13. Chen, Y.; Cai, X.; Li, J.; Zhang, W.; Liu, Z. The values and barriers of Building Information Modeling (BIM) implementation combination evaluation in smart building energy and efficiency. *Energy Reports* **2022**, *8*, 96–111. [CrossRef]
14. Alaloul, W.S.; Saad, S.; Qureshi, A.H. Construction Sector: IR 4.0 Applications. In *Handbook of Smart Materials, Technologies, and Devices*; Springer International Publishing: Cham, Switzerland, 2021; pp. 1–50. ISBN 9783030586751.
15. Qureshi, A.H.; Alaloul, W.S.; Alzubi, K.M. Internet of Things (IoT) for Construction Cyber-Physical Systems. In *Cyber-Physical Systems in the Construction Sector*; CRC Press: Boca Raton, FL, USA, 2022; pp. 60–77. ISBN 1003190138.
16. Charef, R.; Emmitt, S.; Alaka, H.; Fouchal, F. Building Information Modelling adoption in the European Union: An overview. *J. Build. Eng.* **2019**, *25*, 100777. [CrossRef]
17. Shirowzhan, S.; Sepasgozar, S.M.E.; Edwards, D.J.; Li, H.; Wang, C. BIM compatibility and its differentiation with interoperability challenges as an innovation factor. *Autom. Constr.* **2020**, *112*, 103086. [CrossRef]
18. Ahmed, S. Barriers to Implementation of Building Information Modeling (BIM) to the Construction Industry: A Review. *J. Civ. Eng. Constr.* **2018**, *7*, 107. [CrossRef]
19. Hamid, A.B.A.; Embi, M.R. Key factors of bim implementation for interior design firms in Malaysia. *Int. J. Sustain. Constr. Eng. Technol.* **2020**, *11*, 175–184. [CrossRef]
20. Alwee, S.N.A.S.; Salleh, H.; Zulkifli, U.K. Strategic process protocol for building information modeling (Bim) contract administration in Malaysia—A concept paper. *Malaysian Constr. Res. J.* **2021**, *12*, 37.
21. Taat, N.H.M.; Abas, N.H.; Hasmori, M.F. The Barriers of Building Information Modelling (BIM) for Construction Safety. In Proceedings of the Lecture Notes in Civil Engineering, Penang, Malaysia, 9 September 2021.
22. Manzoor, B.; Othman, I.; Gardezi, S.S.S.; Altan, H.; Abdalla, S.B. BIM-Based Research Framework for Sustainable Building Projects: A Strategy for Mitigating BIM Implementation Barriers. *Appl. Sci.* **2021**, *11*, 5397. [CrossRef]
23. Belayutham, S.; Zabidin, N.S.; Ibrahim, C.K.I.C. Dynamic representation of barriers for adopting building information modelling in Malaysian tertiary education. *Constr. Econ. Build.* **2018**, *18*, 24–44. [CrossRef]
24. Ringle, C.M.; Sarstedt, M.; Mitchell, R.; Gudergan, S.P. Partial least squares structural equation modeling in HRM research. *Int. J. Hum. Resour. Manag.* **2018**, *31*, 1617–1643. [CrossRef]
25. Wang, Y.A.; Rhemtulla, M. Power Analysis for Parameter Estimation in Structural Equation Modeling: A Discussion and Tutorial. *Adv. Methods Pract. Psychol. Sci.* **2021**. [CrossRef]
26. Arif, N.K.; Hasmori, M.F.; Deraman, R.; Yasin, M.N.; Mohd Yassin, M.A. Readiness of Malaysian Small and Medium Enterprises Construction Companies for Building Information Modelling Implementation. In Proceedings of the IOP Conference Series: Materials Science and Engineering, Johor, Malaysia, 24 August 2021. [CrossRef]

27. Yaakob, M.; James, J.; Nawi, M.N.M.; Radzuan, K. Study on benefits and barriers of implementing Building Information Modelling (BIM) in Malaysian construction industry. In Proceedings of the International Conference on Industrial Engineering and Operations Management, online. 26 September 2016. [CrossRef]
28. Julián, M.; Bonavia, T. Understanding unethical behaviors at the university level: A multiple regression analysis. *Ethics Behav.* **2021**, *31*, 257–269. [CrossRef]
29. Király, G.; Miskolczi, P. Dynamics of participation: System dynamics and participation—An empirical review. *Syst. Res. Behav. Sci.* **2019**, *36*, 199–210. [CrossRef]
30. Abiodun, O.I.; Jantan, A.; Omolara, A.E.; Dada, K.V.; Mohamed, N.A.E.; Arshad, H. State-of-the-art in artificial neural network applications: A survey. *Heliyon* **2018**, *4*, e00938. [CrossRef]
31. Bialas, F.; Wapelhorst, V.; Brokbals, S.; Čadež, I. Quantitative cross-sectional study of the BIM-application in planning offices—Benefits and barriers of the BIM-implementation. *Bautechnik* **2019**, *96*, 229–238. [CrossRef]
32. Gledson, B.; Bleanch, P.; Henry, D. Does size matter? Experiences and perspectives of BIM implementation from large and SME construction contractors. In Proceedings of the 1st UK Academic Conference on Building Information Management (BIM), Newcastle upon Tyne, UK, 5–7 September 2012.
33. Hosseini, M.R.; Banihashemi, S.; Chileshe, N.; Namzadi, M.O.; Udaeja, C.; Rameezdeen, R.; McCuen, T. BIM adoption within Australian small and medium-sized enterprises (SMEs): An innovation diffusion model. *Constr. Econ. Build.* **2016**, *16*, 71–86. [CrossRef]
34. Al-Yami, A.; Sanni-Anibire, M.O. BIM in the Saudi Arabian construction industry: State of the art, benefit and barriers. *Int. J. Build. Pathol. Adapt.* **2021**, *39*, 33–47. [CrossRef]
35. Babatunde, S.O.; Ekundayo, D. Barriers to the incorporation of BIM into quantity surveying undergraduate curriculum in the Nigerian universities. *J. Eng. Des. Technol.* **2019**, *17*, 629–648. [CrossRef]
36. Waterhouse, R.; Philp, D. *National BIM Report*; NBS National BIM Library: Cookstown, NJ, USA, 2016.
37. NBS Enterprises Ltd. *10th Annual UK's National Building Specification Report 2020*; NBS Enterprises Ltd.: Newcastle upon Tyne, UK, 2020.
38. Shibani, A.; Awwad, K.A.; Ghostin, M.; Siddiqui, K.; Farji, O. Adopting building information modelling in small and medium enterprises of Iraq's construction industry. In Proceedings of the International Conference on Industrial Engineering and Operations Management, Dubai, United Arab Emirates, 10–12 March 2020.
39. Arayici, Y.; Coates, P.; Koskela, L.; Kagioglou, M.; Usher, C.; O'Reilly, K. Technology adoption in the BIM implementation for lean architectural practice. *Autom. Constr.* **2011**, *20*, 189–195. [CrossRef]
40. Saka, A.B.; Chan, D.W.M.; Siu, F.M.F. Adoption of Building Information Modelling in Small and Medium-Sized Enterprises in Developing Countries: A System Dynamics Approach. In Proceedings of the CIB World Building Congress, Hong Kong, China, 17–21 June 2019.
41. Pretti, S.M.; Vieira, D.R. Implementation of a BIM solution in a small construction company. *J. Mod. Proj. Manag.* **2016**, *3*, 164.
42. Yan, T.W.; Kah, K.S. Building Information Modelling (BIM) in Small and Medium Enterprises (SMEs) within Malaysian Construction Sector: Implementation, Barriers, and Solutions. *INTI J.* **2018**, *2*. Available online: http://eprints.intimal.edu.my/1333/ (accessed on 21 December 2022).
43. Georgiadou, M.C. An overview of benefits and challenges of building information modelling (BIM) adoption in UK residential projects. *Constr. Innov.* **2019**, *19*, 298–320. [CrossRef]
44. Vidalakis, C.; Abanda, F.H.; Oti, A.H. BIM adoption and implementation: Focusing on SMEs. *Constr. Innov.* **2020**, *20*, 128–147. [CrossRef]
45. Al-Zwainy, F.; Mohammed, I.A.; Al-Shaikhli, K.A.K. Diagnostic and Assessment Benefits and Barriers of BIM in Construction Project Management. *Civ. Eng. J.* **2017**, *3*, 63–77. [CrossRef]
46. Aredah, A.S.; Baraka, M.A.; Elkhafif, M. Project Scheduling Techniques Within a Building Information Modeling (BIM) Environment: A Survey Study. *IEEE Eng. Manag. Rev.* **2019**, *47*, 133–143. [CrossRef]
47. Girginkaya Akdag, S.; Maqsood, U. A roadmap for BIM adoption and implementation in developing countries: The Pakistan case. *Archnet-IJAR* **2020**, *14*, 112–132. [CrossRef]
48. Alshdiefat Ala'a; Aziz Zeeshan Crucial barriers of building information modelling (BIM) in the Jordanian construction industry. *Glob. J. Eng. Technol. Adv.* **2020**, *3*, 20–30. [CrossRef]
49. Abd Hamid, A.B.; Mohd Taib, M.Z.; Abdul Razak, A.H.N.; Embi, M.R. Building Information Modelling: Challenges and Barriers in Implement of BIM for Interior Design Industry in Malaysia. In Proceedings of the IOP Conference Series: Earth and Environmental Science, Langkawi, Malaysia, 4–5 December 2017. [CrossRef]
50. Ruthankoon, R. Barriers of BIM Implementation: Experience in Thailand. Sustainable Construction, Engineering and Infrastructure Management. 2015. Available online: https://adoc.pub/proceedings-of-narotama-international-conference-on-civil-en.html (accessed on 21 December 2022).
51. Dalui, P.; Elghaish, F.; Brooks, T.; McIlwaine, S. Integrated project delivery with BIM: A methodical approach within the UK consulting sector. *J. Inf. Technol. Constr.* **2021**, *26*, 922–935. [CrossRef]
52. Singh, S.; Ashuri, B. Leveraging Blockchain Technology in AEC Industry during Design Development Phase. In Proceedings of the Computing in Civil Engineering 2019: Visualization, Information Modeling, and Simulation—Selected Papers from the ASCE International Conference on Computing in Civil Engineering 2019, Atlanta, Georgia, 17–19 June 2019. [CrossRef]

53. Abanda, F.H.; Sibilla, M.; Garstecki, P.; Anteneh, B.M. A literature review on BIM for cities Distributed Renewable and Interactive Energy Systems. *Int. J. Urban Sustain. Dev.* **2021**, *13*, 214–232. [CrossRef]
54. Zhang, Q.; Guo, B. Discussion on the Development Barriers of BIM Construction Costs in China. *Am. J. Civ. Eng.* **2019**, *7*, 133. [CrossRef]
55. Shahid, F.; Ahmed, Z.; Hussain Ali, T.; Ali Moriyani, M.; Hussain Khahro, S. A Stepped Wise Approach and Barriers towards Implementation of BIM Toolkits of Infrastructure Project in Pakistan. *Muet* **2019**. Available online: https://docplayer.net/169324891-Icsdc-th-07-th-december-2019.html (accessed on 21 December 2022).
56. Ibrahim, F.S.; Shariff, N.D.; Esa, M.; Rahman, R.A. The barriers factors and driving forces forbimimplementationin Malaysian AEC Companies. *J. Adv. Res. Dyn. Control Syst.* **2019**, *11*, 275–284.
57. Huang, B.; Lei, J.; Ren, F.; Chen, Y.; Zhao, Q.; Li, S.; Lin, Y. Contribution and obstacle analysis of applying BIM in promoting green buildings. *J. Clean. Prod.* **2021**, *278*, 123946. [CrossRef]
58. Elhendawi, A.; Omar, H.; Elbeltagi, E.; Smith, A. Practical approach for paving the way to motivate BIM non-users to adopt BIM. *Int. J. BIM Eng. Sci.* **2019**, *2*, 1–22. [CrossRef]
59. Bahar, Y.N.; Pere, C.; Landrieu, J.; Nicolle, C. A thermal simulation tool for building and its interoperability through the Building Information Modeling (BIM) platform. *Buildings* **2013**, *3*, 380–398. [CrossRef]
60. von Both, P. Potentials and Barriers for Implementing BIM in the German AEC Market. *Digit. Appl. Constr.* **2012**. Available online: https://www.google.com.hk/url?sa=t&rct=j&q=&esrc=s&source=web&cd=&cad=rja&uact=8&ved=2ahUKEwiK8bTa5r78AhXNGuwKHcRuAi0QFnoECAwQAQ&url=https%3A%2F%2Fwww.irbnet.de%2Fdaten%2Fkbf%2Fkbf_e_F_2844.pdf&usg=AOvVaw3RiIHo-lUYFdsPFtEy8xcK (accessed on 21 December 2022).
61. Azhar, S.; Khalfan, M.; Maqsood, T. Building information modeling (BIM): Now and beyond. *Australas. J. Constr. Econ. Build.* **2012**, *12*, 15–28. [CrossRef]
62. Alreshidi, E.; Mourshed, M.; Rezgui, Y. Factors for effective BIM governance. *J. Build. Eng.* **2017**, *10*, 89–101. [CrossRef]
63. Sardroud, J.M.; Mehdizadehtavasani, M.; Khorramabadi, A.; Ranjbardar, A. Barriers analysis to effective implementation of BIM in the construction industry. In Proceedings of the ISARC 2018—35th International Symposium on Automation and Robotics in Construction and International AEC/FM Hackathon: The Future of Building Things, Berlin, Germany, 20–25 July 2018. Available online: https://www.google.com.hk/url?sa=t&rct=j&q=&esrc=s&source=web&cd=&cad=rja&uact=8&ved=2ahUKEwjy2Ifx5r78AhXKyaQKHWvoC7kQFnoECAwQAQ&url=http%3A%2F%2Ftoc.proceedings.com%2F40759webtoc.pdf&usg=AOvVaw0y48GvhM7FPVXZMFY0FpqV (accessed on 21 December 2022).
64. Criminale, A.; Langar, S. 53 rd ASC Annual International Conference Proceedings Challenges with BIM Implementation: A Review of Literature. In Proceedings of the 53rd Associated School of Construction International Conference, Seattle, WA, USA, 5–8 April 2017.
65. Loveday, J.; Kouider, T.; Scott, J. The Big BIM battle: BIM adoption in the UK for large and small companies. In Proceedings of the Conference Proceedings of the 6th International Congress of Architectural Technology, Budapest, Hungary, 6 March 2020.
66. Bain, D. UK BIM Survey 2019 Findings. *Natl. BIM Rep.* **2019**. Available online: https://www.scribd.com/document/410774756/BIM-Report-2019 (accessed on 21 December 2022).
67. McNamara, A.; Sepasgozar, S.M.E. Barriers and drivers of Intelligent Contract implementation in construction. In Proceedings of the 42nd AUBEA Conference 2018: Educating Building Professionals for the Future in the Globalised World, Singapore, 5 January 2018.
68. McPartland, R. *NBS National BIM Report for Manufacturers 2017*; NBS National BIM Library: Cookstown, NJ, USA, 2017.
69. Zhou, Y.; Yang, Y.; Yang, J. Bin Barriers to BIM implementation strategies in China. *Eng. Constr. Archit. Manag.* **2019**, *26*, 554–574. [CrossRef]
70. Tan, T.; Chen, K.; Xue, F.; Lu, W. Barriers to Building Information Modeling (BIM) implementation in China's prefabricated construction: An interpretive structural modeling (ISM) approach. *J. Clean. Prod.* **2019**, *219*, 949–959. [CrossRef]
71. Chan, D.W.M.; Olawumi, T.O.; Ho, A.M.L. Perceived benefits of and barriers to Building Information Modelling (BIM) implementation in construction: The case of Hong Kong. *J. Build. Eng.* **2019**, *25*, 100764. [CrossRef]
72. Tran-Hoang-Minh, H.; Nguyen, T.Q.; Nguyen, D.P.; Pham, Q.T. Barriers of BIM adoption in Vietnamese contractors. *AIP Conf. Proc.* **2021**, *2428*, 020004. [CrossRef]
73. Kassem, M.; Brogden, T.; Dawood, N. BIM and 4D planning: A holistic study of the barriers and drivers to widespread adoption. *J. Constr. Eng. Proj. Manag.* **2012**, *2*, 1–10. [CrossRef]
74. Barqawi, M.; Chong, H.Y.; Jonescu, E. A Review of Employer-Caused Delay Factors in Traditional and Building Information Modeling (BIM)-Enabled Projects: Research Framework. *Adv. Civ. Eng.* **2021**, *2021*, 6696203. [CrossRef]
75. Bouhmoud, H.; Loudyi, D. Building information modeling (BIM) barriers in Africa versus global challenges. In Proceedings of the Colloquium in Information Science and Technology, CIST, Agadir-Essaouira, Morocco, 5–12 June 2021.
76. Sriyolja, Z.; Harwin, N.; Yahya, K. Barriers to Implement Building Information Modeling (BIM) in Construction Industry: A Critical Review. In Proceedings of the IOP Conference Series: Earth and Environmental Science, Chennai, India, 27 November 2021.
77. Gibbs, D.J.; Lord, W.; Emmitt, S.B.A.; Ruikar, K. BIM and construction contracts—CPC 2013's approach. *Proc. Inst. Civ. Eng. Manag. Procure. Law* **2015**, *168*, 285–293. [CrossRef]

78. Hedayati, A.; Mohandes, S.R.; Preece, C. Studying the Obstacles to Implementing BIM in Educational System and Making Some Recommendations. *J. Basic. Appl. Sci. Res* **2015**, *5*, 29–35.
79. Babatunde, S.O.; Perera, S.; Ekundayo, D.; Adeleke, D.S. An investigation into BIM uptake among contracting firms: An empirical study in Nigeria. *J. Financ. Manag. Prop. Constr.* **2020**, *26*, 23–48. [CrossRef]
80. Saka, A.B.; Chan, D.W.M. Profound barriers to building information modelling (BIM) adoption in construction small and medium-sized enterprises (SMEs): An interpretive structural modelling approach. *Constr. Innov.* **2020**, *20*, 261–284. [CrossRef]
81. Nasila, M.; Cloete, C. Adoption of Building Information Modelling in the construction industry in Kenya. *Acta Structilia* **2018**, *25*, 1–38. [CrossRef]
82. Halttula, H.; Haapasalo, H.; Herva, M. Barriers to Achieving the Benefits of BIM. *Int. J. 3-D Inf. Model.* **2015**, *4*, 16–33. [CrossRef]
83. Olawumi, T.O.; Chan, D.W.M.; Wong, J.K.W.; Chan, A.P.C. Barriers to the integration of BIM and sustainability practices in construction projects: A Delphi survey of international experts. *J. Build. Eng.* **2018**, *20*, 60–71. [CrossRef]
84. Hossain, M.A.; Yeoh, J.K.W. BIM for Existing Buildings: Potential Opportunities and Barriers. In Proceedings of the IOP Conference Series: Materials Science and Engineering, Nha Trang, Vietnam, 23–25 February 2018.
85. Al-Hammadi, M.A.; Tian, W. Challenges and Barriers of Building Information Modeling Adoption in the Saudi Arabian Construction Industry. *Open Constr. Build. Technol. J.* **2020**, *14*, 98–110. [CrossRef]
86. Alemayehu, S.; Nejat, A.; Ghebrab, T.; Ghosh, S. A multivariate regression approach toward prioritizing BIM adoption barriers in the Ethiopian construction industry. *Eng. Constr. Archit. Manag.* **2022**, *29*, 2635–2664. [CrossRef]
87. Deng, Y.; Li, J.; Wu, Q.; Pei, S.; Xu, N.; Ni, G. Using network theory to explore bim application barriers for BIM sustainable development in China. *Sustainability* **2020**, *12*, 3190. [CrossRef]
88. El Hajj, C.; Martínez Montes, G.; Jawad, D. An overview of BIM adoption barriers in the Middle East and North Africa developing countries. *Eng. Constr. Archit. Manag.* **2021**. [CrossRef]
89. Enshassi, A.; Ayyash, A.; Choudhry, R.M. BIM for construction safety improvement in Gaza strip: Awareness, applications and barriers. *Int. J. Constr. Manag.* **2016**, *16*, 249–265. [CrossRef]
90. Hatem, W.A.; Abd, A.M.; Abbas, N.N. Barriers of adoption building information modeling (BIM) in construction projects of Iraq. *Eng. J.* **2018**, *22*, 59–81. [CrossRef]
91. Doan, D.T.; GhaffarianHoseini, A.; Naismith, N.; Ghaffarianhoseini, A.; Zhang, T.; Tookey, J. Examining critical perspectives on Building Information Modelling (BIM) adoption in New Zealand. *Smart Sustain. Built Environ.* **2021**, *10*, 594–615. [CrossRef]
92. Reza Hosseini, M.; Pärn, E.A.; Edwards, D.J.; Papadonikolaki, E.; Oraee, M. Roadmap to Mature BIM Use in Australian SMEs: Competitive Dynamics Perspective. *J. Manag. Eng.* **2018**, *34*, 05018008. [CrossRef]
93. Saka, A.B.; Chan, D.W.M. BIM divide: An international comparative analysis of perceived barriers to implementation of BIM in the construction industry. *J. Eng. Des. Technol.* **2021**. [CrossRef]
94. Sodangi, M.; Salman, A.F.; Saleem, M. Building Information Modeling: Awareness Across the Subcontracting Sector of Saudi Arabian Construction Industry. *Arab. J. Sci. Eng.* **2018**, *43*, 1807–1816. [CrossRef]
95. Zhan, Z.; Tang, Y.; Wang, C.; Yap, J.B.H.; Lim, Y.S. System Dynamics Outlook on BIM and LEAN Interaction in Construction Quantity Surveying. *Iran. J. Sci. Technol.—Trans. Civ. Eng.* **2022**, *46*, 3947–3962. [CrossRef]
96. Ullah, K.; Lill, I.; Witt, E. An overview of BIM adoption in the construction industry: Benefits and barriers. *Emerald Reach Proc. Ser.* **2019**, *2*, 297–303. [CrossRef]
97. Marefat, A.; Toosi, H.; Mahmoudi Hasankhanlo, R. A BIM approach for construction safety: Applications, barriers and solutions. *Eng. Constr. Archit. Manag.* **2019**, *26*, 1855–1877. [CrossRef]
98. Belay, S.; Goedert, J.; Woldesenbet, A.; Rokooei, S. Enhancing BIM implementation in the Ethiopian public construction sector: An empirical study. *Cogent Eng.* **2021**, *8*, 1886476. [CrossRef]
99. Abu Aisheh, Y.I.; Tayeh, B.A.; Alaloul, W.S.; Jouda, A.F. Barriers of Occupational Safety Implementation in Infrastructure Projects: Gaza Strip Case. *Int. J. Environ. Res. Public Health* **2021**, *18*, 3553. [CrossRef] [PubMed]
100. Leśniak, A.; Górka, M.; Skrzypczak, I. Barriers to bim implementation in architecture, construction, and engineering projects—The Polish study. *Energies* **2021**, *14*, 2090. [CrossRef]
101. Olanrewaju, O.I.; Chileshe, N.; Babarinde, S.A.; Sandanayake, M. Investigating the barriers to building information modeling (BIM) implementation within the Nigerian construction industry. *Eng. Constr. Archit. Manag.* **2020**, *27*, 2931–2958. [CrossRef]
102. Oraee, M.; Hosseini, M.R.; Edwards, D.J.; Li, H.; Papadonikolaki, E.; Cao, D. Collaboration barriers in BIM-based construction networks: A conceptual model. *Int. J. Proj. Manag.* **2019**, *37*, 839–854. [CrossRef]
103. Saka, A.B.; Chan, D.W.M. A scientometric review and metasynthesis of building information modelling (BIM) research in Africa. *Buildings* **2019**, *9*, 85. [CrossRef]
104. El Hajj, C.; Jawad, D.; Montes, G.M. Analysis of a Construction Innovative Solution from the Perspective of an Information System Theory. *J. Constr. Eng. Manag.* **2021**, *147*, 03121003. [CrossRef]
105. Franco, J.; Mahdi, F.; Abaza, H. Using Building Information Modeling (BIM) for Estimating and Scheduling, Adoption Barriers. *Univers. J. Manag.* **2015**, *3*, 376–384. [CrossRef]
106. Stanley, R.; Thurnell, D. The benefits of, and barriers to, implementation of 5D BIM for quantity surveying in New Zealand. *Australas. J. Constr. Econ. Build.* **2014**, *14*, 105–117. [CrossRef]
107. Elagiry, M.; Marino, V.; Lasarte, N.; Elguezabal, P.; Messervey, T. BIM4Ren: Barriers to BIM implementation in renovation processes in the Italian market. *Buildings* **2019**, *9*, 200. [CrossRef]

108. Kim, K.P.; Freda, R.; Nguyen, T.H.D. Building information modelling feasibility study for building surveying. *Sustainability* **2020**, *12*, 4791. [CrossRef]
109. Ding, L.; Jiang, W.; Zhou, C. IoT sensor-based BIM system for smart safety barriers of hazardous energy in petrochemical construction. *Front. Eng. Manag.* **2022**, *9*, 1–15. [CrossRef]
110. Kekana, G.; Aigbavboa, C.; Thwala, W. *Overcoming Barriers That Hinders the Adoption and Implementation of Building Information Modeling in the South African Construction Industry*; University of Johannesburg: Johannesburg, South Africa, 2015.
111. Huong, Q.T.T.; Lou, E.C.W.; Hoai, N. Le Enhancing bim diffusion through pilot projects in Vietnam. *Eng. J.* **2021**, *25*, 167–176. [CrossRef]
112. Bialas, F.; Wapelhorst, V.; Brokbals, S.; Čadež, I. Quantitative Querschnittsstudie zur BIM-Anwendung in Planungsbüros: Vorteile und Hemmnisse bei der Implementierung der BIM-Methodik. *Bautechnik* **2019**, *96*, 229–238. [CrossRef]
113. Dao, T.N.; Chen, P.H.; Nguyen, T.Q. Critical Success Factors and a Contractual Framework for Construction Projects Adopting Building Information Modeling in Vietnam. *Int. J. Civ. Eng.* **2021**, *19*, 85–102. [CrossRef]
114. Salleh, H.; Fung, W.P. Building information modelling application: Focus-group discussion. *Gradjevinar* **2014**, *66*, 705–714. [CrossRef]
115. Wong, S.Y.; Gray, J. Barriers to implementing Building Information Modelling (BIM) in the Malaysian construction industry. In Proceedings of the IOP Conference Series: Materials Science and Engineering, Sarawak, Malaysia, 26–28 November 2018.
116. Ibrahim, F.S.B.; Esa, M.B.; Kamal, E.B.M. Towards construction 4.0: Empowering bim skilled talents in malaysia. *Int. J. Sci. Technol. Res.* **2019**, *8*, 1694–1700.
117. CREAM Malaysian Construction Research Journal (MCRJ). *Malaysian Constr. Res. J.* **2018**. Available online: https://www.cream.my/usr/product.aspx?pgid=40&id=17&lang=en (accessed on 21 December 2022).
118. Evans, M.; Farrell, P. Barriers to integrating building information modelling (BIM) and lean construction practices on construction mega-projects: A Delphi study. *Benchmarking* **2021**, *28*, 652–669. [CrossRef]
119. Jasiński, A. Impact of BIM implementation on architectural practice. *Archit. Eng. Des. Manag.* **2021**, *17*, 447–457. [CrossRef]
120. Karampour, B.; Mohamed, S.; Karampour, H.; Spagnolo, S.L. Formulating a Strategic Plan for BIM Diffusion within the AEC Italian Industry: The Application of Diffusion of Innovation Theory. *J. Constr. Dev. Ctries.* **2021**, *26*, 161–184. [CrossRef]
121. King, L.S.; Wei, D.K.H.; Kamarzaly, M.A.; Yaakob, A.M.; Xiong, Y.B. The Implementation of Bim Learning Module in Quantity Surveying Degree. *J. Built Environ. Technol. Eng.* **2018**. Available online: https://wenku.baidu.com/view/1d77964115888486876 2caaedd3383c4bb4cb4f6.html?_wkts_=1673416041223&bdQuery (accessed on 21 December 2022).
122. Wu, P.; Jin, R.; Xu, Y.; Lin, F.; Dong, Y.; Pan, Z. The analysis of barriers to bim implementation for industrialized building construction: A China study. *J. Civ. Eng. Manag.* **2021**, *27*, 1–13. [CrossRef]
123. Nguyen, T.Q.; Nguyen, D.P. Barriers in bim adoption and the legal considerations in Vietnam. *Int. J. Sustain. Constr. Eng. Technol.* **2021**, *12*, 283–295. [CrossRef]
124. Olugboyega, O.; Windapo, A.O. Structural equation model of the barriers to preliminary and sustained BIM adoption in a developing country. *Constr. Innov.* **2021**. [CrossRef]
125. Farooq, U.; Ur Rehman, S.K.; Javed, M.F.; Jameel, M.; Aslam, F.; Alyousef, R. Investigating bim implementation barriers and issues in Pakistan using ism approach. *Appl. Sci.* **2020**, *10*, 7250. [CrossRef]
126. Arrotéia, A.V.; Freitas, R.C.; Melhado, S.B. Barriers to BIM Adoption in Brazil. *Front. Built Environ.* **2021**, *7*, 520154. [CrossRef]
127. Babatunde, S.O.; Udeaja, C.; Adekunle, A.O. Barriers to BIM implementation and ways forward to improve its adoption in the Nigerian AEC firms. *Int. J. Build. Pathol. Adapt.* **2021**, *39*, 48–71. [CrossRef]
128. Olugboyega, O.; Windapo, A. Modelling the indicators of a reduction in BIM adoption barriers in a developing country. *Int. J. Constr. Manag.* **2021**, 1–11. [CrossRef]
129. Siebelink, S.; Voordijk, H.; Endedijk, M.; Adriaanse, A. Understanding barriers to BIM implementation: Their impact across organizational levels in relation to BIM maturity. *Front. Eng. Manag.* **2021**, *8*, 236–257. [CrossRef]
130. Olatunji, O.A.; Lee, J.J.S.; Chong, H.Y.; Akanmu, A.A. Building information modelling (BIM) penetration in quantity surveying (QS) practice. *Built Environ. Proj. Asset Manag.* **2021**, *11*, 888–902. [CrossRef]
131. Fountain, J.; Langar, S. Building Information Modeling (BIM) outsourcing among general contractors. *Autom. Constr.* **2018**, *95*, 107–117. [CrossRef]
132. Abu Aisheh, Y.I.; Alaloul, W.S.; Alhammadi, S.A.; Tayeh, B.A. Safety Management Implementation Drivers for Construction Projects: A Structural Equation Modelling Approach. *Int. J. Occup. Saf. Ergon.* **2022**, 1–10. [CrossRef] [PubMed]
133. Fitriani, H.; Budiarto, A.; Ajayi, S.; Idris, Y. Implementing BIM in architecture, engineering and construction companies: Perceived benefits and barriers among local contractors in palembang, Indonesia. *Int. J. Constr. Supply Chain Manag.* **2019**, *9*, 20–34. [CrossRef]
134. Durdyev, S.; Mbachu, J.; Thurnell, D.; Zhao, L.; Reza Hosseini, M. BIM adoption in the cambodian construction industry: Key drivers and barriers. *ISPRS Int. J. Geo-Inf.* **2021**, *10*, 215. [CrossRef]
135. Cao, Y.; Zhang, L.H.; McCabe, B.; Shahi, A. The benefits of and barriers to BIM adoption in Canada. In Proceedings of the 36th International Symposium on Automation and Robotics in Construction, ISARC 2019, Banff, AB, Canada, 24 May 2019.
136. Ademci, E.; Gundes, S. Individual and Organisational Level Drivers and Barriers to Building Information Modelling. *J. Constr. Dev. Ctries.* **2021**, *26*, 89–109. [CrossRef]

137. Zakari, Z.; Ali, N.M.A.; Haron, A.T.; Marshall Ponting, A.; Hamid, Z.A. Exploring the Barriers and Driving Factors in Implementing Building Information Modelling (BIM) in the Malaysian Construction Industry: A Preliminary Study. *J. Inst. Eng. Malaysia* **2014**, 75. [CrossRef]
138. Alhumayn, S.; Chinyio, E.; Ndekugri, I. The barriers and strategies of implementing BIM in Saudi Arabia. In *WIT Transactions on the Built Environment*; WIT Press: Ashurst, UK, 2017.
139. NBS; NBS. Jenny Archer National BIM Report 2019—The Definitive Industry Update. *Natl. BIM Rep. 2019 Defin. Ind. Updat.* **2019**. Available online: https://www.google.com.hk/url?sa=t&rct=j&q=&esrc=s&source=web&cd=&cad=rja&uact=8&ved=2ahUKEwi-odbu6L78AhUqMewKHcYACxUQFnoECA8QAQ&url=https%3A%2F%2Fwww.thenbs.com%2Fknowledge%2Fnational-bim-report-2019&usg=AOvVaw0JAb6Cnpq9QNy76MtywG0G (accessed on 21 December 2022).
140. Tayeh, B.A.; Salem, T.J.; Abu Aisheh, Y.I.; Alaloul, W.S. Risk Factors Affecting the Performance of Construction Projects in Gaza Strip. *Open Civ. Eng. J.* **2020**, *14*, 94–104. [CrossRef]
141. Mellon, S.; Kouider, T. SME's and Level 2, the way forward. In Proceedings of the 6th International Congress of Architectural Technology (ICAT 2016): Healthy Buildings: Innovation, Design and Technology, Alicante, Spain, 12–14 May 2016; pp. 121–135.
142. Alhammadi, S.A.; Tayeh, B.A.; Alaloul, W.S.; Jouda, A.F. Occupational Health and Safety Practice in Infrastructure Projects. *Int. J. Occup. Saf. Ergon.* **2022**, *28*, 2631–2644. [CrossRef]
143. Saka, A.B.; Chan, D.W.M. Adoption and implementation of building information modelling (BIM) in small and medium-sized enterprises (SMEs): A review and conceptualization. *Eng. Constr. Archit. Manag.* **2020**, *28*, 1829–1862. [CrossRef]
144. Alzubi, K.M.; Alaloul, W.S.; Qureshi, A.H. Applications of Cyber-Physical Systems in Construction Projects. In *Cyber-Physical Systems in the Construction Sector*; CRC Press: Boca Raton, FL, USA, 2022; pp. 153–171. [CrossRef]
145. Jones, B.I. A study of building information modeling (BIM) uptake and proposed evaluation framework. *J. Inf. Technol. Constr.* **2020**, *25*, 452–468. [CrossRef]
146. Whitlock, K.; Abanda, F.H.; Manjia, M.B.; Pettang, C.; Nkeng, G.E. 4D BIM for Construction Logistics Management. *CivilEng* **2021**, *2*, 325–348. [CrossRef]
147. Gerrish, T.; Ruikar, K.; Cook, M.; Johnson, M.; Phillip, M.; Lowry, C. BIM application to building energy performance visualisation and managementChallenges and potential. *Energy Build.* **2017**, *144*, 218–228. [CrossRef]
148. Al Amin, A. Factors Affecting as Barrier for Adaptation of Building Information Modelling in Architecture Practice in Bangladesh. *SEU J. Sci. Eng.* **2020**, *14*. Available online: https://www.google.com.hk/url?sa=t&rct=j&q=&esrc=s&source=web&cd=&cad=rja&uact=8&ved=2ahUKEwjg2PW26b78AhVS3aQKHdMVAasQFnoECAoQAQ&url=https%3A%2F%2Fseu.edu.bd%2Fseujse%2Fdownloads%2Fvol_14_no_1_Jun_2020%2FSEUJSE-Vol14No1-08.pdf&usg=AOvVaw3rSTO9XTXj0YIR4S9g4uq3 (accessed on 21 December 2022).
149. Eadie, R.; Odeyinka, H.; Browne, M.; Mckeown, C.; Yohanis, M. Building Information Modelling Adoption: An Analysis of the Barriers to Implementation. *J. Eng. Archit.* **2014**, *2*, 77–101. [CrossRef]
150. Yao, Y.; Qin, H.; Wang, J. Barriers of 5D BIM implementation in prefabrication construction of buildings. In Proceedings of the IOP Conference Series: Earth and Environmental Science, Xi'an, China, 14–16 August 2020.
151. Ting, T.Y.; Taib, M. The awareness of building information modeling in Malaysia construction industry from contractor perspective. *Malaysian Constr. Res. J.* **2018**, *3*, 75–81.
152. Liao, L.; Zhou, K.; Fan, C.; Ma, Y. Evaluation of Complexity Issues in Building Information Modeling Diffusion Research. *Sustainability* **2022**, *14*, 3005. [CrossRef]
153. Jiang, H.J.; Cui, Z.P.; Yin, H.; Yang, Z.B. BIM Performance, Project Complexity, and User Satisfaction: A QCA Study of 39 Cases. *Adv. Civ. Eng.* **2021**, *2021*, 1–10. [CrossRef]
154. Charlton, J.; Kelly, K.; Greenwood, D.; Moreton, L. The complexities of managing historic buildings with BIM. *Eng. Constr. Archit. Manag.* **2021**, *28*, 570–583. [CrossRef]
155. Rice, L. Healthy BIM: The feasibility of integrating architecture health indicators using a building information model (BIM) computer system. *Archnet-IJAR* **2021**, *15*, 252–265. [CrossRef]
156. Crowther, J.; Ajayi, S.O. Impacts of 4D BIM on construction project performance. *Int. J. Constr. Manag.* **2021**, *21*, 724–737. [CrossRef]
157. Joo, S.U.; Kim, C.-K.; Kim, S.-U.; Noh, J.-O. BIM-Based Quantity Takeoff and Cost Estimation Guidelines for Reinforced Concrete Structures. *J. Comput. Struct. Eng. Inst. Korea* **2017**, *30*, 567–576. [CrossRef]
158. Monazan, N.H.; Hamidimonazan, H.; Hosseini, M.R.; Zaeri, F. Barriers to adopting building information modelling (bim) within south australian small and medium sized enterprises. In Proceedings of the Fifth International Scientific Conference on Project management in the Baltic Countries, Riga, Latvia, 4–15 April 2016.
159. Bonanomi, M.; Paganin, G.; Talamo, C. Bim Implementation in Design Firms. Risk-Response. In Proceedings of the 41st IAHS WORLD CONGRESS. Sustainability and Innovation for the Future, Albufeira, Portugal, 13–16 September 2016.
160. Abu Aisheh, Y.I.; Tayeh, B.A.; Alaloul, W.S.; Almalki, A. Health and Safety Improvement in Construction Projects: A Lean Construction Approach. *Int. J. Occup. Saf. Ergon.* **2022**, *28*, 1981–1993. [CrossRef]
161. Burt, J.; Purver, K. Building information modelling for small-scale residential projects. In Proceedings of the Institution of Civil Engineers: Management, Procurement and Law, London, UK, 14 March 2014.
162. Utomo, F.R.; Rohman, M.A. The Barrier and Driver Factors of Building Information Modelling (BIM) Adoption in Indonesia: A Preliminary Survey. *IPTEK J. Proc. Ser.* **2019**, *5*, 133–139. [CrossRef]

163. Piroozfar, P.; Farr, E.R.P.; Zadeh, A.H.M.; Timoteo Inacio, S.; Kilgallon, S.; Jin, R. Facilitating Building Information Modelling (BIM) using Integrated Project Delivery (IPD): A UK perspective. *J. Build. Eng.* **2019**, *26*, 100907. [CrossRef]
164. Kushwaha, V. Contribution Of Building Information Modeling (BIM) To Solve Problems In Architecture, Engineering and Construction (AEC) Industry and Addressing Barriers to Implementation of BIM. *Int. Res. J. Eng. Technol.* **2016**, *3*, 100–105.
165. Stojanovska-Georgievska, L.; Sandeva, I.; Krleski, A.; Spasevska, H.; Ginovska, M.; Panchevski, I.; Ivanov, R.; Arnal, I.P.; Cerovsek, T.; Funtik, T. BIM in the Center of Digital Transformation of the Construction Sector—The Status of BIM Adoption in North Macedonia†. *Buildings* **2022**, *12*, 218. [CrossRef]
166. Braila, N.; Panchenko, N.; Kankhva, V. Building Information Modeling for Existing Sustainable Buildings. E3S Web of Conferences, 27 May 2021, France, 2021. Available online: https://ui.adsabs.harvard.edu/abs/2021E3SWC.24405024B/abstract (accessed on 21 December 2022).
167. Mehran, D. Exploring the Adoption of BIM in the UAE Construction Industry for AEC Firms. In Proceedings of the Procedia Engineering, Tirgu Mures, Romania, 7 October 2016.
168. Zhang, L.; Chu, Z.; He, Q.; Zhai, P. Investigating the constraints to buidling information modeling (BIM) applications for sustainable building projects: A case of China. *Sustainability* **2019**, *11*, 1896. [CrossRef]
169. Whitlock, K.; Abanda, F.H.; Manjia, M.B.; Pettang, C.; Nkeng, G.E. BIM for Construction Site Logistics Management. *J. Eng. Proj. Prod. Manag.* **2018**, *8*, 47–55. [CrossRef]
170. Arokiaprakash, A.; Aparna, K. Study on Adoption of BIM in Indian Construction Industry. *J. Adv. Res. Dyn. Control Syst.* **2018**. Available online: https://books.google.co.kr/books?id=g61jEAAAQBAJ&pg=PA96&lpg=PA96&dq=Arokiaprakash,+A.%3B+Aparna,+K&source=bl&ots=UDFKgvrnWU&sig=ACfU3U15Xfryfx-9Xb8xRsqWjKbPzu-QNA&hl=en&sa=X&redir_esc=y#v=onepage&q=Arokiaprakash%2C%20A.%3B%20Aparna%2C%20KStudy%20on%20adoption%20&f=false (accessed on 21 December 2022).
171. Harrison, C.; Thurnell, D. BIM implementation in a new zealand consulting quantity surveying practice. *Int. J. Constr. Supply Chain Manag.* **2015**, *5*, 1–15. [CrossRef]
172. Ma, X.; Darko, A.; Chan, A.P.C.; Wang, R.; Zhang, B. An empirical analysis of barriers to building information modelling (BIM) implementation in construction projects: Evidence from the Chinese context. *Int. J. Constr. Manag.* **2020**, *22*, 3119–3127. [CrossRef]
173. Casasayas, O.; Hosseini, M.R.; Edwards, D.J.; Shuchi, S.; Chowdhury, M. Integrating BIM in Higher Education Programs: Barriers and Remedial Solutions in Australia. *J. Archit. Eng.* **2021**, *27*, 05020010. [CrossRef]
174. Liu, S.; Xie, B.; Tivendal, L.; Liu, C. Critical barriers to BIM implementation in the AEC industry. *Int. J. Mark. Stud.* **2015**, *7*, 162. [CrossRef]
175. Hamma-adama, M.H.; Kouider, T.; Salman, H. Analysis of Barriers and Drivers for BIM Adoption. *Int. J. BIM Eng. Sci.* **2020**, *3*, 18–41. [CrossRef]
176. Demirdöğen, G.; Diren, N.S.; Aladağ, H.; Işık, Z. Lean based maturity framework integrating value, BIM and big data analytics: Evidence from AEC industry. *Sustainability* **2021**, *13*, 10029. [CrossRef]
177. Barbosa, M.J.; Pauwels, P.; Ferreira, V.; Mateus, L. Towards increased BIM usage for existing building interventions. *Struct. Surv.* **2016**, *34*, 168–190. [CrossRef]
178. Tushar, Q.; Bhuiyan, M.A.; Zhang, G.; Maqsood, T. An integrated approach of BIM-enabled LCA and energy simulation: The optimized solution towards sustainable development. *J. Clean. Prod.* **2021**, *289*, 125622. [CrossRef]
179. Bruno, N.; Roncella, R. HBIM for conservation: A new proposal for information modeling. *Remote Sens.* **2019**, *11*, 1751. [CrossRef]
180. de Almeida Pereira Santana, C.C.; de Souza Freitas, A.T.V.; Barreto, G.O.; de Avelar, I.S.; Mazaro-Costa, R.; Bueno, G.N.; Ribeiro, D.C.; Silva, G.D.; Naghettini, A.V. Serious game on a smartphone for adolescents undergoing hemodialysis: Development and evaluation. *JMIR Serious Games* **2020**, *8*, e17979. [CrossRef]
181. Peters, A.; Thon, A. Best practices and first steps of implementing bim in landscape architecture and its reflection of necessary workflows and working processes. *J. Digit. Landsc. Archit.* **2019**, 106–113. [CrossRef]
182. Tarandi, V. A BIM Collaboration Lab for Improved through Life Support. *Procedia Econ. Financ.* **2015**, *21*, 383–390. [CrossRef]
183. Krämer, M.; Besenyoi, Z. Towards Digitalization of Building Operations with BIM. *IOP Conf. Ser. Mater. Sci. Eng.* **2018**, *365*, 022067. [CrossRef]
184. Saleeb, N.; Marzouk, M.; Atteya, U. A comparative suitability study between classification systems for bim in heritage. *Int. J. Sustain. Dev. Plan.* **2018**, 130–138. [CrossRef]
185. Mostafa, S.; Kim, K.P.; Tam, V.W.Y.; Rahnamayiezekavat, P. Exploring the status, benefits, barriers and opportunities of using BIM for advancing prefabrication practice. *Int. J. Constr. Manag.* **2020**, *20*, 146–156. [CrossRef]
186. Memon, A.H.; Rahman, I.A.; Memon, I.; Azman, N.I.A. BIM in Malaysian construction industry: Status, advantages, barriers and strategies to enhance the implementation level. *Res. J. Appl. Sci. Eng. Technol.* **2014**, *8*, 606–614. [CrossRef]
187. Alreshidi, E.; Mourshed, M.; Rezgui, Y. Requirements for cloud-based BIM governance solutions to facilitate team collaboration in construction projects. *Requir. Eng.* **2018**, *23*, 1–31. [CrossRef]
188. Aitbayeva, D.; Hossain, M.A. Building Information Model (BIM) Implementation in Perspective of Kazakhstan: Opportunities and Barriers. *J. Eng. Res. Reports* **2020**, *14*, 13–24. [CrossRef]
189. Vishnu Vardan, M.; Raj Prasad, J. Developing a strategic model to improve the reuse of construction material by integrating CBM and BIM. *Int. J. Recent Technol. Eng.* **2019**, *8*, 630–634.

190. Oduyemi, O.; Okoroh, M.I.; Fajana, O.S. The application and barriers of BIM in sustainable building design. *J. Facil. Manag.* **2017**, *15*, 15–34. [CrossRef]
191. Onososen, A.; Musonda, I. Barriers to BIM-Based Life Cycle Sustainability Assessment for Buildings: An Interpretive Structural Modelling Approach. *Buildings* **2022**, *12*, 324. [CrossRef]
192. Hamada, H.M.; Haron, A.; Zakiria, Z.; Humada, A.M. Benefits and Barriers of BIM Adoption in the Iraqi Construction Firms. *Int. J. Innov. Res. Adv. Eng.* **2016**, *3*, 76–84.
193. Lindblad, H. Study of the Implementation Process of BIM in Construction Projects: Analysis of the Barriers Limiting BIM Adopotion in the AEC Industry. *Unpubl. MSc Thesis* **2013**, 633132. Available online: http://kth.diva-portal.org/smash/get/diva2 (accessed on 21 December 2022).
194. Chan, C.T.W. Barriers of Implementing BIM in Construction Industry from the Designers' Perspective: A Hong Kong Experience. *ISSN J. Syst. Manag. Sci. J. Syst. Manag. Sci.* **2014**, *4*, 24–40.
195. Dworkin, S.L. Sample Size Policy for Qualitative Studies Using In-Depth Interviews. *Arch. Sex. Behav.* **2012**, *41*, 1319–1320. [CrossRef] [PubMed]
196. Hesse-Biber, S. Qualitative approaches to mixed methods practice. *Qual. Inq.* **2010**, *16*, 455–468. [CrossRef]
197. Halim, E.; Mohamed, A.; Fathi, M.S. Building Information Modelling (BIM) Implementation for Highway Project from Consultant's Perspectives in Malaysia. In Proceedings of the IOP Conference Series: Earth and Environmental Science, Bangkok, Thailand, 16 October 2022.
198. Al-Aidrous, A.H.M.H.; Rahmawati, Y.; Wan Yusof, K.; Omar Baarimah, A.; Alawag, A.M. Review of Industrialized Buildings Experience in Malaysia: An Example of a Developing Country. In Proceedings of the IOP Conference Series: Earth and Environmental Science, Putrajaya, Malaysia, 16–17 October 2020.
199. Noor, S.S.M.; Esa, M. Relationship between project managers personality and small public construction project success in Malaysia. *Int. J. Sustain. Constr. Eng. Technol.* **2021**, *12*, 18–30. [CrossRef]
200. Vasudevan, G.; Wei, C.C. *Implementation of BIM with Integrated E-Procurement System in Malaysian Construction Industry*; Springer: Berlin/Heidelberg, Germany, 2021. [CrossRef]

Disclaimer/Publisher's Note: The statements, opinions and data contained in all publications are solely those of the individual author(s) and contributor(s) and not of MDPI and/or the editor(s). MDPI and/or the editor(s) disclaim responsibility for any injury to people or property resulting from any ideas, methods, instructions or products referred to in the content.

Article

Health and Safety Improvement through Industrial Revolution 4.0: Malaysian Construction Industry Case

Muhammad Ali Musarat [1,2], Wesam Salah Alaloul [1,*], Muhammad Irfan [3], Pravin Sreenivasan [1] and Muhammad Babar Ali Rabbani [4]

1 Department of Civil and Environmental Engineering, Universiti Teknologi PETRONAS, Bandar Seri Iskandar 32610, Perak, Malaysia
2 Offshore Engineering Centre, Institute of Autonomous System, Universiti Teknologi PETRONAS, Bandar Seri Iskandar 32610, Perak, Malaysia
3 Civil Engineering Department, HITEC University, Taxila 47080, Pakistan
4 Faculty of Engineering, University of New Brunswick, Fredericton, NB E3B 5A3, Canada
* Correspondence: wesam.alaloul@utp.edu.my

Abstract: Safety on construction sites is now a top priority for the construction industry all around the world. Construction labor is often seen as hazardous, putting employees at risk of serious accidents and diseases. The use of Industrial Revolution (IR) 4.0 advanced technologies such as robotics and automation, building information modelling (BIM), augmented reality and virtualization, and wireless monitoring and sensors are seen to be an effective way to improve the health and safety of construction workers at the job site, as well as to ensure construction safety management in general. The main aim of this research was to analyze the IR-4.0-related technologies for improving the health and safety problems in the construction industry of Malaysia by utilizing the analytical hierarchy process (AHP) technique. IR-4.0-related technologies show great potential in addressing the construction industry's existing health and safety problems from the perspective of civil engineering practitioners and industry experts. This research adopted the analytical hierarchy process (AHP) for quantitative analysis of data collected through the survey questionnaire approach. The findings of the study indicate that from matrix multiplication, the highest importance among the criteria and the alternatives was for BIM with a score of 0.3855, followed by wireless monitoring and sensors (0.3509). This research suggests that building information modelling (BIM) and integrated systems had the greatest potential as advanced technology and should be prioritized when it comes to introducing it to the construction industry to improve the current health and safety performances.

Keywords: IR 4.0; health and safety; BIM; AHP; construction industry

1. Introduction

Evolution in science and technology has always been of paramount significance since the advent of the Industrial Revolution [1]. This hunger and appetite of humans to evolve further have assisted mankind to go the extra mile [2]. Since the initiation of the industrial revolution in the 1700s, each industrial revolution has played a key part in the growth of today's development [3]. Mechanical looms, powered by water and steam on mechanical equipment, were first introduced in the First Industrial Revolution (IR 1.0) in the 1700s, and they replaced agricultural sectors, further increasing the economic structure. Furthermore, the Second Industrial Revolution (IR 2.0) between 1870 and 1914 was marked by dense innovation based on valuable knowledge being mapped onto technology that propelled the industry forward with low-cost, high-efficiency mass production of steel, telegraphs, and railways [4]. Similarly, electrical energy was also introduced in the 1870s, resulting in the formation of a large system known as mass production. The Internet, information technology (IT), and the widespread access to personal computers in the late 1950s ushered in a new digital revolution in which mechanical and analogue procedures were digitized, and mass manufacturing gave way to mass customization [5]. Consequently, with the emergence of electronics in the 1970s, the Third Industrial Revolution (IR 3.0) began. This changed the fate of mankind in the sense that microchips and supercomputers revolutionized almost every industry. The human hunger for evolution in science and technology did not stop here. In the late 19th or early 20th century, large amounts of research in science and technology gave birth to the fourth industrial revolution. Thus, the Fourth Industrial Revolution (IR 4.0) is based on the Digital Revolution, which connects technology and people. Research on the Fourth Industrial Revolution is relatively new and includes robotics and automation, smart factories, augmented and virtual reality, artificial intelligence, integrated systems, BIM, and cloud computing [6].

Significantly, in Malaysia, the Ministry of Works, in conjunction with parties with interests in the construction industry and through Construction Industry Development Board (CIDB) is preparing a Construction Strategy Plan 4.0 (2021–2050) to help the industry adapt to the changes [7]. In all fields of industry, notably in the construction industry, health and safety issues are of great concern [8]. Even though the construction industry is constantly changing due to new methods, equipment, and machinery, it is never without safety issues, including fatalities [9]. Over the years, most countries, including Malaysia, have created safety and health legislation. The Occupational Safety and Health Act [10] regulates safety and health issues in the construction industry in Malaysia [11]. The Social Security Organization's data, on the other hand, revealed that there had been a wave of construction site accidents. In 2016, 7338 accidents were reported, up from 4330 in 2011, indicating a 69.47% increase [12]. Furthermore, the construction industry continues to be the leading cause of fatal accidents, with a fatality rate of 14.57 per 10,000 persons [13].

However, Industry 4.0 has been around for a while in the construction industry, and the technologies are at various stages of development [14]. Industry 4.0 or IR 4.0 is the fourth industrial revolution in which human–machine interaction is achieved. Key technologies in IR 4.0 include: building information modelling (BIM), cloud computing, artificial intelligence, robotics, 3D printing, and modularization, which have advanced greatly, while other technologies such as augmented, virtual, and mixed reality are still being improved and may have an impact on the industry's long-term viability [15]. These advanced technologies can make the construction site much safer and more productive for the project teams, thus avoiding hazardous incidents on sites. Despite having access to these technologies, the application of IR 4.0 in the construction industry is still woefully weak. Within the construction industry, IR 4.0 methods have been applied, and the procedures have exhibited considerable impacts across different platforms [16]. Nonetheless, all parties involved must resolve the issues that arose to assure a successful implementation. The social factor has been recognized as the most important factor influencing successful implementation; nevertheless, the other contributing factors imply that these factors are interconnected and should be addressed at the same time.

The construction industry had the greatest number of occupational fatalities examined out of any Malaysian industry sector in 2018 [17]. In addition, the construction industry was responsible for 118 deaths (45.4%) in total. Poor safety performance in the construction industry is one of the leading causes of work-related injuries and death. In many nations, including Malaysia, the construction industry continues to be one of the most dangerous places to work when compared to other industries, with one of the highest rates of fatal occupational accidents [18]. Because of the complicated nature of the construction site and the activities that take place there, safety work becomes more difficult. Occupational Safety and Health (OSH) in the construction sector has been recognized as a top priority topic since it is the most dangerous industry with complicated and aggressive methods, resulting in large numbers of accidents and deaths among construction employees and the general public [19].

Furthermore, the utilization of a multidisciplinary workforce adds to the complexity, as does the problem of controlling the borders between disciplines and the characteristics of worker behavior, which are not as uniform as in manufacturing industries [20]. Accidents can occur as a result of workers' dangerous actions, which are difficult to monitor and manage. Therefore, safety should be prioritized throughout the design process, as well as downstream operations such as the construction phases by utilizing IR-4.0-related methods and tools [21]. While other economic sectors have experienced significant technological transformations, the construction industry continues to fall behind in several areas, including occupational health and safety [22]. However, in the view of published literature and given the official statistics and figures on the safety performance in the Malaysian construction industry [12,17,18], it is clear to see that a necessary upgrade in technology is needed to replace existing and long-standing safety technologies that are outdated and to mitigate the worrying upward trend of accidents in the recent years [8]. The published literature justifies the urgency of the utilization of the latest tools and methods for safe construction industry practices [23].

Therefore, to bridge this research gap, the current study was undertaken. The main aim of this research was to analyze the IR-4.0-related technologies for improving the health and safety problems in the construction industry of Malaysia by utilizing the analytical hierarchy process (AHP) technique. The findings of the study will assist in examining how the technology application in Industrial Revolution 4.0 may overcome the limitations and unsatisfactory safety performances that are typically encountered in the construction industry. The significant findings of the study will expand the body of knowledge by proposing a framework through AHP, which will guide top management, all stakeholders, and key decisionmakers to establish an effective and safe environment in the organizations and assist policymakers to strive for the incorporation of safe work practices on construction sites.

2. Literature Review

Health and safety are a common necessity for all branches of industry, business, and commerce including information technology businesses, traditional industries, the National Health Service, care homes, schools, universities, leisure facilities, and offices. Globally, many legislation acts or regulations have been crafted throughout the years to help solidify the health and safety standards in all branches of industry. Occupational Health and Safety (OHS) is a multidisciplinary field concerned with the prevention of occupational risks inherent to each work activity or job. The fundamental goal is the promotion and maintenance of the highest degree of safety and health at work, therefore creating conditions to avoid the occurrence of work accidents and illness [8]. As a result, OHS is more than just preventing work accidents or occupational diseases, but is also the consequence of taking actions to identify their causes (hazards existent in the workplace) and putting in place suitable preventive OHS control measures. According to Zid et al. (2018) [24], Occupational Health and Safety (OHS) is concerned primarily with safeguarding employees in the workplace against accidents, injuries, and harmful exposure [25]. While accidents can occur at any given moment, it is the employer's responsibility to ensure that precautions are taken to decrease the likelihood of mishaps and to maintain a safe working environment.

According to the writers' scientific fields and nationalities, there are a variety of definitions used in occupational health and safety literature [26]. However, in the current study context, the definition of the accident as outlined by OSHA, i.e., an unforeseen incident that causes property damage or personal injury is referred to as an accident [27], will be used. The construction industry involves high risk due to its production processes, labor-intensive characteristics, and financial losses on a large scale in the event of occupational accidents due to risks and hazards [18]. One in five worker deaths in the private sector occurred in construction in 2019, accounting for almost 20% (1061) of all worker fatalities [28]. Moreover, construction-related illnesses and injuries cost Great Britain's economy approximately GBP 16.2 billion in 2018–2019. The majority of these expenses—59%, or GBP 9.56 billion—were borne by the hurt or ill people themselves. Compared to 2018, the overall cost of injuries in the construction industry increased by 34% in 2020, and the fatal injuries rate in the construction industry as compared to other sectors is four times more [29]. Some of the most common hazards in the construction industry include falls from height, falling objects, exposure to dangerous substances, dust inhalation, working in confined spaces, and being hit in vehicle accidents [30–32]. Safety is a vital part of finishing a project on or under budget. Downtime is costly for companies, as is finding replacement workers when someone cannot do their job after an accident [20]. Worker's compensation claims and lawsuits can drive up a company's insurance costs. Focusing on safety helps keep your costs low. Safety and health should always be strictly adhered to at construction sites because it protects the public, reduces work-related accidents, and decreases time and money lost after an accident [33].

2.1. Health and Safety in the Construction Industry in Malaysia

The construction industry has generally met the obstacles of rapid physical expansion and development throughout this time. According to Abdul-Aziz and Hussin (2003) [34], even though other economic sectors have experienced significant transformations [35], the construction industry continues to lag in many areas including occupational health and safety. A survey conducted by Saifullah and Ismail (2012) [36] about research priorities in occupational health and safety in Malaysia highlighted that construction is one of two economic prospects that should be given top priority (the other being plantation). In contrast, the Malaysian construction industry's safety record has remained stubbornly low, despite many efforts.

The Construction Industry Development Board (CIDB), which has been in existence since 1995, was established to ensure that Malaysia's construction industry develops in a manner that is consistent with national aspirations [37]. CIDB has undertaken several projects, including the "green card" program, which was launched in May 2000 with a Malaysian ringgit (MYR) 16 million budget. All CIDB-registered site personnel are required to attend a one-day safety training led by CIDB staff or CDIB-accredited independent trainers, after which they are awarded a green registration card (as opposed to a white registration card for those who have not). Green card holders are automatically covered against industrial accidents through a CIDB-arranged insurance policy. Before beginning any construction job, contractors must obtain Workmen's Compensation Insurance (WCI) coverage. Every worker on the job, including subcontractors, is covered by the policy. Industrial accidents, occupational illness, and commuting accidents are covered by the Employment Injury Insurance Scheme, whereas invalidity or death from any cause is covered by the Invalidity Pension Scheme [38].

The construction industry is regarded as the most hazardous in terms of Occupational Health and Safety (OHS) due to the nature of the process involved [22]. According to the Department of Safety and Health Malaysia's (DOSH) statistics on fatal accidents, the Malaysian construction industry had the highest number of fatalities during the study period, making it the most critical sector that requires effective OHS management to reduce the significant number of fatalities on construction sites [39]. In 2018 alone, there was a record high of 118 worker fatalities in the Malaysian construction industry.

2.2. Technological Application in Industrial Revolution 4.0

Since the days of hard helmets and safety glasses, digital technology has been used in construction. The use of big data, technological brilliance, and construction has improved worker safety on construction sites in recent years [40]. Several studies have highlighted digital technologies such as BIM, VR and AR, drones, GIS, automation and robots, unmanned equipment, sensing and warning systems, and 4D CAD as useful solutions for accident prevention and project delivery [41–46]. The next subsections go through these technologies in depth.

2.2.1. Robotics and Automation

Construction is a labor-intensive sector of the economy. Being one of the least mechanized businesses, the construction industry is far behind in embracing robots, automation, and digital technology [47]. The newest innovations in the building construction sector, including robots and artificial intelligence, are now the talk of the town. By using such technology, the building may be completed more quickly and accurately while also conserving time, money, and other resources. Robotics technology in construction enables construction professionals with quality-assured outputs and decreased human mistakes in a fast-paced building process [48].

Thus, numerous studies have suggested that robots and automation can alleviate construction-related health and safety concerns. Robotic technologies are currently being employed to complete risky and difficult tasks [49]. Robotics and automation, according to Zanchettin et al. (2018) [50], are most successful in operations that need speed and repetition and are carried out in adverse settings. Construction workers are particularly vulnerable to overexertion diseases and injuries as a result of these activities. The use of robotic devices to speed up and automate building operations is a recent theme right now [51]. Different kinds of robotic systems have been created to improve construction project quality, productivity, worker health, and other safety issues [42]. Robots are used in the autonomous installation and collection of heavy construction materials, which typically demand huge manpower, as well as the building of structures such as skyscraper towers, which are controlled by computers on site and rely on improved recognition and control [50]. Exoskeletons, welding robots, and forklift robots are just a few of the new robotic technologies that have recently been created and can be used in construction. Wearable robotics, such as exoskeletons for decreasing lower back stress, are used while lifting and moving large things and are often used by elderly employees. The AWN-03 Suit detects worker movement and offers back, shoulder, and thigh support. Signals are delivered to the motors, which rotate the gears when the hip and spine are moved, reducing the amount of energy required. Another wearable robotics gear that improves users' strength and endurance is the FORTIS Exoskeleton. Robotic arms are made out of aluminum servo brackets and resemble human arms in appearance. Its ergonomic tool arms are made up of radial distance infrared sensors and a USB camera that records angles and gives the arms input on whether or not they can lift an object [43]. Robotic arms have been discovered to be useful for improving worker comfort and safety by stabilizing and bracing, reaching and transferring, and decreasing effort in repetitive tasks.

2.2.2. Augmented Reality and Virtualization

Despite being one of the largest and oldest sectors in the world, construction struggles frequently to adopt new tools. Nevertheless, innovation is a vital resource for the building sector. Another innovative development in the construction sector is augmented reality, which is a priceless tool for all facets of this sector. It makes use of cutting-edge camera and sensor technologies that blend digital components, audio, and sensory input. They all support the real-time presentation of the data [52].

Unlike virtual reality, which employs computer-generated images to imitate real-life events, augmented reality uses technology to improve reality [53]. However, if such interactions are integrated into programs and mobile devices, improvements may be seen [54]. As people walk around building sites with mobile devices or special helmets, augmented reality projects 3D images on their physical environment, employing GPS and cameras to show real-time data geospatially and provide updated user feedback [46]. This technology might be used to perform safety training by allowing employees to wear augmented reality headsets to receive virtual exercises, instructions, and safety situations with minimum training expenses and downtime.

A virtual reality experience is a computer-generated simulation of a real-life scenario or setting. It creates realistic visuals and sound, giving the user the impression that they are experiencing the simulated world [55]. Virtual reality has been used by the safety team in construction to review safety tie-off points and coordinate major crane picks over occupied facilities that cannot be disrupted, allowing for an effective means of visualizing and communicating the impact of major construction activities in existing facilities that may be overlooked when viewing through traditional techniques [56]. Virtual reality also offers a realistic work environment that may be used for construction safety instruction [43]. These advantages include the ability to conduct safety exercises in the absence of a certified safety administrator by merely replicating the training environment on a computer. Traditional paper-based handouts, video cassettes, or slide displays, according to [57], do not adequately communicate electric dangers to trainees and do not give enough opportunities for trainees to interact in activities. This type of interactive training incorporates a real-world scenario into the training in the form of an "it might happen to you" scenario, allowing the learner to link these rules and circumstances to real-life scenarios involving life and death.

2.2.3. Building Information Modelling (BIM)

Building information modeling (BIM) is a critical and sometimes required procedure used all over the world to guarantee that the planning, design, and construction of buildings are extremely efficient and collaborative [58]. It has clearly stated aims and objectives that are advantageous to everyone who progresses through the stages. Undoubtedly, the building will become much more digital and collaborative in the future. Four-dimensional, five-dimensional, and even six-dimensional BIM will start to participate in the process as BIM grows more complex [59].

Sun et al. (2017) [60] suggested that BIM might be used to efficiently complete construction tasks to solve difficulties originating from the perplexing nature of construction sites and projects. The BIM system is used as a supporting technology for health and safety planning and management on construction sites [57]. BIM allows for a visual evaluation of the construction site and the identification of possible risks [44]. Site workers may have a better grasp of the actual site circumstances by using the BIM model to perform visual safety training. Before carrying out construction operations, construction workers are given enough time and knowledge to plan and manage their safety. By verifying data-collecting procedures using sensors, the BIM can effectively minimize the risk of site accidents (Fargnoli and Lombardi, 2020) [61]. One of the primary benefits of BIM adoption is its ability to represent and manage visuals as well as automatically analyze designs, create drawings, reports, design schedules, and manage facilities. Furthermore, BIM technology is successful not only in the design phase but also in the operation phases, notably for construction process simulation [62].

2.2.4. Wireless Monitoring and Sensors

The acceptance of new technology in the construction sector is a long process, much like new standards and laws. In reality, the industry's failure to stay up with technology is one of the key reasons why adopting new standards and regulations may be so difficult. Construction decisionmakers are reluctant to abandon conventional design, construction, and testing techniques. The use of smart technology in the building is crucial in a world where everything is becoming more linked and new systems are being built for every aspect of life [63]. More precisely, the use of sensors and other tools is crucial for keeping track of and evaluating the structural [47] and material characteristics of concrete.

Through real-time monitoring of structures or building components, sensors play an important role in establishing construction safety [56]. By monitoring the whole environment on construction projects, sensor-based technologies have been deployed to avoid incidents and worker–equipment collisions. Sensor-integrated location, vision-oriented sensing, and wireless sensor networks are some of the sensor technologies used in construction safety management [45]. Wireless sensor networks improve and facilitate information flow among design teams on building sites. Because of the complexity of the building environment, network circulation is challenging; nevertheless, wireless networks offer answers to this challenge. Visual inspection methods used to monitor bridge building projects do not offer thorough and trustworthy information [64].

2.3. Industrial Revolution 4.0 as a Solution to Overcoming Current Health and Safety Issues

IR 4.0 is a concept that aims to digitize industrial processes to create a flexible yet vast production and service network without any employee's safety being jeopardized. The introduction of IR 4.0 creates an environment in which all mechanized automation will be networked via technological advancements to function and share information with reduced risks directed at workers, hence increasing the efficiency and safety of workers [16]. The advantages are obvious because of its introduction, since it enhances product quality, reduces time to market, improves operational performance, and enhances health and safety.

The use of various safety technologies on construction sites, such as 3D and 4D CAD, RFID, augmented reality, virtual reality, building information modelling, smart sensor and wireless technology, online databases, robotics, and automation, has significantly improved the effectiveness of health and safety management [46]. By keeping people away from hazardous work environments, new technology can be leveraged to establish a safe workplace [65]. For instance, it is possible to track chemical leaks or employee mishaps in real time by using a deep-learning algorithm to detect human behavior patterns using surveillance cameras. The appropriate system can promptly notify the operator, safety officer, or responsible department of a potentially hazardous situation to help avoid an accident [66].

However, collaborative robots are one of the numerous technologies that Industry 4.0 has adopted. Using them has many benefits, including the reduction of the need for physical labor in many tasks that were previously handled by people. However, the psychological costs of increasing worker monitoring and demands must also be thoroughly assessed. If human factors issues are not carefully considered throughout the process, the decision-making needs for human–robot interaction may potentially have certain hazards [67].

3. Methodology

The defined objectives were attained using a thorough research technique that included three qualitative stages, as indicated in Figure 1.

A thorough review of the published literature was carefully obtained from the literature that helped to identify the advanced technological applications of IR 4.0 towards safe construction practices in the first phase, and furthermore, to replace out-of-date and long-standing traditional health and safety technologies, tools, and procedures to improve their health and safety performance. The study was supported by a semi-structured interview and a feedback questionnaire to collect feedback and opinions from industry experts.

The next step involved creating an AHP and industry feedback questionnaire that had been narrowed down for use in determining the extent of advanced technological applications of IR 4.0 deployment in enterprises and on construction sites. In the second phase, the developed questionnaires were circulated to the industry experts (who included general civil engineering practitioners) to obtain their input on the adoption of advanced technological applications of IR 4.0 for safe construction practices.

In the third and final phase, the gathered data were analyzed using the analytical hierarchy process (AHP), followed by its application and pertinent discussions. The details of the adopted methodology are explained further in the subsequent sections.

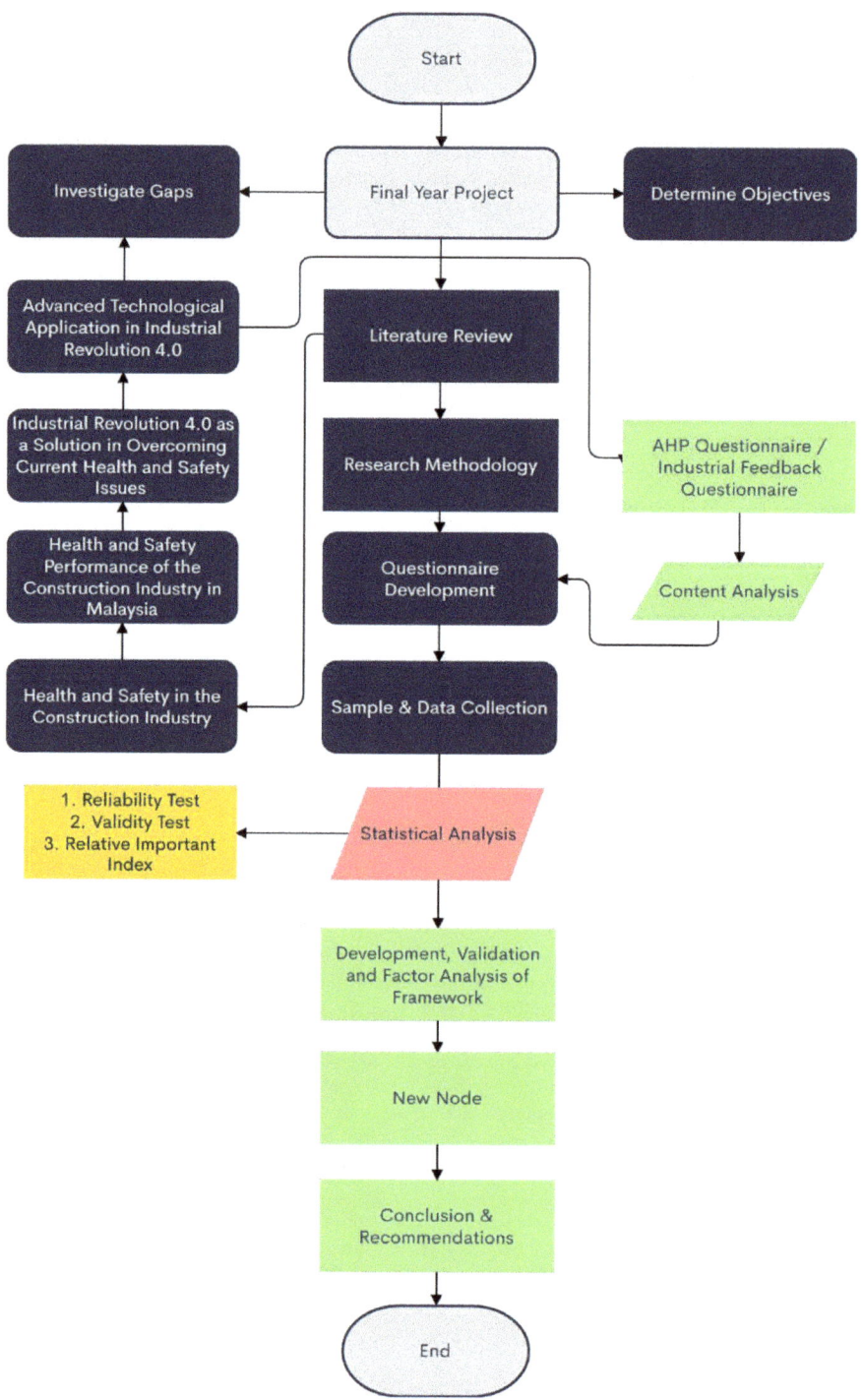

Figure 1. Methodology flowchart.

3.1. Industrial Expert Feedback Analysis Questionnaire Design

The semi-structured interview and industrial feedback questionnaire approach were utilized to achieve the study objectives. Three key respondents were interviewed in a semi-structured interview to collect relevant feedback and comments. Lecturers with an academic background, an engineer from a contractor company, and a consulting company's engineer were the three main acquired interviewees. Interviewees were chosen on the basis of slightly different criteria than the targeted survey respondents, for instance, they work in the construction industry and have extensive civil engineering teaching experience. This approach was chosen to ensure that data collected from the interviews were based on accurate information and knowledge through the company's work in the field. Interview invites were issued to the author's networks, who were also asked to recommend additional people for the interview. A virtual meeting was organized due to the current COVID-19 pandemic to obtain important input from these experts, including feedback and recommendations. These suggestions and opinions are critical to the success of this research project. A total of 18 open-ended questions were developed and were divided into four main sections, allowing respondents to offer comprehensive examples from their own experiences about the poor safety performance experienced in the Malaysian construction industry and how we can use technologies from Industrial Revolution 4.0 as a potential solution to the problem. This would aid in the discovery of new obstacles and technological applications in Industrial Revolution 4.0 that were not covered in the literature study and are relevant to the Malaysian construction industry.

Moreover, the feedback questionnaire survey was conducted online through the Microsoft Forms platform whereby the survey was properly administered and distributed personally to targeted respondents via email; current connections; and LinkedIn, a professional networking platform. Although this method consumes a large amount of time rather than just posting the questionnaire link via a public online platform, this approach was adopted to ensure the quality and reliability of the data collected. Of 60 private messages sent through LinkedIn, 120 emails, and WhatsApp messages sent to connections, there were only 15 responses, which meant a response rate of 8.33%. The lack of response was closely related to two factors, namely, the COVID-19 pandemic and the busy schedule of industrial experts.

The questions were designed in the view of a literature review to identify whether the findings from those studies are still applicable and remain relevant to the industry. Consequently, the collected data are expected to be in descriptive, continuous, and free-form responses, whereby the descriptive data are obtained from multiple choice questions, the continuous data are based on a 5-point Likert scale, and the free-form response was collected from open-ended questions.

3.2. Analytical Hierarchy Process (AHP) Questionnaire Design

The analytic hierarchy process questionnaire survey technique was carried out in the current study. The AHP method is a mathematical decision-based tool [68] for solving complex and ambiguous issues in policy and decision making. AHP helps break down the complex problem into a hierarchy of simple criteria and sub-criteria and makes analysis easier with the help of a relative analysis. One of the key significance of the AHP method is the subjective- and objective-based problem evaluation [69]. This method is primarily composed of breaking down a multi-criteria into discrete levels of hierarchy with the top hierarchy as the goal, mid-level as criteria, sub-criteria, and lower levels as an alternative [70]. Therefore, the AHP is considered the most suitable for the current study.

The analytic hierarchy process (AHP) is a math- and psychology-based approach for organizing and evaluating complicated choices. It consists of three parts: the aim or problem you are attempting to address, all viable solutions (referred to as alternatives), and the criteria you will use to evaluate the alternatives. The AHP provides a solid basis for a necessary conclusion by outlining its criteria and alternate alternatives and connecting those elements to the larger goal. The numerical priority for each of the alternative alternatives is computed in the last phase of the procedure. On the basis of the values of all users, these figures reflect the most requested solutions. When it comes to making decisions on complicated problems with high stakes, the AHP comes in handy. It differs from previous decision-making approaches in that it quantifies criteria and alternatives that are impossible to quantify with concrete statistics in the past. Rather than prescribing a "right" solution, the AHP assists decisionmakers in identifying the option that best fits their values and understanding of the situation. The AHP differs from a traditional poll or meeting in that it removes bias from the decision-making process. An analytic hierarchy process (AHP) survey had been performed to obtain feedback and insights from a limited group of people. The analytic hierarchy process (AHP) questionnaire survey is made up of 25 questions and it was further divided into six sections. The six sections contain varying pairwise comparisons between each alternative (advanced technologies). The alternatives were evaluated against four main constraints or criteria to output the alternatives that show the highest popularity in a ranked order. This was then circulated among 250 respondents. This section is critical for any survey of this type since it helps to improve the understanding of the overall results and provides extra important feedback and suggestions. For the analytic hierarchy process (AHP) questionnaire survey, online questionnaire survey platforms such as Google Form had been used to conduct the survey questions and collect feedback from respondents. Moreover, there were no precise figures for any organization, role, gender, etc., in the sample used for this study, which consisted of Malaysian civil engineers who were picked at random. A total of 250 full responses were obtained from 300 invites, for an overall response percentage of 83.3%. Because this sample size was more than the required minimum of 96, it is considered representative and significant for further analysis.

The analytical hierarchy process (AHP) was conducted on the first questionnaire that was targeted toward civil engineering practitioners to evaluate their responses. The analytical hierarchy process (AHP) began by defining the alternatives that needed to be evaluated. The alternatives in this study were represented by four types of advanced technologies, namely, "robotics and automation", "augmented reality and virtualization", "building information modelling (BIM)", and "wireless monitoring and sensors". These alternatives are the different criteria that solutions must be evaluated against.

The next step taken was defining the problem and the criteria. According to the analytical hierarchy process (AHP) methodology, a problem is a related set of sub-problems. Therefore, dividing the problem into a hierarchy of smaller problems is the foundation of the AHP technique. In the process of breaking down the sub-problem, criteria to evaluate the solutions emerge. The problem, in this case, was identified to be the deteriorating health and safety performances in the Malaysian construction industry while the criteria or constraints that the four advanced technologies were bounded by were "cost", "innovation", "ecosystem", and "interoperability".

The third step taken was establishing a priority amongst criteria using pairwise comparison. This priority scale was set with a scale of numbers ranging from 1 to 10, where 1 represents the two alternatives having an equal amount of importance and 10 verifies that the first alternative has extremely high importance when compared to the second alternative. The priority scale used is shown in Table 1.

Table 1. Priority scale table.

Priority Level	Number
Equally preferred	1
Moderately preferred	3
Strongly preferred	5
Very strongly preferred	7
Extremely preferred	9
Intermediate judgment values	2, 4, 6, 8

The questions for the analytical hierarchy process (AHP) questionnaire directed toward civil engineering practitioners follow an order of four questions for each of the six sections. In this order, all alternatives will be able to be compared with each other once. The four questions contain a criterion each for which the pairwise comparison is to be compared concerning selected criteria. Each alternative is compared against the remaining three alternatives only once concerning the given criteria.

3.3. Sample Selection and Data Analysis

The semi-structured interview, industrial experts' feedback analysis, and the analytic hierarchy process (AHP) were utilized to gather information for the study. The first method, a semi-structured interview, was performed with 15 interviewees, who included academic professionals, contractors, and consultants. Academicians offered input from a theoretical standpoint, while construction industry stakeholders provided feedback from a practical standpoint. The analytic hierarchy process (AHP), the second technique, primarily targeted general civil engineering practitioners as respondents. A series of questionnaires was sent to them to obtain important feedback and insight for this project.

The semi-structured interview was performed with 15 industrial experts, and the feedback and remarks were compiled. A total of 50 completed surveys were given to civil engineering practitioners in the industry. We utilized the analytical hierarchy process (AHP) to analyze the information collected through the questionnaire crafted for its purpose. The AHP is a systematic approach for organizing and analyzing complicated choices that are based on mathematics and psychology. This method will provide detailed and thorough conclusions on the study and other key results. Throughout this research project, the industrial expert's feedback questionnaire was used to analyze data and was subsequently verified using the Cronbach's alpha test, validity test, and AHP.

3.4. Reliability and Validity Test

The Cronbach's alpha coefficient, which ranges from 0 to 1, was used to assess data dependability. The higher the number, the more reliable the data. The examination of the coefficient of internal accuracy of data used in statistics is a measure of reliability. The presence of a "high" alpha score does not imply that the test is one-dimensional [71]. Aside from that, rather than a statistical test using SPSS, this test is a coefficient of reliability or consistency. If the coefficient value surpasses 0.6 throughout this examination, the calculating procedure is considered trustworthy.

In most cases, the validity test is used to compute the degree to be assessed. The criterion that distinguishes it from the durability test is different. As a result, when the validity and reliability of the data measure have been validated, further analysis can be carried out. The square root of the measure's reliability establishes an upper bound for its resemblance to any other example; for example, a measure with a reliability score of 0.65 will never correlate higher than 0.81 with another test.

4. Results

The industrial expert's feedback questionnaire managed to obtain 15 responses from a wide range of states in Malaysia, which included four in the Federal Territories, three in Selangor, one in Perak, one in Penang, one in Pahang, one in Negeri Sembilan, one in Kedah, and two in Johor. The designation of respondents had project engineers and consultants as its majority, which gave this survey higher reliability in terms of input and experience. These respondents had a minimum qualification of a BSc, which included five of them, alongside four MScs, three Ph.D. holders, and three remaining others. The working experience of the respondents was also beneficial to this survey as seven of them had experience of 1–5 years, four of them had experience of 6–10 years, three of them had experience of 11–15 years, and one of them had experience of 16–20 years. These respondents had been involved in various project types, which included residential, non-residential, infrastructure, industrial, commercial, and others. The key outcome from this section was that the experts suggested USD 25,000–60,000 to be used annually on advanced technology to improve health and safety performances at project and construction sites with proven results that are well above the cost of implementation for the four advanced technologies under study.

Moreover, the majority of industrial experts suggested that accidents in the construction industry occur often and that the health and safety performances of the construction industry have deteriorated. This feedback is significant as it portrays the alarming occurrence frequency of accidents in the Malaysian construction industry. Apart from that, these experts also suggest that traditional health and safety procedures, methods, and personal protective equipment (PPE) are not as effective. It is highly suggested that more effort should be made in the industry to improve traditional health and safety technologies. Moreover, most respondents disagreed with the statement, suggesting that the replacement of old health and safety methods, procedures, and personal protective equipment (PPE) is too costly or risky. These data can be highly related to the construction industry's incompetence in reducing the health and safety risks of workers on site.

Thus, it was verified that the respondents chosen for this survey knew the topic of advanced technologies. It was also found that two respondents attended high-level training courses to better understand advanced technologies. The other respondents included five who had medium-level training, six who had basic-level training, and two who did not attend any sort of training on advanced technologies. It is also suggested that some forms of advanced technologies were used occasionally in projects during their tenure in their company. A list of 12 statements portraying the capabilities of the four advanced technologies was provided to respondents in this section. The 12 statements were divided into four parts, and three of the statements were related to one of the types of advanced technology which included robotics and automation, building information modelling (BIM), augmented reality and virtualization, and wireless monitoring and sensors. All 12 questions obtained strong agreements; however, the statement designated for robotics had a 6.7% disagreement from the respondents in terms of using it to reduce the need for workers to lift or carry heavy objects. This feedback is strongly believed to be based on worker employment rates with the potential emergence of robotics and automation into the industry.

Consequently, the appropriate level of importance was selected for each criterion of advanced technologies to be used in the construction industry. The four criteria included cost, innovativeness, interoperability, and ecosystem. It was highlighted that cost is the criteria that are most favored or given the most importance by industrial experts when selecting advanced technology, which could potentially improve the health and safety performance of the construction industry. This level of importance was followed by innovativeness, interoperability, and ecosystem, consecutively. This outcome was highly expected as the cost is a crucial factor in the construction industry. The industrial experts were also asked to rank the four advanced technologies in the order which shows the most promise in improving the health and safety performance in the construction industry. Building information modelling obtained the highest rank, followed by wireless monitoring and sensors, robotics and automation, and augmented reality and virtualization.

A list of 12 statements portraying the capabilities of the four advanced technologies was provided to respondents in this section. The 12 statements were divided into four parts, and three of the statements were related to one of the types of advanced technology, which included robotics and automation, building information modelling (BIM), augmented reality and virtualization, and wireless monitoring and sensors. All 12 questions obtained strong agreements; however, two statements designated for robotics and automation picked up disagreements by 6.7% each. The disagreements were in terms of using automated systems to respond faster to emergencies and using exoskeletons for workers to complete tasks. These disagreements are highly believed to be based on worker employment rates with the potential emergence of robotics and automation into the industry.

4.1. Reliability and Validity Test

The Cronbach's alpha value, which assesses the internal consistency and reliability of data, was 0.74, as shown in Table 2. For additional analysis, values between 0.7 and 0.95 are acceptable [72].

Table 2. Cronbach's alpha value.

Variables	Description	Values	Internal Consistency
K	No. of items	32	
$\sum S^2 y$	Sum of the item variance	11.93	Acceptable
$\sum S^2 x$	The variance of the total score	41.53	
α	Cronbach's alpha	0.74	

The validity test, which is used to compute the degree to be assessed and will indicate how useful it is in a certain circumstance, was 0.86. The value implies that the test, survey, or measuring method includes all pertinent aspects of the subject it seeks to assess.

4.2. Analytical Hierarchy Process (AHP)

At the very start of pairwise comparison, criteria are compared in relation to the main goal, followed by a sub-criteria comparison with project health and safety criteria, consecutively. To compare the importance of one criterion to the other, each cluster is compared in pairs with all its links or connections. The reliability of subjective input can be evaluated by the consistency ratio in the pairwise comparison matrix. It should be less than 0.1, otherwise it should be re-evaluated [73]. The value of 0.046, as shown in Table 3, for the comparison of criteria for reliable and consistent data is well within the acceptable range.

Table 3. Consistency ratio calculation.

Consistency (0.046)	Cost	Innovation	Ecosystem	Interoperability	Priority Vector
Cost	1	4	1/3	3	0.29
Innovation	1/4	1	1/4	1/2	0.09
Ecosystem	3	4	1	3	0.48
Interoperability	1/3	2	1/3	1	0.14

Furthermore, in the analytical hierarchy process (AHP) survey given in Appendix A, responders were allowed to identify which criteria are the most important when evaluating a new technology for industrial application and which type of advanced technology should be prioritized for use in the construction industry for the improvement of health and safety of workers on site. Table 4 represents the calculated level of importance among each of the four criteria.

Table 4. Level of importance among the criteria.

Criteria	Cost	Innovation	Ecosystem	Interoperability
Cost	1.0000	8.0000	5.0000	7.0000
Innovation	0.1250	1.0000	0.3333	0.5000
Ecosystem	0.2000	3.0000	1.0000	0.3333
Interoperability	0.1429	2.0000	3.0000	1.0000
Total	1.4679	14.0000	9.3333	8.8333

The results presented in Tables 4–14 need to be read from two different perspectives i.e., horizontally and vertically. In the horizontal direction, each variable on the leftmost column is compared with each variable horizontally. Conversely, their vertical scores were totaled and are presented in the last rows. The variable with the highest total score (score calculated after adding vertical scores of each variable) is considered more significant as compared to other variables with lesser scores. The same sense was adopted in all subsequent tables, and due to replication of information, it is explained at this location.

Table 5. Level of importance of alternatives for cost.

Cost	Robotics and Automation	Augmented Reality and Virtualisation	Building Information Modelling (BIM)	Wireless Monitoring and Sensors
Robotics and automation	1.0000	2.0000	0.1250	0.1250
Augmented reality and virtualisation	0.5000	1.0000	0.5000	0.1429
Building information modelling (BIM)	8.0000	2.0000	1.0000	3.0000
Wireless monitoring and sensors	8.0000	7.0000	0.3333	1.0000
Total	17.5000	12.0000	1.9583	4.2679

Table 6. Level of importance of alternatives for innovation.

Innovation	Robotics and Automation	Augmented Reality and Virtualisation	Building Information Modelling (BIM)	Wireless Monitoring and Sensors
Robotics and automation	1.0000	0.1429	2.0000	3.0000
Augmented reality and virtualisation	7.0000	1.0000	9.0000	8.0000
Building information modelling (BIM)	0.5000	0.1111	1.0000	2.0000
Wireless monitoring and sensors	0.3333	0.1250	0.5000	1.0000
Total	8.8333	1.3790	12.5000	14.0000

Table 7. Level of importance of alternatives for the ecosystem.

Ecosystem	Robotics and Automation	Augmented Reality and Virtualisation	Building Information Modelling (BIM)	Wireless Monitoring and Sensors
Robotics and automation	1.0000	1.0000	0.3333	0.3333
Augmented reality and virtualisation	7.0000	1.0000	1.0000	0.3333
Building information modelling (BIM)	3.0000	1.0000	1.0000	8.0000
Wireless monitoring and sensors	3.0000	3.0000	0.1250	1.0000
Total	14.0000	6.0000	2.4583	9.6667

Table 8. Level of importance of alternatives for interoperability.

Interoperability	Robotics and Automation	Augmented Reality and Virtualisation	Building Information Modelling (BIM)	Wireless Monitoring and Sensors
Robotics and automation	1.0000	7.0000	0.3333	0.1250
Augmented reality and virtualisation	0.1429	1.0000	0.5000	0.3333
Building information modelling (BIM)	3.0000	2.0000	1.0000	0.3333
Wireless monitoring and sensors	8.0000	3.0000	3.0000	1.0000
Total	12.1429	13.0000	4.8333	1.7917

Table 9. The average level of importance among the criteria.

Criteria	Cost	Innovation	Ecosystem	Interoperability	Average
Cost	0.6813	0.5714	0.5357	0.7925	0.6452
Innovation	0.0852	0.0714	0.0357	0.0566	0.0622
Ecosystem	0.1363	0.2143	0.1071	0.0377	0.1239
Interoperability	0.0973	0.1429	0.3214	0.1132	0.1687
Total	1.0000	1.0000	1.0000	1.0000	

Table 10. The average level of importance of alternatives for cost.

Cost	Robotics and Automation	Augmented Reality and Virtualisation	Building Information Modelling (BIM)	Wireless Monitoring and Sensors	Average
Robotics and automation	0.0571	0.1667	0.0638	0.0293	0.0792
Augmented reality and virtualisation	0.0286	0.0833	0.2553	0.0335	0.1002
Building information modelling (BIM)	0.4571	0.1667	0.5106	0.7029	0.4593
Wireless monitoring and sensors	0.4571	0.5833	0.1702	0.2343	0.3612
Total	1.0000	1.0000	1.0000	1.0000	

Table 11. The average level of importance of alternatives for innovation.

Innovation	Robotics and Automation	Augmented Reality and Virtualisation	Building Information Modelling (BIM)	Wireless Monitoring and Sensors	Average
Robotics and automation	0.1132	0.1036	0.1600	0.2143	0.1478
Augmented reality and virtualisation	0.7925	0.7252	0.7200	0.5714	0.7023
Building information modelling (BIM)	0.0566	0.0806	0.0800	0.1429	0.0900
Wireless monitoring and sensors	0.0377	0.0906	0.0400	0.0714	0.0600
Total	1.0000	1.0000	1.0000	1.0000	

Table 12. The average level of importance of alternatives for the ecosystem.

Ecosystem	Robotics and Automation	Augmented Reality and Virtualisation	Building Information Modelling (BIM)	Wireless Monitoring and Sensors	Average
Robotics and automation	0.0714	0.1667	0.1356	0.0345	0.1020
Augmented reality and virtualisation	0.5000	0.1667	0.4068	0.0345	0.2770
Building information modelling (BIM)	0.2143	0.1667	0.4068	0.8276	0.4038
Wireless monitoring and sensors	0.2143	0.5000	0.0508	0.1034	0.2171
Total	1.0000	1.0000	1.0000	1.0000	

Table 13. The average level of importance of alternatives for interoperability.

Interoperability	Robotics and Automation	Augmented Reality and Virtualisation	Building Information Modelling (BIM)	Wireless Monitoring and Sensors	Average
Robotics and automation	0.0824	0.5385	0.0690	0.0698	0.1899
Augmented reality and virtualisation	0.0118	0.0769	0.1034	0.1860	0.0945
Building information modelling (BIM)	0.2471	0.1538	0.2069	0.1860	0.1985
Wireless monitoring and sensors	0.6588	0.2308	0.6207	0.5581	0.5171
Total	1.0000	1.0000	1.0000	1.0000	

Table 14. Matrix multiplication of the average level of importance among the criteria and the alternatives.

	Cost	Innovation	Ecosystem	Interoperability	Criteria	Analytical Hierarchy
Robotics and automation	0.0792	0.1478	0.1020	0.1899	0.6452	0.1050
Augmented reality and virtualisation	0.1002	0.7023	0.2770	0.0945	0.0622	0.1586
Building information modelling (BIM)	0.4593	0.0900	0.4038	0.1985	0.1239	0.3855
Wireless monitoring and sensors	0.3612	0.0600	0.2171	0.5171	0.1687	0.3509

The results from Table 4 indicate that the criteria of innovation had the highest total score of 14, followed by ecosystem (9.33) and interoperability (8.83). The cost has the lowest total value of 1.47 when compared vertically. Similarly, cost criteria showed the highest values when compared horizontally with other criteria, i.e., innovation (8), ecosystem (5), and interoperability (7). Moreover, Tables 5–8 represent the level of importance of each alternative to each of the four criteria.

As shown in Table 5, robotics and automation had the highest level of importance of alternatives for the cost (17.5), followed by augmented reality and virtualization (12), and then by wireless monitoring and sensors (4.27). The lowest importance was shown for BIM (1.96).

As shown in Table 6, wireless monitoring and sensors had the highest level of importance of alternatives for innovation (14), followed by BIM (12.5), and then by robotics and automation (8.83). The lowest importance for innovation was shown by augmented reality and virtualization (1.38).

Virtual and augmented reality help with safety on the job site as well, as they call attention to important features in the environment, such as temperatures or potentially unsafe conditions. BIM allows construction managers to create better designs and spot problems before they occur, which optimizes the workforce by helping them perform tasks smoothly with reduced safety risks and avoiding delays.

As seen in Table 7, robotics and automation had the highest level of importance as alternatives for the ecosystem (14). The second highest alternative was wireless monitoring and sensors (9.67), followed by augmented reality and virtualization (6). The lowest importance for innovation was shown by BIM (2.46). There was not much emphasis given to having proper R&D sectors in the construction industry, resulting in the old methods persisting, while it takes a long time to implement newer methods such as BIM. Adopting safety sensors to monitor rigidity or structural stability can help reduce work-related incidents that may harm the public such as in LRT construction sites. Advanced technologies are currently underutilized in the Malaysian construction industry due to the high cost involved to invest, and companies are unsure about the returns they will be getting in terms of their investment in advanced technology, despite the many advantages advanced technology holds.

As shown in Table 8, augmented reality and virtualization had the highest level of importance of alternatives for interoperability with a value of 13. The second highest alternative was robotics and automation (12.143), followed by BIM (4.833). The lowest importance for innovation was shown by augmented reality and virtualization (1.792).

Table 9 represents the calculations in obtaining the average value of the level of importance for each of the criteria. The highest average level of importance among the criteria was for cost, with a value of 0.6452, followed by interoperability (0.169).

Conversely, Tables 10–13 represent the calculations in obtaining the average value of the level of importance for each alternative for each of the four criteria. The final calculations were performed by conducting matrix multiplication between the average values of the level of importance obtained from the criteria and the subsequent alternatives.

Falls, electric shocks, being struck by an object, and being caught between objects are the top safety risks and have remained constant for some time. However, people are now turning to technology for new and innovative answers to make the industry safer. Despite a large proportion of construction managers saying that more needs to be done to improve safety levels, the adoption of technology-led initiatives has remained relatively conservative. The argument is that new technologies have always accelerated the construction business, and therefore it is strange that so many businesses are reluctant to accept them.

The highest average level of importance of alternatives for cost was for wireless monitoring and sensors, with a value of 0.36, as shown in Table 10. This highlights the significance of integrated systems during the execution of construction projects for maintaining the safety and adoption of a safe work environment.

As is shown in Table 11, the highest average level of importance of alternatives for innovation was for augmented reality and virtualization with a value of 0.7023. The construction industry runs on conventional tools, methods, and practices that need to change. Therefore, the utilization of augmented reality and virtualization can play a key role during the inception and planning phases of the project.

As seen in Table 12, the highest average level of importance of alternatives for the ecosystem was for BIM, with a value of 0.4038. For establishing a sound and productive ecosystem, it is very important to use BIM during the execution of construction projects.

Site sensors that can be deployed across a construction site to monitor things such as temperature, noise levels, dust particulates, and volatile organic compounds to help limit exposure to workers can be implemented. Safety training and equipment operator training are two areas where virtual reality (VR) could have a strong impact on the construction industry. With VR, workers could have exposure to environments such as confined spaces or working at height in a safe, controlled environment. Wearables can be used to monitor workers and their environment to make job sites safer. Wearable tech in construction is embedded into apparel and personal protective equipment (PPE) and is already common on construction sites. The apparels and personal protective equipment (PPE) that are already being experimented on include hard hats, gloves, safety vests, and work boots.

Table 13 shows the average level of importance of alternatives for interoperability. The highest value was obtained by wireless monitoring and sensors (0.5171), followed by BIM (0.1985). These results indicate the significance of BIM in construction projects. With the usage of BIM, communication between project teams can be improved, planning can be done effectively, and execution of projects can be undertaken with minimum errors.

Moreover, the results obtained from the analytical hierarchy process (AHP) suggest that civil engineering practitioners opted for building information modelling (BIM) as the advanced technology that provides the most promise in improving the health and safety of the Malaysian construction industry in terms of the four main criteria. Wireless monitoring and sensors, augmented reality and virtualization, and robotics and automation were ranked second, third, and fourth, respectively. The calculated results are shown in Table 14.

5. Discussion

The IR 4.0 in the construction industry system focuses on the physical-to-digital transition and then the digital-to-physical transition to help coordinate, design, and execute built environment infrastructure more effectively and efficiently [74]. IR 4.0 integrates organizations, procedures, and information to effectively plan, build, and operate assets using cyber-physical structures, the internet of things, data, and services to associate the digital layer, comprising BIM and CDE, with the physical layer, comprising the asset, over its entire existence in order to establish an integrated environment incorporating organizations, processes, and information. According to Rastogi, the primary objective of IR 4.0 in construction is to build a digital construction site that uses various techniques to track progression during a project's life cycle. IR 4.0 implementation in construction would transform not only the construction process, but also the company and project frameworks, transforming the fragmented construction industry into an integrated one [75]. Although it is becoming one of the most profitable industries, the construction industry has among the poorest research and development initiatives. Likewise, employment growth in the AEC has decreased over time, although it has nearly doubled in other sectors [76]. The function of human resources in an IR 4.0 world is shifting from machine operator to strategic decision creator. Robots support humans in dangerous, stressful, and time-consuming tasks, for which humans must be adequately prepared for successful human–machine interaction. Since it is a labor-intensive sector, the construction industry has a significant opportunity to increase productivity through technical advancement (e.g., robot use), particularly for potentially hazardous human labor. Robots are used in restricted ways in the digital building platform, such as 3D printing, structuring walls, placing rebar, welding, and drones [40]. Furthermore, a discussion on IR 4.0 technologies has been presented in detail in the subsequent sections in light of our findings.

5.1. Robotics and Automation

Robotics and automation offer huge potential to enhance productivity, efficiency, and manufacturing flexibility throughout the construction industry, including automating the fabrication of modular homes and building components off site, robotic welding and material handling on building sites, and robot 3D printing of houses and customized structures. As well as making the industry safer and more cost effective, robots are improving sustainability and reducing environmental impact by enhancing the quality and cutting waste. With so few construction businesses using automation today, there is a huge potential for us to transform the industry through robotics [77]. Robots on construction sites can be very useful during high-risk-oriented activities. They can be operated from distant operating rooms. This can significantly reduce the on-site safety incidents, and thus the health and safety of the workers can be considerably ensured.

5.2. Augmented Reality and Virtualisation

Augmented reality (AR) and virtualization have proven their value across multiple industries and have demonstrated that they can be leveraged beyond the realms of gaming and entertainment. As such, the AR/VR market is expected to see a 77% compound annual growth rate (CAGR) from 2019 to 2023. In retail, brands are adopting augmented reality to enrich the online shopping experience, allowing shoppers to see what an item (e.g., a piece of furniture or appliance) might look like in their environment [78]. AR has also been making waves in the construction industry. When leveraged properly, augmented reality can help you win more projects, collaborate with team members, and even improve safety [79]. Taking advantage of augmented reality in construction requires a solid understanding of the technology's capabilities and technical requirements.

Using augmented reality, designers can place digital models in a real-world setting. This provides a chance to visualize how a building project's components will look from the outside. As a result, a building or construction site that is now unoccupied might be made to appear to be in use. Before beginning construction, project teams can identify potential problems and avoid errors by using 3D plans instead of traditional paper ones [80]. Additionally, project teams can see exactly where ductwork, pipes, columns, windows, and access points will be installed. It is also simpler to make changes before, during, and after the project. Unable to physically be on the job site, virtual reality offers an innovative alternative that places collaborating teams there. By producing an exact depiction on a computer, virtual reality (VR) simulates an environment. Stakeholders can study a building's layout by stepping inside the simulation using a VR head-sets. This makes it possible for different teams to organize and plan how to carry out the job successfully and efficiently far in advance of the building project getting underway. Using a realistic and immersive environment, architects may precisely measure the spacing, layout, and materials throughout the entire project [81]. Making knowledgeable selections with confidence is made easier when the project manager can observe a project while it is being completed without a physical structure being there.

5.3. Building Information Modelling (BIM)

Moreover, the use of various safety technologies on construction sites, such as building information modelling (BIM), Wireless monitoring and sensors, robotics and automation, and augmented reality and virtualization have gained strong feedback from civil engineering practitioners and industrial experts in the field of construction and have significantly improved the effectiveness of health and safety management [82]. It was apparent that building information modelling (BIM) was ranked first in being the preferred advanced technology to be introduced to the construction industry to improve health and safety performances. Furthermore, it was discovered that combining traditional safety practices and standards with the use of digital health and safety tools and techniques may help site managers, supervisors, and coordinators assure the efficiency of their construction projects. Moreover, worker fatalities and injuries would be significantly reduced. As a result, investing in cutting-edge technologies shows great potential in improving construction site safety. Building information modelling (BIM) should be prioritized when it comes to introducing advanced technologies into the construction industry to improve health and safety performances [83]. Research possibilities are discovered on the basis of an evaluation of the research literature, an analysis of the industrial requirements, and the researcher's experiences in the relevant fields [82].

The building sector is still in the early stages of adopting the building information modelling (BIM) methodology. Throughout every stage of a building's life cycle, 2D AutoCAD drawings are still heavily utilized. A significant drive to transfer the entire process into 3D models is being led by architects. Many pilot programs have shown significant time and cost reductions for construction projects [84]. Through BIM, planning the layout of the site, visualizing and simulating construction risks and countermeasures, making decisions quickly and effectively, taking immediate action to stop the use of hazardous chemicals and equipment, and monitoring and controlling site safety can be done effectively. Additionally, it has been noted that automatic safety checking has been rated as the least advantageous feature, which reflects a lack of knowledge about the true advantages of adopting BIM [85]. BIM is one of the most acceptable ways for safety checking with a time- and money-saving quality, and it is applicable for automatic safety checking.

5.4. Wireless Monitoring and Sensors (Cloud Computing)

Construction cloud computing is a developing field that offers several prospects. Even if this list is not all inclusive, some advantages gained from the deployment of cloud computing technology in the construction industry include financial advantages, on-demand scalability of computing resources, secure platforms, vast storage, and facilitation of collaborative practices [86]. Construction companies are starting to employ cloud computing, and there are many prospects for them. For SMEs in the sector to use BIM-enabled software to digitize their operations, cloud services are essential [87]. Cloud computing can be very effective in maintaining safe work environments because of integrated systems. The detection of emergencies and abrupt response can be established with the utilization of cloud computing tools during the life cycle of any construction project, especially in difficult terrains and geographic locations.

6. Conclusions

Despite being the heart of any economy, the construction sector may, like the industrial sector, improve its performance by utilizing IR 4.0 for maintaining a safe work environment. The goal of this study is to investigate the development of IR 4.0 and digitalization in the construction sector for improving safety practices. IR 4.0 is being pushed forward by the creation and adoption of additional technologies that have contributed to its success thus far. The findings of the current study reaffirm that the construction sector is contributing more than its fair share by developing and promoting the usage of digital methodologies such as robotics and automation, augmented reality and virtualization, integrated systems, and BIM. Moreover, IR 4.0 also brings about some repercussions for the construction sector. The use of digital techniques in design and construction has evolved significantly over the last three years and is now the norm for many businesses. The performance gap—a term used to describe the discrepancy between how building designs are evaluated to demonstrate conformity in the virtual world and how structures operate in the real world—has become abundantly obvious. To address this challenge, it will be necessary to reconsider what is needed to train the next generation for the jobs of the future.

According to the findings of the current study, building information modelling (BIM), wireless monitoring and sensors, and robotics and automation have advanced substantially over the recent years, while other technologies such as augmented reality and virtualization are still being improved and may have an impact on the industry's long-term viability. Despite having access to these technologies, the application of IR 4.0 in the construction industry is still woefully insufficient. Nonetheless, all parties involved must resolve the issues that arose to assure a successful implementation. The social component has been recognized as the most important factor influencing effective implementation; nevertheless, the other contributing variables show that these aspects are interconnected and should be addressed at the same time.

The application of IR 4.0 in the construction sector would boost the industry's performance to equal that of its counterparts in other industries, such as the manufacturing and automotive industries. The use of various safety technologies on construction sites, such as building information modelling (BIM), wireless monitoring and sensors, robotics and automation, and augmented reality and virtualization have gained strong feedback from civil engineering practitioners and industrial experts in the field of construction and have significantly improved the effectiveness of health and safety management. It was apparent that building information modelling (BIM) was ranked first, and Wireless monitoring and sensors second, followed by augmented reality and virtualization, with robotics and automation being the preferred advanced technology to be introduced into the construction industry to improve health and safety performances. Furthermore, it was discovered that combining traditional safety practices and standards with the use of digital health and safety tools and techniques may help site managers, supervisors, and coordinators assure the efficiency of their construction projects. Moreover, worker fatalities and injuries would be significantly reduced. As a result, investing in cutting-edge technologies improves construction site safety.

The study was limited to the Malaysian construction industry in order to assess its health and safety performances and identify IR 4.0 advanced technologies that can be used to improve it. Because this study was limited to the Malaysian construction industry, a similar study is needed in Southeast Asia to assess the suitability and compatibility of the advanced technologies being studied. This study can be further extended globally by having widespread dissemination of the questionnaires, which can provide an in-depth insight into the subject in other regions that can be contrasted. The comparison will aid in the development of a unified approach that would result in reforms in the construction industry all around the world, on the basis of which updated guidelines can be drafted.

Author Contributions: All authors contributed equally to this research. All authors have read and agreed to the published version of the manuscript.

Funding: This research received no external funding.

Institutional Review Board Statement: Not Applicable.

Informed Consent Statement: Not Applicable.

Data Availability Statement: Not Applicable.

Acknowledgments: The authors would like to appreciate the YUTP-FRG 1/2021 project (cost center # 015LC0-369) and the URIF project (cost center # 015LB0-051) in Universiti Teknologi PETRONAS (UTP) awarded to Wesam Alaloul for the support.

Conflicts of Interest: The authors declare no conflict of interest.

Appendix A. AHP Survey Questions

Appendix A.1. Health and Safety Performances in the Construction Industry

8. How often do you hear about accidents in the construction industry lately?
   ○ Never
   ○ Rarely
   ○ Sometimes
   ○ Often
   ○ Always

9. How often do accidents happen at project/construction sites during your tenure?
- ○ Never
- ○ Rarely
- ○ Sometimes
- ○ Often
- ○ Always

10. How much do you agree with the statements below?

1 = Strongly disagree
2 = Disagree
3 = Maybe
4 = Agree
5 = Strongly agree

	1	2	3	4	5
The health and safety performances of the construction industry have deteriorated.	○	○	○	○	○
Traditional health and safety procedures, methods, and PPE are not as effective.	○	○	○	○	○
More effort should be made in the industry to improve traditional health and safety technologies.	○	○	○	○	○
The replacement of old health and safety methods, procedures, and PPE is too costly or risky.	○	○	○	○	○

Appendix A.2. Advanced Technology Application during the Industrial Revolution 4.0 (IR 4.0)

11. Your knowledge of advanced technologies in the construction industry
- ○ No knowledge
- ○ Very weak
- ○ Weak
- ○ Well
- ○ Very well

12. Have you attended any training or course to better understand advanced industrial technologies?
- ○ Not attended
- ○ Basic level
- ○ Medium level
- ○ High level
- ○ Advanced level

13. Did you or your organization use advanced technologies in any project during your tenure?
- ○ Never
- ○ Rarely
- ○ Sometimes
- ○ Often
- ○ Always

14. How much do you agree with the statements below?

1 = Strongly disagree
2 = Disagree
3 = Maybe
4 = Agree
5 = Strongly agree

	1	2	3	4	5
VR provides the opportunity to experience situations that you wouldn't be able to easily construct for training in a real-life situation.	○	○	○	○	○
VR can simulate a construction site module, giving a worker first-hand experience of the job before they even start.	○	○	○	○	○
With augmented training, you can also introduce the realistic sensations of heights, distractions, stress, and environmental hazards.	○	○	○	○	○
Automation reduces the risk of injury at the operational level by removing workers from dangerous work procedures.	○	○	○	○	○
Safety automation software makes it easy for companies to update safety protocols and create a plan of action in the case of an emergency.	○	○	○	○	○
Robotics can also be used to reduce the need for workers to lift or carry heavy objects.	○	○	○	○	○
BIM allows you to consider and think about hazards and risks earlier within the design as information is shared and coordinated among the project team.	○	○	○	○	○
Using 4D scheduling and sequencing with site logistics planning can easily help identify traffic considerations and potential hazards around the site before even breaking ground.	○	○	○	○	○
BIM allows doing more prefabrication of materials off site in a safe and controlled environment.	○	○	○	○	○
Smart sensors can help site managers detect and make accommodations for work as it begins to shift away from known benchmarks or tolerances.	○	○	○	○	○
A sensor embedded in a critical machine lets operators and maintenance personnel identify vibrations, sounds, and other variables that might indicate that the machine, or even a single part within it, is about to fail.	○	○	○	○	○
Sensors can be integrated into a BIM model once a building is up to provide data about the structure.	○	○	○	○	○

Appendix A.3. Potential of Advanced Technology Application during IR 4.0 to Improve Health and Safety Performances in the Construction Industry

15. Please select the appropriate level of importance of each criterion for advanced technologies to be used in the construction industry.

	1	2	3	4	5
Cost	○	○	○	○	○
Innovativeness	○	○	○	○	○
Interoperability	○	○	○	○	○
Ecosystem	○	○	○	○	○

16. On the basis of your analysis, rank these advanced technologies in the order of which shows the most promise in improving health and safety performances in the construction industry.

1 = first
2 = second
3 = third
4 = fourth

	1	2	3	4
Robotics and automation	○	○	○	○
Building information modelling (BIM)	○	○	○	○
Augmented reality and virtualisation	○	○	○	○
Wireless monitoring and sensors	○	○	○	○

17. How much do you agree with the statements below?

1 = Strongly disagree
2 = Disagree
3 = Maybe
4 = Agree
5 = Strongly agree

	1	2	3	4	5
AR typically respond faster to emergencies by providing real-time monitoring.	○	○	○	○	○
AR can be used to overlay the digital model on the as-built structure in real time, making it possible to identify any discrepancies that could pose a threat to worker safety.	○	○	○	○	○
AR can also provide accurate in-person measurements of physical spaces to make sure everything can fit where it is supposed to be, and changes can be made before mistakes happen.	○	○	○	○	○
Automated systems typically respond faster to emergencies by providing real-time monitoring. Situations that had previously been perceived as unavoidable can be taken care of before they even occur.	○	○	○	○	○
Robotics reduces workers' exposure to mechanical hazards that pose a significant risk of injury or death.	○	○	○	○	○
Exoskeleton robots can reduce the need for workers to perform repetitive motion tasks, which often lead to musculoskeletal disorders (MSDs).	○	○	○	○	○
BIM can focus on each task so workers can better identify the risks; prepare for the work at hand; and, therefore, complete the task more efficiently and safely.	○	○	○	○	○
The ability to communicate information visually can break down any language barriers, aid understanding, and reduce accidents.	○	○	○	○	○
Using tracking and sensing technology, working fatalities and injuries relating to being struck by moving construction vehicles can be dramatically reduced.	○	○	○	○	○
Smart sensors deliver tremendous value because they can wirelessly deliver real-time updates on project status; the location of vehicles, deliveries, and assets; or the condition of various features as they're built.	○	○	○	○	○
Smart sensors can track and monitor the condition of personnel much more easily, read their vitals, and ensure that any emergencies are brought to the attention of all invested parties immediately.	○	○	○	○	○
Adding sensing technology could lower insurance premiums since many insurance companies provide discounts if companies engage in activities aimed at reducing workplace injuries and accidents.	○	○	○	○	○

References

1. Jovane, F.; Yoshikawa, H.; Alting, L.; Boer, C.R.; Westkamper, E.; Williams, D.; Tseng, M.; Seliger, G.; Paci, A.M. The Incoming Global Technological and Industrial Revolution towards Competitive Sustainable Manufacturing. *CIRP Ann.* **2008**, *57*, 641–659. [CrossRef]
2. Rozin, V. From Engineering and Technological Process to Post-Cultural Technology. *Futur. Hum. Image* **2021**, *15*, 99–109. [CrossRef]
3. Deane, P.M.; Deane, P.M. *The First Industrial Revolution*; Cambridge University Press: Cambridge, UK, 1979.
4. Mokyr, J. Eurocentricity Triumphant. *Am. Hist. Rev.* **1999**, *104*, 1241–1246. [CrossRef]
5. Fisher, T. Welcome to the Third Industrial Revolution: The Mass-Customisation of Architecture, Practice and Education. *Archit. Des.* **2015**, *85*, 40–45. [CrossRef]
6. Lasi, H.; Fettke, P.; Kemper, H.G.; Feld, T.; Hoffmann, M. Industry 4.0. *Bus. Inf. Syst. Eng.* **2014**, *6*, 239–242. [CrossRef]
7. CIBD. IR4.0 in Construction; Malaysia, 2020. Available online: https://www.cidb.gov.my/en/construction-info/technology/ir40-construction (accessed on 23 June 2022).
8. Pamidimukkala, A.; Kermanshachi, S. Occupational Health and Safety Challenges in Construction Industry: A Gender-Based Analysis. In Proceedings of the ASCE Construction Research Congress, Arlington, VA, USA, 9–12 March 2022; pp. 9–12.
9. Al-Emad, N. Construction Workers' Issues From Worldwide and Saudi Arabia Studies. *Int. J. Sustain. Constr. Eng. Technol.* **2021**, *12*, 85–100. [CrossRef]
10. OSHA. Occupational Safety and Health Act. 1994. Available online: https://science.utm.my/oshe/files/2017/06/APPENDICES.pdf (accessed on 25 June 2022).
11. Sapuan, S.M.; Ilyas, R.A.; Asyraf, M.R.M. Occupational Safety and Health Administration in Composite Industry. In *Safety and Health in Composite Industry*; Springer: Singapore, 2022; pp. 229–252. ISBN 978-981-16-6136-5.
12. Zid, C.; Kasim, N.; Soomro, A.R. Effective Project Management Approach to Attain Project Success, Based on Cost-Time-Quality. *Int. J. Proj. Organ. Manag.* **2020**, *12*, 149–163. [CrossRef]
13. Wong, S.S.; Soo, A.L. Factors Influencing Safety Performance in the Construction Industry. *J. Soc. Sci. Humanit.* **2019**, *16*, 1–9.
14. Qureshi, A.H.; Alaloul, W.S.; Manzoor, B.; Musarat, M.A.; Saad, S.; Ammad, S. Implications of Machine Learning Integrated Technologies for Construction Progress Detection under Industry 4.0 (IR 4.0). In Proceedings of the Second International Sustainability and Resilience Conference: Technology and Innovation in Building Designs, Sakheer, Bahrain, 11–12 November 2020; pp. 1–6.
15. Lu, Y. A Survey on Technologies, Applications and Open Research Issues. *J. Ind. Inf. Integr.* **2017**, *6*, 1–10.
16. You, Z.; Feng, L. Integration of Industry 4.0 Related Technologies in Construction Industry: A Framework of Cyber-Physical System. *IEEE Access* **2020**, *8*, 122908–122922. [CrossRef]
17. Hamid, A.R.A.; Azmi, M.N.; Aminudin, E.; Jaya, R.P.; Zakaria, R.; Zawawi, A.M.M.; Yahya, K.; Haron, Z.; Yunus, R.; Saar, C.C. Causes of Fatal Construction Accidents in Malaysia. In Proceedings of the IOP Conference Series: Earth and Environmental Science, Moscow, Russia, 27 May–6 June 2019; p. 012044.
18. Halim, N.N.A.A.; Jaafar, M.H.; Anuar, M.O.H.A.M.A.D.; Kamaruddin, N.A.K.; Jamir, P.S. The Causes of Malaysian Construction Fatalities. *J. Sustain. Sci. Manag.* **2020**, *15*, 236–256. [CrossRef]
19. Andolfo, C.; Sadeghpour, F. A Probabilistic Accident Prediction Model for Construction Sites. *Procedia Eng.* **2015**, *123*, 15–23. [CrossRef]

20. Danielsen, D.A.; Torp, O.; Lohne, J. HSE in Civil Engineering Programs and Industry Expectations. *Procedia Eng.* **2017**, *196*, 327–334. [CrossRef]
21. Alaloul, W.S.; Musarat, M.A.; Liew, M.S.; Zawawi, N.A.W.A. Influential Safety Performance and Assessment in Construction Projects: A Review. In Proceedings of the AWAM International Conference on Civil Engineering, Penang, Malaysia, 21–22 August 2019; pp. 719–728.
22. Mohammadi, A.; Tavakolan, M.; Khosravi, Y. Factors Influencing Safety Performance on Construction Projects: A Review. *Saf. Sci.* **2018**, *109*, 382–397. [CrossRef]
23. Oke, A.E.; Kineber, A.F.; Alsolami, B.; Kingsley, C. Adoption of Cloud Computing Tools for Sustainable Construction: A Structural Equation Modelling Approach. *J. Facil. Manag.* **2022**. [CrossRef]
24. Zid, C.; Kasim, N.; Benseghir, H.; Kabir, M.N.; Ibrahim, A.B. Developing an Effective Conceptual Framework for Safety Behaviour in Construction Industry. In Proceedings of the E3S Web of Conferences, Kuala Lumpur, Malaysia, 2–5 October 2018; p. 03006.
25. Wu, J.; Xiao, J.; Li, T.; Li, X.; Sun, H.; Chow, E.P.; Lu, Y.; Tian, T.; Li, X.; Wang, Q.; et al. A Cross-Sectional Survey on the Health Status and the Health-Related Quality of Life of the Elderly after Flood Disaster in Bazhong City, Sichuan, China. *BMC Public Health* **2015**, *15*, 163. [CrossRef]
26. Niza, C.; Silva, S.; Lima, L. Work Accidents in the Empirical Literature: Implications for the Future. *Saf. Reliab. Manag. Risk* **2006**, *1*, 809–814.
27. OSHA Worker's Rights. Available online: https://www.osha.gov/sites/default/files/publications/osha3021.pdf. (accessed on 27 June 2022).
28. USDL Occupational Safety and Health Administration. Available online: https://www.osha.gov/. (accessed on 27 June 2022).
29. HSM Construction Crisis: Data Highlights Concerning Statistics. Available online: https://www.hsmsearch.com/Construction-Data-highlights-injury-concerns (accessed on 27 June 2022).
30. Kowalik, T.; Logoń, D.; Maj, M.; Rybak, J.; Ubysz, A.; Wojtowicz, A. Chemical Hazards in Construction Industry. In Proceedings of the E3S Web of Conferences, Bucharest, Romania, 26–29 May 2019; p. 03032.
31. Parn, E.A.; Edwards, D.; Riaz, Z.; Mehmood, F.; Lai, J. Engineering-out Hazards: Digitising the Management Working Safety in Confined Spaces. *Facilities* **2019**, *37*, 196–215. [CrossRef]
32. Tunji-Olayeni, P.F.; Afolabi, A.O.; Okpalamoka, O.I. Survey Dataset on Occupational Hazards on Construction Sites. *Data Br.* **2018**, *18*, 1365–1371. [CrossRef]
33. Muñoz-La Rivera, F.; Mora-Serrano, J.; Oñate, E. Factors Influencing Safety on Construction Projects (FSCPs): Types and Categories. *Int. J. Environ. Res. Public Health* **2021**, *18*, 10884. [CrossRef]
34. Abdul-Aziz, A.R.; Hussin, A.A. Construction Safety in Malaysia: A Review of Industry Performance and Outlook for the Future. *J. Constr. Res.* **2003**, *4*, 141–153. [CrossRef]
35. Alaloul, W.S.; Musarat, M.A.; Rabbani, M.B.A.; Iqbal, Q.; Maqsoom, A.; Farooq, W. Construction Sector Contribution to Economic Stability: Malaysian GDP Distribution. *Sustainability* **2021**, *13*, 5012. [CrossRef]
36. Saifullah, N.M.; Ismail, F. Integration of Occupational Safety and Health during Pre-Construction Stage in Malaysia. *Procedia-Soc. Behav. Sci.* **2012**, *15*, 603–610. [CrossRef]
37. Hussain, M.; Hadi, A. Corporate Governance, Risky Business and Construction Industry: A Divergence between Bursa and Construction Industry Development Board (CIDB) Klang Valley, Malaysia. *Corp. Gov. Int. J. Bus. Soc.* **2019**, *19*, 438–457. [CrossRef]
38. Hussain, M.A.; Abdul Hadi, R.A. Corporate Governance, Small Medium Enterprises (SMEs) and Firms' Performance: Evidence from Construction Business, Construction Industry Development Board (CIDB) Malaysia. *Int. J. Bus. Manag.* **2018**, *13*, 14–28. [CrossRef]
39. DOSH. Occupational Accident Statistics; Malaysia, 2021. Available online: https://www.dosh.gov.my/index.php/statistic-v/occupational-accident-statistics/occupational-accident-statistic-2021 (accessed on 28 June 2022).
40. Liu, Z.; Xie, K.; Li, L.; Chen, Y. A Paradigm of Safety Management in Industry 4.0. *Syst. Res. Behav. Sci.* **2020**, *37*, 632–645. [CrossRef]
41. Alaloul, W.S.; Liew, M.S.; Zawawi, N.A.W.A.; Mohammed, B.S. Industry Revolution IR 4.0: Future Opportunities and Challenges in Construction Industry. In Proceedings of the MATEC Web of Conferences, Lisbon, Portugal, 23–25 April 2018; p. 02010.
42. Ryu, J.; McFarland, T.; Banting, B.; Haas, C.T.; Abdel-Rahman, E. Health and Productivity Impact of Semi-Automated Work Systems in Construction. *Autom. Constr.* **2020**, *120*, 103396. [CrossRef]
43. He, W.; Li, Z.; Chen, C.P. A Survey of Human-Centered Intelligent Robots: Issues and Challenges. *J. Autom. Sin.* **2017**, *4*, 602–609. [CrossRef]
44. Azhar, S. Role of Visualization Technologies in Safety Planning and Management at Construction Jobsites. *Procedia Eng.* **2017**, *171*, 215–226. [CrossRef]
45. Rao, A.S.; Radanovic, M.; Liu, Y.; Hu, S.; Fang, Y.; Khoshelham, K.; Palaniswami, M.; Ngo, T. Real-Time Monitoring of Construction Sites: Sensors, Methods, and Applications. *Autom. Constr.* **2022**, *136*, 104099. [CrossRef]
46. Adepoju, O. Construction 4.0. In *Re-Skilling Human Resources for Construction 4.0*; Springer Cham: Cham, Denmark, 2022; pp. 17–39. ISBN 978-3-030-85973-2.
47. Seyrfar, A.; Ataei, H.; Osman, I. Robotics and Automation in Construction (RAC): Priorities and Barriers toward Productivity Improvement in Civil Infrastructure Projects. In Proceedings of the Automation and Robotics in the Architecture, Engineering, and Construction Industry, Bogotá, Colombia, 13–15 July 2022; Springer: Cham, Denmark, 2022; pp. 59–71.

48. Firth, C.; Dunn, K.; Haeusler, M.H.; Sun, Y. Anthropomorphic Soft Robotic End-Effector for Use with Collaborative Robots in the Construction Industry. *Autom. Constr.* **2022**, *138*, 104218. [CrossRef]
49. Akinradewo, O.; Oke, A.; Aigbavboa, C.; Mashangoane, M. Willingness to Adopt Robotics and Construction Automation in the South African Construction Industry. In Proceedings of the International Conference on Industrial Engineering and Operations Management, Pretoria/Johannesburg, South Africa, 29 October–1 November 2018; pp. 1639–1646.
50. Zanchettin, A.M.; Croft, E.; Ding, H.; Li, M. Collaborative Robots in the Workplace. *IEEE Robot. Autom. Mag.* **2018**, *25*, 16–17. [CrossRef]
51. Pan, W.; Hua, R.; Bock, T. A Novel Approach to Develop Vertical City Utilizing Construction Automation and Robotics. In Proceedings of the Creative Construction Conference, Ljubljana, Slovenia, 30 June–3 July 2018.
52. Assaad, R.H.; El-Adaway, I.H.; Hastak, M.; Needy, K.L. Smart and Emerging Technologies: Shaping the Future of the Industry and Offsite Construction. *Comput. Civ. Eng.* **2021**, 787–794.
53. Bouchlaghem, D.; Shang, H.; Whyte, J.; Ganah, A. Visualisation in Architecture, Engineering and Construction. *Autom. Constr.* **2005**, *14*, 287–295. [CrossRef]
54. Patrucco, M.; Bersano, D.; Cigna, C.; Fissore, F. Computer Image Generation for Job Simulation: An Effective Approach to Occupational Risk Analysis. *Saf. Sci.* **2010**, *48*, 508–516. [CrossRef]
55. Ghaffarianhoseini, A.; Doan, D.T.; Zhang, T.; Rehman, A.U.; Naismith, N.; Tookey, J.; Rotimi, J.; Ghaffarianhoseini, M.; Yu, T.; Mu, Z.; et al. Integrating Augmented Reality and Building Information Modelling to Facilitate Construction Site Co-ordination. In Proceedings of the 16th International Conference on Construction Application of Virtual Reality, Kowloon, Hong Kong, 11–13 December 2016.
56. Zhang, M.; Cao, T.; Zhao, X. Applying Sensor-Based Technology to Improve Construction Safety Management. *Sensors* **2017**, *17*, 1841. [CrossRef]
57. Zhou, Z.; Irizarry, J.; Li, Q. Applying Advanced Technology to Improve Safety Management in the Construction Industry: A Literature Review. *Constr. Manag. Econ.* **2013**, *31*, 606–622. [CrossRef]
58. Hire, S.; Sandbhor, S.; Ruikar, K. Bibliometric Survey for Adoption of Building Information Modeling (BIM) in Construction Industry–a Safety Perspective. *Arch. Comput. Methods Eng.* **2022**, *29*, 679–693. [CrossRef]
59. Olanrewaju, O.I.; Kineber, A.F.; Chileshe, N.; Edwards, D.J. Modelling the Relationship between Building Information Modelling (BIM) Implementation Barriers, Usage and Awareness on Building Project Lifecycle. *Build. Environ.* **2022**, *207*, 108556. [CrossRef]
60. Sun, C.; Jiang, S.; Skibniewski, M.J.; Man, Q.; Shen, L. A Literature Review of the Factors Limiting the Application of BIM in the Construction Industry. *Technol. Econ. Dev. Econ.* **2017**, *23*, 764–779. [CrossRef]
61. Fargnoli, M.; Lombardi, M. Building Information Modelling (BIM) to Enhance Occupational Safety in Construction Activities: Research Trends Emerging from One Decade of Studies. *Buildings* **2020**, *10*, 98. [CrossRef]
62. Rodrigues, F.; Baptista, J.S.; Pinto, D. BIM Approach in Construction Safety—A Case Study on Preventing Falls from Height. *Buildings* **2022**, *12*, 73. [CrossRef]
63. Qi, N. Optimization and Simulation of a Dynamic Management System for Building Construction Based on Low-Power Wireless Sensor Networks. *J. Sens.* **2022**, *2022*, 1841066. [CrossRef]
64. Ahsan, S.; El-Hamalawi, A.; Bouchlaghem, D.; Ahmad, S. Mobile Technologies for Improved Collaboration on Construction Sites. *Archit. Eng. Des. Manag.* **2007**, *3*, 257–272. [CrossRef]
65. Min, J.; Kim, Y.; Lee, S.; Jang, T.W.; Kim, I.; Song, J. The Fourth Industrial Revolution and Its Impact on Occupational Health and Safety, Worker's Compensation and Labor Conditions. *Saf. Health Work* **2019**, *10*, 400–408. [CrossRef] [PubMed]
66. Chen, Y.K. Challenges and Opportunities of Internet of Things. In Proceedings of the 17th Asia and South Pacific Design Automation Conference, Sydney, Australia, 30 January–2 February 2012; pp. 383–388.
67. Mendes, J.; Chaves, C. Industry 4.0: Is There Any Impact on Worker's Health and Safety?–A Literature Review. In Proceedings of the XXXIX Encontro Nacional DE Engenharia De Producao, Santos, São Paulo, Brasil, 15–18 October 2019.
68. Saaty, T.L. Decision Making with the Analytic Hierarchy Process. *Int. J. Serv. Sci.* **2008**, *1*, 83–98. [CrossRef]
69. Dewi, N.K.; Putra, A.S. Decision Support System for Head of Warehouse Selection Recommendation Using Analytic Hierarchy Process (AHP) Method. In Proceedings of the International Conference Universitas Pekalongan 2021, Online, 8–9 March 2021; pp. 43–50.
70. Wolnowska, A.E.; Konicki, W. Multi-Criterial Analysis of Oversize Cargo Transport through the City, Using the AHP Method. *Transp. Res. Procedia* **2019**, *39*, 614–623. [CrossRef]
71. Taber, K.S. The Use of Cronbach's Alpha When Developing and Reporting Research Instruments in Science Education. *Res. Sci. Educ.* **2018**, *48*, 1273–1296. [CrossRef]
72. Tavakol, M.; Dennick, R. Making Sense of Cronbach's Alpha. *Int. J. Med. Educ.* **2011**, *2*, 53–55. [CrossRef]
73. Ho, D.; Newell, G.; Walker, A. The Importance of Property-specific Attributes in Assessing CBD Office Building Quality. *J. Prop. Invest. Financ.* **2005**, *23*, 424–444. [CrossRef]
74. Dallasega, P.; Rauch, E.; Linder, C. Industry 4.0 as an Enabler of Proximity for Construction Supply Chains: A Systematic Literature Review. *Comput. Ind.* **2018**, *99*, 205–225. [CrossRef]
75. Hassankhani Dolatabadi, S.; Budinska, I. Systematic Literature Review Predictive Maintenance Solutions for SMEs from the Last Decade. *Machines* **2021**, *9*, 191. [CrossRef]

76. Oesterreich, T.D.; Teuteberg, F. Understanding the Implications of Digitisation and Automation in the Context of Industry 4.0: A Triangulation Approach and Elements of a Research Agenda for the Construction Industry. *Comput. Ind.* **2016**, *83*, 121–139. [CrossRef]
77. Wichmann, R.L.; Eisenbart, B.; Gericke, K. The Direction of Industry: A Literature Review on Industry 4.0. In Proceedings of the Design Society: International Conference on Engineering Design, Delft, The Netherlands, 5–8 August 2019; pp. 2129–2138.
78. Raji, I.O.; Shevtshenko, E.; Rossi, T.; Strozzi, F. Industry 4.0 Technologies as Enablers of Lean and Agile Supply Chain Strategies: An Exploratory Investigation. *Int. J. Logist. Manag.* **2021**, *32*, 1150–1189. [CrossRef]
79. Sawhney, A.; Riley, M.; Irizarry, J. Construction 4.0: Introduction and Overview. In *Construction 4.0: An Innovation Platform for the Built Environment*; Sawhney, A., Riley, M., Irizarry, J., Eds.; Routledge: London, UK, 2020; pp. 3–22. ISBN 978-042-939-810-0.
80. Zhang, M.; Shu, L.; Luo, X.; Yuan, M.; Zheng, X. Virtual Reality Technology in Construction Safety Training: Extended Technology Acceptance Model. *Autom. Constr.* **2022**, *135*, 104113. [CrossRef]
81. Behzadi, A. Using Augmented and Virtual Reality Technology in the Construction Industry. *Am. J. Eng. Res.* **2016**, *5*, 350–353.
82. Gallego-García, S.; Groten, M.; Halstrick, J. Integration of Improvement Strategies and Industry 4.0 Technologies in a Dynamic Evaluation Model for Target-Oriented Optimization. *Appl. Sci.* **2022**, *12*, 1530. [CrossRef]
83. Latiffi, A.A.; Mohd, S.; Kasim, N.; Fathi, M.S. Building Information Modeling (BIM) Application in Malaysian Construction Industry. *Int. J. Constr. Eng. Manag.* **2013**, *2*, 1–6.
84. Lepore, D.; Spigarelli, F. Integrating Industry 4.0 Plans into Regional Innovation Strategies. *Local Econ.* **2020**, *35*, 496–510. [CrossRef]
85. Hasan, A.N.; Rasheed, S.M. The Benefits of and Challenges to Implement 5D BIM in Construction Industry. *Civ. Eng. J.* **2019**, *5*, 412. [CrossRef]
86. Salam, M.A. Analyzing Manufacturing Strategies and Industry 4.0 Supplier Performance Relationships from a Resource-Based Perspective. *Benchmarking Int. J.* **2019**, *28*, 1697–1716. [CrossRef]
87. Vinodh, S.; Antony, J.; Agrawal, R.; Douglas, J.A. Integration of Continuous Improvement Strategies with Industry 4.0: A Systematic Review and Agenda for Further Research. *TQM J.* **2020**, *33*, 441–472. [CrossRef]

Disclaimer/Publisher's Note: The statements, opinions and data contained in all publications are solely those of the individual author(s) and contributor(s) and not of MDPI and/or the editor(s). MDPI and/or the editor(s) disclaim responsibility for any injury to people or property resulting from any ideas, methods, instructions or products referred to in the content.

Article

An Analysis of the Energy Consumption Forecasting Problem in Smart Buildings Using LSTM

Daniela Durand [1], Jose Aguilar [1,2,3,*] and Maria D. R-Moreno [1,4]

1 Escuela Politécnica Superior, University of Alcalá, 28805 Alcalá de Henares, Spain
2 Centro de Microcomputación y Sistemas Distribuidos (CEMISID), University of The Andes, Mérida 5101, Venezuela
3 Grupo de Investigación, Desarrollo e Innovación en Tecnologías de Informacion y Comunicación (GIDITIC), EAFIT University, Medellín 50022, Colombia
4 Intelligent Autonomous Systems Group (IAS), TNO, 2597 AK The Hague, The Netherlands
* Correspondence: jose.aguilar@uah.es

Citation: Durand, D.; Aguilar, J.; R-Moreno, M.D. An Analysis of the Energy Consumption Forecasting Problem in Smart Buildings Using LSTM. *Sustainability* 2022, *14*, 13358. https://doi.org/10.3390/su142013358

Academic Editors: Wesam Salah Alaloul, Bassam A. Tayeh and Muhammad Ali Musarat

Received: 19 September 2022
Accepted: 13 October 2022
Published: 17 October 2022

Publisher's Note: MDPI stays neutral with regard to jurisdictional claims in published maps and institutional affiliations.

Copyright: © 2022 by the authors. Licensee MDPI, Basel, Switzerland. This article is an open access article distributed under the terms and conditions of the Creative Commons Attribution (CC BY) license (https://creativecommons.org/licenses/by/4.0/).

Abstract: This work explores the process of predicting energy consumption in smart buildings based on the consumption of devices and appliances. Particularly, this work studies the process of data analysis and generation of prediction models of energy consumption in Smart Buildings. Specifically, this article defines a feature engineering approach to analyze the energy consumption variables of buildings. Thus, it presents a detailed analysis of the process to build prediction models based on time series, using real energy consumption data. According to this approach, the relationships between variables are analyzed, thanks to techniques such as Pearson and Spearman correlations and Multiple Linear Regression models. From the results obtained with these, an extraction of characteristics is carried out with the Principal Component Analysis (PCA) technique. On the other hand, the relationship of each variable with itself over time is analyzed, with techniques such as autocorrelation (simple and partial), and Autoregressive Integrated Moving Average (ARIMA) models, which help to determine the time window to generate prediction models. Finally, prediction models are generated using the Long Short-Term Memory (LSTM) neural network technique, taking into account that we are working with time series. This technique is useful for generating predictive models due to its architecture and long-term memory, which allow it to handle time series very well. The generation of prediction models is organized into three groups, differentiated by the variables that are considered as descriptors in each of them. Evaluation metrics, RMSE, MAPE, and R^2 are used. Finally, the results of LSTM are compared with other techniques in different datasets.

Keywords: forecasting models; energy consumption; smart buildings; machine learning; time series; LSTM technique

1. Introduction

This section discusses the need for energy consumption forecasting models in the context of building energy management systems. In addition, the novelty and contributions of this research are presented.

1.1. Motivation

One of the main current challenges is the efficient consumption of energy due to economic and environmental reasons, among others. The massive energy consumption entails more economic expenses, impact on the environment, etc. However, thanks to the evolution of technology, it is possible to develop smart energy management systems (SEMSs) that efficiently save energy, without degrading user comfort. In this context, the application of soft computing techniques is necessary.

On the other hand, the building sector consumes more energy than the industry and transportation sectors, which is due mainly to Heating, Ventilation, and Air Conditioning

(HVAC) systems, appliances/devices, and lighting [1,2]. Particularly, it is interesting to analyze the appliances/devices consumption in the context of a SEMS for several reasons, e.g., to minimize the utilization of the energy when the prices are excessive to guarantee the comfort to the users, or to have a sustainable rate of consumption considering the environment.

A smart building is a dynamic system where technology is used to improve its functioning, considering hundreds of elements, such as its HVAC system, appliances and devices, etc. In this context, SEMS must seek energy efficiency, implementing energy management tasks, such as monitoring of energy supply, predicting energy consumption, and anomaly detecting of energy use, among others.

Thus, among the possible reasons to analyze the energy consumption in smart buildings are the following: to determine the electrical load; to detect anomalies in consumption; to estimate energy consumption; to define load profiles using consumption behavior; and to classify the consumers, among others. In this way, to reach optimal management of energy consumption in a smart building, it is necessary to study the consumption, which is precisely the scope of this paper. Thereby, possible energy problems can be detected and solved. At the same time, around a smart grid, the energy is intermittent, distributed, mobile, and able to be stored. For example, renewable energy resources (RES) are characterized by their variability and intermittency, which make the prediction of the generated energy complex [1,2]. These attributes make the implementation of SEMSs more challenging, because more flexibility and stability is needed to secure its normal operation in a building, for which efficient energy consumption forecasting models are required. SEMSs today do not consider these aspects for this highly complex and rapidly changing scenario.

On the other hand, Artificial Intelligence (AI) can build useful knowledge of factors such as the prediction of energy consumption and the prediction of occupancy behavior, among others. AI techniques are already being used in the SEMSs, such as tasks of modeling, learning, and reasoning, among others. The motivation of this work is to analyze the behavior of the energy consumption data of the devices and electrical appliances in a building, in order to build models that allow prediction of their behavior.

1.2. Background

In the literature, there are some works similar to this work. For example, Rodriguez-Mier et al. [3] proposed a knowledge model to define predictive models of energy consumption for smart buildings, and a multi-step prediction model based on a hybrid genetic-fuzzy system, which includes a feature selection method. The authors use a database that stores two types of signals: synchronous signals that record at a constant rate of 10 s (e.g., temperature, sensors, etc.) and asynchronous signals that record when a value changes (e.g., the indoor temperatures, error signals, etc.). In addition, they collect the humidity, solar radiation power, and pressure. Garcia et al. [4] present a comparative study of different forecasting strategies of the energy consumption of smart buildings. Particularly, they determine that strategies based on Machine Learning (ML) approaches are more suitable. Alduailij et al. [5] analyze several statistical and ML techniques to predict energy consumption for five different building types. They especially predict the peak demand that serves to achieve energy efficiency. Hernández et al. [6] present an energy consumption forecasting strategy that allows hourly day-ahead predictions using several ML techniques. Then, they define an ensemble model using the mean of the prediction values of the top five models. In addition, Hernández et al. [7] present a review of energy consumption forecasting for improving energy efficiency in smart buildings. They analyze different forecasting methods in nonresidential and residential buildings in terms of forecasting methods, forecasting objectives, input variables, and prediction horizon.

Moreno et al. [8] define predictive models of energy consumption and save energy for buildings based on the Radial Basis Function (RBF) technique. Nabavi et al. [9] propose a Deep Learning (DL) method that uses a discrete wavelet transformation and the long short-term memory method to forecast building energy demand and energy supply. These methods consider several factors, such as energy consumption patterns in buildings, electricity price, availability of renewable energy sources, and uncertainty in climatic factors. Somu et al. [10] present an energy consumption forecasting model which employs LSTM. The hyperparameter optimization process (learning rate, number of layers, momentum, and weight decay) of the LSTM was optimized using the sine–cosine optimization algorithm.

On the other hand, Le et al. [11] develop a framework for multiple energy consumption forecasting of a smart building based on the use of the Transfer Learning concept. Hadri et al. [12] implement different energy consumption forecasting approaches of appliances by integrating the occupancy and the context-driven control information of buildings. In addition, Gonzalez-Vidal et al. [13] defined a methodology to transform the multivariate time-dependent series to be used by ML algorithms for energy forecasting. Then, González-Vidal et al. [14] proposed ML and grey-box approaches to predict energy consumption based on the physics of the building's heat transfer. Sulo et al. [15] analyzed the ways to improve the efficiency of the energy used by buildings using an LSTM model to predict the energy consumption of the buildings on the campuses of the City University of New York.

In other contexts, Aliberti et al. [16] proposed a predictive model to estimate the indoor air temperature in individual rooms with a prediction window of up to three hours, and for the whole building with a prediction window of four hours. In addition, Lawadi et al. [17] compared several ML algorithms to estimate the indoor temperature in a building, which were evaluated using different metrics, such as accuracy and robustness to weather changes. Siddiqui et al. [18] introduced a DL approach to recommend consumption patterns for the appliances based on Term Frequency–Inverse Document Frequency (TF-IDF) to quantify the energy tags. The aim of the work of Bhatt et al. [19] was to forecast the cost of energy consumption in smart buildings. They proposed a balanced DL algorithm that considers three constraints to solve the price management problem and high-level energy consumption in HVAC systems [20]. Bourhnane et al. [21] used Artificial Neural Networks (ANN) along with Genetic Algorithms (GA) to define an approach for energy consumption prediction and scheduling.

The motivation of the work of Hadri et al. [22] was to determine the forecasting quality and the computational time of the XGBOOST, LSTM, and SARIMA algorithms in the context of forecasting energy consumption. Khan et al. [23] proposed a short-term electric consumption forecasting model based on spatial and temporal ensemble forecasting. The ensemble forecasting model consists of a K-means algorithm to determine energy consumption profiles, and two deep learning models, LSTM and Gated Recurrent Unit (GRU). The model forecasts the energy consumption at three spatial scales (apartment, building, and floor level) for hourly, daily, and weekly forecasting horizons. The work of Keytingan et al. [24] proposed predictive models for energy consumption based on a Support Vector Machine, Artificial Neural Networks, and K-Nearest Neighbour using real-life data of a commercial building from Malaysia. The goal of the work of Son et al. [25] was to study adaptive energy consumption forecasting models in order to follow the dynamics of buildings. They consider active and passive change detection methods, which are integrated into the decision tree and deep learning models. The results showed that constant retraining, in some cases, is not good in performance. Moon et al. [26] proposed an online learning approach to enable fast learning of building energy consumption patterns for unseen data. In addition, Pinto et al. [27] presented three ensemble learning models (XGBOOST, random forests, and an adaptation of Adaboost) for energy consumption forecasting an hour ahead, using real data from an office building. Finally, the work of Somu et al. [28] described a deep learning framework based on CNN (Convolutional Neural Networks)-LSTM to provide building energy consumption forecasts. CNN-LSTM uses K-means to determine the energy consumption pattern/trend, CNN to extract features about energy consumption, and LSTM to handle long-term dependencies.

In summary, the vast majority of recent works have been dedicated to carrying out comparative studies of different building energy consumption forecasting strategies based on statistical and ML techniques for different types of buildings (residential, office, among others), using specific datasets [4–8,16–19,22,24,27]. In some of these comparisons, specific aspects have been analyzed, such as occupation [12], how to follow the dynamics of energy consumption [25], or the use of online learning approaches to follow the consumption pattern in real time [26]. On the other hand, some works analyzed the relationships of temporal dependencies in the time series in order to forecast the energy demand and/or the energy supply of the building [9,10,14,23], and in some cases, use the LSTM model [15,28] or feature selection methods [3,28] to predict energy consumption in buildings. Other works have studied the Transfer Learning concept in the context of forecasting the energy consumption of intelligent buildings [11], or have combined it with other techniques to consider the prediction and programming of energy consumption [20].

In conclusion, there are many works on the prediction of energy consumption in smart buildings, but none of them propose a scheme to carry out an exhaustive feature engineering process to analyze the variables, their dependencies, and their transformations, which allows improvement of the prediction of the forecasting models. Specifically, the great gap in the previous works is that they do not propose strategies to analyze the implicit temporal relationships in the time series that describe the pattern of energy consumption, which affects/degrades the ML and the statistical algorithms used to build the models of forecasting of energy consumption.

1.3. Novelty and Our Contribution

This work studies the process of data analysis and generation of prediction models of energy consumption in smart buildings. The focus of this paper is to estimate energy consumption in smart buildings based on the consumption of the appliances and devices in them. Therefore, thanks to the data collected in smart buildings, the study is carried out on this energy consumption as a function of time, in order to obtain a model capable of estimating total consumption, knowing the consumption of the devices and appliances in the building. The reason for working with time series is that energy consumption can be labeled by times of the year, days of the week, or even hours on the same day. For example, at Christmas, there may be a greater consumption of Christmas lights, and in holiday months, the consumption may be lower if we are traveling. The research question is whether the prediction of energy consumption in a building depends on an exhaustive analysis of time series that describe its behavior, which would imply carrying out a specific feature engineering process for that context.

This work carries out an analysis of these variables and their relationships, thanks to techniques such as the Pearson and Spearman correlations, and Multiple Linear Regression models. With the results obtained, the fusion and extraction of characteristics are carried out with the Principal Component Analysis (PCA) technique. On the other hand, the relationship of each variable with itself over time is analyzed, using techniques such as autocorrelation (simple and partial) and ARIMA models. Finally, several forecasting models are generated with LSTM. We start from the hypothesis that LSTM is an excellent technique for treating time series [29,30], so we have chosen this technique to analyze the results of our feature engineering process. With these results, the generation of prediction models is organized into three groups. The first group consists of prediction models in which only the first Principal Component (PC1) is taken into account. The second group includes PC2. The last group uses the original variables. These groups of models are evaluated using the following metrics: Root Mean Square Error (RMSE), Mean Absolute Percentage Error (MAPE), and R^2. Thus, the main contribution of this article is the definition of a feature engineering methodological approach to analyze the energy consumption variables of buildings. Other specific contributions derived from that contribution are:

- The definition of the phases to study the time series that define energy consumption in buildings;
- The definition of the analysis process of the dependency relations between the variables, especially the temporal ones;

- The definition of the analysis process of the dimensions in the dataset to determine the fusion and extraction of characteristics.
- The utilization of this approach for the definition of forecast models based on time series.

The remainder of this work is organized as follows. A preliminary theoretical framework is described in Section 2. The analysis of the energy forecasting problem is reported in Section 3 based on two aspects. It begins by defining our feature engineering approach for energy consumption time series, then describes its detailed application in a case study, and finally builds a forecast model using LSTM from the results obtained with it. Section 4 presents a comparison of LSTM with other machine learning techniques in different time series on building energy consumption, using our feature engineering approach to define the forecast model. Finally, a discussion about future directions in this domain is pointed out in Section 5.

2. Theoretical Framework

In this section, we formalize the problem of energy consumption forecasting in smart buildings and present the ML technique used for the forecasting task.

2.1. The Energy Consumption Forecasting Problem

Particularly, the energy consumed by the 'n' elements in a building (e.g., an HVAC system, an electrical appliance) has a timestamp t, recorded by the sensors deployed in the building, and can be described as [10]:

$$X_t = \{x_{1t}, x_{1t}, \ldots, x_{1nt}\}$$

where x_{it} is the energy consumption captured by the ith sensor at tth timestamp, and n is the number of sensors. Normally, the data-driven forecasting models use a window-based approach for forecasting, defined by two variables, the input window (I_w) and the forecast window (F_w) sizes. Thus, the total number of inputs and forecasts is defined in [10] as $k = (S_n - I_w - a)/F_w$, where a is the forecast interval and S_n is the total number of samples.

Subsequently, the input (S_I) of size I_w is $S_I = \{X_t, X_{t+1}, \ldots, X_{t+k}\}$. Similarly, the forecast (S_F) of size F_w is $\{S_F = \{\widehat{X_t}, \widehat{X_{t+1}}, \ldots, \widehat{X_{t+k}}\}$.

Therefore, a non-linear approximation function (f) can be defined as $S_F = f(S_I)$. Hence, for a given time window (I_w), the model (f) learns to forecast their values for the time window (F_w) with minimal forecast error *between* the actual and forecasted energy consumption value of the sensors at each timestamp. The input window and forecast window sizes can be adjusted according to the forecast problem (long-term, mid-term, and short-term memory).

2.2. Long Short-Term Memory (LSTM) Technique

Long Short-Term Memory (LSTM) technique is a type of Recurrent Neural Network (RNN). The RNNs [30], unlike traditional neural networks, are capable of processing data sequences, that is, data that need to be interpreted together in a specific order to have meaning. This is possible thanks to the fact that the RNNs have an architecture that allows them, at each instant of time, to receive the input corresponding to that instant, in addition to the value of the activation in the previous instant. These previous instants of time allow a certain "memory", since they retain the information from previous steps. Thus, they have a memory cell that preserves the state over time. In addition, the RNNs apply backpropagation through time (BPTT).

However, the "memory" of traditional RRNs is a short-term memory, since, as the information is transmitted through time, the effect of the state of an instant will be very small at a very distant instant. Given this drawback, the LSTM [29,30] emerged. These networks are capable of maintaining information both in the short and long term, thanks to the incorporation of the cell state (c). The cell state c encodes the information of the inputs (relevant info) that have been observed up to that step. This cell state is updated according to the current state of the cell, the output at time t−1, the input at time t, and a series of gates that are responsible for processing this information. The gates are the following:

A forget gate that, receiving input x at time t and output h at time t−1, decides what information will be discarded; that is, it will not be part of the state for that instant.

An update gate (or input gate) that determines what information in the cell state c to update from the input x at time t and the output h at time t−1.

An output gate, which will generate the output h for the instant t from the information in the cell state c. However, not all of the information from the cell state is dumped in the output; instead, a selection is made of the information of the cell state that has been considered important to use at the next time.

Figure 1 shows the architecture of an LSTM network, with each of the parts explained above [29,30].

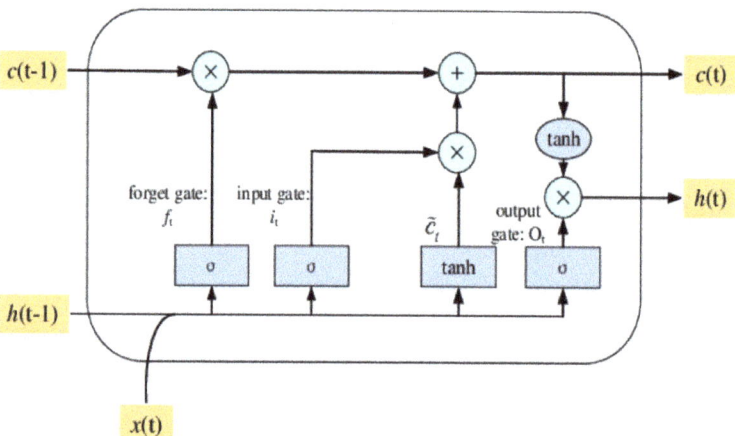

Figure 1. Architecture of a neuron in an LSTM network.

In this way, the behavior of an LSTM neural network tries to replicate human learning. To process information that requires long-term memory, we do not keep all the data, but only the most relevant, which will be useful to understand what we are dealing with. This ability of LSTM networks to retain information in the long term, makes this type of network the ideal option to create the prediction models in our work. This is because the time series of energy consumption are correlated with values in past time intervals, which can be days, but can also be weeks, months, or years. For example, in the context of energy consumption of devices, the following case may occur: if we want to predict the total consumption on Mondays, the forget gate will be in charge of discarding the information on the consumption of the devices of the last Wednesday, because it is not relevant. On the other hand, the update gate will add data on the consumption of the devices for Mondays, since this information is useful for predicting the total consumption of that day. In this way, the output gate could determine the total consumption for Monday by combining the input data with the selected data from the cell state, such as the consumption of last Monday.

3. Analysis of the Energy Consumption Forecasting Problem

This section presents a deep analysis of the Energy Consumption Forecasting Problem. It is divided into three subsections. The first carries out a feature engineering process to obtain the variables to be used in the predictive models. This process analyzes the variables of the dataset to determine their quality, their correlations and autocorrelations, and the selection and fusion of variables, among other things. The next subsection describes the generation of energy consumption prediction models using the LSTM technique. Particularly, an initial general predictive model is generated, from which three groups of models are created according to the descriptive variables selected by the feature engineering process. In the third subsection, a comparison is made between the models of each group.

3.1. Analysis of the Variables (Feature Engineering Process)

This section presents the feature engineering process proposed in this work to analyze the energy consumption in smart buildings (see Figure 2).

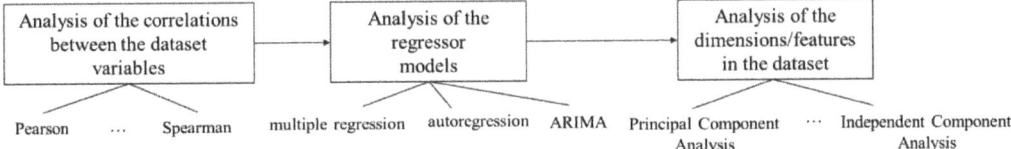

Figure 2. Feature Engineering Process for Energy Datasets of Smart Buildings.

Figure 2 shows that there are 3 major processes: (i) analysis of the correlations between the variables; (ii) analysis of regressive models; (iii) analysis of the dimensions. In the first case, the different types of correlations, linear or not (Pearson and Spearman, among others), between the variables and with the target variable (in our case, the energy consumption) are analyzed. In particular, the variables that are not correlated with the objective variable to be estimated are eliminated, or one of the variables that has a high correlation between them is chosen in order to avoid the collinearity problem. Regarding the second case, since it is a forecast model, its different forms are analyzed by studying the possible combinations between the variable of interest and all of the predictor variables (multiple regression model), or the regression of the variable against itself (autoregression or ARIMA models, among others). Finally, the last phase creates an analysis of the dimensions/characteristics in the dataset to determine if they can be reduced (Principal Component Analysis) or extract information of interest (Independent Component Analysis), among other possible studies.

Next, in this section, we explain this feature engineering process in detail for a dataset about the energy consumption in a building. The dataset used is Raw_Data.csv [31], which is a time-series of the energy consumption of the Research Support Facility (RSF) building of the United States National Renewable Energy Laboratory. It has 34 attributes, some of which are:

- Date: represents the date each sample is taken, with the format: YYYY: MM: DD.
- Month: represents the month in which each sample is taken (integer).
- Day: represents the day of the week (Sunday–Monday) on which each sample is taken (string).
- Time: represents the time of day at which the sample is taken, with the format hh: mm (time).
- Hour and Minute: represent the hour and minute of the sample collection, respectively (integer).
- Skyspark: represents the total energy consumption for each observation, in kilowatts (kW). This will be the target or dependent variable.
- AV.Controller, Coffee.Maker, Copier, Office Computer, Lamp, Laptop, Microwave, Monitor, Phone.Charger, Printer, Projector, Toaster.Oven, TV, Video.Conference.Camera, Water.Boiler, Conference.Podium, Auto.Door.Opener, Treadmill, Refrigerator, Central-Monitoring-Station, TVs, etc. represent the energy consumption of each device in each observation (kilowatts (kW)). These will be our descriptive or independent variables.

This data set has 26,496 observations, collected every 5 min, from 1 October 2019 at 00:00 to 31 December 2019 at 23:55. Additionally, the variables Month, Day, Hour, and Minute are not necessary to keep them, since this information is condensed in the variable Date_Time. This variable will be set as the index (timestamp). It can also be observed that some columns do not provide information, since all their values are 0 (e.g., the variables

Conference.Podium, Central.Monitoring.Station, and Auto.Door.Opener, among others). Therefore, they are eliminated.

3.1.1. Analysis Using Pearson's Correlation

In order to analyze the existence of linear relationships, the Pearson correlation coefficients between each pair of variables are used. Figure 3 shows these results.

	Skyspark	AV.Controller	Coffee.Maker	Copier	Desktop.Server	Headset	Lamp	Laptop	Microwave	Monitor	Phone.Charger	Printer	Projector	Toaster.Oven	TV	Video.Conference.Camera	Water.Boiler
Skyspark	1.00	0.07	0.15	0.46	0.07	-0.01	0.39	0.82	0.38	0.86	0.19	0.53	0.45	0.07	0.47	0.19	0.32
AV.Controller	0.07	1.00	0.00	0.04	0.02	0.04	-0.01	0.07	0.03	0.06	0.00	0.05	0.04	0.00	0.07	0.08	0.02
Coffee.Maker	0.15	0.00	1.00	0.09	-0.01	-0.01	0.03	0.10	0.03	0.10	0.02	0.07	0.06	0.02	0.03	0.00	0.10
Copier	0.46	0.04	0.09	1.00	0.03	0.01	0.20	0.45	0.17	0.46	0.05	0.31	0.24	0.03	0.24	0.03	0.18
Desktop.Server	0.07	0.02	-0.01	0.03	1.00	-0.01	0.10	0.06	0.03	0.07	0.12	0.04	0.03	-0.00	0.01	0.09	-0.00
Headset	-0.01	0.04	-0.01	0.01	-0.01	1.00	0.00	0.07	-0.00	0.01	-0.02	0.01	0.01	0.01	-0.05	-0.06	-0.00
Lamp	0.39	-0.01	0.03	0.20	0.10	0.00	1.00	0.43	0.13	0.43	0.14	0.25	0.24	0.02	0.23	0.03	0.11
Laptop	0.82	0.07	0.10	0.45	0.06	0.07	0.43	1.00	0.35	0.92	0.16	0.50	0.42	0.04	0.44	0.06	0.26
Microwave	0.38	0.03	0.03	0.17	0.03	-0.00	0.13	0.35	1.00	0.35	0.06	0.19	0.16	0.04	0.20	0.06	0.13
Monitor	0.86	0.06	0.10	0.46	0.07	0.01	0.43	0.92	0.35	1.00	0.18	0.53	0.45	0.04	0.50	0.10	0.27
Phone.Charger	0.19	0.00	0.02	0.05	0.12	-0.02	0.14	0.16	0.06	0.18	1.00	0.14	0.16	0.06	0.11	0.04	0.07
Printer	0.53	0.05	0.07	0.31	0.04	0.01	0.25	0.50	0.19	0.53	0.14	1.00	0.24	0.07	0.26	0.04	0.16
Projector	0.45	0.04	0.06	0.24	0.03	0.01	0.24	0.42	0.16	0.45	0.16	0.24	1.00	0.02	0.28	0.06	0.13
Toaster.Oven	0.07	0.00	0.02	0.03	-0.00	0.01	0.02	0.04	0.04	0.04	0.06	0.07	0.02	1.00	0.03	-0.01	0.05
TV	0.47	0.07	0.03	0.24	0.01	-0.05	0.23	0.44	0.20	0.50	0.11	0.26	0.28	0.03	1.00	0.06	0.13
Video.Conference.Camera	0.19	0.08	0.00	0.03	0.09	-0.06	0.03	0.06	0.06	0.10	0.04	0.04	0.06	-0.01	0.06	1.00	0.03
Water.Boiler	0.32	0.02	0.10	0.18	-0.00	-0.00	0.11	0.26	0.13	0.27	0.07	0.16	0.13	0.05	0.13	0.03	1.00

Figure 3. Pearson's correlation coefficient matrix.

According to the literature, Pearson correlations between variables less than 0.2 indicate that there is little relationship between them, while correlations greater than 0.8 indicate that there are high correlations between them [3,13,32]. In Figure 2, the Pearson correlation coefficients between the Skyspark variable and the following descriptor variables are less than 0.2: AV.Controller: 0.07, Coffee.Maker: 0.15, Desktop.Server: 0.07, Headset: −0.01, Phone.Charger: 0.19, Toaster.Oven: 0.07, Video.Conference.Camera: 0.19. This means that these variables are very poorly correlated with the target variable. They do not provide relevant information for the generation of a predictive model of the Skyspark variable. Therefore, they can be deleted.

On the other hand, the following pairs of variables have a coefficient greater than 0.8: Skyspark–Laptop: 0.82, Skyspark–Monitor: 0.86 and Laptop–Monitor: 0.92. The descriptor variables Laptop and Monitor are highly correlated with each other. Furthermore, these same variables are, in turn, highly correlated with the target variable. This can negatively affect the modeling, so it will be analyzed later.

3.1.2. Analysis Using Spearman's Correlation

In addition to analyzing the linear relationships between the variables, it is necessary to analyze the non-linear relationships. Thus, the Spearman correlation coefficients are determined (see Figure 4).

The most significant correlations are between the following variables: Laptop–Monitor: 0.79, Coffee.Maker–Microwave: 0.67, Microwave–Water.Boiler: 0.65. In this case, it is observed that there is a non-linear correlation between two pairs of variables that were not linearly related (Coffee.Maker–Microwave and Microwave–Water.Boiler). However, a high coefficient is also obtained for the Laptop and Monitor variables, thus confirming that these two variables are mutually correlated. Furthermore, the variables that had a weak linear correlation with the target variable (Skyspark) also have a weak non-linear correlation (AV.Controller, Coffee.Maker, Desktop.Server, Headset, Microwave, Phone.Charger, Toaster.Oven, Video.Conference.Camera, and Water.Boiler). The variables that are not correlated with Skyspark can be eliminated. On the other hand, for the descriptive variables that are highly correlated with each other (Laptop and Monitor), there are two options. The first one consists of selecting one of the variables, thus eliminating the redundant information that this correlation supposes. The second option consists of carrying out an extraction/fusion of characteristics, so that we obtain new variables, which have information about each of the original variables, without being correlated between them. In order to do this, we apply a Principal Component Analysis (PCA).

	Skyspark	AV.Controller	Coffee.Maker	Copier	Desktop.Server	Headset	Lamp	Laptop	Microwave	Monitor	Phone.Charger	Printer	Projector	Toaster.Oven	TV	Video.Conference.Camera	Water.Boiler
Skyspark	1.00	0.06	0.15	0.30	0.15	-0.06	0.32	0.56	0.06	0.62	0.17	0.51	0.26	0.15	0.40	0.18	0.04
AV.Controller	0.06	1.00	0.05	-0.03	0.08	0.13	-0.19	0.19	0.08	0.04	0.02	-0.00	0.06	0.01	-0.12	0.12	0.04
Coffee.Maker	0.15	0.05	1.00	0.34	0.09	0.00	0.15	0.09	0.67	0.11	0.28	0.16	0.22	0.04	0.12	-0.00	0.63
Copier	0.30	-0.03	0.34	1.00	0.12	-0.03	0.24	0.15	0.30	0.19	0.19	0.30	0.25	0.06	0.14	-0.00	0.30
Desktop.Server	0.15	0.08	0.09	0.12	1.00	-0.04	0.22	-0.14	0.01	-0.14	0.33	0.04	0.15	0.01	-0.13	-0.02	0.06
Headset	-0.06	0.13	0.00	-0.03	-0.04	1.00	-0.02	0.12	0.03	-0.00	-0.05	0.04	-0.11	0.02	-0.06	-0.06	-0.00
Lamp	0.32	-0.19	0.15	0.24	0.22	-0.02	1.00	0.18	0.05	0.14	0.36	0.31	0.13	0.06	0.07	-0.11	0.10
Laptop	0.56	0.19	0.09	0.15	-0.14	0.12	0.18	1.00	0.09	0.79	-0.16	0.45	0.08	0.14	0.38	0.03	0.01
Microwave	0.06	0.08	0.67	0.30	0.01	0.03	0.05	0.09	1.00	0.08	0.20	0.11	0.18	0.07	0.10	0.04	0.65
Monitor	0.62	0.04	0.11	0.19	-0.14	-0.00	0.14	0.79	0.08	1.00	-0.17	0.46	0.17	0.14	0.54	0.09	0.02
Phone.Charger	0.17	0.02	0.28	0.19	0.33	-0.05	0.36	-0.16	0.20	-0.17	1.00	0.10	0.09	0.02	-0.09	-0.05	0.24
Printer	0.51	-0.00	0.16	0.30	0.04	0.04	0.31	0.45	0.11	0.46	0.10	1.00	0.17	0.10	0.36	0.00	0.08
Projector	0.26	0.06	0.22	0.25	0.15	-0.11	0.13	0.08	0.18	0.17	0.09	0.17	1.00	0.04	0.12	0.21	0.17
Toaster.Oven	0.15	0.01	0.04	0.06	0.01	0.02	0.06	0.14	0.07	0.14	0.02	0.10	0.04	1.00	0.09	0.02	0.03
TV	0.40	-0.12	0.12	0.14	-0.13	-0.06	0.07	0.38	0.10	0.54	-0.09	0.36	0.12	0.09	1.00	0.12	0.07
Video.Conference.Camera	0.18	0.12	-0.00	-0.00	-0.02	-0.06	-0.11	0.03	0.04	0.09	-0.05	0.00	0.21	0.02	0.12	1.00	-0.02
Water.Boiler	0.04	0.04	0.63	0.30	0.06	-0.00	0.10	0.01	0.65	0.02	0.24	0.08	0.17	0.03	0.07	-0.02	1.00

Figure 4. Spearman's correlation coefficient matrix.

3.1.3. Analysis Using Multiple Linear Regression

Another technique used to analyze linear relationships between variables is the generation of Multiple Linear Regression models. In this case, we generate a model for each descriptor variable in order to detect possible collinearities. That is, for each descriptor variable, there is a model in which it is the dependent variable and the other descriptor variables are independent variables. Table 1 shows R^2 and the p-value for each model generated.

Table 1. Multiple Linear Regression Models.

Model	R2	p-Value
AV.Controller	0.019	2.99×10^{99}
Coffee.Maker	0.020	3.62×10^{106}
Copier	0.230	0.00
Desktop.Server	0.031	8.48×10^{174}
Headset	0.031	1.69×10^{173}
Lamp	0.205	0.00
Laptop	0.848	0.00
Microwave	0.132	0.00
Monitor	0.867	0.00
Phone.Charger	0.063	0.00
Printer	0.297	0.00
Projector	0.214	0.00
Toaster.Oven	0.010	2.90×10^{50}
TV	0.262	0.00
Video.Conference.Camera	0.036	8.81×10^{199}
Water.Boiler	0.086	0.00

Table 1 shows that there are two models with a very high R^2 (for the variables Laptop and Monitor), with a value of 0.848 and 0.867, respectively. In addition, the p-value for these models is less than 0.05. These results indicate that these variables have a fairly strong linear relationship with at least one of the other descriptor variables. This coincides with the results of Pearson's correlation, where it was seen that a linear relationship exists between these two variables.

3.1.4. Autocorrelation Analysis

The autocorrelation determines how many previous instants of time affect the energy consumption of a given observation. This number, known as time delay or lag, is required by the LSTM technique. Simple autocorrelation and partial autocorrelation of the target variable (Skyspark) are shown in Figure 5. The x-axis represents the lags, and the y-axis represents the autocorrelation value.

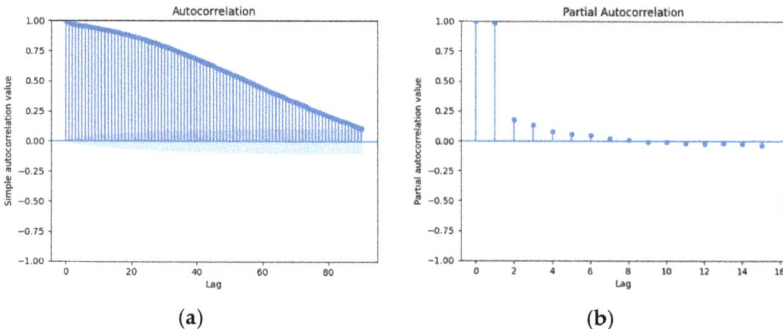

Figure 5. Autocorrelation of Skyspark. (**a**) Simple Skyspark autocorrelation. (**b**) Partial autocorrelation of Skyspark.

According to the simple autocorrelation (Figure 5a), the limit of the confidence interval of 0.05 is at lag 90. This means that the first 90 lags are statistically significant and, therefore, directly or indirectly influence the energy consumption values for each observation. On the other hand, the partial autocorrelation indicates that only seven lags are statistically significant (Figure 5b). That is, only six previous instants produce a direct effect on the values of a given observation. However, the utility of simple and partial correlations is that they are used to obtain ARIMA models. This is because the delay obtained from the simple autocorrelation corresponds to the parameter q of the ARIMA models, while the delay obtained from the partial autocorrelation corresponds to the parameter p (see Section 3.1.6 where there is an introduction to ARIMA models).

In this way, from an initial model created with these values, new models can be adjusted to find the most accurate one. With this last adjusted model, the time window to be used in our LSTM models will be established later. For example, the ARIMA model which starts the fit would be ARIMA (90, 0, 6). The parameter d is 0 because the time series in the Skyspark variable is non-stationary; that is, there is a trend or seasonality. This is shown in Figure 6, where there is a representation of the values of the Skyspark variable over time (first Figure), and their decomposition into trend, seasonality, and residuals, respectively.

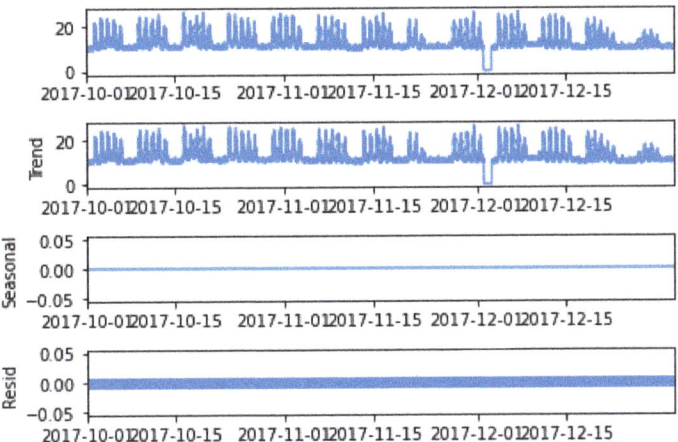

Figure 6. Decomposition of the time series of the Skyspark variable.

3.1.5. Principal Component Analysis (PCA)

This technique can reduce the dimensionality of the time-series, so that in each Principal Component (PC), we have a degree of information about each of the variables. Note that the PCs are not correlated between them. After normalizing the values of the dataset, PCA is applied to the selected descriptor variables (Copier, Lamp, Laptop, Monitor, Printer, Projector, and TV). Figure 7 shows the PCs calculated from these variables.

	PC1	PC2	PC3	PC4	PC5	PC6	PC7
Date_Time							
2017-10-01 00:00:00	-1.149516	-0.291675	0.255529	-0.087660	-0.122129	0.289152	-0.179384
2017-10-01 00:05:00	-1.142055	-0.293963	0.256674	-0.086361	-0.124481	0.281334	-0.169564
2017-10-01 00:10:00	-1.149035	-0.292193	0.256722	-0.087701	-0.121914	0.289727	-0.179343
2017-10-01 00:15:00	-1.139473	-0.288521	0.262416	-0.085559	-0.113571	0.291975	-0.173272
2017-10-01 00:20:00	-1.143439	-0.295435	0.259982	-0.088086	-0.124909	0.284403	-0.172609
...	...	...	...	...	...	...	...
2017-12-31 23:35:00	-1.031105	0.092535	-0.166260	-0.022208	0.106422	-0.264954	0.233805
2017-12-31 23:40:00	-1.032868	0.095190	-0.170062	-0.022025	0.107757	-0.266492	0.233767
2017-12-31 23:45:00	-1.031613	0.093634	-0.167631	-0.022136	0.107004	-0.266037	0.234478
2017-12-31 23:50:00	-1.029545	0.091782	-0.164500	-0.022273	0.106580	-0.264511	0.234629
2017-12-31 23:55:00	-1.028449	0.095251	-0.165527	-0.021345	0.110507	-0.263606	0.235436

Figure 7. PCs for our case study.

Figure 8 shows how much information about each variable is contained in each PC. In addition, the sign indicates whether the relationship between the component and the variable is direct or inverse.

	Copier	Lamp	Laptop	Monitor	Printer	Projector	TV
PC1	0.318164	0.298390	0.486168	0.499520	0.350511	0.309250	0.324595
PC2	0.611761	-0.538207	0.033347	0.024401	0.366971	-0.391302	-0.215846
PC3	-0.048093	0.674872	0.055905	0.009641	0.256955	-0.492015	-0.480533
PC4	-0.320911	-0.066689	0.030654	0.068862	0.145588	-0.639390	0.675923
PC5	-0.594770	-0.295527	0.046877	0.050451	0.664432	0.237660	-0.237102
PC6	0.255683	0.272132	-0.552809	-0.445708	0.461170	0.208420	0.316544
PC7	0.009936	0.002213	0.671318	-0.737461	0.033247	0.020680	0.062023

Figure 8. Relationship between PCs and variables.

For example, this means that PC1 is calculated as:

$$PC1 = 0.318164 * Copier + 0.298390 * Lamp + 0.486168 * Laptop + 0.499520 * Monitor + 0.350511 \\ *Printer + 0.309250 * Projector + 0.324595 * Tp$$

The other components are calculated analogously.

On the other hand, the PCs are arranged in descending order according to their eigenvalues (that is, those that best explain the variability of the dataset are first). This is shown in Table 2, where it can be seen that PC1 has an explained variability of 3.417665, much higher than the following components.

Table 2. Explained variability of each PC.

PC1	PC2	PC3	PC4	PC5	PC6	PC7
3.4175	0.8202	0.7896	0.7264	0.6735	0.4923	0.0804

However, to get a better idea of the proportion of explained variability of each component, the ratios of explained variability are analyzed. This is more intuitive information to select the PCs. Table 3 shows the accumulated explained variability ratios. One way to choose the components is to select those that reach a certain threshold. In this case, for a threshold of 98% of explained variability, the first 6 PCs can be selected, since they explain 98.85% of the variability.

Table 3. Cumulative Explained Variability Ratios.

PC1	PC2	PC3	PC4	PC5	PC6	PC7
0.4882	0.6053	0.7181	0.8219	0.9181	0.9885	1.0

Another option for selecting the number of PCs is to use the elbow method. In order to do this, the variability ratios explained by each component are graphically displayed, and the PC from which the curve flattens is identified, since the following components do not include relevant information on the initial variables. Figure 9 shows these ratios. It can be seen that the value drops considerably in PC2, confirming that PC1 explains much of the total variability.

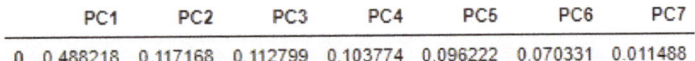

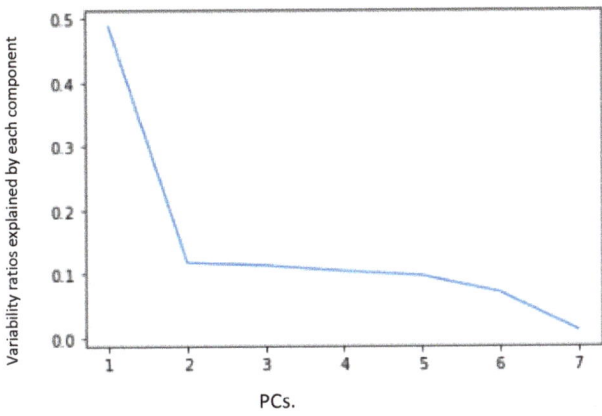

Figure 9. Explained variability ratios by each PC.

Based on the above analysis, different numbers of PCs can be used. The former provides the greatest amount of information according to Table 3 and Figure 9, but up to PC6, there is an important accumulation of information. After this analysis, it is decided that the first six PCs be considered for further analysis.

Now, it is possible to determine the time window to generate the prediction models. In order to do this, the autocorrelations and the ARIMA models are used, both for the target variable and for each component.

3.1.6. Analysis Using ARIMA Models

ARIMA models are an approach to time series forecasting which seeks to describe the autocorrelations in the data. The AR in ARIMA indicates an autoregression model, i.e., it forecasts the variable of interest using a linear combination of past (i.e., prior) values of the variable. The MA in ARIMA indicates a moving average model, i.e., it uses past forecast errors in a regression model (the regression error is a linear combination of errors in the past). Finally, the I in ARIMA indicates that the data values have been replaced with the differences between consecutive observations (the difference between their values and their previous values). Thus, an ARIMA model is defined as ARIMA (p,d,q), where p is the order of the autoregressive model, d is the degree of differencing involved, and q is the order of the moving average model:

$$y'_t = c + \alpha_1 y'_{t-1} + \ldots + \alpha_p y'_{t-p} + \varepsilon_{t-1} + \beta_1 \varepsilon_{t-1} + \ldots + \beta_q \varepsilon_{t-q}$$

where y'_t is the differenced series, c is a constant, α_i are the parameters of the autoregressive model, β_i are the parameters of the moving average model, and ε_i are error terms. Thus, the purpose of an ARIMA model with these features is to make the model fit the data as well as possible.

In this section, the ARIMA models for each variable are generated automatically. Before applying this, it is guaranteed that each variable is non-stationary. In the previous section, this has been performed visually through the trend and seasonality figures, but to perform it systematically, we use the Dickey–Fuller test.

Starting with the target variable (Skyspark), the Dickey–Fuller test returns a p-value of 0.01. This value is below the threshold of statistical significance of 0.05, which indicates that the series is, in effect, non-stationary. Once the non-stationarity has been verified, the ARIMA model for Skyspark is obtained. The model obtained has a value of 5 as a parameter p and a value of 1 as a parameter q. Similarly, this process is repeated for the other PCs. However, it will not be applied to the six components selected, but only to those that are more correlated with Skyspark and have a high variability ratio/explained variability. These components are PC1 and PC2, as is shown in Figure 10 using the Pearson correlation, Table 2, and Figure 9.

	Skyspark	PC1	PC2	PC3	PC4	PC5	PC6
Skyspark	1.000000	8.474843e-01	4.226491e-02	-2.407451e-02	1.475352e-02	4.591142e-02	-1.780296e-01
PC1	0.847484	1.000000e+00	5.289389e-15	-5.612004e-16	-2.389070e-15	-5.023234e-15	9.405892e-16
PC2	0.042265	5.289389e-15	1.000000e+00	-5.431375e-16	-3.181588e-15	-4.501825e-15	2.084218e-15
PC3	-0.024075	-5.612004e-16	-5.431375e-16	1.000000e+00	-5.011040e-16	2.248216e-16	7.851842e-17
PC4	0.014754	-2.389070e-15	-3.181588e-15	-5.011040e-16	1.000000e+00	2.392261e-15	-7.541371e-16
PC5	0.045911	-5.023234e-15	-4.501825e-15	2.248216e-16	2.392261e-15	1.000000e+00	-2.460730e-15
PC6	-0.178030	9.405892e-16	2.084218e-15	7.851842e-17	-7.541371e-16	-2.460730e-15	1.000000e+00

Figure 10. Pearson's Correlation of the PCs.

The Dickey–Fuller test indicates that the time series in both components is stationary. After this, it is possible to obtain their respective ARIMA models. For PC1, the parameter p has a value of 5 and q of 0, while for PC2, the parameter p has a value of 5 and q has a value of 1. Table 4 shows the p and q values of the adjusted models for each variable.

Table 4. Parameters p and q of the ARIMA models.

Model	p	q
Skyspark	5	1
PC1	5	0
PC2	5	1

3.2. Generation and Evaluation of the Forecasting Models Using LSTM

For the modeling phase, it has been decided that LSTM neural networks would be used due to their advantages. In order to accomplish this, it is necessary to determine the time intervals to be used for the predictions. According to the feature engineering process results, it was decided that three groups of models be formed.

The first group consists of prediction models, in which only PC1 is used, since it is the variable most correlated with the target variable. The second group includes, in addition to PC1, PC2, because it is the second-most correlated variable with Skyspark. Finally, in the third group, the Skyspark estimation is carried out from the original variables after the feature engineering process, in order to determine if the extraction/fusion of characteristics has provided any advantage.

Given that, for the three variables, a value of 5 was obtained as the parameter p of the ARIMA models, this being greater than the value of the parameter q in the three cases, the three groups of models were tested with lags of t−5. The modeling process follows the following steps:

First, the values are normalized and the predictor variables are separated from the target variable. LSTM neural networks require that the input dataset be a three-dimensional array: number of observations, number of time intervals, and number of variables. In this way, for each observation, there is an array for each time interval to take into account, including the current instant and the established delays.

Once the dataset has been prepared according to the time window, it is necessary to divide the total observations into training data (70%) and test data (30%).

The predictive models are generated with the training data. The number of neurons and epochs are adjusted according to the results. As the loss function, the Least Squared Errors metric is used.

The quality of each model is determined using RMSE, MAPE, and R^2 metrics.

Before creating the three groups of models, it is decided that an initial prediction model be generated using the variables selected by the feature engineering process before reaching the dimension analysis process, which will serve as a starting point to adjust both the number of neurons and epochs in the following groups of models. In order to produce this first model, the established time window is related to the number of neurons and training epochs of the LSTM network. Therefore, the model is generated with a network composed of five neurons in the hidden layer and trained in five epochs. Figure 11 shows the loss function for the training and test datasets. It can be seen that the values for the test dataset do not reach the values of the training dataset, remaining higher at all points.

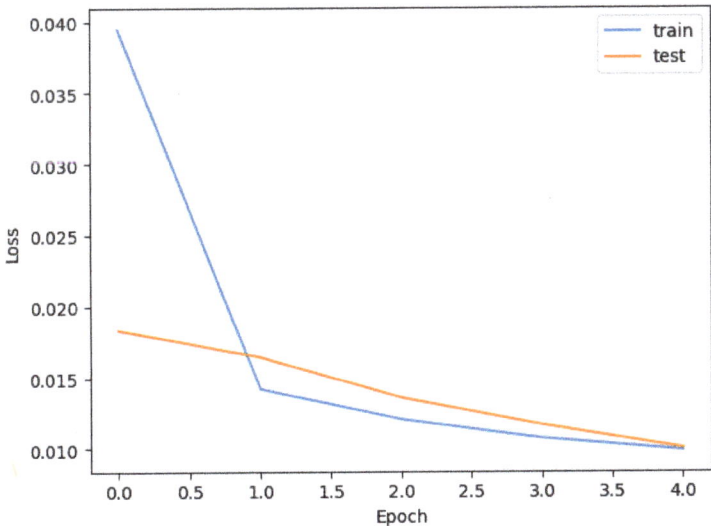

Figure 11. The loss function for 5 neuron models and 5 training epochs.

On the other hand, the prediction quality metrics are not optimal enough. Regarding the errors obtained, RMSE is 0.10, MAPE is 0.17, and R^2 is 0.40. Given these inaccurate results, it was decided that a greater number of neurons and training periods be tested in each of the groups, with the aim of obtaining better prediction models.

3.2.1. Group 1: PC1 and Skyspark

After a process of hyperparameterization of the LSTM for the number of neurons and training epochs, the best model was with 100 neurons and 100 training epochs. Figure 12 shows that the loss function for the test data reaches the loss values of the training data before epoch 20. Thereafter, it undergoes small fluctuations, but these fluctuations gradually disappear. The quality metrics are: RMSE is 0.07, MAPE is 0.10, and R^2 is 0.74.

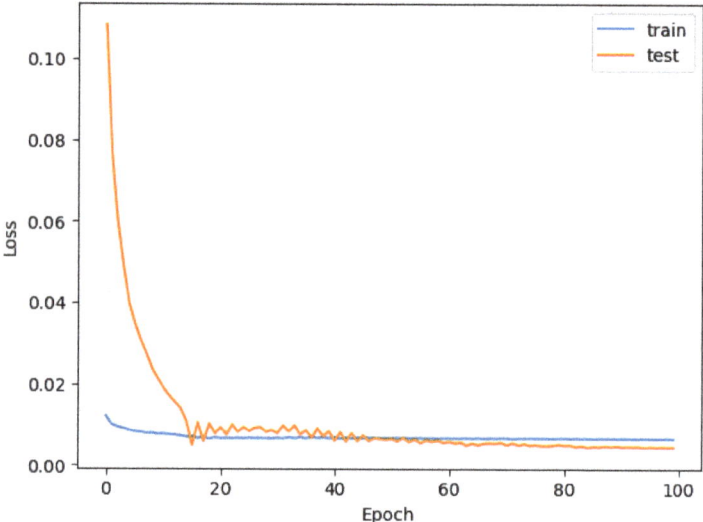

Figure 12. Best results of the loss functions for Group 1.

3.2.2. Model Group 2: PC1, PC2 and Skyspark

In a similar way to the first group, after an hyperparameterization process for the number of neurons and training periods, the best model was for 50 neurons and 100 training epochs. In this case, the loss function of the test data set reaches the values of the training data set shortly after epoch 20, as seen in Figure 13. Although small variations occur from that point, the function tends eventually to stabilize. Regarding the quality metrics, RMSE is 0.07, MAPE is 0.11, and R^2 is 0.72.

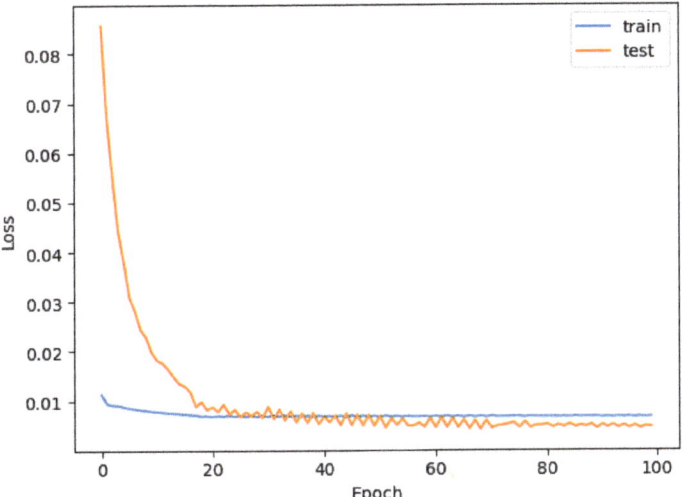

Figure 13. Best results of the loss functions for Group 2.

3.2.3. Model Group 3: Original Variables and Skyspark

Unlike the previous groups, these models are not generated with the PCs as input data. Instead, the input data set for the models in this group corresponds to the original variables before applying PCA, determined by our feature engineering process.

In this case, after the hyperparameterization process, the best LSTM model was with 50 neurons and 100 epochs. In this case, the loss function for the test data set appears stable, descending until reaching the loss values of the training data set around epoch 20 (see Figure 14). Regarding the prediction quality metrics, RMSE is 0.07, MAPE is 0.12, and R^2 is 0.74.

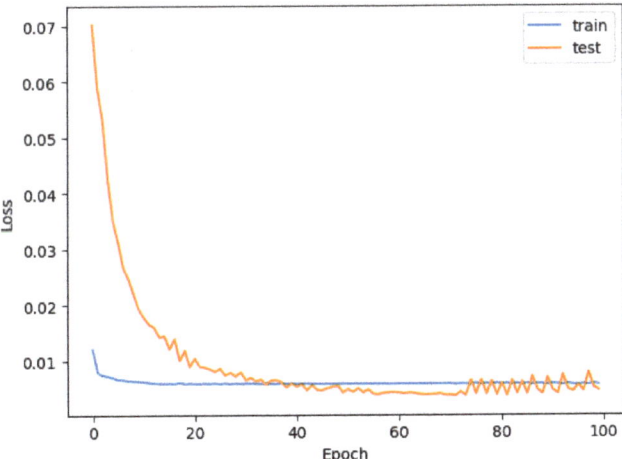

Figure 14. Loss functions for Group 3.

3.3. Comparison of the Forecasting Models of Each Group

According to [21,22], the relevant LSTM hyperparameters to be optimized are the number of neurons and epochs (number of times each training example is passed through the network). In the previous subsections, we have optimized them for each different set of input data. In this section, we compare the best LSTMs obtained for each input dataset.

Comparing the results of the groups, we can see they are very similar. In some cases, the errors or precision are better or worse. In addition, the optimal number of neurons and training epochs is very variable in each group. Starting from the best models obtained in each group, summarized in Table 5, we make a comparison between them. In the first group, better results were obtained with a model of 100 neurons and 100 training periods, while in the second group, the best results were obtained with a model of 50 neurons and 100 training periods.

According to the results, the inclusion of PC2 as a descriptive variable does not provide advantages to the generation of an optimal model. It can even be said that it negatively affects the learning process, since it causes the neural network to take into account a greater amount of input data, which does not provide relevant information, since PC2 does not have a significant correlation with the target variable (Skyspark). Therefore, from this first comparison, the model from the first group can be selected as the best option.

On the other hand, in the third group are obtained predictions as accurate as those of the first group. In particular, in the third group, the variables used are the ones selected by the initial feature engineering process, before reaching the dimensional analysis phase, which is where PCA is used (see Figure 2). Specifically, the results indicate that the phase of analysis of the correlations to determine the descriptive variables to use (the first phase of our feature engineering process, see Figure 2) is quite good, since similar results are obtained when using PC1. These variables are sufficiently correlated with Skyspark to provide relevant information in the learning process. In addition, including a greater number of variables requires more neurons and training cycles to achieve a model with a performance similar to the first group. However, increasing the number of neurons and epochs can lead to overfitting, causing less accurate predictions. This is reflected in the cases in which the loss function for the test data set presents values below those of the loss function for the training data set (see Figure 14) [33].

Thus, this third group model seems a viable option. However, if we review the loss function for this model, we see that it undergoes variations for the test data set, which means that it is not truly a stable model. That is, in certain cases, it can give good predictions, and in other cases, it cannot. From this, we can say that, although considering the initial variables as descriptors can give good results, it does not assure us that this happens in all cases. Once the best models of each group have been compared, the model of the first group is the most appropriate.

Table 5. Summary of best models of each group.

Group	No. of Neurons	No. of Epochs	RMSE	MAPE	R^2
1	100	100	0.07	0.10	0.74
2	50	100	0.07	0.11	0.72
3	50	100	0.07	0.12	0.74

In general, in the case of very few epochs, the network does not learn enough, producing underfitting, as in the case of the initial model of 5 neurons and 5 training epochs, where the values of the loss function for the test data set are very low [33]. On the contrary, if there are many epochs, the model begins to memorize and stops learning, producing overfitting, as observed in the cases in which the loss function for the test data set goes below that of the training data set [33].

Finally, some possible extensions to this study concern the use of other concepts in the feature engineering process [32,34,35], such as the evaluation of the temporal dependence relationship of the descriptor variables using regressors combined with the autoregression of the objective variable [34,35], the use of a descriptor variable selection algorithm that uses different criteria in real-time for said selection [36,37], and their effects on the behavior of predictive models. In addition, the use of techniques that allow the construction of explanatory predictive models for time series is important in the energy field, such as cognitive maps [38,39], and will require analysis of variables and parameters similar to the one proposed in this work.

4. Comparison of LSTM with Other Techniques

In order to show the feasibility of the feature engineering process proposed in this work, several datasets are used in this section. Specifically, for each dataset, a feature engineering process is carried out according to the steps shown in Section 3, which:

- Studies the temporal relationship between the variable to be predicted and the rest of the variables;
- Performs a feature reduction analysis using PCA.

With the results of the feature engineering process, prediction models are built using LSTM and other techniques in order to compare them (see Table 6). For each model used, as for the LSTM model, an hyperparameter optimization process is performed. For example, for the LSTM model and dataset [40], the best results were with 30 neurons and 25 epochs; for dataset [41], 40 neurons and 40 epochs; for dataset [42], 120 neurons and 60 epochs; and for dataset [43], 100 neurons and 40 epochs.

Table 6. Results of the predictive models in different datasets and techniques.

Dataset	Technique	RMSE	MAPE	R^2
[40]	Gradient boosting	0.0249	75.7002	0.9937
	Random forest	0.0244	75.4282	0.9928
	LSTM	**0.0101**	**75.3260**	0.9920
	L-BFGS	0.0220	76.7360	**0.9939**
	CNN	0.0229	75.7002	0.9936
[41]	Gradient boosting	0.0331	21.6070	0.9750
	Random forest	0.0351	21.5721	0.9600
	LSTM	**0.0310**	**19.4190**	**0.9710**
	L-BFGS	0.0320	19.4860	0.9570
	CNN	0.0663	21.6816	0.9019
[42]	Gradient boosting	**0.0395**	17.1182	0.9309
	Random forest	0.0457	17.4153	0.9336
	LSTM	0.0417	**15.0250**	**0.9497**
	L-BFGS	0.0487	15.1240	0.9393
	CNN	0.0608	17.0162	0.9401
[43]	Gradient boosting	0.0645	17.5819	**0.9352**
	Random forest	0.0914	18.0262	0.8930
	LSTM	**0.0625**	**17.3260**	0.9147
	L-BFGS	0.0910	22.7140	0.9126
	CNN	0.1318	19.8812	0.8966
[44]	Gradient boosting	0.1415	21.7193	0.6254
	Random forest	0.1177	21.9066	0.6694
	LSTM	0.1351	21.0200	0.7300
	L-BFGS	0.1313	21.0180	0.7430
	CNN	**0.0973**	**21.0040**	**0.7671**

According to the results, we see that our approach to pre-toning the LSTM with our feature engineering process to define the backward sequence of the technique makes it a very robust method. In particular, in the different datasets, it obtains the best result (see colors in bold), or it is always among the best. Of the other techniques evaluated, some of their metrics are never among the best (for example, random forest), or they are good in some cases and not in others (for example, CNN).

Thus, we can see that our feature engineering process to establish the temporal relationships of the time series that describe energy consumption is necessary. In addition, this indicates the need for such a process, and for techniques such as LSTM, in energy consumption prediction tasks. Particularly, this process gives a lot of robustness to the LSTM technique, regardless of the energy consumption dataset (time series).

5. Conclusions

This research presents an analysis of the energy consumption forecasting problem. The paper carries out an analysis proposing a feature engineering process to obtain the variables to be used in the predictive models. The results of this process are used to build different energy consumption prediction models using the LSTM technique. In this context, different groups are defined in order to build the predictive models and to experiment with them.

The main contribution is the feature engineering methodological approach created to analyze the energy consumption variables of buildings. This approach defines several phases to study the time series that define energy consumption in buildings. Particularly, it proposes an analysis phase of the dependency relations between the variables (correlations), as well as the temporal ones (regressors). In addition, it proposes an analysis phase of the dimensions in the dataset to fuse/extract characteristics. Thus, in the feature engineering process, thanks to the correlation coefficients and linear regression models, we analyze the relationships between the variables, and with these results, we perform the feature extraction/fusion by applying PCA. Another aspect to be considered is the temporal relationship of the variables with themselves. For this, we rely on the models of autocorrelation and ARIMA, thanks to which we obtain the optimal time window to make the predictions.

Finally, we have compared LSTM with other machine learning techniques, using our feature engineering process to analyze the time series of various energy consumption datasets (temporal series). Based on the results, we see that this feature engineering process helps LSTM to obtain an excellent fit of the time sequences to be considered, in order to build a predictive model. This quality is the best in the group of datasets tested, since their metrics are the best, or are consistently among the best.

The next steps in our work are: (i) to analyze the impact of other feature engineering processes that can be used with time series; (ii) to define an adaptation mechanism of the predictive model in real-time; (iii) to analyze the effect of inclusion of new variables like climatological variables in our dataset. Climatic factors affect energy consumption since, depending on these values, devices such as heating or air conditioning are used to a greater or lesser extent. In addition, these variables have a marked temporality.

Author Contributions: Conceptualization, J.A.; Data curation, D.D.; Formal analysis, D.D., J.A. and M.D.R.-M.; Funding acquisition, J.A. and M.D.R.-M.; Investigation, D.D., J.A. and M.D.R.-M.; Methodology, J.A.; Writing—original draft, D.D.; Writing—review & editing, J.A. and M.D.R.-M. All authors have read and agreed to the published version of the manuscript.

Funding: J. Aguilar is supported by the European Union's Horizon 2020 research and innovation program under the Marie Skłodowska-Curie grant agreement No 754382 GOT ENERGY TALENT. M.D. R-Moreno is supported by the JCLM project SBPLY/19/180501/000024 and the Spanish Ministry of Science and Innovation project PID2019−109891RB-I00, both under the European Regional Development Fund (FEDER).

Institutional Review Board Statement: Not applicable.

Informed Consent Statement: Not applicable.

Data Availability Statement: The data is open.

Acknowledgments: This project has received funding from the European Union's Horizon 2020 research and innovation program under the Marie Skłodowska-Curie grant agreement No 754382 GOT ENERGY TALENT.

Conflicts of Interest: The authors declare no conflict of interest. The content of this publication does not reflect the official opinion of the European Union. Responsibility for the information and views expressed herein lies entirely with the authors. The funders had no role in the design of the study; in the collection, analyses, or interpretation of data; in the writing of the manuscript, or in the decision to publish the results.

References

1. Aguilar, J.; Garces-Jimenez, A.; R-Moreno, M.D.; García, R. A systematic literature review on the use of artificial intelligence in energy self-management in smart buildings. *Renew. Sustain. Energy Rev.* **2021**, *151*, 111530. [CrossRef]
2. Minoli, D.; Sohraby, K.; Occhiogrosso, B. IoT considerations, requirements, and architectures for smart buildings—Energy optimization and next-generation building management systems. *IEEE Internet Things J.* **2017**, *4*, 269–283. [CrossRef]
3. Rodriguez-Mier, P.; Mucientes, M.; Bugarín, A. Feature Selection and Evolutionary Rule Learning for Big Data in Smart Building Energy Management. *Cogn. Comput.* **2019**, *11*, 418–433. [CrossRef]
4. García, D.; Goméz, M.; Noguera, F.V. A Comparative Study of Time Series Forecasting Methods for Short Term Electric Energy Consumption Prediction in Smart Buildings. *Energies* **2019**, *12*, 1934.
5. Alduailij, M.; Petri, I.; Rana, O.; Alduailij, M.; Aldawood, A. Forecasting peak energy demand for smart buildings. *J. Supercomput.* **2021**, *77*, 6356–6380. [CrossRef]
6. Hernández, M.; Hernández-Callejo, L.; Solís, M.; Zorita-Lamadrid, A.; Duque-Perez, O.; Gonzalez-Morales, L.; Santos-García, F. Data-Driven Forecasting Strategy to Predict Continuous Hourly Energy Demand in Smart Buildings. *Appl. Sci.* **2021**, *11*, 7886. [CrossRef]
7. Hernández, M.; Hernández-Callejo, L.; García, F.; Duque-Perez, O.; Zorita-Lamadrid, A. A Review of Energy Consumption Forecasting in Smart Buildings: Methods, Input Variables, Forecasting Horizon and Metrics. *Appl. Sci.* **2020**, *10*, 8323. [CrossRef]
8. Moreno, M.; Dufour, L.; Skarmeta, A. Big data: The key to energy efficiency in smart buildings. *Soft Comput.* **2016**, *20*, 1749–1762. [CrossRef]
9. Nabavi, S.; Motlagh, N.; Zaidan, M.; Aslani, A.; Zakeri, B. Deep Learning in Energy Modeling: Application in Smart Buildings with Distributed Energy Generation. *IEEE Access* **2021**, *9*, 125439–125461. [CrossRef]
10. Somu, N.; Raman, G.; Ramamritham, K. A hybrid model for building energy consumption forecasting using long short term memory networks. *Appl. Energy* **2020**, *261*, 114131. [CrossRef]
11. Le, T.; Vo, M.; Kieu, T.; Hwang, E.; Rho, S.; Baik, W. Multiple Electric Energy Consumption Forecasting Using a Cluster-Based Strategy for Transfer Learning in Smart Building. *Sensors* **2020**, *20*, 2668. [CrossRef] [PubMed]
12. Hadri, S.; Naitmalek, Y.; Najib, M.; Bakhouya, M.; Fakhri, Y.; Elaroussi, M. A Comparative Study of Predictive Approaches for Load Forecasting in Smart Buildings. *Procedia Comput. Sci.* **2019**, *160*, 173–180. [CrossRef]
13. González-Vidal, A.; Jiménez, F.; Gómez-Skarmeta, A.F. A methodology for energy multivariate time series forecasting in smart buildings based on feature selection. *Energy Build.* **2019**, *196*, 71–82. [CrossRef]
14. González-Vidal, A.; Ramallo-González, A.; Terroso-Sáenz, F.; Skarmeta, A. Data driven modeling for energy consumption prediction in smart buildings. In Proceedings of the IEEE International Conference on Big Data, Boston, MA, USA, 11–14 December 2017; pp. 4562–4569.
15. Sülo, S.; Keskin, G.; Dogan, T.; Brown, T. Energy Efficient Smart Buildings: LSTM Neural Networks for Time Series Prediction. In Proceedings of the International Conference on Deep Learning and Machine Learning in Emerging Applications, Istanbul, Turkey, 26–28 August 2019; pp. 18–22.
16. Aliberti, A.; Bottaccioli, L.; Macii, E.; di Cataldo, S.; Acquaviva, A.; Patti, E. A Non-Linear Autoregressive Model for Indoor Air-Temperature Predictions in Smart Buildings. *Electronics* **2019**, *8*, 979. [CrossRef]
17. Alawadi, S.; Mera, D.; Fernández-Delgado, M. A comparison of Machine Learning algorithms for forecasting indoor temperature in smart buildings. *Energy Syst.* **2020**, *13*, 689–705. [CrossRef]
18. Siddiqui, A.; Sibal, A. Energy Disaggregation in Smart Home Appliances: A Deep Learning Approach. *Energy*, **2021**, *in press*.
19. Bhatt, D.; Hariharasudan, A.; Lis, M.; Grabowska, M. Forecasting of Energy Demands for Smart Home Applications. *Energies* **2021**, *14*, 1045. [CrossRef]
20. Escobar, L.M.; Aguilar, J.; Garcés-Jiménez, A.; de Mesa, J.A.G.; Gomez-Pulido, J.M. Advanced Fuzzy-Logic-Based Context-Driven Control for HVAC Management Systems in Buildings. *IEEE Access* **2020**, *8*, 16111–16126. [CrossRef]
21. Bourhnane, S.; Abid, M.; Lghoul, R.; Zine-Dine, K.; Elkamoun, N.; Benhaddou, D. Machine Learning for energy consumption prediction and scheduling in smart buildings. *SN Appl. Sci.* **2020**, *2*, 297. [CrossRef]

22. Hadri, S.; Najib, M.; Bakhouya, M.; Fakhri, Y.; el Arroussi, M. Performance Evaluation of Forecasting Strategies for Electricity Consumption in Buildings. *Energies* **2021**, *14*, 5831. [CrossRef]
23. Khan, A.-N.; Iqbal, N.; Rizwan, A.; Ahmad, R.; Kim, D. An Ensemble Energy Consumption Forecasting Model Based on Spatial-Temporal Clustering Analysis in Residential Buildings. *Energies* **2021**, *14*, 3020. [CrossRef]
24. Keytingan, M.; Shapi, M.; Ramli, N.; Awalin, L. Energy consumption prediction by using machine learning for smart building: Case study in Malaysia. *Dev. Built Environ.* **2021**, *5*, 100037.
25. Son, N.; Yang, S.; Na, J. Deep Neural Network and Long Short-Term Memory for Electric Power Load Forecasting. *Appl. Sci.* **2020**, *10*, 6489. [CrossRef]
26. Moon, J.; Park, S.; Rho, S.; Hwang, E. Robust building energy consumption forecasting using an online learning approach with R ranger. *J. Build. Eng.* **2022**, *47*, 103851. [CrossRef]
27. Pinto, T.; Praça, I.; Vale, Z.; Silva, J. Ensemble learning for electricity consumption forecasting in office buildings. *Neurocomputing* **2021**, *423*, 747–755. [CrossRef]
28. Somu, N.; Raman, G.; Ramamritham, K. A deep learning framework for building energy consumption forecast. *Renew. Sustain. Energy Rev.* **2021**, *137*, 110591. [CrossRef]
29. Amidi, S. Recurrent Neural Networks Cheatsheet. Available online: https://stanford.edu/~{}shervine/teaching/cs-230/cheatsheet-recurrent-neural-networks (accessed on 15 April 2021).
30. Yuan, X.; Li, L.; Wang, Y. Nonlinear Dynamic Soft Sensor Modeling with Supervised Long Short-Term Memory Network. *IEEE Trans. Ind. Inform.* **2021**, *16*, 3168–3176.
31. Doherty, K. Trenbath, Raw_Data, CO, USA: Mendeley Data. 2019. Available online: https://data.mendeley.com/datasets/g392vt7db9/1 (accessed on 15 April 2021).
32. Aguilar, J.; Terán, O. Modelo del proceso de Influencia de los Medios de Comunicación Social en la Opinión Pública. *Educere* **2018**, *22*, 179–191.
33. Brownlee, J. Multivariate Time Series Forecasting with LSTMs in Keras. Available online: https://machinelearningmastery.com/multivariate-time-series-forecasting-lstms-keras/ (accessed on 15 April 2021).
34. Quintero, Y.; Ardila, D.; Camargo, E.; Rivas, F.; Aguilar, J. Machine learning models for the prediction of the SEIRD variables for the COVID-19 pandemic based on a deep dependence analysis of variables. *Comput. Biol. Med.* **2021**, *134*, 104500. [CrossRef]
35. Cavaleiro, J.; Neves, M.; Hewlins, M.; Jackson, A. The photo-oxidation of *meso*-tetraphenylporphyrins. *J. Chem. Soc.* **1990**, *7*, 1937–1943, Perkin Transactions 1. [CrossRef]
36. Jiménez, M.; Aguilar, J.; Monsalve-Pulido, J.; Montoya, E. An automatic approach of audio feature engineering for the extraction, analysis and selection of descriptors. *Int. J. Multimed. Info. Retr.* **2021**, *10*, 33–42. [CrossRef]
37. Aguilar, J.; Salazar, C.; Velasco, H.; Monsalve-Pulido, J.; Montoya, E. Comparison and Evaluation of Different Methods for the Feature Extraction from Educational Contents. *Computation* **2020**, *8*, 30. [CrossRef]
38. Aguilar, J. A Fuzzy Cognitive Map Based on the Random Neural Model. *Lect. Notes Comput. Sci.* **2001**, *2070*, 333–338.
39. Sánchez, H.; Aguilar, J.; Terán, O.; de Mesa, J.G. Modeling the process of shaping the public opinion through Multilevel Fuzzy Cognitive Maps. *Appl. Soft Comput.* **2019**, *85*, 105756. [CrossRef]
40. Papaioannou, T.; Stamoulis, G. Teaming and competition for demand-side management in office buildings. In Proceedings of the IEEE International Conference on Smart Grid Communications (SmartGridComm), Dresden, Germany, 23–26 October 2017; pp. 332–337.
41. Power Consumption Data of a Hotel Building. Available online: https://ieee-dataport.org/documents/power-consumption-data-hotel-building (accessed on 15 April 2021).
42. Zhang, L.; Wen, J. A systematic feature selection procedure for short-term data-driven building energy forecasting model development. *Energy Build.* **2019**, *183*, 428–442. [CrossRef]
43. Pipattanasomporn, M.; Chitalia, G.; Songsiri, J.; Aswakul, C.; Pora, W.; Suwankawin, S.; Audomvongseree, K.; Hooncharoen, N. CU-BEMS, smart building electricity consumption and indoor environmental sensor datasets. *Sci. Data* **2020**, *7*, 241. [CrossRef]
44. Long-Term Energy. Consumption & Outdoor Air Temperature for 11 Commercial Buildings. Available online: https://trynthink.github.io/buildingsdatasets/show.html?title_id=long-term-energy-consumption-outdoor-air-temperature-for-11-commercial-buildings (accessed on 15 April 2021).

Article

Key Enablers of Resilient and Sustainable Construction Supply Chains: A Systems Thinking Approach

Maria Ghufran [1], Khurram Iqbal Ahmad Khan [1,*], Fahim Ullah [2], Wesam Salah Alaloul [3,*] and Muhammad Ali Musarat [3]

1. Department of Construction Engineering and Management, National University of Sciences and Technology (N.U.S.T.), Islamabad 44000, Pakistan
2. School of Surveying and Built Environment, University of Southern Queensland, Springfield, QLD 4300, Australia
3. Department of Civil and Environmental Engineering, Universiti Teknologi PETRONAS, Bandar Seri Iskandar 32610, Malaysia
* Correspondence: khurramiqbal@nit.nust.edu.pk (K.I.A.K.); wesam.alaloul@utp.edu.my (W.S.A.)

Abstract: In the globalized world, one significant challenge for organizations is minimizing risk by building resilient supply chains (SCs). This is important to achieve a competitive advantage in an unpredictable and ever-changing environment. However, the key enablers of such resilient and sustainable supply chain management are less explored in construction projects. Therefore, the present research aims to determine the causality among the crucial drivers of resilient and sustainable supply chain management (RSSCM) in construction projects. Based on the literature review, 12 enablers of RSSCM were shortlisted. Using the systems thinking (ST) approach, this article portrays the interrelation between the 12 shortlisted resilience enablers crucial for sustainability in construction projects. The causality and interrelationships among identified enablers in the developed causal loop diagram (CLD) show their dynamic interactions and impacts within the RSSCM system. Based on the results of this study, agility, information sharing, strategic risk planning, corporate social responsibility, and visibility are the key enablers for the RSSCM. The findings of this research will enable the construction managers to compare different SCs while understanding how supply chain characteristics increase or decrease the durability and ultimately affect the exposure to risk in the construction SCs.

Keywords: causal loop diagram; construction management; resilient supply chain; sustainable supply chain; supply chain management; systems thinking

Citation: Ghufran, M.; Khan, K.I.A.; Ullah, F.; Alaloul, W.S.; Musarat, M.A. Key Enablers of Resilient and Sustainable Construction Supply Chains: A Systems Thinking Approach. *Sustainability* 2022, 14, 11815. https://doi.org/10.3390/su141911815

Academic Editors: Vincenzo Torretta and Roberto Cerchione

Received: 10 May 2022
Accepted: 28 August 2022
Published: 20 September 2022

Publisher's Note: MDPI stays neutral with regard to jurisdictional claims in published maps and institutional affiliations.

Copyright: © 2022 by the authors. Licensee MDPI, Basel, Switzerland. This article is an open access article distributed under the terms and conditions of the Creative Commons Attribution (CC BY) license (https://creativecommons.org/licenses/by/4.0/).

1. Introduction

A supply chain (SC) consists of a network of organizations involved in different processes and activities for delivering services to users. An SC produces value through upstream and downstream linkages in products and services delivered to the end-user [1]. Thus, an SC consists of several entities: upstream (supply), downstream (distribution), and the final consumer [2]. In line with the global sustainability drive, academic researchers have recently focused on designing sustainable SC (SSC) networks. Such SSCs can potentially impact the efficiency of the global SCs [3,4]. A balance between economic, social, and environmental factors has become increasingly crucial for SSCs as consumers demand sustainable products [4–7]. However, as world businesses have become intensely competitive and unpredictable, sustainability in the SC is often threatened [8]. Unforeseen circumstances frequently disrupt businesses and their SC, questioning the continuity of the SC [9,10]. Sustainability is hard to achieve when there are persistent SC disruptions. Therefore, to achieve reliable SSCs, the resilience capabilities of the organizations must be developed and improved. Thus, it is essential to investigate whether the SCs need resilience to be sustainable [11,12].

While the terms SC and sustainability have been explored by various researchers, resilient and sustainable supply chain management (RSSCM) has not been explored holistically. Resilience in supply chains is the ability to anticipate and withstand disruptions, respond to them, and effectively recover from disruptions [13]. RSSCM is defined as the management of resources toward satisfying stakeholder expectations to create high resilience and sustainability in an organization's supply chain [14]. The literature on sustainable supply chain management (SSCM) and SC resilience highlight that no systematic study has been performed to date that incorporates SC resilience and sustainability, particularly in developing countries [15]. This is in line with the general dearth of research in such countries [16,17].

Nevertheless, among the relevant studies, Pettit et al. [18] mentioned that SC resilience is a prerequisite for SC sustainability that increases system complexity. Chowdhury et al. [19] emphasized the development of the systems thinking (ST) approach to address the increasing complexity. ST is the ability to see the world as a dynamic system; everything is related to everything else, and an individual item may not be achieved in isolation [20–22]. Accordingly, RSSCM cannot be achieved independently, and a holistic assessment of the system is needed. This presents a gap in the existing literature that is targeted in the current study. The elementary idea of this research is to demonstrate the relationship between SC resilience and SC sustainability through the causal loop diagram (CLD). The developed CLD considers the RSSCM a holistic system and comprises its key enablers and linkages. Based on the above, this paper has the following objectives:

To identify the key resilience enablers for sustainable SCs.

To determine the causality among the identified key resilience enablers for sustainable SCs.

To achieve these objectives, this study uses the ST approach, a holistic method focused on the interrelationship of the constituent parts of a system and addressing the inherent complexity. ST is a conceptual problem-solving methodology that considers issues in their entirety (at the systems level). The findings of this study will help achieve a competitive advantage in an unpredictable construction environment where change is imperative. Moreover, this will lower the organizational risk by enabling real-time insights into all operations across the SC networks.

It is expected that the construction organizations would be empowered to optimize and adjust their processes and logistics and move towards an RSSCM. Further, the results of this study will help make SCs more resilient and sustainable, resulting in lower costs, enhanced manufacturing efficiency and flexibility, and consequently higher profits for construction organizations. The associated RSSCM can handle disruptive events, respond quickly, and resume normal operations after the disruption. This study is a novel attempt to determine the causality among the identified enablers of resilience in SCs using the ST approach. It utilizes Vensim® for developing the CLDs of RSSCM in developing economies.

The paper is organized as follows. Firstly, the background and introduction are presented in Section 1. Secondly, in the Section 2, SSCM and resilient SC management are presented, followed by RSSCM and ST approaches. In this step, the key enablers of resilience in developing countries' sustainable construction SCs are identified. Thirdly, the Section 3 is described, articulating the data collection and data analysis process. In the Section 4, findings and outcomes are deliberated, and a CLD is developed. Finally, the paper is concluded, and limitations, recommendations, and directions for further research are presented.

2. Literature Review

2.1. Sustainable Supply Chain Management (SSCM)

An SC is a network that connects all the people, organizations, resources, and activities involved in producing and distributing a product [2]. It encompasses everything from delivering source materials from the supplier to the manufacturer and eventual delivery to the end-user [1]. It is the process of managing how goods and services evolve from concept to finished product [3]. Modern SCs are complicated systems where various players work together in distinct steps to deliver various products to customers [23]. In order to decrease the uncertainties and disruption risks and increase the SCs' resilience and flexibility, independent businesses must collaborate [24]. SCM encompasses all aspects of an organization's operation integrated into one system [4].

SSCM includes all three pillars of sustainability, i.e., environmental, social, and financial, throughout the production lifecycle. The lifecycle includes product design and production to material sourcing, processing, packaging, shipping, warehousing, distribution, consumption, return, and disposal [25,26]. The SSCM effectively and efficiently manages interrelated environmental, social, and economic aspects in the global supply chains [27]. In sustainable SCs, the participants must meet environmental, economic, and social requirements [28]. The assumption is that competition would be preserved by fulfilling consumer demands and associated economic criteria.

SSCM has gained significant recognition with a surge in scholarly publications over the past few years. Such sustainable SCs lead to Value Management (VM) [29]. Value engineering (VE) or VM is a systematic process to increase the value of a product. It is a strategy that examines and optimizes the function of each item and its associated cost to increase the value of the project or product [30]. When it comes to construction projects, VE can be very beneficial. Using VE early in the project can save time and money in the long run, resulting in a higher return on investment and more cost savings. VE encourages substituting less expensive materials and technologies without affecting the product's functionality. VE helps improve the performance of construction SCs by cutting costs through supply chain integration while maintaining a high quality of service, thus making them more sustainable [31].

In the SCM, the social aspect of sustainability has been less addressed than the environmental and economic dimensions [32]. SC sustainability aims to include environmental, economic, and social efforts into traditional, cost-oriented SCM strategies [3,24]. A sustainable SC is described as an interaction among organizations in an SC that provides holistic environmental and social benefits to all SC partners [33,34]. It encompasses businesses' attempts to address the environmental and human impact of their products' path throughout the SCs, i.e., from raw material sourcing to production, storage, and delivery [32,35].

2.2. Resilient Supply Chain Management

Resilience is the supply chain's adaptive capability to plan for unanticipated events and respond to and recover from disruptions by maintaining operational stability at the optimal level of connectivity and control over the structure and function [36,37]. Resilience, in simpler terms, is the ability to recover from adversity [37]. A resilient SC can withstand or avoid the consequences of an SC disruption and recover from one quickly. Resilience is at the core of current thinking regarding SCM [38,39].

A resilient SC can resist or prevent the consequences of an SC disruption and recover from it in an economical and timely manner [40]. Resilience has always been a key factor in ensuring organizational success. Supply chain resilience no longer refers solely to risk management [37]. It is now recognized that managing risk encompasses being better positioned than competitors to deal with disruptions in the SCs. Further resilient SCs provide an advantage to organizations through competitive gains [39].

It is necessary to consider the measurement of resilience to build a resilient system. The level of resilience needed by the system is context-dependent [40]. SC resilience is impacted by the antecedents of capability, vulnerability, SC orientation, and SC design [41,42]. SC disruptions are unexpected events interrupting the usual operation and flow between the SC players: products, components, and materials [43]. Disruptions in SCs are characterized by a high degree of uncertainty that may occur from several sources, such as physical hazards, personal events, information disruptions, environmental disasters, acts of terrorism, and political upheaval [44].

Organizations are more likely to experience a wide range of unforeseen vulnerabilities, producing minor to large disruptions throughout their SCs [45]. Accordingly, these organizations must recognize and focus on their inherent component of the SC, while policymakers should reevaluate methods for making global SCs more resilient [46]. For example, digital technologies have disrupted the construction industry and associated fields [47,48]. Accordingly, construction managers have been focused on creating more resilient SCs to mitigate the effects of disruptions [45]. A resilient SC can tackle the adverse effects of disturbances and substantially reduce the recovery period necessary for construction organizations to return to normal operation [46].

2.3. Resilient Sustainable Supply Chain Management (RSSCM)

RSSCM is the management of resources to meet the needs of stakeholders to attain high resilience and sustainability in the SC [49,50]. Risk management is a key feature of RSSCM. According to Kamalahmadi and Parast [46], SC resilience is a core element of SC management that helps quicker recovery from disruptions. Various methodologies are used to achieve RSSCM. These include Transaction Cost Analysis, Network Perspective, Total Quality Management, and the ST approach [51,52].

At the strategic network design stage, there are linkages between SC resilience and sustainability performance [53]. Fahimnia and Jabbarzadeh [54] elucidate how variations in the resilience level affect the economic, environmental, and social sustainability of an SC. Similarly, the simulation-based model suggested by Ivanov [55] shows how sustainability factors can be linked to SC resilience in multiple ways. Jabbarzadeh et al. [53] considered a situation in which the aims of sustainability and resilience are in contradiction. Nevertheless, facility protection must simultaneously promote sustainability and resilience [56].

Based on the key concepts of SCM and RSSCM in construction projects, the current paper sheds light on the key enablers of resilience in SSCM. The focus is on RSSCM in the construction industries of developing countries.

3. Research Methodology

Research methodology defines how research is to be carried out to achieve its objectives [57]. Accordingly, this research has been divided into three stages to achieve the predefined objectives, as presented in Figure 1, below. These stages are subsequently explained.

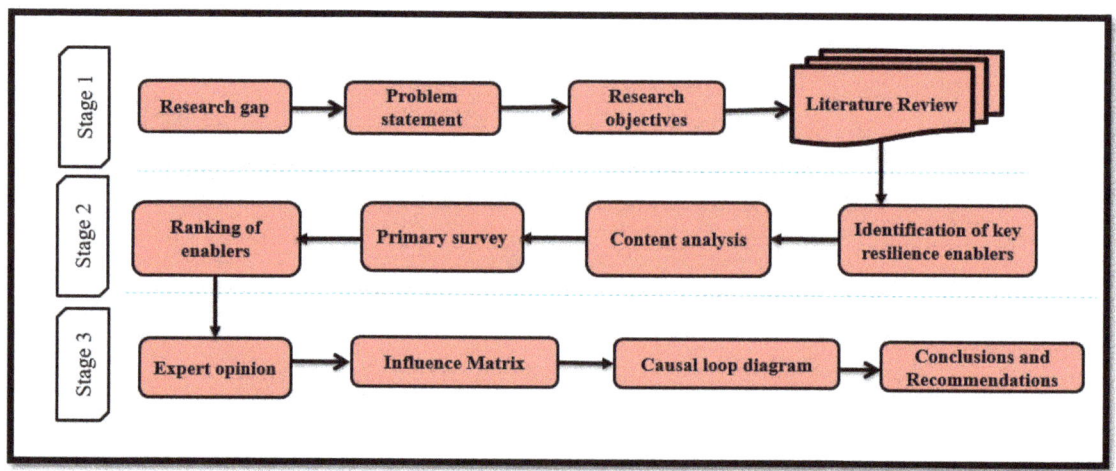

Figure 1. Methodology chart.

3.1. Stage 1: Initial Study

The first stage of the method of the current study comprises the initial study. The initial study was conducted to identify the research gap, draft the problem statement, and formulate the research objectives of the current study. Then, a detailed literature review was conducted to identify the key resilience enablers for sustainable SCs. Following recent studies, the four major databases selected for paper collection include Science Direct, Scopus, Web of Science, and IEEE Xplore [58,59]. The inclusion and exclusion criteria of the referred study were adopted to ensure that the literature review was exhaustive and comprehensive.

3.2. Stage 2: Factors Shortlisting

The second stage of the current study method deals with the shortlisting of key RSSCM factors. Due to the literature review, key resilience enablers for sustainable SCs were identified. A total of 55 articles were scrutinized using the keywords "enablers of resilient construction supply chain" and "enablers of sustainable construction supply chains". The keywords were joined using boolean operators "AND" or "OR", resulting in a total of 26 relevant articles. Initially, 32 enablers were identified from 26 papers published in the last decade that focused on SCs in developing countries. The identified enablers include top management support, adaptability, agility, transparency, leadership, tenacity, resource efficiency, and others, as shown in Table 1.

A quantitative number was assigned to each enabler according to its influence (high as 5, medium as 3, and low as 1) following Rasul et al. [60]. This led to calculating the literature score (LS) using Equation 1, where W is the product of frequency (repetition of enablers in papers) and assigned impact score (5,3,1) following the referred study. A is the highest possible score, and N is the total number of papers considered for enabler identification [60]. The scores are normalized to have a uniform scale.

Normalization is the process of converting values measured on various scales to a theoretically common scale (out of 1). It is a data-shifting and rescaling technique in which data points are shifted and rescaled till they are in the 0 to 1 range. Normalization is required to ensure that the data directly related to the database is considered. Further, each data field contains only one data element, which removes redundant (unnecessary) data. The normalized literature score (NLS) was computed by dividing each enabler's LS by the sum of the literature scores, as shown in Equation 2. The identified 32 enablers from the literature, along with references and the respective NLS, are shown in Table 1.

$$\text{RII} = (\sum W)/(A \times N) \quad (1)$$

$$\text{NLS} = (\text{LS})/(\sum \text{LS}) \quad (2)$$

A primary survey was conducted to calculate the field scores with a response rate of 106. The final ranking of enablers was based on the combined field and literature data score, with a weightage of 60/40 (60 percent of the respondent's normalized score and 40% of the literature's normalized score). Factors having a 50% impact score were then shortlisted [16,61].

Statistical tools are used to check the reliability of the data. The IBM® SPSS® Statistics software platform is a robust statistical software platform. This software is one of the most widely used statistical packages, capable of handling and analyzing large amounts of data [4,60]. Accordingly, it was used to check the normality and reliability of the data in the current study by applying basic statistical tests (Cronbach's alpha). The threshold value for Cronbach's alpha is 0.7. Any value of the data above 0.7 shows its reliability. A Cronbach alpha value of 0.92 was obtained in this study, showing that the data are highly reliable for further analysis [4]. Table 1 represents the collective NLS and ranks of the 32 enablers. Moreover, the classification categories of the papers are also elaborated in Table 1, where "S" represents the classification of the factors into the category of "sustainability" and "R" represents the "resilience" category.

Table 1. Enablers identification via the literature review.

Sr.#	Enablers	References	Category	NLS	Sr.#	Enablers	References	Category	NLS
1	Top Management Support	[62–66]	S	0.035	17	Information Security	[43,62,63,65,67,68]	S	0.025
2	Adaptability	[11,18,62,67,69,70]	R, S	0.042	18	Strategic Risk Planning	[18,43,62,64,67,68,71,72]	S	0.034
3	Visibility	[11,18,43,62,66,67,69–75]	R, S	0.106	19	Corporate Social Responsibility	[43,67,72]	S	0.035
4	Health	[18,62]	S	0.014	20	Contingency Planning	[43,62,65,66,71,76]	S, R	0.030
5	Compatibility	[49,62–66,69]	S	0.056	21	Safety Stock	[62,69,73,76]	S, R	0.035
6	Quality Awareness	[77]	R	0.007	22	Flexible Transportation	[43,49,62,68,69,71,73]	S, R	0.030
7	Responsiveness	[76,77]	R	0.014	23	Resource Efficiency	[49,63,68,72]	S	0.013
8	Technological Capability	[70,77]	S, R	0.021	24	Transparency	[49,65,68,72]	S	0.014
9	Agility	[63,66,67,69–72,77,78]	S	0.092	25	Self-Regulation	[75,78]	S	0.013
10	Supply Chain Security	[64,69]	S	0.021	26	Market Sensitivity	[66,68,70,71]	R, S	0.007
11	Collaboration	[18,43,49,62,65–67,69–71,79]	R, S	0.092	27	Tenacity	[73]	R	0.021
12	Swift Trust	[11,64,66,67,70,71]	R, S	0.030	28	Leadership	[66,71,79]	S	0.035
13	Risk and Revenue Sharing	[67,70]	R, S	0.014	29	Just in Time	[73,76,77]	R, S	0.001
14	Information Sharing	[11,43,64,65,67,70,71,73,79]	R, S	0.085	30	Proper Scheduling	[64]	S	0.007
15	Flexible Structure	[62,63,67,69–71,73,76]	R, S	0.034	31	Composure	[71]	S	0.003
16	Risk Management Culture	[67,70]	R, S	0.014	32	Reasoning	[62,78]	S	

3.3. Stage 3: Systems Thinking Approach

The third stage of the current study method deals with the ST approach. ST is a cognitive endeavor that is more systematic, abstract, and planned [80]. Although the hierarchical thinking process is complex, not all processes and cognition are always complicated [20]. ST is a conceptual problem-solving methodology that considers issues in their entirety rather than dealing with them individually [81]. A CLD is used to ascertain the relationship between variables and balance and reinforce feedback loops in a complex environment [82]. Polarities are assigned to the loops to show their reinforcing or balancing impact. Polarities within links merely anticipate what will happen if something changes and do not demonstrate how variables behave [21,83]. The polarity of a variable is determined by tracing its effects as they propagate around the loop [80].

The ST approach focuses on how the integral parts of a system interact and operate over time in complex networks. Accordingly, it has been used in this research to deal with the complexities of the RSSCM. The ST approach would make it easier for SC managers to overcome disruptions and vulnerabilities and build an RSSCM. The global business has become increasingly volatile, and uncertainties frequently interrupt the functions of the SC. Accordingly, SC managers can use ST to get ideas about disruption mitigation [21]. Furthermore, managers can know the association amongst different variables in the CLD, how the variables are linked, and the antecedents via the ST approach.

In this stage of the study, expert opinion was acquired to determine the polarity and interrelationships among the enablers, which resulted in the development of an influence matrix. Stella Professional, AnyLogic, Vensim ®PLE, and iThink are some of the software packages used to design CLDs and associated system dynamics models. This research utilized Vensim® PLE for CLD development based upon the shortlisted enabler's interrelationships. This is because Vensim® is the most powerful package in terms of computing speed, capabilities, and flexibility [58]. In terms of capacity, performance, and functionality, Vensim® PLE is unrivaled. The optimization possibilities are powerful and the simulation speed is rapid. Thus, it has been used to develop the CLDs that represent the causality among the key enablers of RSSCM. The CLD provides a snapshot of all the important relationships in the RSSCM system [82]. In addition, it visualizes key variables and their relationships, composed of balancing and reinforcing loops [83,84].

3.4. Data Collection and Analysis

Demographics of Primary Survey Respondents

After the content analysis, a primary survey was conducted to shortlist the key resilience enablers. Due to the lack of research on developing economies, these countries were identified following Samans et al. [85]. The questionnaire was floated to over 2000 respondents via LinkedIn®, ResearchGate®, Facebook®, and organizational emails. A total of 106 responses were received, including those from Pakistan (37%), South Africa (14%), Malaysia (9%), Turkey (8%), UAE (7%), India (6%), Saudi Arabia (5%), Iran (4%), and from other developing countries (10%), as shown in Figure 2. The respondents' profiles are shown in Table 2 below.

As shown in Table 2, 12% of respondents had 0–1 year of experience, 9% had 2–5, 18% had 6–10, 10% had 11–15, 7% had 16–20, and 24% had experience of more than 20 years. Regarding qualification, 6% of respondents were diploma holders, 52% had a graduate degree, 36% had a post-graduate degree, and 6% were Ph.D. holders. In addition, 33% of respondents were from the government sector, whereas 53% and 14% were from private and semi-government sectors. To check the level of knowledge of the respondents about the understanding of the topic, respondents were asked to rank their level of knowledge of the topic as no understanding at all, slight, moderate, and high, respectively. Accordingly, 55% of the respondents had a moderate level of knowledge about RSSCM, 28% of respondents had a high level of knowledge, and 17% had slight to no knowledge of the research topic.

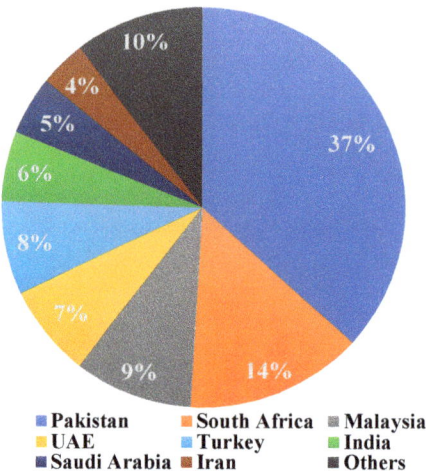

Figure 2. Regional distribution of respondents.

Table 2. Frequency distribution of responses.

Profile	Frequency	Percentage
Total responses = 106		
Job title		
CEO	4	4%
Construction Manager	5	5%
Assistant Manager	14	13%
Site Manager	11	10%
Architect/Designer	7	7%
Planning Engineer	14	13%
Project Manager	16	15%
Project Director	10	9%
Academician	12	12%
Others	13	12%
Years of Professional Experience		
0–1	13	12%
2–5	31	29%
6–10	19	18%
11–15	11	10%
16–20	7	7%
>20	25	24%
Education		
Diploma Holder	6	6%
Graduation	55	52%
Post-Graduation	39	36%
PhD	6	6%
Organization type		
Government	35	33%
Semi-Government	15	14%
Private	56	53%
Understanding of resilience and risk management in supply chains		
No understanding at all	8	8%
Slight	10	9%
Moderate	58	55%
High	30	28%

4. Results and Discussions

4.1. Factors Shortlisting

Table 3 represents the collective scores and ranks of the 32 enablers. The normalized literature score (40%) and normalized field score (60%) were selected to calculate the collective score to rank the enablers. After arranging factors in descending order with respect to their collective score, enablers with a cumulative percentage normalized score up to 51 percent were shortlisted for further analysis.

Table 3. Ranking via collective score.

Sr.No	Enablers	Normalized Literature Score (40%)	Normalized Field Score (60%)	Collective Score	Rank
1	Top Management Support	0.014	0.023	0.038	6
2	Adaptability	0.017	0.019	0.036	8
3	Visibility	0.042	0.019	0.061	1
4	Health	0.006	0.023	0.029	15
5	Compatibility	0.023	0.019	0.041	5
6	Quality Awareness	0.003	0.023	0.026	21
7	Responsiveness	0.006	0.023	0.029	17
8	Technological Capability	0.008	0.019	0.027	20
9	Agility	0.037	0.023	0.060	2
10	Supply Chain Security	0.008	0.019	0.027	19
11	Collaboration	0.037	0.019	0.055	3
12	Swift Trust	0.012	0.019	0.031	13
13	Risk and Revenue Sharing	0.006	0.019	0.024	22
14	Information Sharing	0.034	0.019	0.053	4
15	Flexible Structure	0.014	0.019	0.032	11
16	Risk Management Culture	0.006	0.019	0.024	23
17	Information Security	0.010	0.014	0.024	24
18	Strategic Risk Planning	0.014	0.019	0.032	10
19	Corporate Social Responsibility	0.014	0.019	0.033	9
20	Contingency Planning	0.012	0.019	0.031	15
21	Safety Stock	0.014	0.014	0.028	18
22	Flexible Transportation	0.012	0.019	0.031	14
23	Resource Efficiency	0.005	0.019	0.024	25
24	Transparency	0.006	0.019	0.024	26
25	Self-Regulation	0.005	0.019	0.024	27
26	Market Sensitivity	0.008	0.014	0.023	28
27	Tenacity	0.003	0.019	0.022	29
28	Leadership	0.008	0.023	0.032	12
29	Just in Time	0.014	0.023	0.038	7
30	Proper Scheduling	0.001	0.019	0.019	30
31	Composure	0.003	0.009	0.012	31
32	Reasoning	0.001	0.009	0.011	32

A Pareto Chart was used in this study to show the cut-off point for key enablers, as shown in Figure 3. It is a bar graph showing the variables and their ordered percentages.

In addition, it shows the ordered frequency counts of values for the different levels of a variable [86]. A Pareto chart aims to separate the significant aspects of a problem from the trivial ones [4]. In this case, the cut-off point for variable selection was set at 51 percent for cumulative normalized scores [4,86]. The total number of elements under this score was 12, identified as the key enablers. These include visibility, agility, collaboration, information sharing, compatibility, top management support, just in time, adaptability, corporate social responsibility, flexible structure, strategic risk planning, and leadership. The x-axis of Figure 3 represents the variables, and the y-axis displays the combined score and cumulative percentages of the enablers obtained from Table 3.

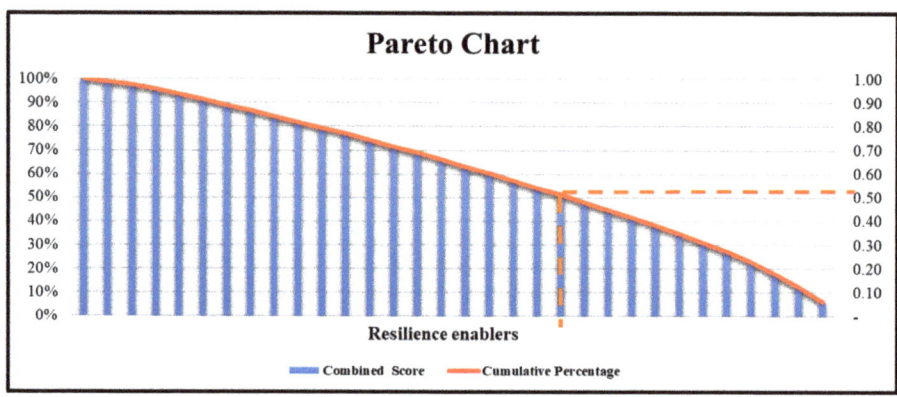

Figure 3. Shortlisting of enablers using 50% impact.

4.2. Influence Matrix

The Influence Matrix (IM) for the CLD was developed based on expert opinion. The IM shows interrelationships and polarities of influence (positive or negative) among the variables. In this case, IM shows 16 relationships among 12 enablers where a value of +1 indicates a direct relationship and −1 indicates an indirect relationship, as shown in Figure 4.

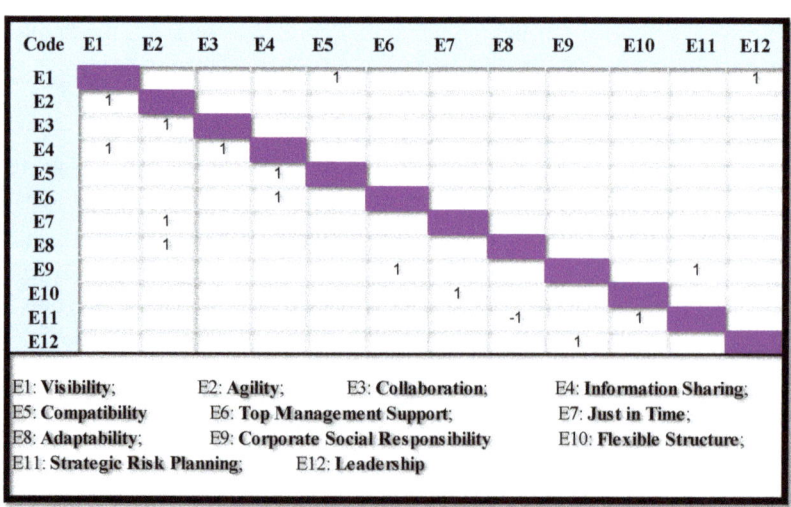

Figure 4. Influence matrix.

4.3. Causal Loop Diagram (CLD)

The CLD was constructed to show the loops, polarities, and images of variables. Vensim® was used for the construction of the CLD. A total of 16 substantial interrelationships were addressed by the CLD, of which one was indirect and the other 15 were direct in terms of polarity. The CLD was developed based on the opinions of 15 construction personnel with over 20 years of experience in developing countries. In addition, a wider experience of the respondents helped confirm the CLDs' significance and applicability to the construction industry. Figure 5 is a consolidated CLD developed in the current study. It comprises five loops, i.e., four reinforcing and one balancing loop. The explanation of each loop is given below.

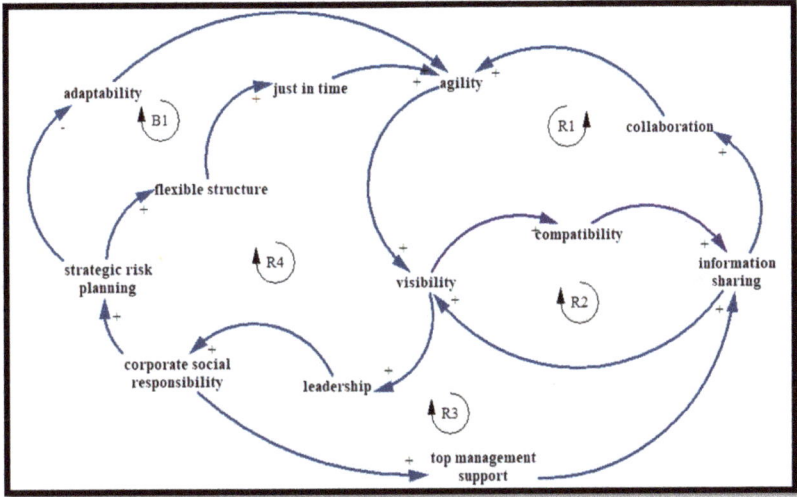

Figure 5. Causal loop diagram.

4.3.1. Reinforcing Loop R1

Reinforcing loop R1 demonstrates that an increase in visibility increases compatibility, leading to increased information sharing. Furthermore, an increase in information sharing promotes collaboration, which increases agility. This further increases visibility, as shown in Figure 6. Hence, this loop clarifies that if there is a visible SC, there would be a more amicable relationship among SC partners, leading to information sharing and cooperation, and ultimately the SC would be faster and more resilient.

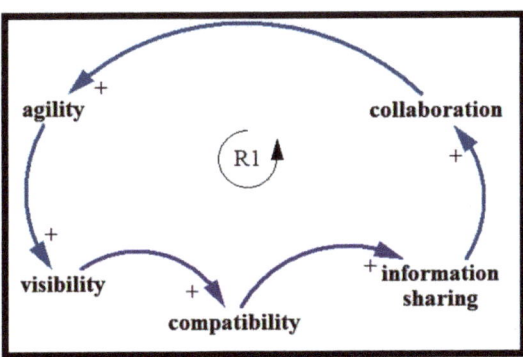

Figure 6. Reinforcing loop R1.

4.3.2. Reinforcing Loop R2

R2, as presented in Figure 7, illustrates that an increase in visibility leads to an increase in SC compatibility. Furthermore, this increase leads to increased information sharing that promotes visibility. This loop elucidates that a more visible SC will lead to an amicable relationship among the partners and more information sharing.

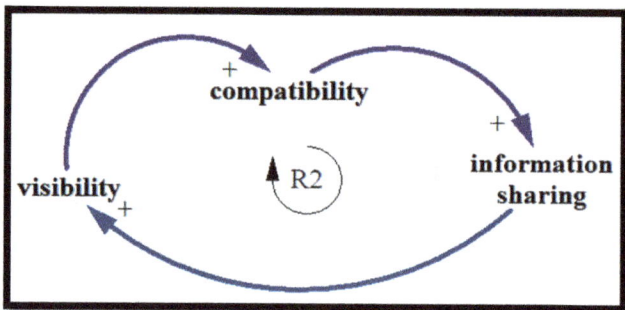

Figure 7. Reinforcing loop R2.

4.3.3. Reinforcing Loop R3

Reinforcing loop R3 shows that increased visibility promotes leadership, leading to increased corporate social responsibility. This, in turn, promotes top management support, leading to increased information sharing, which again leads to increased visibility, as displayed in Figure 8. This loop explains how leadership reinforces corporate social responsibility and top management support and leads to a more visible SC with increased information sharing among the SC partners.

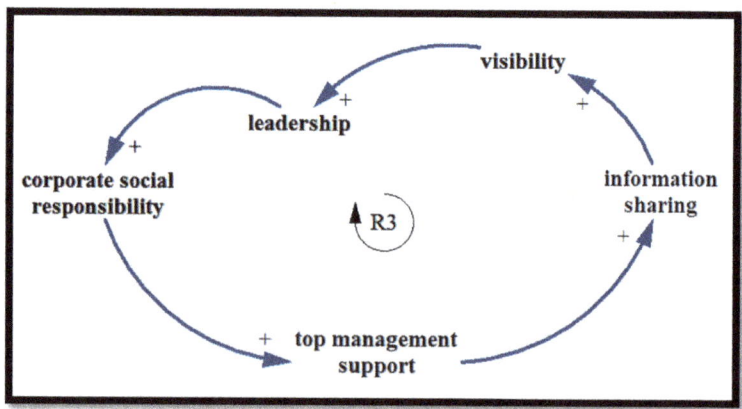

Figure 8. Reinforcing loop R3.

4.3.4. Reinforcing Loop R4

Reinforcing loop R4 illustrates that increased visibility promotes leadership, increasing corporate social responsibility and strategic risk planning. An increase in strategic risk planning promotes a flexible structure that leads to a just-in-time approach. This, in turn, promotes agility, which will increase visibility, as shown in Figure 9. This loop clarifies that when leadership reinforces corporate social responsibility, there would be a flexible SC structure leading to a just-in-time approach that will make the SC faster and more visible. This is due to the strategic risk planning process.

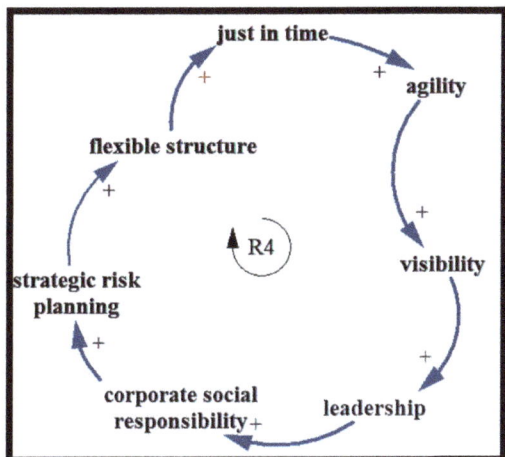

Figure 9. Reinforcing loop R4.

4.3.5. Balancing Loop B1

Balancing loop B1 depicts that an increase in strategic risk planning leads to decreased adaptability, leading to a decrease in agility. A decrease in agility leads to a decrease in visibility which decreases leadership. A decrease in leadership will decrease corporate social responsibility, leading to a decrease in strategic risk planning, as shown in Figure 10. This loop explains the balancing effect of strategic risk planning on adaptability.

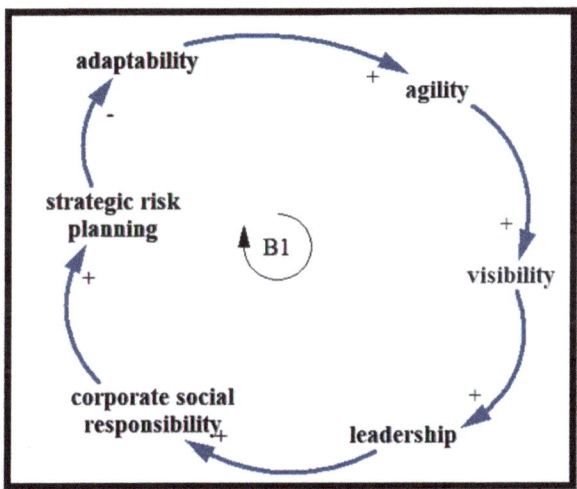

Figure 10. Balancing loop B1.

4.4. Loop Analysis

The magnitude and speed of influence on system outputs serves as a thorough criterion for loop classification [14,71]. Table 4 summarizes the results for each feedback loop. It predicts the speed, strength, and nature of the influence of the loop [87]. The four reinforcing loops, R1, R2, R3, and R4, have a strong influence with a low speed. This indicates that these loops hold great potential but will take time and be long-lasting.

Table 4. Overall loop analysis results.

Loop	Speed of Influence	Strength of Influence	Nature of Influence
R1	Slow	Strong	Reinforcing
R2	Slow	Strong	Reinforcing
R3	Slow	Strong	Reinforcing
R4	Slow	Strong	Reinforcing
B1	Fast	Strong	Balancing

On the contrary, B1 is fast, having a balancing effect. Reinforcing loops have a resonant effect that lasts for a long period, whereas balancing loops have a fading impact that lasts for a short time. The CLD's validity was qualitatively assured and verified through expert opinion [88]. All four reinforcing loops have a strong influence with a slow speed. On the contrary, the balancing loop has a fast speed and strong influence [88]. The results of this study can enable organizations to acclimate to disruptions by sourcing their inputs from a versatile or redundant supply base that allows a business to move suppliers when production is at risk.

4.5. Discussion

In this study, 32 resilience enablers were selected based on a literature review, as shown in Table 1. These enablers were reduced to 12 key enablers of RSSCM. The shortlisting was achieved through a field survey where the 12 enablers with cumulative normalized scores of up to 51% were selected. These key enablers include visibility, agility, collaboration, information sharing, compatibility, top management support, just in time, adaptability, corporate social responsibility, flexible structure, strategic risk planning, and leadership. The IM, as presented in Figure 4, was developed based on these key enablers. The IM has 16 interrelationships between the 12 key enablers. Finally, the CLD was developed based on the IM, as shown in Figure 5.

The CLD developed in this study comprises five loops: four reinforcing and one balancing loop. Figure 6 clarifies that a more visible and established SC would create a more amicable relationship among SC partners. Such a relationship leads to information sharing and cooperation; ultimately, the SC would be faster and more resilient. Figure 7 shows that if an organization's SC is agile, visible, and has a compatible infrastructure, with proper collaboration and information sharing, it will ultimately make it more resilient to avoid disruptions. This finding is in line with [87].

Moreover, top management support, corporate social responsibility, and strong strategic risk planning can reinforce the resilience of any SC, as shown in Figure 8. The same has been concluded by [14]. Figure 9 highlights that through information sharing, exchange, and integration among SC partners, the RSSCM will increase. This is in line with [89] and clarifies that when leadership reinforces corporate social responsibility, then, due to strategic risk planning, there would be a flexible SC structure, leading to a just-in-time approach. Such an approach will make the SC both faster and more visible. Figure 10 explains the balancing effect of strategic risk planning on adaptability. Overall, adaptability and a just-in-time approach play a key role in enabling RSSCM as they promote the use of minimal raw materials, leading to enhanced sustainability [88].

Table 4 shows the loop analysis of the study. Accordingly, the four reinforcing loops, R1, R2, R3, and R4, strongly influence at lower speeds. This indicates that these loops hold great potential but take some time to materialize. This is in line with [48]. On the contrary, B1 is fast, having a balancing effect. Therefore, the impacts of B1, which may not be that significant, have more chances and speed of occurrence. This encourages the SCM managers to be proactive and take timely measures. Furthermore, reinforcing loops have a resonant effect that lasts for a long period, whereas balancing loops have a fading impact that lasts for a short time. Finally, the CLD's validity was qualitatively assured through expert opinion for verification [88].

The outcomes of the study will help firms acclimatize to disturbances in their SCs. It is the first study of the complexity of resilient and sustainable construction SCs. This study has added to the existing body of knowledge by identifying the enablers that aid in developing a more resilient SC network, bridging the research gap identified by Chowdhury et al. [19], Nguyen and Bosch [20], and Sapiri et al. [21]. These authors emphasized demonstrating the relationship between SC resilience and SC sustainability for developing an RSSCM.

5. Conclusions

Resilience is a key organizational capability for achieving sustainability in the current tempestuous global situation. To develop more resilient and sustainable SC networks, this paper illustrates the crucial enablers of resilience for RSSCM. A total of 32 enablers were extracted from the body of knowledge using a literature review. Data were later collected from the respondents in the construction industry of developing countries. Two types of normalized scoring were used to shortlist the key enablers: industry and the literature. After combining the industry and literature scores, the 32 enablers were reduced to 12. Finally, the top 12 enablers were added to the IM, involved in creating a CLD that showed the relationships between the identified enablers. The CLD show four reinforcing and one balancing loop.

Based on the results of this study, agility, information sharing, strategic risk planning, corporate social responsibility, and visibility are the key resilience enablers for RSSCM in developing countries. These enablers serve as significant tools for organizations to plan for and adapt to disruptions in SCs in construction projects. The causality and interrelationships among these enablers in the developed CLD show their dynamicity and impact within the construction SC system.

The findings of the study will assist organizations in adapting to SC disruptions by acquiring inputs from a flexible supply base that allows them to switch providers when production is threatened. There has not been any published work utilizing the ST methodology for similar purposes. As a result, this study's methodology is innovative, and it is the first to address complexity in the construction sector of developing countries for moving towards an RSSCM.

The limitation of this study consists of the inclusion of respondents only from developing countries. In addition, this study utilized an ST approach for constructing CLDs and did not perform system dynamics modeling. Moreover, this study only considered limited enablers based on the literature review, which may not be exhaustive in the future.

A further study involving participants from developed countries would be more beneficial. Future research can explore the application of the developed CLD to real-time projects. A follow-up study could focus on developing a system dynamics model to explore the constructs of sustainability and resilience in the RSSCM.

Author Contributions: Conceptualization, M.G. and K.I.A.K.; methodology, M.G. and K.I.A.K.; software, M.G. and K.I.A.K.; validation, M.G., K.I.A.K. and F.U.; formal analysis, M.G. and K.I.A.K.; investigation, M.G. and K.I.A.K.; resources, F.U., M.A.M. and W.S.A.; data curation, M.G., K.I.A.K. and F.U.; writing—original draft preparation, M.G. and K.I.A.K.; writing—review and editing, K.I.A.K. and F.U.; visualization, M.G. and K.I.A.K.; supervision, K.I.A.K. and F.U.; project administration, K.I.A.K., F.U., M.A.M. and W.S.A.; funding acquisition, M.A.M. and W.S.A. All authors have read and agreed to the published version of the manuscript.

Funding: This research received no external funding.

Institutional Review Board Statement: Not applicable.

Informed Consent Statement: Informed consent was obtained from all subjects involved in the study.

Data Availability Statement: Data are available from the first author and can be shared upon reasonable request.

Acknowledgments: The authors would like to appreciate the YUTP-FRG 1/2021 project (cost center # 015LC0-369) in Universiti Teknologi PETRONAS (UTP) awarded to Wesam Alaloul for the support.

Conflicts of Interest: The authors declare no conflict of interest.

References

1. Touboulic, A.; Walker, H. Theories in sustainable supply chain management: A structured literature review. *Int. J. Phys. Distrib. Logist. Manag.* **2015**, *45*, 16–42. [CrossRef]
2. Naslund, D.; Williamson, S. What is management in supply chain management?-a critical review of definitions, frameworks and terminology. *J. Manag. Policy Pract.* **2010**, *11*, 11–28.
3. Reefke, H.; Sundaram, D. Key themes and research opportunities in sustainable supply chain management–identification and evaluation. *Omega* **2017**, *66*, 195–211. [CrossRef]
4. Ghufran, M.; Khan, K.I.A.; Thaheem, M.J.; Nasir, A.R.; Ullah, F. Adoption of Sustainable Supply Chain Management for Performance Improvement in the Construction Industry: A System Dynamics Approach. *Architecture* **2021**, *1*, 161–182. [CrossRef]
5. Cetinkaya, B.; Cuthbertson, R.; Ewer, G.; Klaas-Wissing, T.; Piotrowicz, W.; Tyssen, C. *Sustainable Supply Chain Management: Practical Ideas for Moving towards Best Practice*; Springer Science & Business Media: Berlin/Heidelberg, Germany, 2011.
6. Janvier-James, A.M. A new introduction to supply chains and supply chain management: Definitions and theories perspective. *Int. Bus. Res.* **2012**, *5*, 194–207. [CrossRef]
7. Sutrisna, M.; Kumaraswamy, M.M. Advanced ICT and smart systems for innovative "engineering, construction and architectural management". *Eng. Constr. Archit. Manag.* **2015**, *22*, 5. [CrossRef]
8. Gattorna, J. *Dynamic Supply Chain Alignment: A New Business Model for Peak Performance in Enterprise Supply Chains Across all Geographies*; CRC Press: Boca Raton, FL, USA, 2016.
9. Marley, K.A.; Ward, P.T.; Hill, J.A. Mitigating supply chain disruptions—A normal accident perspective. *Supply Chain. Manag. Int. J.* **2014**, *19*, 142–152. [CrossRef]
10. Revilla, E.; Saenz, M.J. The impact of risk management on the frequency of supply chain disruptions: A configurational approach. *Int. J. Oper. Prod. Manag.* **2017**, *37*, 557–576. [CrossRef]
11. Papadopoulos, T.; Gunasekaran, A.; Dubey, R.; Altay, N.; Childe, S.J.; Fosso-Wamba, S. The role of Big Data in explaining disaster resilience in supply chains for sustainability. *J. Clean. Prod.* **2017**, *142*, 1108–1118. [CrossRef]
12. Lohmer, J.; Bugert, N.; Lasch, R. Analysis of resilience strategies and ripple effect in blockchain-coordinated supply chains: An agent-based simulation study. *Int. J. Prod. Econ.* **2020**, *228*, 107882. [CrossRef]
13. Sawyerr, E.; Harrison, C. Developing resilient supply chains: Lessons from high-reliability organisations. *Supply Chain. Manag. Int. J.* **2019**, *25*, 77–100. [CrossRef]
14. Katsaliaki, K.; Galetsi, P.; Kumar, S. Supply chain disruptions and resilience: A major review and future research agenda. *Ann. Oper. Res.* **2021**, 1–38. [CrossRef] [PubMed]
15. Petit-Boix, A.; Leipold, S. Circular economy in cities: Reviewing how environmental research aligns with local practices. *J. Clean. Prod.* **2018**, *195*, 1270–1281. [CrossRef]
16. Ullah, F.; Thaheem, M.J. Concession period of public private partnership projects: Industry–academia gap analysis. *Int. J. Constr. Manag.* **2018**, *18*, 418–429. [CrossRef]
17. Ullah, F.; Siddiqui, S. An investigation of real estate technology utilization in technologically advanced marketplace. In Proceedings of the 9th International Civil Engineering Congress (ICEC-2017),"Striving Towards Resilient Built Environment", Karachi, Pakistan, 22–23 December 2017.
18. Pettit, T.J.; Croxton, K.L.; Fiksel, J. The evolution of resilience in supply chain management: A retrospective on ensuring supply chain resilience. *J. Bus. Logist.* **2019**, *40*, 56–65. [CrossRef]
19. Chowdhury, M.M.H.; Quaddus, M.; Agarwal, R. Supply chain resilience for performance: Role of relational practices and network complexities. *Supply Chain. Manag. Int. J.* **2019**, *24*, 5. [CrossRef]
20. Nguyen, N.C.; Bosch, O.J. A systems thinking approach to identify leverage points for sustainability: A case study in the Cat Ba Biosphere Reserve, Vietnam. *Syst. Res. Behav. Sci.* **2013**, *30*, 104–115. [CrossRef]
21. Sapiri, H.; Zulkepli, J.; Ahmad, N.; Abidin, N.Z.; Hawari, N.N. *Introduction to System Dynamic Modelling and Vensim Software*; UUM Press; UUM Press: Sintok Kedah, Malaysia, 2017.
22. Ullah, F.; Sepasgozar, S.M. A study of information technology adoption for real-estate management: A system dynamic model. In *Innovative Production And Construction: Transforming Construction Through Emerging Technologies*; World Scientific: Perth, Australia, 2019; pp. 469–486.
23. Cavone, G.; Dotoli, M.; Epicoco, N.; Morelli, D.; Seatzu, C. Design of modern supply chain networks using fuzzy bargaining game and data envelopment analysis. *IEEE Trans. Autom. Sci. Eng.* **2020**, *17*, 1221–1236. [CrossRef]
24. Cavone, G.; Dotoli, M.; Epicoco, N.; Morelli, D.; Seatzu, C. A game-theoretical design technique for multi-stage supply chains under uncertainty. In Proceedings of the 2018 IEEE 14th International Conference on Automation Science and Engineering (CASE), Munich, Germany, 20–24 August 2018; pp. 528–533.
25. Seuring, S.; Müller, M. From a literature review to a conceptual framework for sustainable supply chain management. *J. Clean. Prod.* **2008**, *16*, 1699–1710. [CrossRef]
26. Stindt, D. A generic planning approach for sustainable supply chain management-How to integrate concepts and methods to address the issues of sustainability? *J. Clean. Prod.* **2017**, *153*, 146–163. [CrossRef]

27. Parsa, S.; Roper, I.; Muller-Camen, M.; Szigetvari, E. Have labour practices and human rights disclosures enhanced corporate accountability? The case of the GRI framework. *Account. Forum* **2018**, *42*, 47–64. [CrossRef]
28. Sabri, Y.; Micheli, G.J.; Cagno, E. Supplier selection and supply chain configuration in the projects environment. *Prod. Plan. Control.* **2020**, *33*, 1–19. [CrossRef]
29. Rajeev, A.; Pati, R.K.; Padhi, S.S.; Govindan, K. Evolution of sustainability in supply chain management: A literature review. *J. Clean. Prod.* **2017**, *162*, 299–314. [CrossRef]
30. Daraei, A.; H Sherwani, A.F.; Faraj, R.H.; Kalhor, Q.; Zare, S.; Mahmoodzadeh, A. Optimization of the outlet portal of Heybat Sultan twin tunnels based on the value engineering methodology. *SN Appl. Sci.* **2019**, *1*, 270. [CrossRef]
31. Rachwan, R.; Abotaleb, I.; Elgazouli, M. The influence of value engineering and sustainability considerations on the project value. *Procedia Environ. Sci.* **2016**, *34*, 431–438. [CrossRef]
32. Seuring, S. Supply chain management for sustainable products–insights from research applying mixed methodologies. *Bus. Strategy Environ.* **2011**, *20*, 471–484. [CrossRef]
33. Ashby, A.; Leat, M.; Hudson-Smith, M. Making connections: A review of supply chain management and sustainability literature. *Supply Chain. Manag. Int. J.* **2012**, *17*, 497–516. [CrossRef]
34. Kshetri, N. 1 Blockchain's roles in meeting key supply chain management objectives. *Int. J. Inf. Manag.* **2018**, *39*, 80–89. [CrossRef]
35. Parmigiani, A.; Klassen, R.D.; Russo, M.V. Efficiency meets accountability: Performance implications of supply chain configuration, control, and capabilities. *J. Oper. Manag.* **2011**, *29*, 212–223. [CrossRef]
36. Ribeiro, J.P.; Barbosa-Povoa, A. Supply Chain Resilience: Definitions and quantitative modelling approaches–A literature review. *Comput. Ind. Eng.* **2018**, *115*, 109–122. [CrossRef]
37. Golan, M.S.; Jernegan, L.H.; Linkov, I. Trends and applications of resilience analytics in supply chain modeling: Systematic literature review in the context of the COVID-19 pandemic. *Environ. Syst. Decis.* **2020**, *40*, 222–243. [CrossRef] [PubMed]
38. Waters, D. *Supply Chain Risk Management: Vulnerability and Resilience in Logistics*; Kogan Page Publishers: Philadelphia, PA, USA, 2011.
39. Adobor, H.; McMullen, R.S. Supply chain resilience: A dynamic and multidimensional approach. *Int. J. Logist. Manag.* **2018**, *29*, 1451–1471. [CrossRef]
40. Quinlan, A.E.; Berbés-Blázquez, M.; Haider, L.J.; Peterson, G.D. Measuring and assessing resilience: Broadening understanding through multiple disciplinary perspectives. *J. Appl. Ecol.* **2016**, *53*, 677–687. [CrossRef]
41. Altay, N.; Gunasekaran, A.; Dubey, R.; Childe, S.J. Agility and resilience as antecedents of supply chain performance under moderating effects of organizational culture within the humanitarian setting: A dynamic capability view. *Prod. Plan. Control.* **2018**, *29*, 1158–1174. [CrossRef]
42. Scholten, K.; Scott, P.S.; Fynes, B. Building routines for non-routine events: Supply chain resilience learning mechanisms and their antecedents. *Supply Chain. Manag. Int. J.* **2019**, *24*, 3. [CrossRef]
43. Ruiz-Benítez, R.; López, C.; Real, J.C. The lean and resilient management of the supply chain and its impact on performance. *Int. J. Prod. Econ.* **2018**, *203*, 190–202. [CrossRef]
44. McAllister, T.; McAllister, T. *Developing Guidelines and Standards for Disaster Resilience of the Built Environment: A Research Needs Assessment*; US Department of Commerce, National Institute of Standards and Technology: Gaithersburg, MD, USA, 2013.
45. Ivanov, D.; Dolgui, A.; Sokolov, B. The impact of digital technology and Industry 4.0 on the ripple effect and supply chain risk analytics. *Int. J. Prod. Res.* **2019**, *57*, 829–846. [CrossRef]
46. Kamalahmadi, M.; Parast, M.M. A review of the literature on the principles of enterprise and supply chain resilience: Major findings and directions for future research. *Int. J. Prod. Econ.* **2016**, *171*, 116–133. [CrossRef]
47. Ullah, F.; Sepasgozar, S.M.; Shirowzhan, S.; Davis, S. Modelling users' perception of the online real estate platforms in a digitally disruptive environment: An integrated KANO-SISQual approach. *Telemat. Inform.* **2021**, *63*, 101660. [CrossRef]
48. Ullah, F.; Sepasgozar, S.M.; Thaheem, M.J.; Wang, C.C.; Imran, M. It's all about perceptions: A DEMATEL approach to exploring user perceptions of real estate online platforms. *Ain Shams Eng. J.* **2021**, *12*, 4297–4317. [CrossRef]
49. Karutz, R.; Riedner, L.; Vega, L.R.; Stumpf, L.; Damert, M. Compromise or complement? Exploring the interactions between sustainable and resilient supply chain management. *Int. J. Supply Chain. Oper. Resil.* **2018**, *3*, 117–142. [CrossRef]
50. Mohamadi Zanjiran, D.; Hashemkhani Zolfani, S.; Prentkovskis, O. LARG supplier selection based on integrating house of quality, Taguchi loss function and MOPA. *Econ. Res. Ekon. Istraživanja* **2019**, *32*, 1944–1964. [CrossRef]
51. Fritz, M.M.; Sustainable Supply Chain Management. *Responsible Consumption and Production. Encyclopedia of the UN Sustainable Development Goals*; Springer: Cham, Switzerland, 2019.
52. Kun, A. Social Dialogue and Corporate Social Responsibility (CSR) in the EU. In *EU Collective Labour Law*; Edward Elgar Publishing: Cheltenham, UK, 2021.
53. Jabbarzadeh, A.; Fahimnia, B.; Sabouhi, F. Resilient and sustainable supply chain design: Sustainability analysis under disruption risks. *Int. J. Prod. Res.* **2018**, *56*, 5945–5968. [CrossRef]
54. Fahimnia, B.; Jabbarzadeh, A. Marrying supply chain sustainability and resilience: A match made in heaven. *Transp. Res. Part E Logist. Transp. Rev.* **2016**, *91*, 306–324. [CrossRef]
55. Ivanov, D. Simulation-based ripple effect modelling in the supply chain. *Int. J. Prod. Res.* **2017**, *55*, 2083–2101. [CrossRef]
56. Ivanov, D. Revealing interfaces of supply chain resilience and sustainability: A simulation study. *Int. J. Prod. Res.* **2018**, *56*, 3507–3523. [CrossRef]

57. Rojon, C.; McDowall, A.; Saunders, M.N. On the experience of conducting a systematic review in industrial, work, and organizational psychology: Yes, it is worthwhile. *J. Pers. Psychol.* **2011**, *10*, 133. [CrossRef]
58. Jahan, S.; Khan, K.I.A.; Thaheem, M.J.; Ullah, F.; Alqurashi, M.; Alsulami, B.T. Modeling Profitability-Influencing Risk Factors for Construction Projects: A System Dynamics Approach. *Buildings* **2022**, *12*, 701. [CrossRef]
59. Ullah, F. A beginner's guide to developing review-based conceptual frameworks in the built environment. *Architecture* **2021**, *1*, 5–24. [CrossRef]
60. Rasul, N.; Malik, M.S.A.; Bakhtawar, B.; Thaheem, M.J. Risk assessment of fast-track projects: A systems-based approach. *Int. J. Constr. Manag.* **2019**, *21*, 1099–1114. [CrossRef]
61. Ullah, F.; Ayub, B.; Siddiqui, S.Q.; Thaheem, M.J. A review of public-private partnership: Critical factors of concession period. *J. Financ. Manag. Prop. Constr.* **2016**, *21*, 3. [CrossRef]
62. Chowdhury, M.H.; Dewan, M.N.A.; Quaddus, M.A. Resilient Sustainable Supply Chain Management-A Conceptual Framework. In Proceedings of the International Conference on e-Business, Hangzhou, China, 9–11 September 2012; pp. 165–173.
63. Ivanov, D. New drivers for supply chain structural dynamics and resilience: Sustainability, industry 4.0, self-adaptation. In *Structural Dynamics and Resilience in Supply Chain Risk Management*; Springer: Berlin/Heidelberg, Germany, 2018; pp. 293–313.
64. Rosič, H.; Bauer, G.; Jammernegg, W. A framework for economic and environmental sustainability and resilience of supply chains. In *Rapid Modelling for Increasing Competitiveness*; Springer: Berlin/Heidelberg, Germany, 2009; pp. 91–104.
65. Badurdeen, F.; Wijekoon, K.; Shuaib, M.; Goldsby, T.J.; Iyengar, D.; Jawahir, I.S. Integrated modeling to enhance resilience in sustainable supply chains. In Proceedings of the 2010 IEEE International Conference on Automation Science and Engineering, Toronto, ON, Canada, 21–24 August 2010; pp. 130–135.
66. Brady, M. Realising supply chain resilience: An exploratory study of Irish firms' priorities in the wake of Brexit. *Contin. Resil. Rev.* **2020**, *3*, 1. [CrossRef]
67. Soni, U.; Jain, V.; Kumar, S. Measuring supply chain resilience using a deterministic modeling approach. *Comput. Ind. Eng.* **2014**, *74*, 11–25. [CrossRef]
68. Mari, S.I.; Lee, Y.H.; Memon, M.S. Sustainable and resilient supply chain network design under disruption risks. *Sustainability* **2014**, *6*, 6666–6686. [CrossRef]
69. Ralston, P.; Blackhurst, J. Industry 4.0 and resilience in the supply chain: A driver of capability enhancement or capability loss? *Int. J. Prod. Res.* **2020**, *58*, 5006–5019. [CrossRef]
70. Jain, V.; Kumar, S.; Soni, U.; Chandra, C. Supply chain resilience: Model development and empirical analysis. *Int. J. Prod. Res.* **2017**, *55*, 6779–6800. [CrossRef]
71. Zavala-Alcívar, A.; Verdecho, M.-J.; Alfaro-Saíz, J.-J. A conceptual framework to manage resilience and increase sustainability in the supply chain. *Sustainability* **2020**, *12*, 6300. [CrossRef]
72. Mangla, S.K.; Kumar, P.; Barua, M.K. Flexible decision approach for analysing performance of sustainable supply chains under risks/uncertainty. *Glob. J. Flex. Syst. Manag.* **2014**, *15*, 113–130. [CrossRef]
73. Govindan, K.; Azevedo, S.G.; Carvalho, H.; Cruz-Machado, V. Lean, green and resilient practices influence on supply chain performance: Interpretive structural modeling approach. *Int. J. Environ. Sci. Technol.* **2015**, *12*, 15–34. [CrossRef]
74. Rha, J.S. Trends of Research on Supply Chain Resilience: A Systematic Review Using Network Analysis. *Sustainability* **2020**, *12*, 4343. [CrossRef]
75. Cabral, I.; Espadinha-Cruz, P.; Grilo, A.; Puga-Leal, R.; Cruz-Machado, V. Decision-Making Models for Interoperable Lean, Agile, Resilient and Green Supply Chains. In Proceedings of the International Symposium on the Analytic Hierarchy Process, Sorrento, Italy, 15–18 June 2011; pp. 1–6.
76. Ivanov, D.; Sokolov, B.; Dolgui, A. The Ripple effect in supply chains: Trade-off 'efficiency-flexibility-resilience'in disruption management. *Int. J. Prod. Res.* **2014**, *52*, 2154–2172. [CrossRef]
77. Christopher, M.; Rutherford, C. Creating supply chain resilience through agile six sigma. *Crit. Eye* **2004**, *7*, 24–28.
78. Rajesh, R. On sustainability, resilience, and the sustainable–resilient supply networks. *Sustain. Prod. Consum.* **2018**, *15*, 74–88. [CrossRef]
79. Parast, M.M.; Sabahi, S.; Kamalahmadi, M. The relationship between firm resilience to supply chain disruptions and firm innovation. In *Revisiting Supply Chain Risk*; Springer: Berlin/Heidelberg, Germany, 2019; pp. 279–298.
80. Arnold, R.D.; Wade, J.P. A definition of systems thinking: A systems approach. *Procedia Comput. Sci.* **2015**, *44*, 669–678. [CrossRef]
81. Mohammadi, A.; Abbasi, A.; Alimohammadlou, M.; Eghtesadifard, M.; Khalifeh, M. Optimal design of a multi-echelon supply chain in a system thinking framework: An integrated financial-operational approach. *Comput. Ind. Eng.* **2017**, *114*, 297–315. [CrossRef]
82. Dhirasasna, N.; Sahin, O. A multi-methodology approach to creating a causal loop diagram. *Systems* **2019**, *7*, 42. [CrossRef]
83. Giannakidou, A. Negative and positive polarity items: Variation, licensing, and compositionality. *Semant. Int. Handb. Nat. Lang. Mean.* **2011**, *2*, 1660–1712.
84. Felli, F.; Liu, C.; Ullah, F.; Sepasgozar, S. Implementation of 360 videos and mobile laser measurement technologies for immersive visualisation of real estate & properties. In Proceedings of the 42nd AUBEA Conference, Singapore, 26–28 September 2018.
85. Samans, R.; Blanke, J.; Drzeniek, M.; Corrigan, G. The inclusive development index 2018 summary and data highlights. In Proceedings of the World Economic Forum, Geneva, Switzerland, 21 June 2018.

86. Heiberger, R.; Robbins, N. Design of diverging stacked bar charts for Likert scales and other applications. *J. Stat. Softw.* **2014**, *57*, 1–32. [CrossRef]
87. Hsueh, C.-F. Improving corporate social responsibility in a supply chain through a new revenue sharing contract. *Int. J. Prod. Econ.* **2014**, *151*, 214–222. [CrossRef]
88. Bhushan, U.; Aserkar, R.; Kumar, K.N.; Seetharaman, A. Effectiveness of Just In Time Manufacturing Practices. *Int. J. Bus. Manag. Econ. Res. (IJBMER)* **2017**, *8*, 1109–1114.
89. Lam, J.S.L.; Bai, X. A quality function deployment approach to improve maritime supply chain resilience. *Transp. Res. Part E Logist. Transp. Rev.* **2016**, *92*, 16–27. [CrossRef]

Article

Error Management Climate and Job Stress in Project-Based Organizations: An Empirical Evidence from Pakistani Aircraft Manufacturing Industry

Hassan Ashraf [1], Ahsen Maqsoom [1], Tayyab Tahir Jajja [1], Rana Faisal Tufail [1], Rashid Farooq [2] and Muhammad Atiq Ur Rehman Tariq [3,4,5,*]

1 Department of Civil Engineering, COMSATS University Islamabad, Wah Campus, Wah Cantt 47040, Pakistan
2 Department of Civil Engineering, International Islamic University, Islamabad 44000, Pakistan
3 College of Engineering, IT & Environment, Charles Darwin University, Darwin, NT 0810, Australia
4 Institute for Sustainable Industries & Liveable Cities, Victoria University, P.O. Box 14428, Melbourne, VIC 8001, Australia
5 Centre of Excellence in Water Resources Engineering, University of Engineering and Technology, Lahore 54890, Pakistan
* Correspondence: atiq.tariq@yahoo.com

Abstract: Drawing on the JD-R model, this study examines the influence of error management climate (EMC) on the job stress of frontline aeronautical employees. It also analyzes the moderating role of psychological capital (PsyCap) dimensions (i.e., hope, optimism, self-efficacy, and resilience) for the relationship between error management climate and job stress. The data was collected from 208 individuals through a questionnaire survey and was analyzed using a partial least squares structural equation modeling (PLS-SEM) approach. The results revealed that employees' perceptions of error management climate have a significant negative impact on job stress. PsyCap optimism and PsyCap self-efficacy were found to have a negative moderating influence on the relationship between EMC and job stress. The other two dimensions of hope and resilience were found to have a moderating influence in the same direction as expected, but not at statistically significant levels. The findings of this study provide a unique perspective in realizing the part national and organizational cultures could play in either enhancing or attenuating the influence of an individual's psychological resources such as psychological capital.

Keywords: error management climate; psychological capital; job stress; aeronautical industry; structural equation modeling

1. Introduction

Occupational accidents are a tremendous burden on organizations and result in substantial pain and suffering [1]. Understanding that organizational environment impinges on workers' performance and safety, researchers have been increasingly interested in identifying variables that are fundamental in creating havoc for individuals and organizations. A number of studies have found that occupational stress has negative consequences and has rapidly affected organizational members' productivity, particularly within complex systems such as aeronautical organizations, construction firms, and the hospital industry [2–4]. Further, job stress is a cause of turnover intention and a poor level of employee well-being [5]. In a recent study conducted by Wang et al. [6], safety-related stress was found to have a negative effect on safety participation, thereby compromising the overall safety performance of individuals. Job stress and its link with safety is further established by the fact that Dupont's [7] Human Performance Model considers stress as one of the twelve precursors to accidents. Project-based organizations operate in an extremely competitive environment, where projects are designed, executed, and are required to be delivered within the stipulated time and cost. Working in these organizations is emotionally and psychologically challenging and stressful [8]. In project-based organizations, job stress mainly depends on a demanding work environment characterized by peak work loads, complex tasks, and high uncertainty [9,10], and on interpersonal and role conflict [11].

The aeronautical industry has a complex organizational structure and the technology used in this industry has changed remarkably over the past few years [12]. In the aeronautical industry, the human factor is very important in handling these complexities and advancements. In human activity, errors and mistakes are natural consequences, particularly in complex systems which lead to job stress [13]. Total elimination of errors is a difficult task as it is nearly impossible to fully eliminate errors from an organization. Where one stream of organizational and management literature connotes error with a negative event that can be life-threatening, inefficient, and costly in some cases [14], the other stream considers errors to be helpful in learning, decision making, and system improvement [15]. Within the latter stream of error management, error management climate refers to shared perceptions of individuals about organizational procedures and practices related to support that individuals provide others in error situations, communicating about errors, sharing error knowledge, and quick detection and handling of errors [15]. A strong error management climate in an organization encourages employees to communicate about errors openly and in a well-coordinated manner. Owing to the error management climate, individuals are more likely to communicate about error occurrence as they feel confident that they will not be blamed, leading to mutual trust and respect [16]. Organizational members who have psychological strengths or personal resources such as psychological capital are more confident in handling negative events [17].

Although previous research has identified that error management climate is negatively related to stress, there is scant work explaining the relationship between error management climate (EMC) and job stress [16] as a function of individuals' predisposition to manage challenges and adversities. Personal resources are theorized to have positive behavioral outcomes such as dedication, job commitment, and work engagement [18]. According to Luthans, Youssef and Avolio [17], psychological capital is a positive psychological state that is reflective of: (1) an individual's confidence in his/her abilities in relation to the successful execution of a task at hand (self-efficacy); (2) the individual's ability to set goals and strategize alternative pathways to surmount challenges in a bid to achieve goals successfully (hope); (3) the individual's tendency to realistically appreciate one's control of life events in order to succeed now and in future (optimism); and (4) the individual's capacity to keep one's mission alive despite challenges and to remain steadfast in the face of adversities (resilience). Therefore, psychological capital is a psychological resource that provides a basis for individuals to succeed at work as they find themselves better equipped to manage daily stressors of work-life. Psychological capital as a psychological resource invokes positive emotions which in turn play their role in influencing positive attitudes such as work engagement [19]. Conversely, empirical studies in the general management literature suggest that psychological resources such as self-efficacy can negatively moderate the relationship between organizational-level variables and individual-level outcomes. For example, Kacmar et al. [20] found that the negative relationship between perceived organizational politics and an individual's job performance is exacerbated by core self-evaluations such as self-efficacy. In another study conducted by Bozeman et al. [21], self-efficacy was found to intensify the negative effects of perceived politics on job satisfaction. Therefore, besides investigating the relationship between EMC and job stress, this study also aims to contribute to psychological capital theory by determining the role PsyCap dimensions play in moderating the relationship between EMC and job stress.

In the extant research, there is ample empirical evidence which suggests that the non-implementation of work-related policies or plans provides the breeding ground for job-related stress [22]. Another stream of research indicates that error also leads to the development of stress in large projects [23]. It is, therefore, important not only to have a climate that promotes the implementation of safety practices but also an environment or climate that provides the basis for error to be managed productively.

Using the job demands-resource (JD-R) model, the present study investigates the impact of error management climate on job stress. It further investigates the moderating role of psychological capital dimensions (hope, optimism, self-efficacy, resilience) for the relationship between error management climate and job stress. Based on the JD-R theory [18], error management climate (EMC) is conceptualized as a potential job resource and psychological capital (PsyCap) as a potential personal resource for the mitigation of employees' job stress.

2. Literature Review

2.1. Theoretical Foundation of Variables

2.1.1. Job Stress

In the past few decades, stress has been a critical problem for organizations [4]. Stress can be categorized as either a stimulus or a response [24]. Job stress refers to psychological strain that leads to tension, anxiety, frustration, job-related hardness, and worry that have roots in one's work [25]. Stress literature points out a lot of key factors, such as workload, management support, psychological support, and work environment, that can affect employees' mental health and psychological emotions [26]. The notion of job stress has gained traction in industrial and organizational management as stress has been found to have a negative influence on the health of working people [11] and to have a role in the impairment of their work performance [27,28].

In organizations, when stress is a result of occupational factors such as required expectations mismatching employees' capabilities, resources, needs, and job demands, it is known as occupational or job stress [23]. Stress exists in every organization either small or big and the place of work becomes complex due to the presence of stress [29].

2.1.2. Error Management Climate

Organizations that follow the "learning from errors" approach have more productive and innovative opportunities [15] and improved safety behavior [30]. Van Dyck, Frese, Baer and Sonnentag [15] argue that error management is comparatively a suitable and supportive approach for an organization as it allows quick error detection, damage control, and learning. Capitalizing on the concept of climate, error management climate is a concept that refers to the shared perception of individuals with regard to error management practices and procedures such as quick error detection and handling of errors, communicating about errors, sharing error knowledge, and helping others in error situations [15].

Error management climate deals with stress and reduces it through reporting, communicating, and sharing with management and other colleagues [16]. A strong error management climate is based on organizational resources such as error communication, error analysis, error competency, and learning from errors [15]. Such resources not only allow employees to improve on their tasks but also provide a basis for handling problems effectively and rendering help when needed. Therefore, the mentioned outcomes of error management climate reduce the employee's turnover intention and job stress [16].

2.1.3. Psychological Capital (PsyCap)

Psychological capital (PsyCap) as a positive psychological state comprises personal resources of hope, efficacy, resilience, and optimism [31]. According to Luthans, Youssef and Avolio [17], Psychological capital is an individual's positive psychological state reflective of the individual's ability to: (1) bounce back from adversity (resilience); (2) strategize alternative pathways with the aim of achieving goals (hope); (3) attribute the reasons for success in a just manner (optimism); and (4) to execute tasks with confidence (self-efficacy).

Psychological capital recognizes the individual's capital and refers to an individual's psychological character development, measurement, and effective management [31,32]. PsyCap has recently received more attention from organizational scholars due to its role in fostering positive behavior and its beneficial effects for an organization [33]. There is a wide range of research in which the relationship between some desirable variables and PsyCap has been examined [6,19,32,34]. The results gathered from the surveys and panel data describe the direct relationship between employees' well-being and psychological capital [35]. Combining the results of different studies into a single study, the coherent analysis showed that there is a strong and direct relationship between PsyCap and workers' behavior, including a worker's psychological well-being, organizational commitment, and job satisfaction [34].

2.2. Research Model and Development of Hypothesis

In this section, the research framework and theoretical basis for hypotheses development are presented. This section presents the relationship between research variables. The current study argues that error management climate (EMC) reduces job stress and that psychological capital (PsyCap) plays a moderating role in the relationship between EMC and job stress.

2.2.1. Job Demands-Resource (JD-R) Model

The job demands-resource (JD-R) model [36] posits that the additive effect of job demands and job resources drives individuals toward either positive or negative behavioral outcomes. Schaufeli and Taris [37] argue that the JD-R model assumes that employee wellbeing and stress are based on the balance between demands (negative) and resources (positive).

Based on the JD-R model, Demerouti et al. [38] argue that every job includes demands as well as resources. Job demands are reflective of elements of a working environment that can lead to stress whereas job resources facilitate work, growth, and learning, and decrease stress levels and stressors of the job [36,39]. Job demands refer to "those physical, social, or organizational aspects of the job that require sustained physical or mental effort and are therefore associated with certain physiological and psychological costs (e.g., exhaustion)" (p. 501). Generally these are energy-consuming efforts at work such as job insecurity, work overload, conflicts, a tense environment, and error-free work requirements. Job resources refer to "those physical, psychological, social, or organizational aspects of the job that may do any of the following: (a) be functional in achieving work goals; (b) reduce job demands at the associated physiological and psychological costs; (c) stimulate personal growth and development"(p. 501) [38]. Job resources are the helping factors in achieving work goals and meeting job demands positively such as social support, performance feedback (which may enhance learning), and job control (which might reduce job demands). Hence, by increasing resources such as job autonomy, job control, social support, climate, a positive workplace, and coworker support, two birds are killed with one stone: stress and negative events are decreased or prevented and positive events are increased [40]. These resources are helpful and stimulate personal growth, development, and learning [38]. The research model is presented in Figure 1.

2.2.2. Error Management Climate and Job Stress

According to Demerouti, Bakker, Nachreiner and Schaufeli [38], resources are helpful in work engagement and decreasing negative events such as stress, burnout, and turnover intentions. Error management climate provides an environment and resources and policies to members so that they can handle and deal with errors more effectively. An error management climate can provide job resources for organizational employees to work in an environment in which they share errors willingly with coworkers and others and seek help and advice from coworkers. At organizations in which strong error management is applied, employees feel more confident and manage errors effectively [15,41]. Error management climate provides a positive organizational environment in which employees help others, gain knowledge about causes of errors, and openly communicate and share their experience about errors. This error-related behavior is helpful for safety compliance [42] and safety citizenship behavior [43].

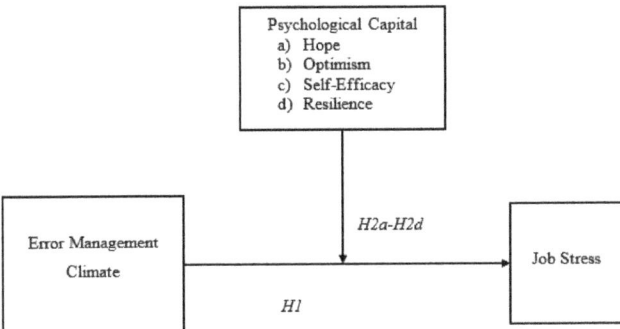

Figure 1. Research model.

Guchait, Paşamehmetoğlu and Madera [16] studied the service industry and noted that strong error management may reduce employees' stress and turnover intention. In a similar vein, Hodges and Gardner [44] have shown that error management climate is negatively related to stress. Error management climate does not remove the errors but instead focuses on changing employees' responses to errors and dealing with an error after its occurrence [16]. When an individual perceives that job demands are high and beyond his perceived ability and resources are not available to achieve goals then the individual b stressed [45]. According to the JD-R model, when job resources are available then organizational members experience less job stress [46]. Thus, a supportive environment enables organizational members to cope with stress. Empirical evidence found that a supportive environment is negatively related to exhaustion, burnout, anxiety, and stress [47]. Given the theoretical reasoning and empirical evidence, it is hypothesized that:

Hypothesis 1. Error management climate is negatively related to job stress.

2.2.3. Psychological Capital Dimensions (Hope, Optimism, Efficacy, Resilience) as Moderators

Credible empirical evidence points out that PsyCap as a higher-order construct plays a significant role in suppressing stress and anxiety. For example, Avey et al.'s [34] meta-analysis and other studies indicate that PsyCap as a personal psychological resource plays an important role in suppressing stress and anxiety and that it is negatively related to undesirable attitudes such as cynicism, turnover intentions, stress, and anxiety. However, there is emerging evidence that suggests that PsyCap's influence as a potential psychological resource becomes diluted under different aspects of organizational and national cultures. For example, in their seminal study, Kacmar, Collins, Harris and Judge [20] found that when perceived organizational politics are combined with core self-evaluations (CSE) such as self-efficacy and locus of control, the deleterious effects of perceived politics on job performance are intensified. Similarly, Rego et al.'s [48] study points out numerous aspects of national culture as potential neutralizers of PsyCap as a resource. They note that organizational cultural aspects such as the absence of performance feedback and lack of clarity on goals could neutralize the positive influence of PsyCap as a resource. Similarly, Rego, Marques, Leal, Sousa and Pina e Cunha [48] note that national cultures characterized by high power distance do not promote proactive and assertive individuals and thus highly self-efficacious individuals find it suitable to be obedient and less assertive.

Referring to Hofstede's [49] insights on national cultures, developing countries such as Pakistan score high on the dimensions of power distance, uncertainty avoidance, and collectivism. People from these cultures are likely to find politics to be high in organizations owing to unequal distribution of power, ambiguity, and chaos, and strong in-groups [50–52]. Hofstede's (2001) insights on the culture of developing countries provide reasonable ground to consider organizational politics to be an inevitable part of organizations working in these countries. This context, therefore, holds a fundamental importance for hypothesizing the moderating role of PsyCap dimensions for the relationship between EMC and job stress.

The concept of locus of control provides a meaningful theoretical distinction between the two similar yet different constructs of hope and optimism [53]. Hope is theorized to be driven by an internal locus of control as opposed to the outer locus of control that feeds optimism. Individuals with an internal locus of control (agency and pathway approach) expect the turn of events as a function of their agency and pathway approach [54]; and hence, they are less susceptible to forces emanating from organizational contexts. It is therefore expected that individuals with high PsyCap hope and PsyCap optimism will yield to negative organizational contexts differently. Hope signifying an individual's ability to strategize alternative pathways in the face of adversities [55,56] and its connection with an internal locus of control [53] is expected to allow individuals to fare better even when the organizational politics impede their expectations to achieve goals and achievements. Therefore, hopeful individuals are expected to take advantage of the prevailing error management climate, resulting in effective management of job stress. In contrast, PsyCap optimism as a function of external locus of control [53] may not be of value to individuals as expectations attached to significant others are compromised in an environment rife with organizational politics [57]. Therefore, optimistic individuals are expected to remain insulated from the theorized benefits of EMC, resulting in poor management of job stress.

Self-efficacy is reflective of an individual's confidence in him/herself to succeed at work [17]. Organizational politics interfering with an individual's chances of succeeding at work is likely to lead an individual to find alternative opportunities where one could employ skills and abilities in the advancement of professional goals. For example, Allen and Griffeth [58] note that high performing individuals are more likely to quit when they find salaries not commensurate with the promotion policies and practices; with this line of reasoning, it is plausible to argue that self-efficacious individuals find organizational politics a hindrance for the advancement of professional goals and so are not expected to capitalize on the benefits of EMC, resulting in the poor management of job stress. Lastly, PsyCap resilience reflective of an individual's capacity to bounce back from adversity [59,60] is expected to provide the basis for individuals to carry on even in a politicized organizational environment. Furthermore, because that resilience plays an important role in replenishing the energy levels of employees and rendering them able to find solutions in difficult organizational circumstances [60], the odds that resilient individuals perceive organizational politics as an obstacle to their work are less [40]. It is therefore expected that individuals with high PsyCap resilience are expected to fare better in cultures characterized by high power distance, uncertainty avoidance, and collectivism. With this line of reasoning, PsyCap resilience is argued to provide the basis for individuals to harness the benefits EMC offers, resulting in the effective management of job stress.

Based on the above theoretical reasoning and empirical evidence it is hypothesized that:

Hypothesis 2a. Hope positively moderates the relationship between error management climate and job stress.

Hypothesis 2b. Optimism negatively moderates the relationship between error management climate and job stress.

Hypothesis 2c. Self-efficacy negatively moderates the relationship between error management climate and job stress.

Hypothesis 2d. Resilience positively moderates the relationship between error management climate and job stress.

3. Research Methodology

3.1. Research Participants

This study analyzes the effect of error management climate on the job-related stress of employees employed in industries related to the development of aerospace and avionics engineering works. The respondents of the current study work in all departments of aerospace and avionics, such as manufacturing, production, support, and light aircraft group.

3.2. Sample and Data Collection Procedure

In this study, the sample is drawn from the employees of the Pakistan aeronautical complex. A sample size of 260 respondents was drawn. The questionnaire was developed with the help of past literatures and empirical studies. Items of the questionnaire were adapted from already developed scales used in the previous researches. The questionnaire was translated into Urdu using the standard translation-back translation procedure [61], as the respondents included frontline workers.

A cross-sectional survey method has been used for data collection in the current study. A total of 250 questionnaires were floated among aeronautical employees, out of which 208 were returned that reflects an 84.8% response rate. Out of 208 responses, 141 respondents were workers (67.8%), 52 respondents were supervisors (25%), and only 14 engineers participated in responding to the questionnaire survey. The demographic characteristics of the sample are summarized in Table 1. After the data was collected from these employees, it was coded into numeric form.

Table 1. Demographic characteristics.

Demographics Category	Frequency	Percentage	Demographics Category	Frequency	Percentage
Age			Designation		
Less than 20 years	4	1.9	Worker	141	67.8
20–30 years	54	26.0	Supervisor	52	25.0
31–40 years	82	39.4	Engineer/Manager	15	7.2
41–50 years	28	13.5	Total Job Experience		
Above 50 years	40	19.2	Less than 1 year	18	8.7
Education			1–5 years	51	24.5
Matric	41	19.7	6–10 years	50	24.0
Intermediate	51	24.5	11–15 years	50	24.0
Bachelor	54	26.0	Above 15 years	39	18.8
Master	61	29.3	Tenure in Current Department		
MS/M.Phil.	1	0.5	Less than 1 year	34	16.3
Employment Status			1–5 years	62	29.8
Permanent	182	87.5	6–10 years	51	24.5
Contractual	20	9.6	11–15 years	44	21.2
Temporary	6	2.9	Above 15 years	17	8.2

Sample size (N) = 208.

3.3. Measures

The questionnaire developed for this study was divided into four parts. The first part included the demographic factors of respondents. It included age, education, total job experience, tenure in the current department, employment status, and designation. The second part included elements of error management climate (EMC), which is the independent variable of this study. The third part included questions related to psychological capital (PsyCap) which is the moderator. The last part included items of job-related stress, which is the dependent variable of this study. All the questions except those of part one were based on a 5-point Likert scale ranging from 1 to 5, where 1 represented "strongly disagree" and 5 represented "strongly agree". Items are scaled because they help the respondent to give an appropriate response by consuming less time [62]. Questionnaires in English as well as in Urdu are reported as Appendices A and B respectively.

3.4. Error Management Climate

In this study, error management climate (EMC), being the independent variable of the study, was measured by sixteen items adapted from the previous study [15,30]. In this scale, one item–"For us, errors are very useful for improving the work process"—was omitted due to a lower internal consistency threshold value (0.6). The Cronbach alpha was 0.976 for 15 items-based EMC in this study (Table 2).

Table 2. Results Summary of measurement model.

Latent Variable	Indicator Codes	Outer Loadings	Cronbach's Alpha (CA)	Composite Reliability (CR)	Average Variance Extracted (AVE)
Error Management Climate	EMC2	0.932	0.976	0.979	0.755
	EMC3	0.922			
	EMC4	0.872			
	EMC5	0.933			
	EMC6	0.857			
	EMC7	0.861			
	EMC8	0.813			
	EMC9	0.664			
	EMC10	0.828			
	EMC11	0.895			
	EMC12	0.927			
	EMC13	0.89			
	EMC14	0.898			
	EMC15	0.854			
	EMC16	0.849			
Efficacy	EFF1	0.879	0.844	0.906	0.764
	EFF2	0.911			
	EFF3	0.83			
Hope	HOP1	0.93	0.937	0.96	0.888
	HOP2	0.961			
	HOP3	0.936			
Optimism	OPT1	0.97	0.937	0.965	0.933
	OPT2	0.962			
Resilience	RES1	0.89	0.87	0.919	0.79
	RES2	0.864			
	RES3	0.913			
Job Stress	JS1	0.767	0.835	0.95	0.596
	JS2	0.807			
	JS4	0.748			
	JS5	0.681			
	JS6	0.691			
	JS7	0.787			
	JS8	0.772			
	JS9	0.756			
	JS11	0.728			
	JS12	0.829			
	JS13	0.845			
	JS14	0.771			
	JS15	0.835			

3.5. Job Stress

In this study, job stress, being the dependent variable, was measured by sixteen items adapted from the study by Parker and DeCotiis [63]. This variable measured the short-term psychological state of job stress. This job stress measure has been used in various previous studies, e.g., [64]. Two items—"My job gets to me more than it should" and "I feel relaxed when I take time off from my job"—were omitted due to a lower internal consistency threshold value (0.6). The Cronbach alpha was 0.944 for 13 items-based job stress measure employed in this study.

3.6. Psychological Capital

Psychological capital (PsyCap), playing the moderating role in the current study, consists of four subscales (i.e., optimism, hope, resilience, and self-efficacy). The PsyCap was measured with the shortened version of the psychological capital questionnaire PCQ-12 developed and validated by Luthans, Avolio, Avey and Norman [31]. In this study hope (4 items), optimism (2 items), self-efficacy (3 items), and resilience (3 items) found Cronbach's alpha value of 0.937, 0.929, 0.844, and 0.870 respectively. The four subscales of PsyCap were measured separately in this study model.

3.7. Data Analysis Technique

Partial Least Squares Structural Equation Modeling (PLS-SEM) was adopted, using the Smart PLS 3.0 software package. PLS-SEM has been used successfully in various researches of a similar kind for assessing the interrelationships among the latent variables [65].

The results of PLS-SEM are based on two sets of models. The first is the measurement model that deals with interrelationships between measurement items and latent constructs. The second is the structural model that shows the relationship results among the latent constructs. The measurement model was assessed by internal consistency reliability convergent validity and discriminant validity [66]. For the assessment of the structural model, path coefficients' t-values and p-values were used. Path coefficients were assessed by adopting bootstrapping. Bootstrapping is a resampling procedure in which the original sample serves as the population.

4. Results

4.1. Measurement Model Evaluation

The measurement model is primarily concerned with the assessment of convergent validity, discriminant validity, and the internal consistency reliability of the constructs of the research model. It is to be noted that the two parameters of loadings of indicator variables and the average variance extracted (AVE) are used to evaluate convergent validity [66]. For convergent validity, the average variance extracted (AVE) threshold should be >0.50. Similarly, the two parameters of Fornell and Larcker and cross-loadings of indicator variables are used to evaluate the discriminant validity.

Table 2 shows the summary of the measurement model. The result shows the Cronbach's alpha and composite reliability of this study to be >0.7 threshold value, which shows the high level of internal consistency and reliability of reflective constructs [67]. Additionally, all outer loadings were greater than 0.50 with the t-values greater than 2.3.

Only four reflective measures are omitted, i.e., EMC1, HOP4, JS3, and JS10. Omitting these reflective measures resulted in an increase in AVE and composite reliability (CR) above the suggested threshold value [68]. Most of the items' outer loading in this study is >0.708 whereas the minimum outer loading of measurement items is equal to 0.664. Three items (EMC9, JS5, JS6) were retained because deletion did not increase AVE and CR above the suggested threshold values. Further, the value of AVE is greater than 0.5 for all constructs that indicate the maximum convergent validity of all constructs (Table 2).

For discriminant validity evaluation, values of cross-loadings and Fornell and Larcker criterion correlation were assessed. Table 3 shows that all the diagonal values are high as compared to the off-diagonal elements in the corresponding rows and columns, indicating that Fornell and Larcker criterion is met and the constructs demonstrate discriminant validity [69]. Table 4 shows that all indicators load on their respective constructs, thereby establishing discriminant validity at the indicator variable level.

Table 3. Correlation Matrix and Square Root of AVE Fornell and Larcker Criterion.

Latent Variables	EMC	Efficacy	Hope	Job Stress	Optimism	Resilience
EMC	0.869					
Efficacy	0.520	0.874				
Hope	0.649	0.821	0.943			
Job Stress	−0.539	−0.477	−0.526	0.772		
Optimism	0.581	0.519	0.524	−0.497	0.966	
Resilience	0.507	0.432	0.460	−0.381	0.826	0.889

Table 4. Cross loadings analysis.

	EMC	Efficacy	Hope	Optimism	Resilience	Job Stress
EMC2	0.664	0.516	0.651	0.571	0.483	−0.526
EMC3	0.828	0.526	0.654	0.591	0.472	−0.562
EMC4	0.895	0.411	0.541	0.513	0.473	−0.461
EMC5	0.927	0.514	0.643	0.539	0.420	−0.488
EMC6	0.890	0.464	0.565	0.539	0.471	−0.442
EMC7	0.898	0.389	0.546	0.452	0.405	−0.438
EMC8	0.854	0.451	0.517	0.445	0.402	−0.421
EMC9	0.849	0.308	0.349	0.274	0.314	−0.289
EMC10	0.932	0.409	0.517	0.455	0.422	−0.460
EMC11	0.922	0.446	0.573	0.515	0.464	−0.441
EMC12	0.872	0.486	0.602	0.543	0.476	−0.544
EMC13	0.933	0.442	0.556	0.562	0.511	−0.520
EMC14	0.857	0.482	0.590	0.457	0.414	−0.422
EMC15	0.861	0.428	0.550	0.519	0.444	−0.432
EMC16	0.813	0.460	0.529	0.500	0.405	−0.489
Eff1	0.376	0.879	0.698	0.453	0.367	−0.401
Eff2	0.470	0.911	0.719	0.430	0.355	−0.440
Eff3	0.515	0.830	0.736	0.481	0.412	−0.409
Hop1	0.609	0.796	0.930	0.501	0.427	−0.486
Hop2	0.593	0.768	0.961	0.477	0.431	−0.511
Hop3	0.634	0.758	0.936	0.506	0.444	−0.489
Opt1	0.601	0.527	0.543	0.970	0.835	−0.503
Opt2	0.517	0.473	0.467	0.962	0.757	−0.454
Res1	0.378	0.309	0.313	0.659	0.890	−0.310
Res2	0.374	0.331	0.380	0.668	0.864	−0.264
Res3	0.560	0.478	0.505	0.840	0.913	−0.411
JS1	−0.551	−0.521	−0.518	−0.565	−0.443	0.767
JS2	−0.481	−0.366	−0.396	−0.404	−0.303	0.807
JS4	−0.444	−0.430	−0.477	−0.493	−0.376	0.748
JS5	−0.314	−0.246	−0.278	−0.268	−0.169	0.681
JS6	−0.324	−0.247	−0.279	−0.183	−0.200	0.691
JS7	−0.494	−0.417	−0.473	−0.362	−0.294	0.787
JS8	−0.339	−0.305	−0.340	−0.289	−0.186	0.772
JS9	−0.279	−0.321	−0.326	−0.301	−0.223	0.756
JS11	−0.323	−0.278	−0.351	−0.304	−0.259	0.728
JS12	−0.413	−0.435	−0.441	−0.431	−0.365	0.829
JS13	−0.405	−0.355	−0.395	−0.400	−0.259	0.845
JS14	−0.454	−0.349	−0.420	−0.347	−0.306	0.771
JS15	−0.426	−0.354	−0.431	−0.419	−0.279	0.835

4.2. Structural Model Evaluation

The structural model was assessed by examining the path coefficients. The R^2 value was used to evaluate the model's predictive accuracy, f^2 to assess the substantial impact of the exogenous variable on an endogenous variable, and Q^2 to evaluate the model's predictive relevance [68].

Structural model prediction power is assessed by the value of R^2 (coefficient of determination). Table 5 shows that the R^2 value for this study is 0.383, that is the combined variation of all independent or exogenous variables can cause 38.3% variance in job stress (endogenous variable), and the Q^2 is larger than zero, which shows the predictive relevance of the model (Table 5).

Table 5. R^2 and Q^2 results.

Endogenous Latent Variable	R^2	Adjusted R^2	Q^2 (=1 − SSE/SSO)	Effect Size
Job Stress	0.383	0.368	0.200	Medium

Small: $0.0 < Q^2$ effect size < 0.15; Medium: $0.15 < Q^2$ effect size < 0.35; Large: Q^2 effect size > 0.35.

The path coefficient is used for structural model assessment and is checked by bootstrapping in Smart PLS. Path coefficient explains how strong one variable influences the other variable; its value must be higher than 0.20 [65]. It is found that three paths (EMC → Job Stress, Optimism → Job Stress, and Efficacy → Job Stress) are significant; on the other side, two paths (Hope → Job Stress and Resilience → Job Stress) are insignificant. However, path relevance is determined by the magnitude of the path coefficients. In this study, the highest path coefficient is that of Mod eff of Optimism → Job Stress (−0.418), followed by EMC → Job Stress (−0.328), and Mod eff of Efficacy → Job Stress (−0.242).

Figure 2 shows the relationship between the studied variables (error management climate, job stress, and psychological capital dimensions). As per the bootstrapping procedure, the significance of path coefficient, p-statistics, and t-values of this study model are shown in Table 6.

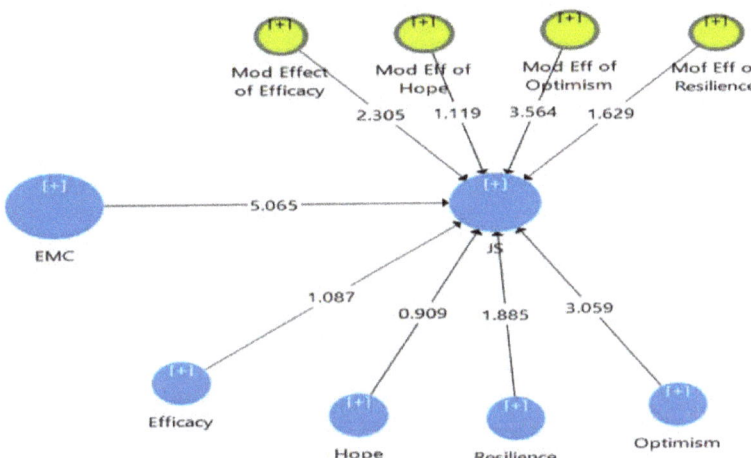

Figure 2. Model constructs relationships.

Table 6. Structural model—Path Coefficients, T-Statistics and Significance of Hypotheses.

Hypotheses		Path Coefficients (β)	T-Values	p-Values	Decision
EMC → Job Stress	H1	−0.328	4.991	0.000	Supported
Mod effect of Hope → Stress	H2a	0.130	1.142	0.254	Not Supported
Mod effect of Optimism → Stress	H2b	−0.418	3.727	0.000	Supported
Mod effect of Self-Efficacy → Stress	H2c	−0.242	2.421	0.016	Supported

Notes: $p < 0.05$ (two tailed); $p < 0.001$ (two tailed).

4.3. Hypothesis Testing

After the validity of the structural model is confirmed, the next step is to assess the paths of the proposed structural model. A total of five hypotheses were proposed in this study. Out of these five hypotheses, one hypothesis is predictive of the direct relationship of the exogenous variable (EMC) on the endogenous variable (Job stress). The other four hypotheses reflect the moderating effect of PsyCap dimensions (hope, optimism, efficacy, and resilience) on the relationship between EMC and the dependent variable (job stress). The hypotheses' results are provided in Table 6 below.

5. Discussion

The purpose of this research was to explore the relationship between error management climate (EMC) and employees' job-related stress. A sample of Pakistani aeronautical employees was used to evaluate error management climate, psychological capital dimensions, and job stress relationships. This study found that error management climate is negatively related to job stress.

Referring to Table 6, the path coefficient for the relationship between EMC and job stress is -0.328, which shows that the individual's perceived organizational error management climate is negatively and significantly associated with job stress [38]. Consistent with the above and in the specific case of the aeronautical employees, it has been found that those who find the organizational climate to be supportive of error management tend to feel low job stress [14]. This study's findings are consistent with the previous study results, e.g., [15,30]. In other words, it could be said that in organizations in which a strong error management climate is provided, employees feel more confident and manage errors effectively [41].

For the moderating role of PsyCap dimensions, Optimism ($\beta = -0.418$, $p = 0.000$) and self-efficacy ($\beta = -0.242$ $p = 0.016$) are found to have a significant negative moderating effect. Therefore, H2b and H2c are accepted. These findings are in line with the findings of Abbas et al.'s [70] study which was also conducted in Pakistan's context. The current study is conducted in the largest and the only aircraft manufacturing facility in Pakistan. This facility operates in the public sector and the personnel's job nature is governed by the Government's policies. Jobs in the public sector at the working-staff level may not appear lucrative owing to tough working environments, continuous pressure to meet deadlines, and almost no incentives on achieving goals and targets. Furthermore, lack of proper feedback and guidance, poor communication, and ambiguous policies and procedures fuel perceived organizational politics [70]. It is possible to argue that organizational politics is a dominant part of Pakistani public sector organizations considering Hofstede's [49] insights on Pakistani culture. Therefore, it could be argued that perceptions of organizational politics when combined with employees' psychological state of self-efficacy and optimism have a role to play in retarding the influence of EMC on job stress.

Results indicate that hope ($\beta = 0.130$, $p = 0.254$) and resilience ($\beta = 0.167$ $p = 0.110$) moderate the relationship between EMC and job stress as hypothesized, but not at statistically significant levels. Therefore, both H2a and H2d are rejected. Results are of significance for understanding that hope and resilience might play a significant role in strengthening the relationship between EMC and job stress provided that organizations are supportive of individuals and provide systemic help in the development and maintenance of psychological resources such as hope and resilience. These results also highlight that the JD-R model in tandem with Hofstede's [49] insights on national cultures holds more relevance in hypothesizing the relationships involving PsyCap dimensions and individual-level outcomes.

6. Conclusions

Current study findings demonstrated that within the context of aeronautical project organizations, error management climate has a direct impact on job stress. This study further suggests that core self-evaluations of individuals in the form of optimism and self-efficacy could have a negative moderating effect on the relationship between EMC and job stress. Thus, it is important to note that the cultivation of an error management climate may not work in combating an individual's stress when an individual's psychological resources are threatened in the wake of organizational politics.

This study's findings are in-line with Kacmar, Collins, Harris and Judge's [20] and Bozeman, Hochwarier, Perrewe and Brymer's [21] findings whereby core self-evaluation in the form of self-efficacy has been found to have counter-productive effects. Furthermore, the results of the study lend support to Avey et al.'s [34] conclusion that industry type and sample base (the US vs non-US) have a significant influence on the effects of PsyCap. This study, nonetheless, provides an alternative perspective on psychological capital which must be investigated further in other countries with similar profiles of power distance and uncertainty avoidance.

The present study has important theoretical implications of error management in several directions. First, it is one of the first studies to investigate the relationship between error management climate (EMC) and job-related stress. Although EMC and stress have been studied independently as important organizational factors [71], their role in the aeronautical industry has been largely neglected. Second, this study is the first to empirically examine error management climate (EMC) in an aeronautical project-based industry context, asserting that EMC principles are relevant to aeronautical employee job stress and need to be applied more extensively. Third, the current study has contributed to the literature on job-related stress by considering the combination of psychological capital (PsyCap) and error management climate (EMC) in the conceptual model.

From a practical perspective, this study's results suggest that interventions can be made from the perspective of error management climate in job-related stress. Considering the negative effect of errors on employee stress, managers should be aware of the benefits error management provides and the effects employees may experience, allowing them to take measures to reduce the errors. In complex organizations, managers should handle error as an event that can provide knowledge and learning, rather than blaming or punishing anyone. Additionally, organizations should promote an environment in which rewards for excellent error recoveries, sharing information, and assisting situations are provided. Where it is important to develop procedures and norms that would be fundamental in cultivating perceptions of error management, it is equally important for management to introduce structural changes in a system for the cultivation of a just culture. Adhering to important elements of justice such as substantive justice, procedural justice, and restorative justice could prove critical in aligning management's efforts to cultivate error management climate. For example, substantive justice underscores the importance of morality and the legitimacy of rules' content [72]. Rules made in isolation and neglecting the requirements of reality may induce pressure on workers to get the job done, paving way for errors that may lead to serious accidents. In a similar vein, procedural justice is what individuals witness and internalize in their subconscious. This internalization later provides a guide for individuals' actions. The cultivation of procedural justice is thought to have a significant role in the successful cultivation of error management climate.. Individuals should be able to witness the investigations in relation to error occurrence through impartial mechanisms. For example, the appointment of objective judges [72] may go a long way in allowing workers to have faith in the procedural justice of the organization, thereby allowing individuals to develop attitudes considered optimum for error management. Lastly, an accountability system based on restorative justice could potentially provide a strong basis for error management climate to develop and thrive. Restorative justice deals with the idea of healing whereby the victims of accidents and those being alleged in accident causation are provided with the opportunity to have their voices heard. Organizations have a crucial role in demonstrating that organizations are not focused on holding individuals responsible for the errors or accidents, rather that their main concern is to understand the principal practices, norms, and work routines that have led to such procedural lapses, errors or accidents. Such an all-inclusive approach is expected to provide firm foundations for EMC to take hold in the organization.

Limitations and Future Directions

The findings of this study like any other research study are not without limitations. The hypothesized moderating influence of hope and resilience did not find support from the data at the statistically significant levels. Although the sample size of this study was determined following the guidelines provided by [73], the relationships must be studied with a larger sample size. Furthermore, this study conducted in the air crafts manufacturing industry may have been influenced by peculiar job routines which may be uncommon in the service industry. Therefore, a similar study in the service industry is recommended to broaden our perspective in understanding the role PsyCap plays in reducing job stress.

Author Contributions: Conceptualization, H.A. and A.M.; methodology, T.T.J.; software, H.A.; validation, R.F.T.; formal analysis, A.M.; investigation, H.A.; resources, A.M.; data curation, T.T.J.; writing—original draft preparation, T.T.J., R.F. and H.A.; writing—review and editing, A.M, R.F., M.A.U.R.T. and R.F.T.; supervision, H.A. and M.A.U.R.T.; project administration, H.A. All authors have read and agreed to the published version of the manuscript.

Funding: This research received no external funding.

Data Availability Statement: Not applicable.

Conflicts of Interest: The authors declare no conflict of interest.

Appendix A Measures Used in the Study (English Version)

Table A1. Error Management Climate.

S.#	Please, Indicate How Strongly You Disagree or Agree with the Following Statements.	Strongly Disagree	Disagree	Neutral	Agree	Strongly Agree
1.	For us, errors are very useful for improving the work process.	1	2	3	4	5
2.	An error provides important information for the continuation of the work.	1	2	3	4	5
3.	Our errors point us at what we can improve.	1	2	3	4	5
4.	When mastering a task, people can learn a lot from their mistakes.	1	2	3	4	5
5.	After an error, people think through how to correct it.	1	2	3	4	5
6.	After an error has occurred, it is analyzed thoroughly.	1	2	3	4	5
7.	If something went wrong, people take the time to think it through.	1	2	3	4	5
8.	After making a mistake, people try to analyze what caused it.	1	2	3	4	5
9.	While working with this organization, people think a lot about how an error could have been avoided.	1	2	3	4	5
10.	Although we make mistakes, we don't let go of the final goal.	1	2	3	4	5
11.	When an error is made, it is corrected right away.	1	2	3	4	5
12.	When an error has occurred, we usually know how to rectify it.	1	2	3	4	5
13.	When people are unable to correct an error by themselves, they turn to their co-workers.	1	2	3	4	5
14.	When people make an error, they can ask others for advice on how to continue.	1	2	3	4	5
15.	If people are unable to continue their work after an error, they can rely on others.	1	2	3	4	5
16.	When someone makes an error, he shares it with others so they don't make the same mistake.	1	2	3	4	5

Table A2. Job Stress.

S.#	How Do You Feel about Your Job? Please Rate the Extent to Which You Agree with the Following Statements by Circling a Number from 1 to 5.	Strongly Disagree	Disagree	Neutral	Agree	Strongly Agree
1.	I have felt fidgety or nervous as a result of my job.	1	2	3	4	5
2.	Working here makes it hard to spend enough time with my family.	1	2	3	4	5
3.	My job gets to me more than it should.	1	2	3	4	5
4.	I spend so much time at work, I can't see the forest for the trees.	1	2	3	4	5
5.	There are lots of times when my job drives me right up the wall.	1	2	3	4	5
6.	Working here leaves little time for other activities.	1	2	3	4	5
7.	Sometimes when I think about my job I get a tight feeling in my chest.	1	2	3	4	5
8.	I frequently get the feeling I am married to the company.	1	2	3	4	5
9.	I have too much work and too little time to do it in.	1	2	3	4	5
10.	I feel relaxed when I take time off from job.	1	2	3	4	5
11.	I sometimes dread the telephone ringing at home because the call might be job-related.	1	2	3	4	5
12.	I feel like I never have a day off.	1	2	3	4	5
13.	Too many people at my level in the company get burned out by job demands.	1	2	3	4	5
14.	I don't have enough time to develop my people.	1	2	3	4	5
15.	People find this place of work uncomfortable.	1	2	3	4	5

Appendix B Measures Used in the Study (Urdu Version)

Table A3. Error Management Climate.

بالکل درست ہے	کچھ حد تک درست ہے	کچھ نہیں کہہ سکتا	کچھ حد تک درست نہیں	بالکل درست نہیں	برائے مہربانی، نشاندہی کریں کہ آپ اس ادارے میں رہتے ہوئے مندرجہ ذیل کیفیات سے کتنا متفق ہیں۔ کسی ایک نمبر کے گرد دائرہ لگائیں۔	نمبر
5	4	3	2	1	کام کرنے کے مراحل کو بہتر بنانے کے لیے غلطیاں ہماری لیے مفید ثابت ہوتی ہیں	1
5	4	3	2	1	کام کو متسلسل برقرار رکھنے کے لیے غلطی میں ہمیں اہم معلومات فراہم کرتی ہے	2
5	4	3	2	1	بھاری غلطیاں اس بات کی نشاندہی کرتی ہیں کہ ہم کیا بہتر کر سکتے ہیں	3
5	4	3	2	1	کسی کام میں مہارت حاصل کرنے کے لیے لوگ اپنی غلطیوں سے بہت زیادہ سیکھ سکتے ہیں	4
5	4	3	2	1	غلطی کے بعد لوگ تفصیلی طور پر سوچتے ہیں کہ اس کو کیسے درست کیا جائے	5
5	4	3	2	1	غلطی بوجھائی کے بعد اس کا بخوبی تجزیہ کیا جاتا ہے	6
5	4	3	2	1	اگر کچھ غلط ہو جائے تو لوگ اس کو مکمل طور پر سمجھنے کے لیے وقت لیتے ہیں	7
5	4	3	2	1	غلطی کرنے کے بعد لوگ اس بات کا تجزیہ کرنے کی کوشش کرتے ہیں کہ اس غلطی کی وجہ کیا تھی	8
5	4	3	2	1	اس اداری کے ساتھ کام کے دوران لوگ اس بارے میں بہت زیادہ سوچتے ہیں کہ غلطیوں سے بہتر طرح بچا جا سکتا ہے	9
5	4	3	2	1	اگرچہ ہم غلطیاں کرتے ہیں، لیکن ہم ہم پھر بھی اپنے اصل ہدف کو نہیں بھولتے	10
5	4	3	2	1	جب غلطی کی جاتی ہے تواس کو درست طریقے میں ٹھیک کیا جاتا ہے	11
5	4	3	2	1	جب کوئی غلطی سرزد ہوتی ہے تو عام طور پر ہم جانتے ہیں کہ اس کو کیسے درست کرنا ہے	12
5	4	3	2	1	جب لوگ خود سے اپنی غلطی درست نہیں کر سکتے تو وہ اپنے ساتھ کام کرنے والے لوگوں سے مدد لیتے ہیں	13
5	4	3	2	1	جب لوگ کوئی غلطی کرتے ہیں تو وہ دوسروں سے مشورہ لے سکتے ہیں کہ کیسے سے آگے بڑھا جا سکتا ہے	14
5	4	3	2	1	اگر لوگوں میں غلطی کرنے کے بعد اپنے کام کو جاری رکھنے کی صلاحیت ہے تو وہ پوری طرح دوسروں پر انحصار کرسکتے ہیں۔	15
5	4	3	2	1	جب کوئی بھی غلطی کرتا ہے تو وہ اپنے دوسروں کو بتاتا ہے، تاکہ دوسری بھی وہی غلطی نہ کرے	16

Table A4. Job Stress.

نمبر		1	2	3	4	5
	آپ اپنی موجودہ ملازمت کی متعلق کیا سوچتی ہیں کسی ایک نمبر کی گرد دائرہ لگائیں	بالکل درست نہیں ہے	کچھ حدتکدرست نہیں ہے	کچھ نہیں کہہ سکتا	کچھ حدتکدرست ہے	بالکل درست ہے
1	مجھی اپنی ملازمت کی وجہ سی گھبراہٹ یا پریشانی محسوس ہوتی ہے	1	2	3	4	5
2	خاندان کی ساتھ مناسب وقت گزاری سی روکتا ہے/ہاں پر کام کرنی کا معمول مجھی اپنی فملی	1	2	3	4	5
3	میری ملازمت مجھی ضرورت سی زیادہ فکر دی ہی	1	2	3	4	5
4	میں کام میں بہت زیادہ وقت گزارتا ہوں، اورکام کوبہت زیادہ باریکی سی دیکھنی کی وجہ سی میں کام کی اہم اصولوں کو نظر انداز کر جاتا ہوں	1	2	3	4	5
5	ایسا وقت اکثر آتا ہی جب میں اپنی کام سی شدید تنگ آجاتا ہوں	1	2	3	4	5
6	ہاں کام کرنی کی بعد باقی کاموں کیلئی کم وقت بچتا ہے	1	2	3	4	5
7	کبھی کبھار جب میں اپنی کام کی باری میں سوچتا ہوں تومیں سنی میں گھٹن محسوس کرتا ہوں	1	2	3	4	5
8	کام کی نوعت کی وجہ سی مجھی اکثر ایسا محسوس ہوتا ہی جیسی میری اس کمپنی سی شادی ہوگئی ہے	1	2	3	4	5
9	مجھی بہت کم وقت میں بہت زیادہ کام کرنا پڑتا ہے	1	2	3	4	5
10	کام سی چھٹی کرنی پر مجھی سکون کا احساس ہوتا ہے	1	2	3	4	5
11	گھریں کبھی کبھار فون کی گھنٹی سن کر مجھی خوف محسوس ہوتا ہی کہ نہ ملازمت سی متعلق ہوگی	1	2	3	4	5
12	مجھی ایسا محسوس ہوتا ہی جیسی مجھیکبھی کام ہی چھٹی نہیں ہوتی	1	2	3	4	5
13	کمپنی میں میری رہنی کی بہت ساری لوگ ملازمت کیمطالبوں سی اکتا گئی ہیں	1	2	3	4	5
14	میری پاس اپنی لوگوں کی تربیت کی لئی مناسب وقت نہیں ہے	1	2	3	4	5
15	لوگ اس جگہ پر کام کرنی کو غیر آرام دہ سمجھتی ہیں	1	2	3	4	5

References

1. Peng, L.; Chan, A.H. Adjusting work conditions to meet the declined health and functional capacity of older construction workers in Hong Kong. *Saf. Sci.* **2020**, *127*, 104711. [CrossRef]
2. Lee, C.; Huang, G.-H.; Ashford, S.J. Job Insecurity and the Changing Workplace: Recent Developments and the Future Trends in Job Insecurity Research. *Annu. Rev. Organ. Psychol. Organ. Behav.* **2018**, *5*, 335–359. [CrossRef]
3. Santos, L.; Melicio, R. Stress, Pressure and Fatigue on Aircraft Maintenance Personal. *Int. Rev. Aerosp. Eng.* **2019**, *12*, 35–45. [CrossRef]
4. Wang, W.; Sakata, K.; Komiya, A.; Li, Y. What Makes Employees' Work So Stressful? Effects of Vertical Leadership and Horizontal Management on Employees' Stress. *Front. Psychol.* **2020**, *11*, 340. [CrossRef] [PubMed]
5. Vui-Yee, K.; Yen-Hwa, T. When does ostracism lead to turnover intention? The moderated mediation model of job stress and job autonomy. *IIMB Manag. Rev.* **2020**, *32*, 238–248. [CrossRef]
6. Wang, D.; Wang, X.; Xia, N. How safety-related stress affects workers' safety behavior: The moderating role of psychological capital. *Saf. Sci.* **2018**, *103*, 247–259. [CrossRef]
7. Dupont, G. The dirty dozen errors in aviation maintenance. In Proceedings of the Eleventh Federal Aviation Administration Meeting on Human Factors Issues in Aircraft Maintenance and Inspection "Human Error in Aviation Maintenance", San Diego, CA, USA, 12–13 March 1997; pp. 45–49.
8. Gemünden, H.G.; Lehner, P.; Kock, A. The project-oriented organization and its contribution to innovation. *Int. J. Proj. Manag.* **2018**, *36*, 147–160. [CrossRef]
9. Turner, R.; Huemann, M.; Keegan, A. Human resource management in the project-oriented organization: Employee well-being and ethical treatment. *Int. J. Proj. Manag.* **2008**, *26*, 577–585. [CrossRef]
10. Liu, J.Y.; Low, S.P. Work–family conflicts experienced by project managers in the Chinese construction industry. *Int. J. Proj. Manag.* **2011**, *29*, 117–128. [CrossRef]
11. Bowen, P.; Edwards, P.; Lingard, H.; Cattell, K. Occupational stress and job demand, control and support factors among construction project consultants. *Int. J. Proj. Manag.* **2014**, *32*, 1273–1284. [CrossRef]
12. Ciampa, P.D.; Nagel, B. AGILE Paradigm: The next generation collaborative MDO for the development of aeronautical systems. *Prog. Aerosp. Sci.* **2020**, *119*, 100643. [CrossRef]
13. Silva, A.V.; Trabasso, L.G. Design for Automation within the aeronautical domain. *J. Braz. Soc. Mech. Sci. Eng.* **2019**, *41*, 292. [CrossRef]
14. Guchait, P.; Paşamehmetoğlu, A.; Dawson, M. Perceived supervisor and co-worker support for error management: Impact on perceived psychological safety and service recovery performance. *Int. J. Hosp. Manag.* **2014**, *41*, 28–37. [CrossRef]
15. Van Dyck, C.; Frese, M.; Baer, M.; Sonnentag, S. Organizational Error Management Culture and its Impact on Performance: A Two-Study Replication. *J. Appl. Psychol.* **2005**, *90*, 1228–1240. [CrossRef] [PubMed]
16. Guchait, P.; Paşamehmetoğlu, A.; Madera, J. Error management culture: Impact on cohesion, stress, and turnover intentions. *Serv. Ind. J.* **2016**, *36*, 124–141. [CrossRef]
17. Luthans, F.; Youssef, C.M.; Avolio, B.J. *Psychological Capital: Developing the Human Competitive Edge*; Oxford University Press: Oxford, UK, 2007.

18. Bakker, A.B.; Demerouti, E. Job demands–resources theory. In *Wellbeing: A Complete Reference Guide*; Cooper, C.L., Ed.; John Wiley & Sons, Ltd.: Hoboken, NJ, USA, 2014; pp. 1–28.
19. Avey, J.B.; Wernsing, T.S.; Luthans, F. Can positive employees help positive organizational change? Impact of psychological capital and emotions on relevant attitudes and behaviors. *J. Appl. Behav. Sci.* 2008, 44, 48–70. [CrossRef]
20. Kacmar, K.M.; Collins, B.J.; Harris, K.J.; Judge, T.A. Core self-evaluations and job performance: The role of the perceived work environment. *J. Appl. Psychol.* 2009, 94, 1572–1580. [CrossRef]
21. Bozeman, D.P.; Hochwarter, W.A.; Perrewe, P.L.; Brymer, R.A. Organizational politics, perceived control, and work outcomes: Boundary conditions on the effects of politics. *J. Appl. Soc. Psychol.* 2001, 31, 486–503. [CrossRef]
22. Casey, T.W.; Krauss, A.D. The role of effective error management practices in increasing miners' safety performance. *Saf. Sci.* 2013, 60, 131–141. [CrossRef]
23. De Silva, N.; Samanmali, R.; De Silva, H.L. Managing occupational stress of professionals in large construction projects. *J. Eng. Des. Technol.* 2017, 15, 488–504. [CrossRef]
24. Matteson, M.T.; Ivancevich, J.M. *Controlling Work Stress: Effective Human Resource and Management Strategies*; Jossey-Bass: San Francisco, CA, USA, 1987.
25. Misis, M.; Kim, B.; Cheeseman, K.; Hogan, N.L.; Lambert, E.G. The impact of correctional officer perceptions of inmates on job stress. *Sage Open* 2013, 3, 2158244013489695. [CrossRef]
26. George, E.; Zakkariya, K.A. Job related stress and job satisfaction: A comparative study among bank employees. *J. Manag. Dev.* 2015, 34, 316–329. [CrossRef]
27. Leung, M.-Y.; Chan, I.Y.S.; Yu, J. Preventing construction worker injury incidents through the management of personal stress and organizational stressors. *Accid. Anal. Prev.* 2012, 48, 156–166. [CrossRef] [PubMed]
28. Wei, W.; Guo, M.; Ye, L.; Liao, G.; Yang, Z. Work-family conflict and safety participation of high-speed railway drivers: Job satisfaction as a mediator. *Accid. Anal. Prev.* 2016, 95, 97–103. [CrossRef] [PubMed]
29. Tiyce, M.; Hing, N.; Cairncross, G.; Breen, H. Employee Stress and Stressors in Gambling and Hospitality Workplaces. *J. Hum. Resour. Hosp. Tour.* 2013, 12, 126–154. [CrossRef]
30. Cigularov, K.P.; Chen, P.Y.; Rosecrance, J. The effects of error management climate and safety communication on safety: A multi-level study. *Accid. Anal. Prev.* 2010, 42, 1498–1506. [CrossRef]
31. Luthans, F.; Avolio, B.J.; Avey, J.B.; Norman, S.M. Positive psychological capital: Measurement and relationship with performance and satisfaction. *Pers. Psychol.* 2007, 60, 541–572. [CrossRef]
32. Sweetman, D.; Luthans, F.; Avey, J.B.; Luthans, B.C. Relationship between positive psychological capital and creative performance. *Can. J. Adm. Sci./Rev. Can. Des Sci. De L'adm.* 2011, 28, 4–13. [CrossRef]
33. Kang, H.J.; Busser, J.A. Impact of service climate and psychological capital on employee engagement: The role of organizational hierarchy. *Int. J. Hosp. Manag.* 2018, 75, 1–9. [CrossRef]
34. Avey, J.B.; Reichard, R.J.; Luthans, F.; Mhatre, K.H. Meta-analysis of the impact of positive psychological capital on employee attitudes, behaviors, and performance. *Hum. Resour. Dev. Q.* 2011, 22, 127–152. [CrossRef]
35. Wu, C.-M.; Chen, T.-J. Collective psychological capital: Linking shared leadership, organizational commitment, and creativity. *Int. J. Hosp. Manag.* 2018, 74, 75–84. [CrossRef]
36. Bakker, A.B.; Demerouti, E. The job demands-resources model: State of the art. *J. Manag. Psychol.* 2007, 22, 309–328. [CrossRef]
37. Schaufeli, W.B.; Taris, T.W. A critical review of the Job Demands-Resources Model: Implications for improving work and health. In *Bridging Occupational, Organizational and Public Health*; Springer: Dordrecht, Germany, 2014; pp. 43–68.
38. Demerouti, E.; Bakker, A.B.; Nachreiner, F.; Schaufeli, W.B. The job demands-resources model of burnout. *J. Appl. Psychol.* 2001, 86, 499–512. [CrossRef] [PubMed]
39. Bakker, A.B.; Demerouti, E.; Verbeke, W. Using the job demands-resources model to predict burnout and performance. *Hum. Resour. Manag. Adv. Hum. Resour. Res. Pract.* 2004, 43, 83–104. [CrossRef]
40. Crawford, E.R.; LePine, J.A.; Rich, B.L. Linking job demands and resources to employee engagement and burnout: A theoretical extension and meta-analytic test. *J. Appl. Psychol.* 2010, 95, 834–848. [CrossRef]
41. Schaufeli, W.B. Applying the job demands-resources model: A 'how to' guide to measuring and tackling work engagement and burnout. *Organ. Dyn.* 2017, 2, 120–132. [CrossRef]
42. Burke, M.J.; Sarpy, S.A.; Tesluk, P.E.; Smith-crowe, K. General Safety Performance: A Test of a Grounded Theoretical Model. *Pers. Psychol.* 2002, 55, 429–457. [CrossRef]
43. Nahrgang, J.D.; Morgeson, F.P.; Hofmann, D.A. Safety at work: A meta-analytic investigation of the link between job demands, job resources, burnout, engagement, and safety outcomes. *J. Appl. Psychol.* 2011, 96, 71–94. [CrossRef]
44. Hodges, M.E.; Gardner, D. Examining the influence of error climate on aviation maintenance performance. *Australas. J. Organ. Psychol.* 2014, 7, E1. [CrossRef]
45. Liu, C.; Li, H. Stressors and stressor appraisals: The moderating effect of task efficacy. *J. Bus. Psychol.* 2018, 33, 141–154. [CrossRef]
46. Steinhardt, M.A.; Dolbier, C.L.; Gottlieb, N.H.; McCalister, K.T. The relationship between hardiness, supervisor support, group cohesion, and job stress as predictors of job satisfaction. *Am. J. Health Promot.* 2003, 17, 382–389. [CrossRef] [PubMed]
47. Halbesleben, J.R. Sources of social support and burnout: A meta-analytic test of the conservation of resources model. *J. Appl. Psychol.* 2011, 96, 182. [CrossRef]

48. Rego, A.; Marques, C.; Leal, S.; Sousa, F.; Pina e Cunha, M. Psychological capital and performance of Portuguese civil servants: Exploring neutralizers in the context of an appraisal system. *Int. J. Hum. Resour. Manag.* **2010**, *21*, 1531–1552. [CrossRef]
49. Hofstede, G. *Culture's Consequences: Comparing Values, Behaviors, Institutions and Organizations Across Nations*, 2nd ed.; Sage Publications: Thousand Oaks, CA, USA, 2001.
50. Romm, T.; Drory, A. Political behavior in organizations—A cross-cultural comparison. *Int. J. Value Based Manag.* **1988**, *1*, 97–113. [CrossRef]
51. Vigoda, E. Reactions to organizational politics: A cross-cultural examination in Israel and Britain. *Hum. Relat.* **2001**, *54*, 1483–1518. [CrossRef]
52. Drory, A.; Vigoda-Gadot, E. Organizational politics and human resource management: A typology and the Israeli experience. *Hum. Resour. Manag. Rev.* **2010**, *20*, 194–202. [CrossRef]
53. Carifio, J.; Rhodes, L. Construct validities and the empirical relationships between optimism, hope, self-efficacy, and locus of control. *Work* **2002**, *19*, 125–136.
54. Gallagher, M.W.; Lopez, S.J. Positive expectancies and mental health: Identifying the unique contributions of hope and optimism. *J. Posit. Psychol.* **2009**, *4*, 548–556. [CrossRef]
55. Snyder, C.R.; Rand, K.L.; Sigmon, D.R. Hope theory. In *Handbook of Positive Psychology*; Snyder, C.R., Rand, K.L., Lopez, S.J., Eds.; Oxford University Press: New York, NY, USA, 2002; pp. 257–276.
56. Luthans, F.; Youssef, C.M.; Avolio, B.J. *Psychological Capital and Beyond*; Oxford University Press: New York, NY, USA, 2015.
57. Poon, J.M. Situational antecedents and outcomes of organizational politics perceptions. *J. Manag. Psychol.* **2003**, *18*, 138–155. [CrossRef]
58. Allen, D.G.; Griffeth, R.W. Job performance and turnover: A review and integrative multi-route model. *Hum. Resour. Manag. Rev.* **1999**, *9*, 525–548. [CrossRef]
59. Luthans, F. The need for and meaning of positive organizational behavior. *J. Organ. Behav.* **2002**, *23*, 695–706. [CrossRef]
60. Youssef, C.M.; Luthans, F. Positive organizational behavior in the workplace the impact of hope, optimism, and resilience. *J. Manag.* **2007**, *33*, 774–800. [CrossRef]
61. Douglas, S.P.; Craig, C.S. Collaborative and iterative translation: An alternative approach to back translation. *J. Int. Mark.* **2007**, *15*, 30–43. [CrossRef]
62. Allen, I.E.; Seaman, C.A. Likert scales and data analyses. *Qual. Prog.* **2007**, *40*, 64–65.
63. Parker, D.F.; DeCotiis, T.A. Organizational determinants of job stress. *Organ. Behav. Hum. Perform.* **1983**, *32*, 160–177. [CrossRef]
64. Jamal, M.; Baba, V.V. Shiftwork and department-type related to job stress, work attitudes and behavioral intentions: A study of nurses. *J. Organ. Behav.* **1992**, *13*, 449–464. [CrossRef]
65. Wong, K.K.-K. Partial least squares structural equation modeling (PLS-SEM) techniques using SmartPLS. *Mark. Bull.* **2013**, *24*, 1–32.
66. Hair, J.F.; Ringle, C.M.; Sarstedt, M. PLS-SEM: Indeed a silver bullet. *J. Mark. Theory Pract.* **2011**, *19*, 139–152. [CrossRef]
67. Hair, J.F.; Sarstedt, M.; Ringle, C.M.; Mena, J.A. An assessment of the use of partial least squares structural equation modeling in marketing research. *J. Acad. Mark. Sci.* **2012**, *40*, 414–433. [CrossRef]
68. Hair, J.F., Jr.; Sarstedt, M.; Hopkins, L.; Kuppelwieser, V.G. Partial least squares structural equation modeling (PLS-SEM) An emerging tool in business research. *Eur. Bus. Rev.* **2014**, *26*, 106–121. [CrossRef]
69. Henseler, J.; Ringle, C.M.; Sarstedt, M. A new criterion for assessing discriminant validity in variance-based structural equation modeling. *J. Acad. Mark. Sci.* **2015**, *43*, 115–135. [CrossRef]
70. Abbas, M.; Raja, U.; Darr, W.; Bouckenooghe, D. Combined effects of perceived politics and psychological capital on job satisfaction, turnover intentions, and performance. *J. Manag.* **2014**, *40*, 1813–1830. [CrossRef]
71. Lei, Z.; Naveh, E.; Novikov, Z. Errors in Organizations: An Integrative Review via Level of Analysis, Temporal Dynamism, and Priority Lenses. *J. Manag.* **2016**, *42*, 1315–1343. [CrossRef]
72. Dekker, S.W.; Breakey, H. 'Just culture:'Improving safety by achieving substantive, procedural and restorative justice. *Saf. Sci.* **2016**, *85*, 187–193. [CrossRef]
73. Cohen, J. A power primer. *Psychol. Bull.* **1992**, *112*, 155–159. [CrossRef]

Article

Intrinsic Workforce Diversity and Construction Worker Productivity in Pakistan: Impact of Employee Age and Industry Experience

Ahsen Maqsoom [1], Hasnain Mubbasit [1], Muwaffaq Alqurashi [2], Iram Shaheen [3], Wesam Salah Alaloul [4,*], Muhammad Ali Musarat [4], Alaa Salman [5], Bilal Aslam [6], Bilel Zerouali [7] and Enas E. Hussein [8,*]

1. Department of Civil Engineering, COMSATS University Islamabad Wah Campus, Wah Cantt 47040, Pakistan; ahsen.maqsoom@ciitwah.edu.pk (A.M.); HasnainM_11@hotmail.com (H.M.)
2. Department of Civil Engineering, College of Engineering, Taif University, P.O. Box 11099, Taif 21944, Saudi Arabia; m.gourashi@tu.edu.sa
3. Department of Management Sciences, COMSATS University Islamabad Wah Campus, Wah Cantt 47040, Pakistan; iramshaheen8765@gmail.com
4. Department of Civil and Environmental Engineering, Universiti Teknologi PETRONAS, Seri Iskandar 32610, Perak, Malaysia; muhammad_19000316@utp.edu.my
5. Department of Civil and Construction Engineering, Imam Abdulrahman Bin Faisal University, Dammam 34212, Saudi Arabia; akalobaidi@iau.edu.sa
6. Department of Earth Sciences, Quaid-i-Azam University, Islamabad 45320, Pakistan; bilalaslam45@gmail.com
7. Vegetable Chemistry-Water-Energy Research Laboratory, Faculty of Civil Engineering and Architecture, Hassiba Benbouali, University of Chlef, B.P. 78C, Ouled Fares, Chlef 02180, Algeria; b.zerouali@univ-chlef.dz
8. National Water Research Center, P.O. Box 74, Shubra El-Kheima 13411, Egypt
* Correspondence: wesam.alaloul@utp.edu.my (W.S.A.); enas_el-sayed@nwrc.gov.eg (E.E.H.)

Abstract: Worker productivity is critical within construction projects as it is the measure of the rate at which work is performed and, more importantly, helps to know how to motivate them to perform at high levels. This research aimed to examine the impact of employee age and industry experience on the intrinsic workforce diversity factors influencing construction worker productivity. Sieving through the previous research and models and theories of analysis, the intrinsic workforce diversity was modeled into the following set of factors, i.e., income, motivation, psychosocial factors, and technical skills. The data were collected by means of a questionnaire survey and examined for the employees having different ages and experiences using the Mann–Whitney U test through SPSS. The results show that employees of varied ages do not concur over motivation-, psychosocial, and technical skills-related workforce diversity factors, whereas employees of varied industrial experiences are in disagreement over some income and motivation related workforce diversity factors. In order to overcome intrinsic workforce diversity, firm support is direly needed for old and mature employees in terms of financial incentives leading to motivation, less supervised scheduling, opportunities for firm advancement, and reporting back every time work is completed. Furthermore, support is required for young employees who are more susceptible due to psychosocial stresses like unevenly distributed work, communication gaps, and technical skills like knowledge of technological equipment and advancement in construction technology which has reduced the skills of workers.

Keywords: workforce diversity; technical skills; motivation; psychosocial; construction worker; productivity

Citation: Maqsoom, A.; Mubbasit, H.; Alqurashi, M.; Shaheen, I.; Alaloul, W.S.; Musarat, M.A.; Salman, A.; Aslam, B.; Zerouali, B.; Hussein, E.E. Intrinsic Workforce Diversity and Construction Worker Productivity in Pakistan: Impact of Employee Age and Industry Experience. *Sustainability* 2022, 14, 232. https://doi.org/10.3390/su14010232

Academic Editor: Elena Cristina Rada

Received: 7 October 2021
Accepted: 22 December 2021
Published: 27 December 2021

Publisher's Note: MDPI stays neutral with regard to jurisdictional claims in published maps and institutional affiliations.

Copyright: © 2021 by the authors. Licensee MDPI, Basel, Switzerland. This article is an open access article distributed under the terms and conditions of the Creative Commons Attribution (CC BY) license (https://creativecommons.org/licenses/by/4.0/).

1. Introduction

With escalating globalization, managing diversity in organizations has turned into a necessarily important issue [1]. There is a broad consensus among scholars about the need to keenly deal with workforce diversity in firms and discover the advantages and drawbacks for the various factors involved in diversity [1–3]. Diversity literature inspects how differences among employees affect team performance [4]. The major effects of those differences yield mixed empirical results on different performance indicators. To make their effects on the performance better understandable, researchers propose a combined analysis of diversity, organizational practices, and different task characteristics.

According to Roberge and Van Dick [5], the foremost vital quality of any organization is a heterogeneous workforce that not only helps mitigate problems, but also supplies different and creative ideas with a competitive advantage to the organization, but diversity can also bring negative outcomes. It has often been acknowledged that heterogeneity can reduce intragroup coherency and lead to disagreements and misunderstandings which, in turn, can lower employee satisfaction, citizenship behaviors, and increase turnover [6], whereas earlier consensus showed that to have strong positive associations with the perceived group and group satisfaction, openness to linguistic, value, and information diversity is essential. Within a team, a homogenous workforce can improve the workflow because there are no social or cultural barriers [7]. In order to manage the growing diversity of the workforce, organizations need to implement systems and practices so that the potential advantages of diversity are maximized and the potential disadvantages are minimized. The influence of demographic differences on work performance needs to be categorized. Particularly on the group level, positive impacts of diversity were shown by Guillaume and Dawson [8] and by Downey and van der Werff [9]. Workforce being a dominant resource in construction, it can be argued that productivity of the construction industry around the globe critically relies on human performance [10,11].

Pakistan is a growing economy with construction being the second largest sector after agriculture. With the China–Pakistan Economic Corridor (CPEC) initiative, workers from both of these demographically diverse countries get employed to work together with national and multinational firms [12]. According to the Planning Commission of Pakistan, the CPEC has employed 30,000 Pakistani engineers and laborers under their Early Harvest projects. These recent steps have led workers from different areas to work and learn together with more interest than ever before. Moreover, teams with diverse strength have subjugated less diverse groups as far as creativity and contentment is concerned [1]. Different perspective about the advantages and disadvantages of diversity in an organization results in different approaches and management ways [13,14]. Therefore, construction worker productivity is a fundamental productivity index of the assignment of manpower to an absolutely specific task.

A few contributions were made by previous studies to the understanding of the effect of workforce diversity factors on employee performance; however, these studies are inconclusive in various dimensions [15–17]. Ibrahim and Brobbey [18] found out that younger workers need more motivation to do an assigned task than older workers; hence, motivation is the most influential factor for employee performance. Van Dalen and Henkens [19] worked on the impact of factors that cause psychosocial stresses and found that employees over the age of 40 face different problems, e.g., tough weather conditions, workspace atmosphere, and long working hours in a firm. Alsuwaiyel [20] examined the consequences of advancement in technology towards unskilled employees in the time to come. Further, he explained the need for old and mature employees to enhance their skill-set with this ongoing technical advancement in order to surpass young age and inexperienced employees. A study held in Belgium proved that the majority of firms in the developing countries have well-documented labor force, overall size, capital used, and productivity, but there is no information on the classification of workforce characteristics that would allow for a more refined breakdown of diversity-defining factors [21].

One of the important issues raised by the ageing society is its impact on productivity, adaptation, and innovation [22]. Age diversity has a definitive negative dependence on performance, but researchers have also expressed that younger and older employees must intermingle to form a coherent and efficient corporate culture and achieve better firm performance. There is a need for different approaches to be developed to study age-based performance evaluation [23]. On the other hand, a meta-analysis by Quińones and Ford [24] revealed a positive relationship between employee experience and performance [19].

A few studies have been able to identify the behavior of experienced and inexperienced employees. By far, no study has examined the association of workforce diversity factors, i.e., income, motivation, psychosocial factors, and technical skills, with the workers' productivity with respect to varied age and experience. This study investigated the impact of workforce diversity factors on worker productivity in the construction industry of Pakistan. Various influencing factors related to income, motivation, psychosocial factors, and technical skills were inspected in this study. Further, the impact of these factors in accordance with age and experience of the employees was analyzed. The outcome of this study will immensely contribute to the literature on labor productivity, where there is a scarcity of literature related to the association between workforce diversity factors, age and experience.

2. Theoretical Framework

The term "diversity" refers to the personnel attributes distributed among codependent members of a work unit [4]. The assorted outcomes of past theories conjointly help to point out the distinction between the theoretical background and the current research.

The job demands–resources (JD–R) model emphasizes the importance of technical skills and psychosocial stress in a workforce. Karasek Jr. [25] proposed that although pressures (both psychological and social) and job demands have an influence on stresses, these demands are not the sole contributors to these stresses. Rather, the amount of stress workers experience in their work majorly depends upon their control over the demands. At the heart of the JD–R model is the assumption that although every occupation may have its own causes of employee well-being, these factors can be classified into two general categories (i.e., job demands and job resources); it thus constitutes an overarching model that may be applied to various occupational settings irrespective of the particular demands and resources involved. In the field of construction, employees from developing countries seem to be losing that control over such demands in sight of comparing themselves with employees in the developed countries who have better technical skills. Factors such as short deadlines, high volume of work, no learning opportunities, etc. are also negative demands that bring down the morale of employees. In this case, the JD–R model postulates job positives/resources that are provided to workers to increase autonomy, organizational rules that strengthen employees, coaching/mentoring, and constructive feedback to generate knowledge/skills in their working environment. JD–R also suggests personal resources such as self-efficacy, optimism, etc. These resources act as a buffer between team members and the demands of their roles.

The equity theory [26] postulates a worker's motivation is a measure of his/her input into the job against the output received from it. The higher the reward, the greater the worker's motivation. Regarding the equity theory, workers who perceive getting better output from their jobs than what they think they put into it feel job satisfaction. These certain aspects vary with every worker's approach toward satisfaction. Tasks that are well-understood and clarified bring up satisfaction since a clear role helps the workforce to be involved and committed. Further, the author identified five characteristics that impact the psychological and motivational state of a worker, namely autonomy, identity and significance of the task, income, and feedback. Further, workers compare their input/output ratio with other workers and if it is found to be fair, it leads to job satisfaction, receiving rewards and motivation. The theory argues that managers should seek to find a balanced way between the inputs that an employee gives and the outputs received. The conceptual framework incorporates technical skills-related and psychosocial workforce diversity factors using the JD–R model. The technical skills related to workforce diversity factors as well as psychosocial factors are examined using the JD–R model, whereas income and motivation related to workforce diversity factors are examined using the equity theory as shown in Table 1.

Table 1. Usage of theoretical models in the conceptual framework of the study.

Theory/Models	Key Components for Diversity	Usage and Framework
Job demands–resources model [25]	Good job positives, e.g., (a) leadership engagement, (b) learning opportunities/skills for advancement, and (c) autonomy, can offset effects of extreme job demands and encourage motivation and engagement	Technical skills, psychosocial factors
Equity theory [26]	Subtle and equitable input and output factors affect each individual's assessment and perception of their work. Dimensions include: (a) personal efforts/satisfaction, (b) rewards (salary, bonuses), and (c) referent to being optimistic.	Income, motivation

3. Workforce Diversity Factors Influencing Worker Productivity

In one stream, there are scholars disputing that a diverse workforce holds a possible market advantage for an organization's success [27–29]. On the other hand, Lauretta McLeod and Lobel [30] stress that a diverse workforce helps in developing cohesion among ideas, turning out to be a success in the end. There are other authors whose position lies within the middle of those flows [31]. However, there seems to be an agreement that if diversity is controlled well, it may enhance productivity, and if unnoticed, it may reduce productivity. Structuring the previous theories to more refined factors, and through the literature support, workforce diversity is classified into the following dimensions: income, motivation, psychosocial factors, and technical skills.

Income has become a subject of serious debate about individuals' economics in the 21st century [32]. Recent developments, such as globalization, have led to vast competition. In order to face this stern competition, firms are continuously looking for ways to improve productivity of their workforce which cannot be achieved without job satisfaction of the working class. Incentives and profit distribution schemes are very common nowadays. In order to handle the problem, a monetary motivation scheme seems to be helpful for both young and old employees [33]. Official labor force statistics derived from administrative data, such as business or taxation records, does not provide a complete picture of trends in the labor market. In lower economic sectors, irregular microdata surveys could be used to have a check upon the work demands and wage policies in that region [34]. Employees with positive attention seem to be much satisfied with the pay relative to people with negative affectivity. Spies [35] studied the differences between the 26 factors of job satisfaction concluding income and the treatment by the parent company of the employees are triggering aspects of the employees' productivity.

Coming to the second factor, motivation is delineated as a force in a person responsible for the meticulousness of efforts spent at work [36]. According to Hofstede [37], managers must verify that the job crew is united and motivated inside the organization through rewards to confirm their continuous commitment. It follows from the top that employment motivation serves two essential purposes: to ensure productivity inside the worksite and list down employees' unity and loyalty through rewards, in particular, for those efforts which will result in higher productivity [38]. It has been observed that young employees need more motivation for their performance as compared to old employees [18]. Clear opportunities for career advancement are an especially powerful employee motivator. Moreover, the impact of job satisfaction and organizational commitment on the performance of employees is more in young employees as compared to old employees.

Stresses causing weak organizational commitment, job dissatisfaction, and demotivation in employees are mentioned as psychosocial stresses. Psychosocial factors include the way work is carried out, i.e., deadlines, workload, work methods, and the context in which work occurs, including relationships and interactions with managers, supervisors, coworkers, and clients or customers [39]. Their study revealed that the impact of job satisfaction, motivation, and organizational commitment on the performance of employees is higher on younger employees as compared to older employees. Another study revealed that psychosocial factors comprising lack of training programs and career mentoring greatly influence the performance of young employees as compared to old employees [40]. Psychological and social stresses have been found to be the most influential factors in carrying out tasks individually and in a group so far [41]. Work discipline, health and safety conditions, work satisfaction, creating competition, relationships with workmates, giving responsibility, sharing problems and their results, social activity opportunities, cultural differences, worker participation in decision-making, distance from home, and distance from population centers have been found to be the key attributes towards psychosocial effects [42–44].

There is ample research evidence suggesting technical skills as a significant factor affecting productivity in the construction industry [45]. For example, Rojas and Aramvareekul [46] reported that skills management and manpower issues are areas of the greatest potential of affecting productivity. Focused on the needs as a result of the globalized market and advantages of workforce diversity, the application of advanced technology has consequences for the relative labor demand [20]. It has been decreasing and will ultimately eliminate the ones without any efficient skills in the near future. Countries with weak economic development consequently exhibit lower levels of employee productivity [47]. A crucial challenge faced by more developed countries is how to make sure that the skills of both experienced and less experienced workers remain consistent throughout their careers. Firms' growth and workers' employability decelerate in the presence of skill gaps [48]. The need to upgrade skills applies not solely to the young generation in schools, universities, and training centers, but also to the current generation of workers following the opportunities to be created in the next two decades [49].

Therefore, a framework for learning the effective management of a diverse workforce and its influence on worker productivity in construction projects is still to be developed. A construction project may be a complicated body that may be planned as a system. However, like Harrison and Price [50] stressed, "people issues" are rather more problematic to unravel than technical problems within the short life of a project. Thus, for assuaging these issues, skills in managing workers are crucial. Therefore, sieving through the previous research and grasping theories, diversity has been carefully molded into the following intrinsic factors: income, motivation, psychosocial factors, and technical skills. The reason why these factors are needed to be studied for this research is that they emphasize the majority part of the difference, enhancements, changes, and processing of the environment that a worker experiences and goes through in their organization. As the JD–R model suggests, increasing work-related resources (autonomy, collaboration with colleagues, mentorship, skills advancement) and selecting staff with proactive personalities and a high level of motivation can increase work engagement and satisfactory employee performance outcomes [51]. The equity theory proposes that communication or status in an organization has an influence on the perceived fairness of pay. Work stresses have an effect on psychosocial strains, but the main pitcher to such strains is the demands that they have to contend with, but fair inputs (income, performance satisfaction, etc.) have a major impact on motivation and psychosocial factors [52]. They are further examined with respect to varied ages and experiences of employees in Pakistani construction contracting firms (CCFs) as shown in Figure 1.

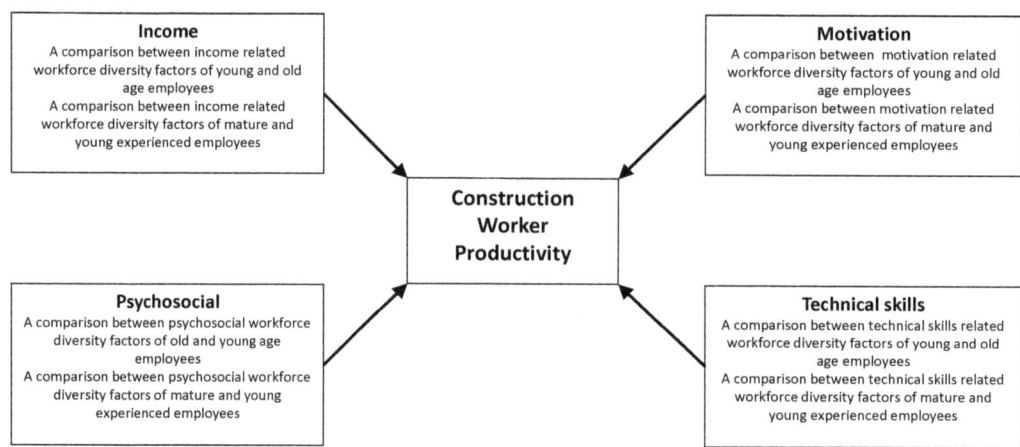

Figure 1. Conceptual framework.

4. Methodology

The computation of the literature reviewed above was used to develop the following research questions:

RQ1. What are the income-related workforce diversity factors that influence the productivity of Pakistani CCFs' workers of varied age and industry experience?

RQ2. What are the motivation-related workforce diversity factors that influence the productivity of Pakistani CCFs' workers of varied age and industry experience?

RQ3. What are the psychosocial workforce diversity factors that influence the productivity of Pakistani CCFs' workers of varied age and industrial experience?

RQ4. What are the technical skills-related workforce diversity factors that influence the productivity of Pakistani CCFs' workers of varied age and industrial experience?

In order to examine the above research questions, a questionnaire was sent to the firms to investigate intrinsic workforce diversity factors and their influence on worker productivity. Pilot testing of the questionnaire was carried out by interviewing eight construction executives. An improved questionnaire was developed by making some essential adjustments to the local industry context based on the recommendations and feedback received during the pilot phase. The questionnaire was divided into three sections incorporating 42 questions in total. The first section consisted of five questions related to the general background of the respondents and the firm. The second section consisted of five questions related to the productivity of employees [15]. The third section consisted of subsections related to four intrinsic factors. For this section, the scales were adapted from well-reputed previous studies, thus ensuring their validity. The first subsection was related to income, comprising eight questions [32], the second subsection consisted of eight questions related to motivation [38], the third subsection consisted of eight questions related to psychosocial stressors [41], and the last subsection related to technical skills consisted of eight questions [45].

To ensure validity of the questionnaire content, the contemporary approach developed by Schriesheim and Powers [53] was used. This approach was also used by Haynes and Weiser [54] in their research regarding surgical safety intervention. The respondents were asked to rate the intrinsic factors (income motivation, psychosocial factors, and technical skills) using a five-point Likert scale ranging from 1 (strongly disagree) to 5 (strongly agree). To reduce the common method variance (CMV) bias, it was conveyed to the correspondents that there were no right or wrong responses. Furthermore, they were assured of the confidentiality of the research so they could respond honestly [55].

Two hundred (200) questionnaires were sent to different engineers working at various construction sites in Pakistan where the profile of the respondents is shown in Table 2. It can be seen that most of the respondents were from firms who were involved in multiple specializations, i.e., involved in more than one business, e.g., civil as well as mechanical, followed by responses from civil and mechanical businesses. All the respondents were registered with the Pakistan Engineering Council (PEC), which is the statutory body responsible for the issuance of licenses to employees working in engineering businesses in Pakistan. Out of the total of 200 questionnaires that were sent to different construction engineers in Pakistan, 131 complete responses were returned, representing a response rate of 66 percent. After receiving the responses, the sample was divided into young vs. old and experienced vs. less experienced employee categories according to age and experience. In order to categorize employees according to their age, various classifications are used by different scholars in their studies, but no clear boundaries have been devised for the criteria. For instance, old employees are considered to be between the age of more than 35 and the age of more than 40 years [56], whereas the World Health Organization classifies old workers as those aged 45 or older. Similarly, the criteria used to distinguish experienced employees from less experienced employees were found to be different in different studies. Chung and Park [57] categorized employees having more than 15 years of work experience as experienced compared to less experienced employees.

Table 2. Profile of the survey respondents.

Specializations	Total Amount of Construction Workers	Responses Received (No.)	Percentage to the Total (%)	Young Workers	Old Workers	Less Experienced Workers	Experienced Workers
Civil	50	41	31	24	17	22	19
Petrochemical	10	5	4	3	2	2	3
Mechanical and electrical	40	31	23	17	14	18	13
Multiple specializations	100	54	42	31	23	28	26
Total	200	131	100	75	56	70	61

In this study, old employees according to age were considered to be those older than 40 years and experienced employees were considered to be those with an experience of more than 10 years [56,58]. The data collected were then analyzed using the Statistical Package for Social Sciences (SPSS), version 21.0. The Shapiro–Wilk normality test was run to assess data normality. The data were found to be not normal, thus necessitating the need for a nonparametric test for the comparison between the studied groups. Hence, the nonparametric Mann–Whitney test was used to compare the scores of workers of different age (young, old) and experience (experienced, less experienced).

5. Findings and Discussion

In this research, only those variables were taken into consideration that recorded statistically significant differences between old vs. young workers and experienced vs. less experienced workers. The other variables having a significant two-tailed p-value more than 0.1 indicated no evidence to reject the null hypothesis. Therefore, the variables which did not record a statistically significant difference in the mean ranks were not documented.

The first stage of the analysis was related to the income-related diversity factors influencing the construction worker productivity. These factors were derived from the previous study based on wages and incremental policies of the workforce [33]. The results obtained indicate a few significant differences in the responses given by workers of varied ages and experiences for the income-related workforce diversity factors influencing the construction worker productivity as shown in Table 3.

Table 3. Income-related workforce diversity factors influencing the construction worker productivity.

Variables	Old Mean Rank	Young Mean Rank	p	Experienced Mean Rank	Less Experienced Mean Rank	p
Basic needs fulfillment	63.27	69.57	0.331	65.39	65.63	0.968
Work demand satisfaction	64.19	67.89	0.531	69.53	60.80	0.124
Satisfactory wage plan	63.39	69.36	0.376	64.49	66.68	0.736
Travel and daily allowances	60.67	74.33	0.043 *	60.00	71.92	0.066 **
Worrying about the living for the next day	62.46	71.05	0.196	59.69	72.28	0.048 *
Annual increments	64.77	66.83	0.759	65.89	65.05	0.896
Wage deductions	63.59	68.99	0.413	61.32	70.38	0.152
Well-paid overtime	64.04	68.17	0.535	62.20	69.35	0.263

Note: * significant at 0.01, ** significant at 0.05.

In terms of age, one variable is regarded more important by younger workers as compared to older workers, i.e., "travel and daily allowances" (mean rank = 74.33 for younger employees and 60.67 for older employees with $p = 0.043$).

Getting extra money for food and fuel is becoming a common routine for employees these days in the developing countries, especially Pakistan, due to high fluctuations in economics in the past decade. Younger employees do not seem to work enthusiastically and productively without such extra allowances compared to past decades, whereas older employees getting decent allowances impart a thought in younger ones. One of the foremost vital elements of running a prosperous business is keeping workers happy and providing them with enough incentives to maximize their productivity. If workers have solid health facilities, insurance plans in place, there is a higher probability that they are going to have a sustainable food supply, regular checkups and take preventative medical steps that ought to facilitate guaranteeing they do not take several sick days [59]. Providing worker advantages might cost a little, and the semipermanent benefits will greatly outweigh those costs and contribute to workers' overall success. Thus, such a business is poised to amass true professionals who are in it for a long time, facilitating a stable workforce.

In terms of experience, two variables are regarded more important by less experienced workers as compared to experienced workers, i.e., "travel and daily allowances" (mean rank = 71.92 for less experienced employees and 60 for experienced employees with $p = 0.066$) and "worrying about the living for the next day" (mean rank = 72.28 for less experienced employees and 59.69 for experienced employees with $p = 0.028$).

Incentives like food, fuel, transport, etc. act as a potential uplift in employees' motivation and, in short, productivity. These incentives definitely appear minor but their impact is quite prominent when analyzed. Experienced employees are always valued; many of them are given attractive salary packages, numerous incentives and securities that keep them motivated to do their work, whereas less experienced employees are not treated in the same way, but such incentives click with their minds [60]. There should be a revision in the policies of firms to add numerous incentives to the packages of their less experienced employees to keep the firm's progress high and the employees motivated.

Worrying about the living for the next day is not just a variable but a condition of many engineers and laborers working in the Pakistani construction industry. At par, 1000–1200 Pak rupees is a wage rate a laborer gets, which is as low as $10–12. Political or injustice strikes hit Pakistan more often, thus leading to the shutdown of businesses, and laborers that survive only on their daily wages do not have food to feed their families. Because of such terrible conditions, many less experienced employees related to the construction industry, primarily laborers, leave for countries abroad, mainly the UAE and other Gulf countries [61]. Wages and job securities of the firms there are far better than those of many of the firms in Pakistan [62]. Wage plans should be revised, and social security should be given to less experienced laborers to keep them motivated. It is not just a matter of keeping them working for their country, but also of the future of their country. If everyone keeps flying to other states like this, Pakistan will soon be deprived of a capable workforce.

The second stage of the analysis was related to the motivation-related diversity factors influencing the construction worker productivity. These factors were derived from the previous study based on the motivation of the workforce [38]. The results obtained indicate several significant differences in terms of age as well as experience of workers for the motivation-related workforce diversity factors influencing construction worker productivity as shown in Table 4.

Table 4. Motivation-related workforce diversity factors influencing the construction worker productivity.

Variables	Old Mean Rank	Young Mean Rank	p	Experienced Mean Rank	Less Experienced Mean Rank	p
Monetary satisfaction	66.42	63.82	0.673	69.80	60.48	0.116
Financial incentives vs. nonfinancial incentives	67.80	61.29	0.310	71.11	58.95	0.048 *
Job security	65.90	64.76	0.857	68.50	62.00	0.284
Responsibility allocation	61.66	72.51	0.097 **	61.28	70.43	0.144
Self-scheduling and less supervised job decisions	69.77	57.70	0.067 **	70.45	59.73	0.089 **
Career advancement	67.29	62.24	0.428	71.39	58.63	0.037 *
Hatred for nosy	72.21	53.24	0.004 *	71.31	58.73	0.047 *
Informal relationships	62.35	71.25	0.166	65.24	65.81	0.926

Note: * significant at 0.01, ** significant at 0.05.

In terms of age, one variable is regarded more important by young workers than by old workers, i.e., "responsibility allocation" (mean rank = 72.51 for younger employees and 61.66 for older employees with $p = 0.097$). On the other hand, two other variables are regarded more important by older employees over younger employees, i.e., "self-scheduling and less supervised job decisions" (mean rank = 69.77 for older employees and 57.7 with $p = 0.067$) and "hatred for nosy" (mean rank = 72.21 for older employees and 53.24 for younger employees with $p = 0.004$).

Responsibility gives a worker confidence over him-/herself, making them learn skills and team-leading qualities. Being a leader and in an authoritative position is always a desire for young employees [63]. In Pakistan, young employees have been given enough responsibilities that the majority of them seem very satisfied and eyeing forward to become more reliable employees. Seeking a satisfactory responsibility in a workplace is necessary for younger employees to keep them motivated towards work more than for the older ones who already have so many responsibilities that many of them are exhausted.

Coming to the second variable, older employees are found to prefer scheduling their work and making job-related decisions with minimum supervision as compared to younger employees [64]. Old workers feel pressed by excessive supervision and tight audits [65]. Over a decade ago, older workers crammed government positions, held places in top management, and the youngest worked on the face lines. Since then, the character of labor is, seemingly, rubbing elbows each day with those in different age groups. Today, Baby Boomers, several of whom have postponed their retirement, have realized themselves working under individuals young enough to be their grandchildren. The older worker/younger boss configuration will produce challenges. Completely different generations have distinctive views on everything, from workplace humor to communication vogue to figure ethics. Baby Boomers might feel awkward taking direction from younger bosses and younger bosses might feel awkward giving directions to them [66].

It is a common fact that an elder is older regardless of whether juniors surpass that person in the organizational race. Being an elder requires that person to be trusted and not just being ordered. Organizations that handle human resources must promote shared values, attitudes, and behaviors while reaching out in ways which are acceptable to every age group. Otherwise, hatred starts building up against young employees and bosses, resulting in an uncomfortable environment. It implies that older employees should be given respect in terms of supervision considering their age and experience, but quality should not be compromised no matter what [66].

In terms of experience, four variables are regarded more important by experienced workers than by less experienced workers, i.e., "financial incentives vs. nonfinancial incentives" (mean rank = 71.11 for experienced employees and 58.95 for less experienced employees with $p = 0.048$), "self-scheduling and less supervised job decisions" (mean rank = 70.45

for experienced employees and 59.73 for less experienced employees with $p = 0.089$), "career advancement" (mean rank = 71.39 for experienced employees and 58.73 for less experienced employees with $p = 0.037$), and "hatred for nosy" (mean rank = 71.31 for experienced employees and 58.73 for less experienced employees with $p = 0.047$).

The analysis depicts that experienced employees emphasize financial incentives over other incentives. Living in a country of struggling economy and economic crises, monetary incentives have always played a productive role in terms of motivation of employees, especially when one has to work at a distant construction site. As an employee becomes more experienced, his/her knowledge becomes more groomed and their observation and decision-making powers become more precise and authentic. Apart from promoting them at a larger scale, incentives should be increased for their dedication to the firm as such a ripe experience cannot be bought with only salaries [67].

"Self-scheduling and less supervised job decisions" was the center of concern for the experienced employees. As the second important variable, it was noted that for an experienced employee, his/her seniority is higher than of those who have just come in the field regardless of how qualified the newcomers are. Therefore, the sense of interference and being supervised makes them resilient [66]. While managing employees who are more experienced than you, it is important to understand how and why they do things their way and only then start implementing changes. While "that is how we have always done" should never be used as an excuse, sometimes things are always done in a certain way for the reasons only an expert would know.

Regarding career advancement, it seems that mature employees got settled in their respective firms and made an identity after working all their lives and moving between different firms, and the firm they are working for now is somehow their final choice where they can see advancements for their career. Comparing to less experienced employees who are yet to touch the peak of their career, they are often seen unsatisfied with their jobs if they do not see advancements coming towards them as seen in the majority of the firms surveyed [68]. Therefore, they are in search of better jobs every day. The last variable, i.e., hatred for nosy, is also very commonly seen in firms with experienced employees. It will be wrong to say that experienced employees should be given full authority to do their work without being asked, but they should be given some authority that makes and keeps them well aware of their place in a workplace and makes them utilize their experience in much better ways [58].

The third stage of the analysis was related to the psychosocial stressors-related diversity factors influencing the construction worker productivity. These factors were derived from a previous study based on work-related psychosocial stressors [41]. The results obtained indicate several significant differences in the responses given by workers of varied ages; however, no significant difference was observed in the responses given by workers of varied experience for the psychosocial stressors-related workforce diversity factors influencing the construction worker productivity, as shown in Table 5.

In terms of age, two variables are regarded as more important by younger workers than by older workers, i.e., "uneven distribution of work" (mean rank = 73.52 for younger employees and 61.11 for older employees with p-value = 0.061) and "workplace satisfaction" (mean rank = 73.7 for younger employees and 61.01 for older employees with $p = 0.049$), whereas, one variable is regarded more important by older workers than by younger workers, i.e., "physically exhausting job" (mean rank = 70.9 for older employees and 55.64 for young workers with $p = 0.019$).

Table 5. Psychosocial workforce diversity factors influencing the construction worker productivity.

Variables	Old Mean Rank	Young Mean Rank	p	Experienced Mean Rank	Less Experienced Mean Rank	p
Uneven distribution of work	61.11	73.52	0.061 **	62.64	68.83	0.330
Relationships with coworkers	63.82	68.57	0.447	67.03	63.72	0.580
Daily task completion	67.23	62.35	0.440	68.28	62.26	0.321
Career advancement	64.63	67.10	0.708	63.88	67.39	0.579
Well-defined tasks	63.32	69.49	0.346	66.74	64.05	0.669
Physically exhausting job	70.90	55.64	0.019 *	66.96	63.80	0.612
Team cooperation	63.45	69.24	0.368	65.05	66.03	0.874
Workplace satisfaction	61.01	73.70	0.049 *	65.14	65.92	0.900

Note: * significant at 0.01, ** significant at 0.05.

Uneven distribution of work is always a problem for a firm or organization where there is a lack of management to utilize proper resources, whether in terms of human resources or machinery; younger employees are always the ones who seem to suffer under such circumstances [41]. Pakistan, since becoming independent, is still in a developing phase. The results show that younger employees do need to be overloaded with work as they are more energetic and enthusiastic, but uneven distribution is never accepted. Creating an inventory of all the work that must be done and then assigning tasks in step with every worker's specific function, position, and strengths is important. This exercise conjointly helps discover any gaps in talent.

Coming to the next variable, "workplace satisfaction", it is a natural fact that every person wants to do respectable work that he/she feels excited and comfortable to tell others about [69]. In the field of construction, younger employees are given enough responsibilities despite uneven distribution of work as discussed earlier that they feel happy to share about their job. In the past 5–8 years, the Pakistani government and the private sector have started many state-of-the-art projects, i.e., metro projects, Bahria Town, and a very recent Orange Line in Lahore, which have created a large space for employment. Numerous engagement surveys have tried to explain the behaviors regarding the workplace that makes workers happy in their jobs. The results found that work values also change as workers grow older [58,70]. Hence, it was concluded that provision of a satisfactory workplace has created a positive impact aside technical abilities in terms of productivity, including employee appreciation and recognition, organizational culture, autonomy, variety learning, and being challenged.

The third and important variable that is highlighted by older employees over younger employees is a "physically exhausting job". Construction is never a white-collar job. Job sites are not always situated in desirable locations with good weather conditions and altitudes, and distances always vary. Due to rapid increases in construction projects in Pakistan, the job demand for older employees has reached very high. Even if younger employees are capable enough to cope with the responsibilities, the experience of older employees can never be taken for granted. Such employees have and always will be in top priority when it comes to decision-making and supervision. In the field of construction, regardless of whether an employee is senior enough, he or she still has to visit distant sites, which might drain off their energies at an older age [44,58]. This can only be minimized by proving good packages and satisfactory job benefits so that older employees feel enthusiastic towards their work [33].

The fourth stage of the analysis was related to the technical skills-related diversity factors influencing the construction worker productivity. These factors were derived from a recent study based on the impact of skills on productivity [47]. The results obtained indicate two significant differences in terms of age and only one significant difference in terms of experience of workers regarding the technical skills-related workforce diversity factors' influence on the construction worker productivity, as shown in Table 6.

Table 6. Technical skills-related workforce diversity factors influencing the construction worker productivity.

Variables	Old Mean Rank	Young Mean Rank	p	Experienced Mean Rank	Less Experienced Mean Rank	p
Technical soundness	62.40	71.16	0.126	65.52	65.48	0.993
Technical training	67.01	62.74	0.477	63.89	67.38	0.546
Importance of vocational training	64.24	67.79	0.583	63.38	67.98	0.459
Vocationally trained employees	62.30	71.34	0.169	61.99	69.60	0.227
Regular interdepartmental trainings	63.48	69.20	0.385	61.99	69.59	0.229
High-technology equipment in the Pakistani construction sector	60.64	74.37	0.034 **	64.16	67.07	0.640
Potential for skills, urban vs. rural	65.47	65.55	0.990	58.31	73.88	0.015 **
Impact of advancements in the construction technology on the skills of workers	61.23	73.30	0.064 ***	62.71	68.75	0.334

Note: ** significant at 0.05, *** significant at 0.1.

In terms of age, two variables are regarded as more important by young workers than by old workers, i.e., "high-technology equipment in the Pakistani construction sector" (mean rank = 74.37 for younger employees and 60.64 for older employees with $p = 0.034$) and "impact of advancements in the construction technology on the skills of workers" (mean rank = 73.3 for younger employees and 61.23 for older employees with $p = 0.064$).

Due to the recent growth of the construction industry of Pakistan, state-of-the-art projects have emerged through collaboration of the government and the private sector. Orange Line in Lahore, Centaurus Mall in Islamabad, Bahria Town, and Neelum Jhelum hydropower are quite prominent among them. Highly advanced technological equipment was used for excavations, pouring, concreting, paving, and in all other phases of construction. This has not only given the construction rapid progress, but also high technical experience to young Pakistani workers and employees as compared to past decades [71].

Coming to the other very important variable, the impact of advancements in the construction technology has weakened the skills of the workers. Change is never an easy thing to cope up with, as the younger employees emphasized, especially when it is related to technologies and skills. Pakistan is not an advanced country in terms of construction, though in the last decade it has seen progressive changes in this industry. The majority of the workers in this industry come from rural areas. Following the basic techniques of masonry, concreting, excavation and other related phases, Pakistani workers are not sufficiently familiar with the latest technologies [71]. Lacking the basic knowledge and skillsets for the use of such equipment, it was found that the majority of young Pakistani workers are resistant to such changes, making them insecure in terms of their job demand.

In terms of experience, only one variable is regarded more important by less experienced workers as compared to experienced workers, i.e., "potential for skills, urban vs. rural" (mean rank = 73.88 for less experienced employees and 58.31 for experienced employees with $p = 0.015$).

Pakistan has been going through its developing phase since becoming independent and is still far behind the world in terms of advancements in technology. There is a very prominent disparity between rural and urban areas here with respect to everything, e.g., food, health, lifestyle, culture, technology, and growth [72]. Old construction techniques are still being followed in the majority of the rural areas, including mud houses, pathways, open sewage, etc. The majority of the less experienced Pakistani employees come from rural areas. When these people come to urban areas for work, they are not very familiar with much of today's construction equipment that is in use. As they have not seen such equipment or built structures or pavements like the ones in urban areas before, they find it very difficult to learn how to use and operate such equipment [71].

6. Conclusions

The stream of the impact of workforce diversity has been examined from a few perspectives by few researchers. Some crosscutting themes on the diverse workforce factors affecting an employee's performance have been provided by the information about the types and roles of diversity in firms. Instead of relying on a sole approach to examine the phenomenon of diversity for Pakistani CCFs, an incorporated framework fetching the key theories on workforce diversity was utilized to gain insight. This research provides a novel insight into the worker productivity literature by showing how workforce diversity is associated with employees' age and experience.

The research analyzed a total of four groups of workforce diversity factors. All these factors were analyzed on the basis of age and experience of an employee, and we conclude that the highlighted factors have a very keen impact on the productivity of workers. The motivation was found to be the artery of a physical workforce. The equity theory of wages depicts that the standard driving force of the performance is motivation and passion. Income and other incentives were found to be of vital importance for younger employees. Job pressures do have an impact on psychosocial strains of young workers, but the main contributor to such strains is control of the employees over the demands they have to deal with, which reduces psychosocial stress. Moving forward, younger employees are found to be more satisfied with the responsibilities they have as compared to older employees. Both the young and the less experienced employees put emphasis on the scheduling of their tasks with less interference.

Conversely, the experienced workers complained about fewer advancement options as a demotivating factor for their careers. Furthermore, the results proposed that older experienced construction workers disliked being asked about the given tasks again and again. When it comes to the uneven distribution of tasks, young workers are found to be the most affected, which results in absenteeism and a negative correlation with the job. That in turn makes such workers reluctant to talk and share about their place of work. Rural areas should be provided with equipment and technologies for construction so that inexperienced rural workers could adapt to such equipment and become more resourceful for the industry.

The verdict of the research has administrative and managerial relevance, with several implications to be observed. The research contributes to worker productivity by showing how employees of varied ages and diverse industry experience are associated with intrinsic workforce diversity factors. This study will help firm owners to reduce the negative relationship of workers with their jobs. It provides insights for contractors and supervisors in Pakistani CCFs on the critical weaknesses associated with management of a well-organized, skillful, and diverse workforce with respect to age and experience of employees which need to be addressed and overcome while undertaking international standards for wages, incentives, opportunities for advancement, health, and sociocultural securities. Another important aspect derived from this research leads to a keen concern about laborers' demotivation, especially in the case of old and experienced workers, and burnout conditions in young workers. The Pakistani government and construction firms need to restructure their labor policies in order to tackle diversity and increase worker productivity.

Author Contributions: All authors contributed equally to this research. All authors have read and agreed to the published version of the manuscript.

Funding: This research was funded by Taif University Researchers Supporting Project (grant No. TURSP-2020/324).

Data Availability Statement: All the data are available within this manuscript.

Acknowledgments: The authors would like to acknowledge the finical support provided by Taif University Researchers Supporting Project (grant No. TURSP-2020/324).

Conflicts of Interest: The authors declare no conflict of interest.

References

1. Podsiadlowski, A.D.; Gröschke, M.; Kogler, C.; Springer, C.; Van Der Zee, K. Managing a culturally diverse workforce: Diversity perspectives in organizations. *Int. J. Intercult. Relat.* **2013**, *37*, 159–175. [CrossRef]
2. Stahl, G.K.; Björkman, I.; Morris, S. *Handbook of Research in International Human Resource Management*; Edward Elgar Publishing: Cheltenham, UK, 2012.
3. McKay, P.F.; Avery, D.R.; Morris, M.A. A tale of two climates: Diversity climate from subordinates' and managers' perspectives and their role in store unit sales performance. *Pers. Psychol.* **2009**, *62*, 767–791. [CrossRef]
4. Jackson, S.E.; Joshi, A.; Erhardt, N.L. Recent research on team and organizational diversity: SWOT analysis and implications. *J. Manag.* **2003**, *29*, 801–830.
5. Roberge, M.-É.; Van Dick, R. Recognizing the benefits of diversity: When and how does diversity increase group performance? *Hum. Resour. Manag. Rev.* **2010**, *20*, 295–308. [CrossRef]
6. Paoletti, J.; Gilberto, J.M.; Beier, M.E.; Salas, E. The role of aging, age diversity, and age heterogeneity within teams. In *Current and Emerging Trends in Aging and Work*; Springer: Berlin/Heidelberg, Germany, 2020; pp. 319–336.
7. Lauring, J.; Selmer, J. Multicultural organizations: Does a positive diversity climate promote performance? *Eur. Manag. Rev.* **2011**, *8*, 81–93. [CrossRef]
8. Guillaume, Y.R.; Dawson, J.F.; Otaye-Ebede, L.; Woods, S.A.; West, M.A. Harnessing demographic differences in organizations: What moderates the effects of workplace diversity? *J. Organ. Behav.* **2017**, *38*, 276–303. [CrossRef]
9. Downey, S.N.; van der Werff, L.; Thomas, K.M.; Plaut, V.C. The role of diversity practices and inclusion in promoting trust and employee engagement. *J. Appl. Soc. Psychol.* **2015**, *45*, 35–44. [CrossRef]
10. Jarkas, A.M. Critical investigation into the applicability of the learning curve theory to rebar fixing labor productivity. *J. Constr. Eng. Manag.* **2010**, *136*, 1279–1288. [CrossRef]
11. Musarat, M.A.; Alaloul, W.S.; Liew, M. Impact of inflation rate on construction projects budget: A review. *Ain Shams Eng. J.* **2020**, *12*, 407–414. [CrossRef]
12. Razzaq, A.; Thaheem, M.J.; Maqsoom, A.; Gabriel, H.F. Critical external risks in international joint ventures for construction industry in Pakistan. *Int. J. Civ. Eng.* **2018**, *16*, 189–205. [CrossRef]
13. Hur, Y.; Strickland, R.A. Diversity management practices and understanding their adoption: Examining local governments in North Carolina. *Public Adm. Q.* **2012**, *36*, 380–412.
14. Maqsoom, A.; Bajwa, S.; Zahoor, H.; Thaheem, M.J.; Dawood, M. Optimizing contractor's selection and bid evaluation process in construction industry: Client's perspective. *Rev. Construcción* **2019**, *18*, 445–458. [CrossRef]
15. Garnero, A.; Kampelmann, S.; Rycx, F. The heterogeneous effects of workforce diversity on productivity, wages, and profits. *Ind. Relat. J. Econ. Soc.* **2014**, *53*, 430–477. [CrossRef]
16. Maqsoom, A.; Ashraf, H.; Choudhry, R.M.; Khan, S.Y.; Dawood, M.; Tariq, A. Extrinsic factors influencing the bid/no-bid decision of construction contracting firms: Impact of firm size and experience. *Rev. Construcción* **2020**, *19*, 146–158. [CrossRef]
17. Alaloul, W.S.; Musarat, M.A.; Liew, M.; Qureshi, A.H.; Maqsoom, A. Investigating the impact of inflation on labour wages in Construction Industry of Malaysia. *Ain Shams Eng. J.* **2021**, *12*, 1575–1582. [CrossRef]
18. Ibrahim, M.; Brobbey, V.A. Impact of motivation on employee performance. *Int. J. Econ. Commer. Manag.* **2015**, *3*, 1218–1237.
19. Van Dalen, H.P.; Henkens, K.; Schippers, J. Productivity of older workers: Perceptions of employers and employees. *Popul. Dev. Rev.* **2010**, *36*, 309–330. [CrossRef]
20. Alsuwaiyel, M.H. *Algorithms: Design Techniques and Analysis (Revised Edition)*; World Scientific: Singapore, 2016; Volume 14.
21. Vandenberghe, V. Is workforce diversity good for efficiency? An approach based on the degree of concavity of the technology. *Int. J. Manpow.* **2016**, *37*, 253–267. [CrossRef]
22. Frosch, K.H. Workforce age and innovation: A literature survey. *Int. J. Manag. Rev.* **2011**, *13*, 414–430. [CrossRef]
23. Joshi, A.; Roh, H. Context matters: A multilevel framework forwork team diversity research. In *Research in Personnel and Human Resources Management*; Emerald Group Publishing Limited: Bingley, UK, 2007; pp. 1–48.
24. Quiñones, M.A.; Ford, J.K.; Teachout, M.S. The relationship between work experience and job performance: A conceptual and meta-analytic review. *Pers. Psychol.* **1995**, *48*, 887–910. [CrossRef]
25. Karasek, R.A., Jr. Job demands, job decision latitude, and mental strain: Implications for job redesign. *Adm. Sci. Q.* **1979**, *24*, 285–308. [CrossRef]
26. Adams, J.S.; Hoffman, B. The frequency of self-reference statements as a function of generalized reinforcement. *J. Abnorm. Soc. Psychol.* **1960**, *60*, 384. [CrossRef] [PubMed]
27. Al-Bayati, A.J.; Abudayyeh, O.; Albert, A. Managing active cultural differences in US construction workplaces: Perspectives from non-Hispanic workers. *J. Saf. Res.* **2018**, *66*, 1–8. [CrossRef] [PubMed]
28. Al-Bayati, A.J.; Abudayyeh, O.; Fredericks, T.; Butt, S.E. Managing Cultural Diversity at US Construction Sites: Hispanic Workers' Perspectives. *J. Constr. Eng. Manag.* **2017**, *143*, 04017064. [CrossRef]
29. Mandell, B.; Kohlergray, S. Management development that values diversity. *Personnel* **1990**, *67*, 41–47.
30. Lauretta McLeod, P.; Lobel, S.A. The effects of ethnic diversity on idea generation in small groups. In *Academy of Management Proceedings*; Academy of Management Briarcliff Manor: New York, NY, USA, 1992; pp. 227–231.
31. Adler, N.J.; Gundersen, A. *International Dimensions of Organizational Behavior*; Cengage Learning: Boston, MA, USA, 2007.

32. Bender, S.; Bloom, N.; Card, D.; van Reenen, J.; Wolter, S. Management practices, workforce selection, and productivity. *J. Labor Econ.* **2018**, *36* (Suppl. S1), S371–S409. [CrossRef]
33. Bonhomme, S.; Hospido, L. The cycle of earnings inequality: Evidence from Spanish social security data. *Econ. J.* **2017**, *127*, 1244–1278. [CrossRef]
34. Jones, S.; Tarp, F. Priorities for Boosting Employment in Sub-Saharan Africa: Evidence for Mozambique. *Afr. Dev. Rev.* **2015**, *27* (Suppl. S1), 56–70. [CrossRef]
35. Spies, M. Distance between home and workplace as a factor for job satisfaction in the North-West Russian oil industry. *Fenn. Int. J. Geogr.* **2006**, *184*, 133–149.
36. Wegge, J.; Jeppesen, H.J.; Weber, W.G.; Pearce, C.L.; Silva, S.A.; Pundt, A.; Jonsson, T.; Wolf, S.; Wassenaar, C.L.; Unterrainer, C. Promoting work motivation in organizations. *J. Pers. Psychol.* **2011**, *9*, 154–171. [CrossRef]
37. Hofstede, G. Motivation, leadership, and organization: Do American theories apply abroad? *Organ. Dyn.* **1980**, *9*, 42–63. [CrossRef]
38. Funso, A.; Sammy, L.; Gerryshom, M. Impact of Motivation on Productivity of Craftsmen in Construction Firms in Lagos, Nigeria. *Int. J. Econ. Financ.* **2016**, *8*, 271. [CrossRef]
39. Safdar, U.; Badir, Y.F.; Afsar, B. Who can I ask? How psychological safety affects knowledge sourcing among new product development team members. *J. High Technol. Manag. Res.* **2017**, *28*, 79–92. [CrossRef]
40. Kakui, I.M. Effects of Career Development on Employee Performance In the Public Sector: A Case of National Cereals and Produce Board. *Strateg. J. Bus. Chang. Manag.* **2016**, *3*, 307–324.
41. Eller, N.H.; Netterstrøm, B.; Gyntelberg, F.; Kristensen, T.S.; Nielsen, F.; Steptoe, A.; Theorell, T. Work-related psychosocial factors and the development of ischemic heart disease: A systematic review. *Cardiol. Rev.* **2009**, *17*, 83–97. [CrossRef]
42. Boyas, J.; Wind, L.H. Employment-based social capital, job stress, and employee burnout: A public child welfare employee structural model. *Child. Youth Serv. Rev.* **2010**, *32*, 380–388. [CrossRef]
43. Glazer, S.; Kruse, B. The role of organizational commitment in occupational stress models. *Int. J. Stress Manag.* **2008**, *15*, 329. [CrossRef]
44. Kazaz, A.; Ulubeyli, S. Drivers of productivity among construction workers: A study in a developing country. *Build. Environ.* **2007**, *42*, 2132–2140. [CrossRef]
45. Abdel-Wahab, M.S.; Dainty, A.R.; Ison, S.G.; Bowen, P.; Hazlehurst, G. Trends of skills and productivity in the UK construction industry. *Eng. Constr. Archit. Manag.* **2008**, *15*, 372–382. [CrossRef]
46. Rojas, E.M.; Aramvareekul, P. Labor productivity drivers and opportunities in the construction industry. *J. Manag. Eng.* **2003**, *19*, 78–82. [CrossRef]
47. Deming, D.J. The growing importance of social skills in the labor market. *Q. J. Econ.* **2017**, *132*, 1593–1640. [CrossRef]
48. Sambasivan, M.; Soon, Y.W. Causes and effects of delays in Malaysian construction industry. *Int. J. Proj. Manag.* **2007**, *25*, 517–526. [CrossRef]
49. Commission, E. *New Skills for New Jobs: Action Now. A Report by the Expert Group on New Skills for New Jobs Prepared for the European Commission*; European Commission: Brussels, Belgium, 2010.
50. Harrison, D.A.; Price, K.H.; Bell, M.P. Beyond relational demography: Time and the effects of surface-and deep-level diversity on work group cohesion. *Acad. Manag. J.* **1998**, *41*, 96–107.
51. Borst, R.T.; Kruyen, P.M.; Lako, C.J. Exploring the job demands–resources model of work engagement in government: Bringing in a psychological perspective. *Rev. Public Pers. Adm.* **2019**, *39*, 372–397. [CrossRef]
52. Al-Zawahreh, A.; Al-Madi, F. The utility of equity theory in enhancing organizational effectiveness. *Eur. J. Econ. Financ. Adm. Sci.* **2012**, *46*, 159–169.
53. Schriesheim, C.A.; Powers, K.J.; Scandura, T.A.; Gardiner, C.C.; Lankau, M.J. Improving construct measurement in management research: Comments and a quantitative approach for assessing the theoretical content adequacy of paper-and-pencil survey-type instruments. *J. Manag.* **1993**, *19*, 385–417. [CrossRef]
54. Haynes, A.B.; Weiser, T.G.; Berry, W.R.; Lipsitz, S.R.; Breizat, A.-H.S.; Dellinger, E.P.; Dziekan, G.; Herbosa, T.; Kibatala, P.L.; Lapitan, M.C.M. Changes in safety attitude and relationship to decreased postoperative morbidity and mortality following implementation of a checklist-based surgical safety intervention. *BMJ Qual. Saf.* **2011**, *20*, 102–107. [CrossRef]
55. Lindell, M.K.; Whitney, D.J. Accounting for common method variance in cross-sectional research designs. *J. Appl. Psychol.* **2001**, *86*, 114. [CrossRef]
56. Idrees, M.D.; Hafeez, M.; Kim, J.-Y. Workers' age and the impact of psychological factors on the perception of safety at construction sites. *Sustainability* **2017**, *9*, 745. [CrossRef]
57. Chung, J.; Park, J.; Cho, M.; Park, Y.; Kim, D.; Yang, D.; Yang, Y. A study on the relationships between age, work experience, cognition, and work ability in older employees working in heavy industry. *J. Phys. Ther. Sci.* **2015**, *27*, 155–157. [CrossRef]
58. Maqsoom, A.; Mughees, A.; Zahoor, H.; Nawaz, A.; Mazher, K.M. Extrinsic psychosocial stressors and workers' productivity: Impact of employee age and industry experience. *Appl. Econ.* **2019**, *52*, 2807–2820. [CrossRef]
59. Srour, F.J.; Srour, I.; Lattouf, M.G. A survey of absenteeism on construction sites. *Int. J. Manpow.* **2017**, *38*, 533–547. [CrossRef]
60. Nadeem, M.; Ahmad, N.; Abdullah, M.; Hamad, N. Impact of employee motivation on employee performance (A Case Study of Private firms: Multan District, Pakistan). *Int. Lett. Soc. Humanist. Sci. (ILSHS)* **2014**, *25*, 51–58. [CrossRef]

61. Kapiszewski, A. Arab versus Asian migrant workers in the GCC countries. In *South Asian Migration to Gulf Countries*; Routledge: London, UK, 2017; pp. 66–90.
62. Maqsoom, A.; Charoenngam, C.; Awais, M. Internationalization process of Pakistani contractors: An exploratory study. In *ICCREM 2013: Construction and Operation in the Context of Sustainability*; 2013; pp. 59–72. Available online: https://scholar.google.com/citations?view_op=view_citation&hl=ja&user=IYQk3foAAAAJ&alert_preview_top_rm=2&citation_for_view=IYQk3foAAAAJ:u5HHmVD_uO8C (accessed on 7 October 2021).
63. Sauermann, H.; Cohen, W.M. What makes them tick? Employee motives and firm innovation. *Manag. Sci.* **2010**, *56*, 2134–2153. [CrossRef]
64. Thomas, K.W.; Velthouse, B.A. Cognitive elements of empowerment: An "interpretive" model of intrinsic task motivation. *Acad. Manag. Rev.* **1990**, *15*, 666–681.
65. Kazaz, A.; Manisali, E.; Ulubeyli, S. Effect of basic motivational factors on construction workforce productivity in Turkey. *J. Civ. Eng. Manag.* **2008**, *14*, 95–106. [CrossRef]
66. Rego, A.; Simpson, A.V. The perceived impact of leaders' humility on team effectiveness: An empirical study. *J. Bus. Ethics* **2018**, *148*, 205–218. [CrossRef]
67. Maqsoom, A.; Mughees, A.; Safdar, U.; Afsar, B.; Ali Zeeshan, B.u. Intrinsic psychosocial stressors and construction worker productivity: Impact of employee age and industry experience. *Econ. Res. Ekon. Istraživanja* **2018**, *31*, 1880–1902. [CrossRef]
68. Whitman, D.S.; van Rooy, D.L.; Viswesvaran, C. Satisfaction, citizenship behaviors, and performance in work units: A meta-analysis of collective construct relations. *Pers. Psychol.* **2010**, *63*, 41–81. [CrossRef]
69. Helm, S. Employees' awareness of their impact on corporate reputation. *J. Bus. Res.* **2011**, *64*, 657–663. [CrossRef]
70. Wey Smola, K.; Sutton, C.D. Generational differences: Revisiting generational work values for the new millennium. *J. Organ. Behav. Int. J. Ind. Occup. Organ. Psychol. Behav.* **2002**, *23*, 363–382. [CrossRef]
71. Gardezi, S.S.S.; Manarvi, I.A.; Gardezi, S.J.S. Time extension factors in construction industry of Pakistan. *Procedia Eng.* **2014**, *77*, 196–204. [CrossRef]
72. Ellison, G.; Glaeser, E.L. The geographic concentration of industry: Does natural advantage explain agglomeration? *Am. Econ. Rev.* **1999**, *89*, 311–316. [CrossRef]

Article

Modified Activated Carbon Synthesized from Oil Palm Leaves Waste as a Novel Green Adsorbent for Chemical Oxygen Demand in Produced Water

Hifsa Khurshid [1,*], Muhammad Raza Ul Mustafa [1,2] and Mohamed Hasnain Isa [3]

1. Department of Civil & Environmental Engineering, Universiti Teknologi PETRONAS, Seri Iskandar 32610, Malaysia; raza.mustafa@utp.edu.my
2. Centre for Urban Resource Sustainability, Institute of Self-Sustainable Building, Universiti Teknologi PETRONAS, Seri Iskandar 32610, Malaysia
3. Civil Engineering Programme, Faculty of Engineering, Universiti Teknologi Brunei, Gadong BE1410, Brunei; mohamed.isa@utb.edu.bn
* Correspondence: hifsa_18002187@utp.edu.my; Tel.: +60-164488362

Abstract: Palm tree waste is one of the most widespread forms of agricultural waste, particularly in areas where oil palms are cultivated, and its management is one of the industry's key concerns. To deal with this palm waste, researchers are working hard to work out the ways to convert this plentiful waste into useful material for future beneficial applications. The objective of this study was to employ chemical activation techniques to prepare a new activated carbon (AC) using discarded oil palm leaves (OPL) in Malaysia. Three chemical agents (H_3PO_4, NaOH and $ZnCl_2$), as well as three pyrolysis temperatures (400 °C, 600 °C and 800 °C) and various impregnation ratios (1:0.5–1:3) were used to optimize the preparation process. As a result, the oil palm leaves activated carbon (OPLAC), with prominent surface properties, was obtained by $ZnCl_2$ activations with a 1:1 impregnation ratio and carbonized at a pyrolysis temperature of 800 °C. The OPLAC-ZC had a surface area of 331.153 m^2/g, pore size of 2.494 nm and carbon content of 81.2%. Results showed that the OPLAC-ZC was able to quickly (90 min) remove the chemical oxygen demand (COD) from produced water (PW), through chemical adsorption and an intraparticle diffusion mechanism. The material followed pseudo-second order kinetic and Freundlich isotherm models. The maximum adsorption capacity of organic pollutants forming COD in PW was found to be 4.62 mg/g (59.6 ± 5%). When compared to previous studies, the OPLAC-ZC showed equivalent or better COD removal capability. It is the first detailed study reporting the preparation of AC from OPL and applying it for organic pollutants adsorption forming COD in PW.

Keywords: wastewater; oil palm leaves activated carbon; chemical activation; COD; adsorption; green technology

Citation: Khurshid, H.; Mustafa, M.R.U.; Isa, M.H. Modified Activated Carbon Synthesized from Oil Palm Leaves Waste as a Novel Green Adsorbent for Chemical Oxygen Demand in Produced Water. *Sustainability* 2022, 14, 1986. https://doi.org/10.3390/su14041986

Academic Editors: Wesam Salah Alaloul, Bassam A. Tayeh and Muhammad Ali Musarat

Received: 6 December 2021
Accepted: 21 January 2022
Published: 10 February 2022

Publisher's Note: MDPI stays neutral with regard to jurisdictional claims in published maps and institutional affiliations.

Copyright: © 2022 by the authors. Licensee MDPI, Basel, Switzerland. This article is an open access article distributed under the terms and conditions of the Creative Commons Attribution (CC BY) license (https://creativecommons.org/licenses/by/4.0/).

1. Introduction

Activated carbons (ACs) have been widely used in adsorption studies due to their chemistry, textural qualities, high surface area and, most notably, surface adsorption properties, for a range of organic and inorganic compounds [1–6]. They have been proven to be extremely useful materials for the adsorption of different contaminants, such as pesticides [7], heavy metals [8,9], dyes [10,11] and hydrocarbons [12]. ACs are produced by combining activation processes (chemical and/or physical) with the thermal decomposition of carbonaceous raw materials [13]. There is growing interest to develop synthetic routes for ACs with stable structures and better adsorption properties, using sustainable and less expensive agricultural by-products. Various materials have been used in this regard, such as tea leaves waste [14], rice husk [15], apricot stones [16], avocado peels [17], pecan nutshell [5], banana peels [18], palm shells [19] and date palm stones [20].

Palm tree waste is one of the most common types of agricultural waste, especially in regions where oil palms are grown. The waste is found in deserts, tropical and semi-tropical regions all across the world [21]. A huge amount of palm waste, including palm seeds, leaves, leaflets, fibers, fronds, trunks, shells and fruit bunches is produced every year in these regions. In the worldwide market, Malaysia is a leading exporter of palm oil and generates huge amounts of waste each year [22]. The management of this waste is one of the major issues in the industry. Open burning is the traditional method of disposing of this waste. It creates smoke, lowers soil fertility and has long-term accumulative impacts on the ecosystem of neighboring countries [23–25]. This method has been banned due to its environmentally hazardous impacts. Palm waste is often littered in milling sites, resulting in contamination of the underground water by leaching [26]. It also releases harmful chemicals into the soil, creates insect colonies and pollutes the atmosphere. Researchers have been determined to investigate ways to turn the plentiful oil palm waste into useful material for future beneficial uses, in an attempt to handle the environmentally daunting palm waste.

Over the years, oil palm waste has been investigated to be used as a precursor material for AC because of its high carbon content. It has been used for the adsorption of various organic and inorganic pollutants in wastewater, such as oil palm shell AC being used for the adsorption of methylene blue (MB) [27] and phenol [28]. Oil palm frond AC was used for adsorption of 2, 4-dichlorophenoxyacetic acid [7] and Janus green dye [29]. Palm empty fruit bunch AC was used for the adsorption of dyes [30] and oil palm kernel shell AC was used for the adsorption of chemical oxygen demand (COD) and color [4]. Oil palm fiber has been reported to have been used for the adsorption of chromium (VI) [31]. However, only a few studies have reported the preparation of AC using the abundantly available oil palm leaves (OPL) waste [32]; no study has been found in Malaysia. Hence, this study focused on the preparation of a novel AC using the OPL waste in Malaysia. The characteristics of AC are dependent on the nature of the precursor material and the activating agent's performance [13]. Chemical activation uses chemical agents for activation, such as alkalis, acids and oxidizing agents, whereas physical activation includes gases, such as steam or CO_2. Usually, chemically activated carbons need a lower temperature and time during thermal decomposition, and also have higher porosity obtained in the final products, when compared to physically activated carbons [33].

The objective of this study was to prepare an oil palm leaves AC (OPLAC) with a comparatively higher surface area and carbon content, to be used as an adsorbent for organic pollutants forming COD in produced water (PW). The PW is a large effluent, generated as a result of oil extraction, accounting for around three times the amount of extracted oil [34]. Three chemical agents, viz., H_3PO_4, NaOH and $ZnCl_2$ were used for the activation of OPLs. The OPLACs were prepared initially at 800 °C with a 1:1 impregnation ratio. The OPLAC with the highest surface area was further evaluated for carbon content at three pyrolysis temperatures (400 °C, 600 °C and 800 °C) and chemical agent to material ratios (1:0.5–1:3). The OPLAC with the highest surface area and carbon content was applied for the adsorption of organic pollutants forming COD in PW. The effects of adsorption factors, such as adsorbent dosage, initial pH and contact time on the COD removal percentage were studied. The equilibrium isotherm and kinetic studies were also performed. A comparative analysis with other studies was made to evaluate the performance of the prepared OPLAC. It is the first detailed study reporting the preparation of AC from OPL and applying it for COD removal in PW. The study will help to extend the application of abundantly available OPL waste, to achieve the goals of sustainable development through waste management.

2. Materials and Methods

2.1. Materials

The OPL waste was collected from FELCRA Berhad Kawasan Nasaruddin Belia, Perak in Malaysia. The chemicals including H_3PO_4, NaOH and $ZnCl_2$ were purchased from

Avantis Laboratory Supply. PW was collected from an oil and gas company operating in Southeast Asia.

2.2. Synthesis of OPLAC

The OPLAC was prepared in three steps. The steps included pre-processing of OPL, processing of OPLAC synthesis and post-processing of OPLAC. In a pre-processing step, the OPL were washed, air dried, cut into small pieces, ground and sieved to 0.5 mm particle size.

In the processing step of OPLAC preparation, the leaf mass was activated using three chemicals and carbonized using three impregnation ratios and temperatures. The purpose was to prepare the OPLAC with the highest surface area and carbon content. As activation agent plays major role in forming the well sized pores and increasing the surface area of ACs, the surface area was considered as evaluation criteria for selection of activation agents. The carbonization majorly depends on the pyrolysis temperature and IR of the material. Therefore, it was considered as the evaluation criteria for the carbon content. The processing steps are shown in Figure 1. Briefly, the leaf mass was chemically activated by wet impregnation technique using an acidic (H_3PO_4), basic (NaOH) and oxidizing agent ($ZnCl_2$). The leaf mass was soaked overnight in chemical solutions of 0.25 M H_3PO_4, 0.1 M NaOH and 0.1 M $ZnCl_2$ with impregnation ratio of 1:1 (w/w). The soaked OPL mass was carbonized in a tube furnace (MTI Corp, Model: OTF-1200X-S) at 800 °C for 2 h with a heating rate of 10 °C/min in the presence of N_2 gas purged at 150 cm^3/min. The prepared ACs were named OPLAC-HP, OPLAC-NO and OPLAC-ZC based on activating agents H_3PO_4, NaOH and $ZnCl_2$, respectively. Among three OPLACs, the OPLAC with the highest surface area was further tested for carbon content by varying temperatures from 400–800 °C and impregnation ratio from 1:0.5–1:3. The surface areas of OPLACs were determined using Brunauer–Emmett–Teller (BET) TriStar II 3020 V1.04. The moderate temperature range of 400–800 °C was selected to avoid incomplete carbonization at low temperatures and higher ash content at high temperatures [21]. The carbon content was determined using a scanning electron microscope equipped with energy-dispersive X-ray spectroscopy (SEM-EDX) technique.

In the post-processing step, the prepared OPLACs were washed repeatedly using 0.1 M HCl and dried in an oven at 105 °C for 12 h. The OPLACs were stored in air-tight storage bottles for further use.

The yield of OPLACs can be calculated using the following equation [35]:

$$\text{Yield of OPLAC} = \frac{\text{g of OPLAC}}{\text{g of dried OPL}} \times 100 \tag{1}$$

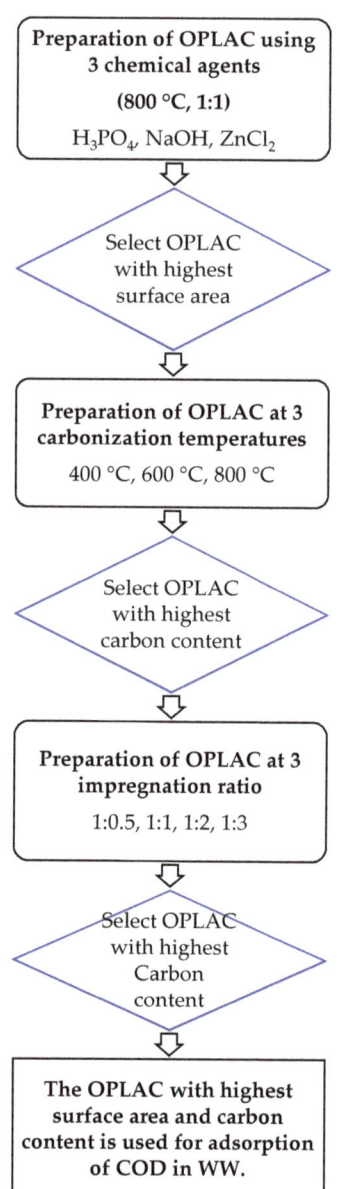

Figure 1. Schematic diagram of OPLAC preparation.

2.3. Characterization of AC and PW

The surface functional groups of OPLAC were analyzed by Fourier transformed infrared spectroscopy (FTIR). The Nicolet Magna 550 spectrometer in KBr was used to obtain FTIR spectra of 5 mg finely ground sample with a scan range of 400–4000 cm^{-1}. Elemental composition was found by Zeiss EVO LS15 SEM-EDX under high vacuum condition and 15 kV EHT. The surface areas and pore volumes of OPLACs were determined using BET analysis on isotherm obtained by nitrogen adsorption-desorption experiment. The sample was first degassed at 300 °C for 240 min with 10 °C/min ramp rate under vacuum environment. Then nitrogen adsorption-desorption experiment was conducted for up to 13 selected equilibrium points. The quantitative and qualitative analysis of the chemical composition of OPLACs was conducted using X-ray photoelectron spectroscopy (XPS) (Micromeritics, Tristar II) with Al K alpha source and background spot of 400 μm. The crystalline properties of OPLAC were determined by X-ray diffraction (XRD) analysis using a Panalytical X'pert3 powder diffractometer. The scanning angle (2θ) was 2–79.99° with a step size of 0.026 s/step.

The water quality parameters of PW are given in supplementary information in Table S1. The PW was mainly analyzed for COD content using HACH DR spectrophotometer and high range COD vials. Briefly, 2 mL of PW sample was poured to the COD vial and kept in a heated digester at 150 °C for 2 h. A blank was also prepared using similar method. The COD concentration (mg/L) was measured after 2 h using USEPA method 800 under programme 430 [36]. The total COD content of PW was 2000 mg/L, whereas the dissolved COD content was 1344 mg/L. Suspended COD was removed using suction filtration unit.

2.4. Adsorption Batch Experiments

To examine the dissolved COD removal capability of OPLAC, batch experiments were conducted at various pH, dosage of AC and contact times. PW was filtered through Whatman filter paper of 45 mm Ø, using suction filtration unit to remove the suspended particles. The experiments were conducted using a hot plate with magnetic stirrer, taking 50 mL of PW in a beaker. Adsorption capacity was evaluated at pH range of 2–12, contact time of 10 min–24 h and adsorbent dosage of 10–3000 mg/L. The samples were placed for stirring at 220 rpm at 25 ± 1 °C. After termination of the stirring time, samples were taken out, filtered with 0.45 μm syringe filter and tested for final COD amount using spectrophotometer.

2.5. Adsorption Isotherm Modelling

To analyze the experimental data from adsorption equilibrium investigations, the Langmuir isotherm model (LIM) and Freundlich isotherm model (FIM) were employed. The experiments were performed for 10–3000 mg/L OPLAC dosages at 25 ± 1 °C.

LIM is the simplest adsorption model and applies to ideal conditions. The model is applicable for monolayer physical adsorption of fluids [37]. Its assumptions include; (i) the adsorption is deemed complete when a monolayer covers all active sites, (ii) each site can hold one molecule that has been adsorbed, the surface is homogeneous and all active sites are identical and, (iii) the occupancy of adjacent sites has no effect on the adsorption of a molecule at a given site [38]. Equation (2), below, represents the linear form of LIM [39]:

$$\frac{C_e}{q_e} = \frac{1}{K_L q_{max}} + \frac{C_e}{q_{max}} \quad (2)$$

where q_e represents the mg of adsorbate adsorbed per gram of adsorbent and C_e represents the equilibrium adsorbate concentration in solution. K_L (L/mg) is the Langmuir constant, and q_{max} is the maximum adsorption capacity of adsorbate (mg/g) [40].

The FIM is mostly applied to characterize the non-ideal adsorption properties of heterogeneous surfaces [41]. It considers that there are various types of accessible sites operating at the same time, each with a different sorption energy [42]. Equation (3) represents the linear form of the FIM [43], shown here:

$$\log q_e = \log K_f + \frac{1}{n} \log C_e \quad (3)$$

where K_f is FIM constant (mg/g) and n is adsorption intensity. When $\frac{1}{n} = 1$, it shows that partitioning between the two phases is unaffected by concentration. When $\frac{1}{n} < 1$, the adsorption is normal. Conversely, a value of $\frac{1}{n} > 1$ implies cooperative adsorption [44].

2.6. Adsorption Kinetic Modelling

Pseudo-first order (PFO), pseudo-second order (PSO) and Weber and Morris (WM) intraparticle diffusion models were used to study the kinetics of COD removal by OPLAC. Batch experiments were conducted from 10 min to 24 h duration. All kinetic models were evaluated based on the R^2 value to compare their performance in describing the kinetics of COD removal by OPLAC.

PFO kinetic model explains the relationship between the rate at which adsorbent sorption sites are occupied and the number of empty sites [45]. The model is based on the premise that adsorption rate is governed by adsorbate diffusion. Equation (4), shown below, is the linear form of the PFO kinetic model:

$$\ln(q_e - q_t) = \ln q_e - K_1 t \qquad (4)$$

where K_1 is rate constant of PFO adsorption (min^{-1}), q_t is adsorption capacity at time t (mg/g) and q_e is equilibrium adsorption capacity (mg/g). The model's parameter K_1 and q_e are calculated by plotting $\ln(q_e - q_t)$ vs. t.

The PSO kinetic model, which is represented by Equation (5), explains the dependence of the adsorbent's adsorption capacity on time [46]. The model assumes that the chemical adsorption is the rate limiting step due to surface adsorption interactions. Equation (5) is shown below:

$$\frac{t}{q_t} = \frac{1}{K_2 q_e^2} + \frac{t}{q_e} \qquad (5)$$

where K_2 is second order rate constant of adsorption (g/mg/min). The model's parameter K_2 and q_e are calculated by plotting $\frac{t}{q_t}$ vs. t.

Weber and Morris (WM) Intraparticle diffusion model is used to understand the intraparticle diffusion process and to find out the rate controlling step [47]. Equation (6) shows the linear form of the model. When intraparticle diffusion is the governing mechanism, the plot of q_t versus $t^{1/2}$ is a straight line that passes through origin. Otherwise multiple processes are involved to regulate the adsorption [48]. Equation (6) is shown below:

$$q_t = K_i t^{1/2} + C \qquad (6)$$

where K_i is rate constant of intraparticle diffusion (mg/g/min$^{1/2}$) and C (mg/g) is intercept of intraparticle diffusion. The model's parameters K_i and C are calculated by plotting q_t vs. $t^{1/2}$.

3. Results and Discussion

3.1. Surface Areas of OPLACS

The three ACs, viz., OPLAC-HP, OPLAC-NO and OPLAC-ZC were characterized for their surface area, total pore volume, pore size and percentage yield (Table 1). The results showed that OPLAC-ZC had the largest BET surface area, pore volume and pore size, compared to the other prepared ACs. A larger surface area indicated the availability of more adsorption sites for OPLAC-ZC, compared to OPLAC-HP and OPLAC-NO.

Table 1. Physical properties of OPLACs.

Activated Carbon	BET Surface Area (m^2/g)	Total Pore Volume (cm^3/g)	Average Pore Size (nm)	Yield (%)
OPLAC-HP	255.840	0.166	2.5	42
OPLAC-NO	267.719	0.0521	0.8	46
OPLAC-ZC	331.153	0.206	2.5	42

3.2. Surface Functional Groups of OPLAC-ZC

OPLAC-ZC not only had dominant surface properties but also surface functional groups, evident from the FTIR spectra. For the analysis of functional groups present at the surfaces, the three ACs were characterized using FTIR. The results are given in Figure 2. The FTIR spectra of OPLAC-HP, OPLAC-NO and OPLAC-ZC showed broad peaks around 3414, 3437 and 3424 cm^{-1}, respectively, suggesting the presence of a −OH stretching vibration in the hydroxyl groups [49]. The peaks for OPLAC-ZC had the lowest transmittance values, showing the presence of higher hydroxyl groups at the surface, compared to OPLAC-HP and OPLAC-NO. An intense peak of C=O stretching compounds including esters, aliphatic ketones and aldehydes was observed for OPLAC-ZC (1725 cm^{-1}), compared to OPLAC-HP (1730 cm^{-1}) and OPLAC-NO (1733 cm^{-1}) [50]. The other peaks, around 1078, 1089, and below 470 cm^{-1} wavenumbers suggested the presence of C-O and C=O stretchings and other complex deformations at the surface of the OPLACs [51]. The transmittance values were lower for OPLAC-ZC, indicating the presence of a higher number of functional groups at the surface. It indicated that OPLAC modified by ZnCl$_2$ had superior surface properties for the adsorption of organic pollutants forming COD, compared to the other two OPLACs.

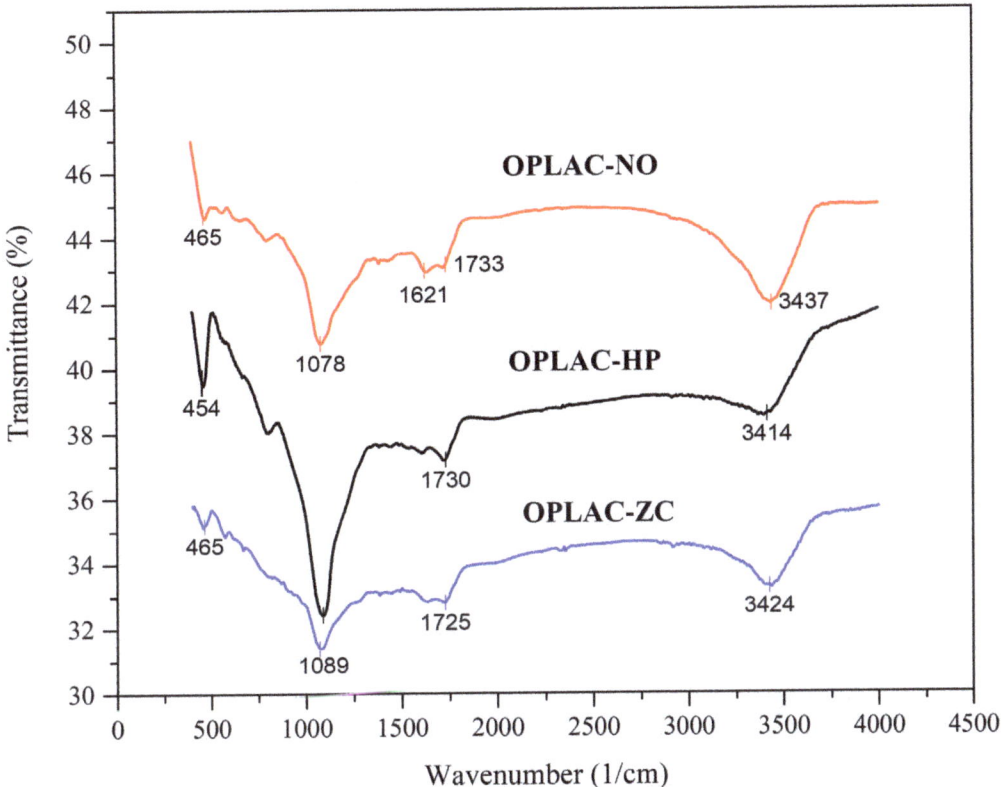

Figure 2. FTIR spectra of OPLACs modified by NaOH, H$_3$PO$_4$ and ZnCl$_2$.

3.3. Elemental Composition of OPLAC-ZC

OPLAC-ZC was further characterized for its carbon content at various temperatures and impregnation ratios. Temperature ranging from 400–800 °C was used for the pyrolysis of ZnCl$_2$ impregnated OPL biomass (1:1). The carbon content was determined using SEM-EDX mapping. The results (Table 2) showed that increasing the temperature, from 400 °C to 800 °C, increased the carbon content of OPLAC-ZC, from 68.3% to 81.2%. The temperature was not increased beyond 800 °C, to avoid higher ash content in OPLAC-ZC. Therefore, 800 °C was considered to be the most suitable temperature for the preparation of OPLAC-ZC.

Table 2. Elemental composition of OPLAC-ZC at various temperatures and impregnation ratios.

Elements	C	O	Zn	Si	Ca
Temperature °C					
400	68.3	20.2	2.8	1.2	0.4
600	74.2	14.29	3.2	2.2	0.9
800	81.2	9.6	4	4.2	1.1
Impregnation ratio					
1:0.5	76.4	11.3	0.8	6.6	3
1:1	81.2	9.6	4	4.2	1.1
1:2	70.9	15.9	1.7	6.1	1.9
1:3	68.3	11.6	7.4	5.4	4.3

The carbon content of OPLAC-ZC was analyzed further using various impregnation ratios at 800 °C. The highest carbon content, of 81.2%, was observed at an impregnation ratio, i.e., OPL biomass to $ZnCl_2$ ratio of 1:1 (Table 2).

Figure 3 shows the SEM images of OPLAC-ZC at the three temperatures and impregnation ratios. The images show that the porous structure was more apparent when the temperature was increased from 400 to 800 °C and at an impregnation ratio of 1:1. Hence, OPLAC-ZC modified by $ZnCl_2$, with an impregnation ratio of 1:1 and carbonized at a pyrolysis temperature of 800 °C, was considered the most suitable adsorbent prepared from OPL.

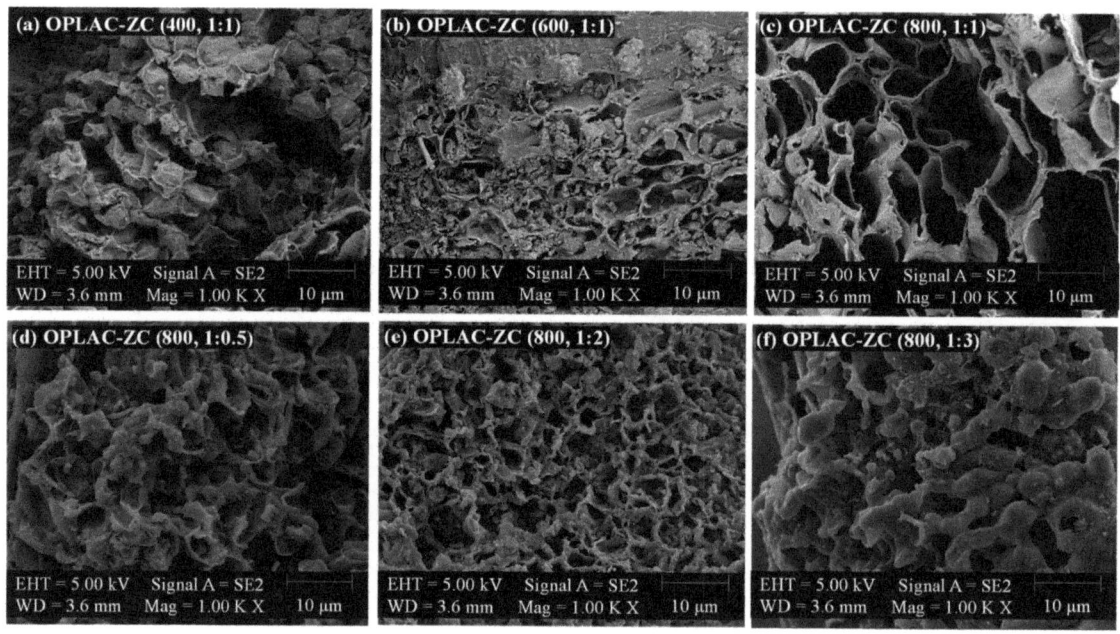

Figure 3. SEM images of OPLAC-ZC at temperatures of (a) 400 °C (1:1); (b) 600 °C (1:1); (c) 800 °C (1:1); impregnation ratios of (d) 1:0.5 (800 °C); (e) 1:2 (800 °C); (f) 1:3 (800 °C).

3.4. Chemical Composition and Crystallinity of OPLAC-ZC

Figure 4a represents the XPS analysis of OPLAC-ZC, where larger peaks of C and O can be observed at the surface. Figure 4b represents the C1s XPS spectra of the OPLAC-ZC, showing binding energies within a range of 284–289 eV. The peaks were assigned to C-C (284.98 eV), C-O/C-COO (286.28 eV), C-N/C-CON (287.28) and C=O/O-C=O

(288.58) functional groups [52]. The O1s spectra of OPLAC-ZC showed peaks at 531.18 eV, 532.18 eV, 533.18 eV and 532.28 eV (Figure 4c). The peaks were assigned to ZnO, C=O, O and C-O functional groups, respectively. The crystallinity of OPLAC-ZC was studied by XRD analysis, which is represented in Figure 4d. The multiple broad peaks in the range of 20–50 °C represented the 0 0 2 and 1 0 0 planes, reflecting the graphitic lattice structure of OPLAC-ZC [53]. The percentage of crystallinity is found by taking the ratio of the area of the crystalline peaks and the area of all the peaks in the XRD graph. The percentage of crystallinity was found to be 64.57%, using OriginPro. The broader peaks showed the presence of amorphous carbon rings at the surface [54].

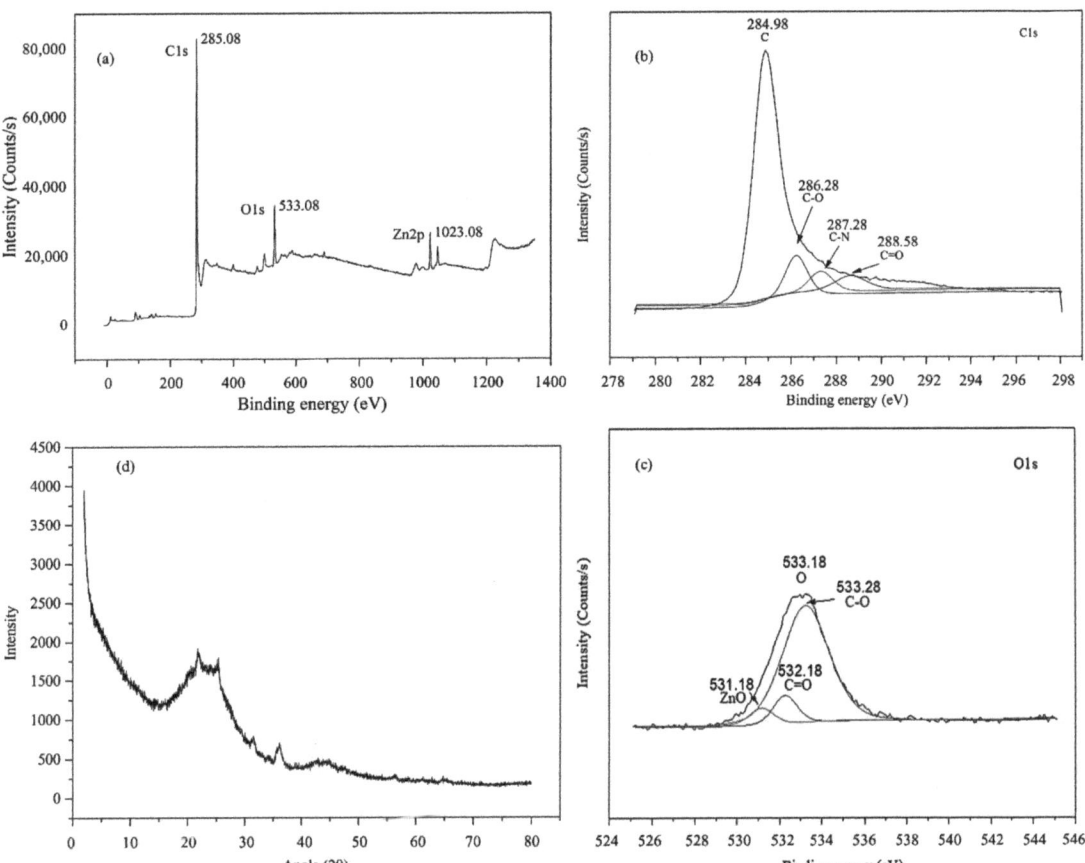

Figure 4. (**a**) XPS spectra of OPLAC-ZC; (**b**) C1s XPS spectra of OPLAC-ZC; (**c**) O1s XPS spectra of OPLAC-ZC and (**d**) XRD spectra of OPLAC-ZC.

3.5. Effect of Initial Solution pH on Removal of COD

The pH of a solution is a key parameter that affects an adsorbent's effectiveness [55]. The adsorbate surface charge and degree of ionization of the adsorbent material are affected by the pH value of the solution and it has a significant impact on the adsorption process [56]. The adsorbent surface will be totally saturated with H^+ ions under acidic conditions and the positively charged organic pollutants will be unable to compete for adsorption sites. The resistance from protons decreases as pH rises, allowing the positively charged organic pollutants to be adsorbed at the negatively charged sites on the adsorbent.

Batch experiments were conducted to study the effect of initial pH levels on adsorptive removal of COD in PW. The initial COD content of PW was 1344 mg/L. It was treated with OPLAC-ZC at various initial pH values, ranging from 2–12, at a dosage of 1000 mg/L and contact time of 60 min at 25 ± 1 °C. The initial pH of PW was adjusted using 0.01 M H_2SO_4 and NaOH. The final COD was measured after each experiment and the impact of pH on adsorptive removal of COD was studied. Figure 5a shows the results of the final COD in PW, with a change in the initial pH value. The results showed that COD in PW reduced from 1344 mg/L to 588 ± 5 mg/L (56.25 ± 5%), with an increase in pH from 2 to 7. A further increase in pH, from 8 to 12, slightly changed the COD amount. Hence, pH 7 was considered as optimum for the adsorption of COD using OPLAC-ZC. The pH value at which the net charge on the adsorbent's surface is zero is called the point of zero charge (pH_{ZPC}). The surface is positively charged below pH_{ZPC} and functional groups are protonated. At a higher pH, the surface gets deprotonated and the negative charge predominates [57]. The pH_{ZPC} of OPLAC-ZC was found as a pH of 9 (Figure 5b). Higher COD adsorptive removal was observed at pH < pH_{ZPC}, when the surface of OPLAC-ZC was positively charged. It indicated that COD was highly favorable to be removed, due to protonated oxygenated functional groups, involving a π–π interaction between OPLAC-ZC and organic pollutants in PW. Many studies have reported that a pH value of 7–8 is the most suitable value for the removal of COD from wastewater [58].

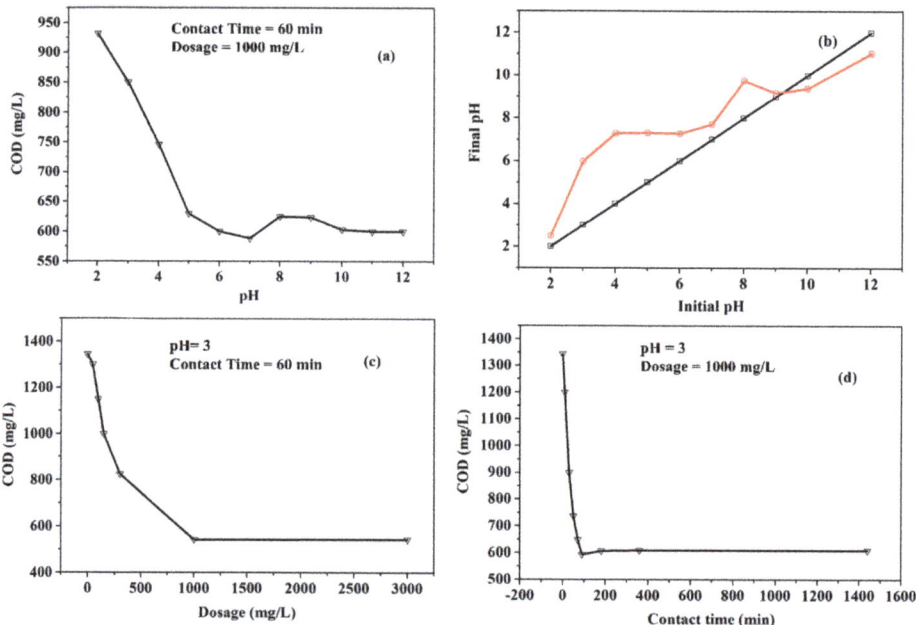

Figure 5. Effect of (**a**) pH; (**b**) pH_{ZPC}; (**c**) dosage of OPLAC-ZC; (**d**) contact time on adsorption of COD in PW.

3.6. Effect of OPLAC-ZC Dosage on Removal of COD

The effect of the dosage of OPLAC-ZC on the removal of COD is shown in Figure 5c. The dosage was varied from 10–3000 mg/L, at a pH of 3 and contact time of 60 min. A significant decrease in the final COD amount occurred with the increase in dosage, up to 1000 mg/L. Increasing the dose further did not result in an increase in COD removal. Therefore, 1000 mg/L dosage of OPLAC-ZC was found to be the optimal dosage for the removal of COD. The final COD concentration was 542 ± 5 mg/L (removal efficiency 59.6 ± 5%) at this dosage.

3.7. Effect of Contact Time on Removal of COD

The effect of contact time on the removal of COD is shown in Figure 5d. The adsorptive removal of COD was considerably increased when the contact time was increased from 5 to 90 min. Increasing the contact time further did not improve the adsorption capacity. The adsorption rate was quicker at first because there were more accessible active sites on the OPLAC-ZC surface; however, the adsorption rate gradually declined as the active sites were gradually occupied until they became saturated. Hence, a contact time of 90 min was the optimal time for the removal of COD. The final COD concentration was 593 ± 5 mg/L (removal efficiency 55.8 ± 5%) at this contact time.

3.8. Adsorption Isotherm Modeling

The study of isotherm models is an important step in choosing the best model to represent the adsorption process. Figure 6 shows the LIM and FIM for the adsorption of COD for OPLAC-ZC. The R^2 values of the linear regression lines produced from the FIM and LIM were 0.92 and 0.707, respectively, suggesting a much better fit for the FIM. The fitting of the adsorption data to the Freundlich model suggested that COD adsorptive removal took place on a heterogeneous surface, and may entail a multilayer adsorption process [59]. The FIM suggested that organic pollutants forming COD adsorbed on active sites, with different activation energies [34]. Other investigations have found similar findings for COD adsorptive removal in wastewaters [60]. Table 3 states the parameters of the isotherm models. A $1/n$ value was observed between 0 and 1, showing a favorable adsorption process.

Table 3. Isothermal adsorption modeling parameters for adsorption of COD onto OPLAC.

Model	Parameters	Units	Values
LIM	q_{max}	mg/g	4.62
	K_L	L/mg	0.0025
	R^2		0.707
FIM	K_F	mg/g	1.67505
	$1/n$		0.98
	R^2		0.92

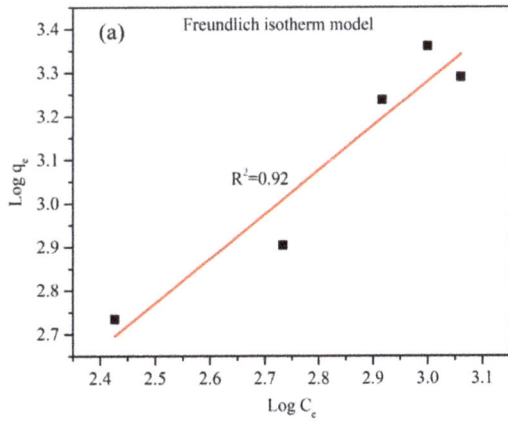
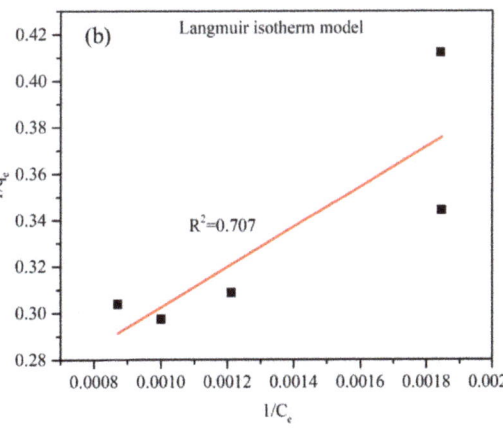

Figure 6. Cont.

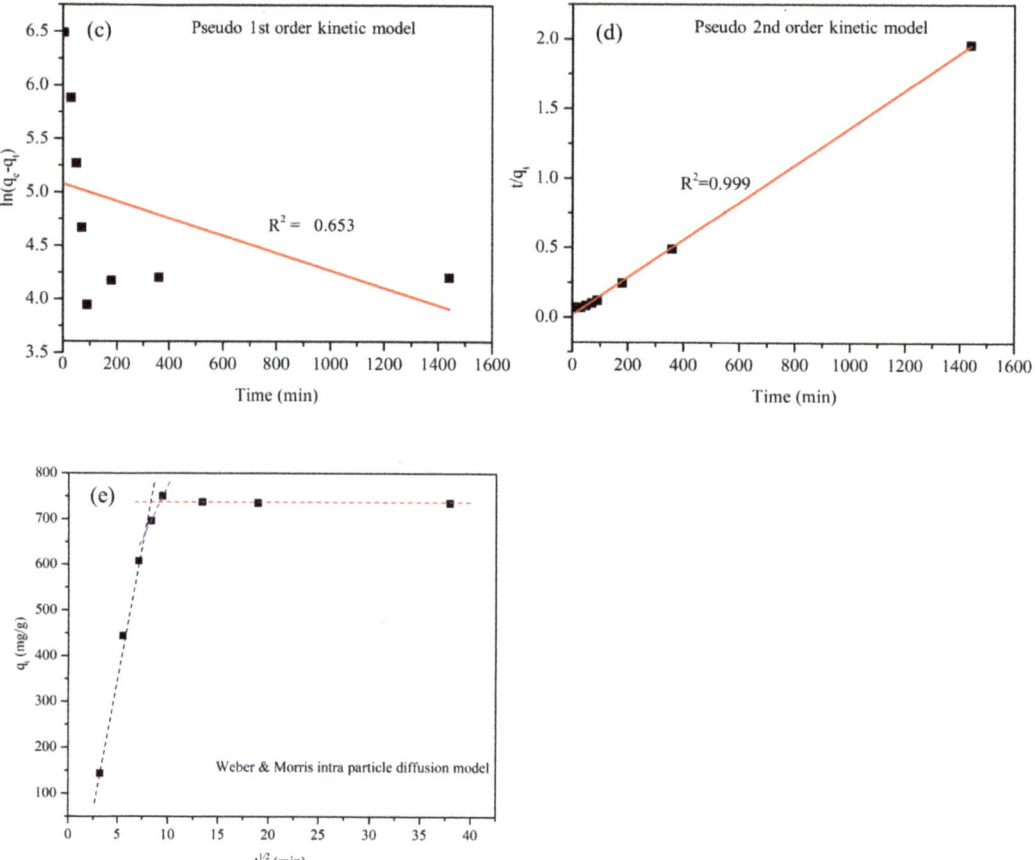

Figure 6. Adsorption isotherm and kinetic models: (**a**) FIM; (**b**) LIM; (**c**) PFO kinetic model; (**d**) PSO kinetic model; (**e**) WM intraparticle diffusion model.

3.9. Adsorption Kinetic Modeling

The study of kinetics is essential in the treatment of wastewaters because it offers significant information about the adsorption mechanism. In 90 min, the COD adsorptive removal process reached equilibrium, and OPLAC-ZC became saturated, as shown in Figure 5c. It shows that one of the advantages of employing OPLAC-ZC as an adsorbent might be the ability to quickly adsorb organic pollutants, measured as COD in PW. The kinetics of COD removal onto OPLAC-ZC were investigated using PFO, PSO and WM intraparticle diffusion models. Figure 6 shows that the PSO model best explained the COD adsorptive removal onto OPLAC-ZC, indicating that chemisorption controlled the process.

Table 4 shows the parameters of the three kinetic models employed in this study. The R^2 value obtained from the PSO kinetic model was higher than that obtained from the PFO kinetic model (Table 4). The adsorption capacity of 746.27 mg/g determined through the PSO model was also closer to the experimentally determined adsorption capacity of 802 mg/g. The model implied that chemical adsorption occurred while forming chemical bonds between the adsorbate and OPLAC-ZC's surface functional groups, as well as that electron transfer between them occurred [61]. The WM intraparticle diffusion model implied that the adsorption on the surface of the OPLAC-ZC was followed by a diffusion step and intraparticle adsorption process. Figure 6e shows that multiple steps were involved in intraparticle diffusion of organic pollutants forming COD on to OPLAC-ZC, using the WM model. If the intraparticle diffusion plot showed a straight line, the intraparticle diffusion would be the single mechanism responsible for the adsorption process [62]. However, as seen in Figure 6e, three linear portions influenced the adsorption process, implying that three stages were involved in the adsorption process. The first, steeply sloping stage was related to surface diffusion, in which a high quantity of COD was immediately adsorbed through OPLAC-ZC's surface active sites. The organic pollutants forming COD adsorbed to the inner surface of OPLAC-ZC pores, when almost all of the surface functional groups of the OPLAC-ZC were occupied, resulting in the second stage, which was the rate-limiting step [43]. The equilibrium state was achieved last, i.e., the third stage. In conclusion, according to the kinetic models, both chemisorption and intraparticle diffusion between the active sites of OPLAC-ZC and organic pollutants forming COD had an influence on the adsorption process.

Table 4. Adsorption kinetic modeling parameters for adsorption of COD.

Model	Parameters	Units	Values
PFO model	q_e, exp	mg/g	802
	q_e	mg/g	160.737
	K_1	min^{-1}	0.0008
	R^2		0.653
PSO model	q_e	mg/g	746.26
	K_2	g/mg/min	0.000107
	R^2		0.999
W&M model	K_i	mg/g/min$^{1/2}$	119.41
	R^2		0.98

4. COD Adsorption Efficiency of Various Materials

Table 5 summaries the COD removal effectiveness and surface areas of several adsorbents, as reported in earlier research. The findings are compared to OPLAC-ZC's adsorption efficiency and surface area. When compared to other adsorbents, OPLAC-ZC was more effective than many in reducing COD. A substantial surface area of AC was also observed using oil palm leaves. As a result, it can be concluded that OPLAC-ZC can be effectively utilized to remove COD and perhaps other contaminants in PW.

Table 5. Comparison of COD adsorption capacity of various materials.

Adsorbent	Surface Area (m^2/g)	Adsorption Capacity	References
Eucalyptus bark	6.1178		[63]
Palm bark powder biochar	1.80	59%	[60]
Zeolite Carbon	60.94	3.23 mg/g	[64]
Palm waste	272		[65]
Oil palm leaves carbon	122		[66]
Activated carbon from red algae	0.8950		[67]
Coal AC	29.7		[68]
Magnetic Coal AC	53.4		
Amygdalus scoparia AC	209.3		[69]
OPLAC	331.153	4.62 mg/g, 59.6 ± 5%	This study
Sludge magnetic biochar		47%	[70]
Sugarcane Bagasse AC		58.73%	[71]
Commercial walnut shell		3.93 mg/g	[72]
Commercial activated carbon		99.02%	[17]

5. Conclusions

The chemical activation technique was used to produce a novel activated carbon (AC), using oil palm leaves (OPL), with various chemical agents to alter its surface properties.

The OPL were chemically activated using three chemical agents, viz., H_3PO_4, NaOH and $ZnCl_2$. The higher surface area (331.153 m^2/g), pore size (2.494 nm) and the chemical structure of OPLAC (C-O and C=O) showed that the synthesis of OPLAC using $ZnCl_2$ was superior in terms of surface properties compared to the others. OPLAC-ZC was synthesized at three pyrolysis temperatures (400 °C, 600 °C and 800 °C) and various OPL to $ZnCl_2$ ratios (1:0.5–1:3). The OPLAC-ZC had a higher carbon content, of 81.2% at 800 °C and 1:1 impregnation ratio. The OPLAC-ZC was used for the removal of COD in PW. The results showed that OPLAC-ZC had good removal efficiency, 4.62 mg/g (59.6 ± 5%), for organic pollutants forming COD in PW. Kinetics showed that the material was able to quickly remove COD (90 min) from PW, following the PSO and WM intraparticle diffusion kinetics. The Freundlich isotherm model best fitted the adsorption data, showing the heterogenous adsorption behavior. Oxygenated functional groups at the surface of OPLAC-ZC were considered responsible for the chemical adsorption of organic pollutants forming COD, through the electron exchange mechanism in PW.

Supplementary Materials: The following supporting information can be downloaded at: https://www.mdpi.com/article/10.3390/su14041986/s1, Table S1: The water quality parameters of PW.

Author Contributions: Conceptualization, M.R.U.M. and H.K.; methodology, M.R.U.M. and H.K.; validation, M.R.U.M. and H.K.; formal analysis, M.R.U.M. and H.K.; investigation, H.K.; resources, M.R.U.M.; data curation, M.R.U.M. and H.K.; writing—original draft preparation, H.K.; writing—review and editing, M.R.U.M., H.K. and M.H.I.; visualization, M.R.U.M. and M.H.I.; supervision, M.R.U.M. and M.H.I.; project administration, M.R.U.M.; funding acquisition, M.R.U.M. All authors have read and agreed to the published version of the manuscript.

Funding: This work was supported by Universiti Teknologi PETRONAS under YUTP, Malaysia grant with cost center 015LCO-190.

Institutional Review Board Statement: Not applicable.

Informed Consent Statement: Not applicable.

Data Availability Statement: Not applicable.

Acknowledgments: The authors acknowledge the financial support provided by Universiti Teknologi PETRONAS under YUTP, Malaysia grant with cost center 015LCO-190. The authors would also like to acknowledge FELCRA Berhad Kawasan Nasaruddin Belia, Perak Malaysia, for providing oil palm leaves waste for this study.

Conflicts of Interest: The authors declare no conflict of interest.

References

1. Borhan, A.; Abdullah, N.A.; Rashidi, N.A.; Taha, M.F. Removal of Cu^{2+} and Zn^{2+} from Single Metal Aqueous Solution Using Rubber-Seed Shell Based Activated Carbon. *Procedia Eng.* **2016**, *148*, 694–701. [CrossRef]
2. Zhou, Y.; Zhang, L.; Cheng, Z. Removal of organic pollutants from aqueous solution using agricultural wastes: A review. *J. Mol. Liq.* **2015**, *212*, 739–762. [CrossRef]
3. Mohan, D.; Singh, K.P.; Singh, V.K. Wastewater treatment using low cost activated carbons derived from agricultural byproducts—A case study. *J. Hazard. Mater.* **2008**, *152*, 1045–1053. [CrossRef]
4. Affam, A.C. Conventional steam activation for conversion of oil palm kernel shell biomass into activated carbon via biochar product. *Glob. J. Environ. Sci. Manag.* **2020**, *6*, 15–30. [CrossRef]
5. Aguayo-Villarreal, I.A.; Bonilla-Petriciolet, A.; Muñiz-Valencia, R. Preparation of activated carbons from pecan nutshell and their application in the antagonistic adsorption of heavy metal ions. *J. Mol. Liq.* **2017**, *230*, 686–695. [CrossRef]
6. Dehghani, M.H.; Karri, R.R.; Yeganeh, Z.T.; Mahvi, A.H.; Nourmoradi, H.; Salari, M.; Zarei, A.; Sillanpää, M. Statistical modelling of endocrine disrupting compounds adsorption onto activated carbon prepared from wood using CCD-RSM and DE hybrid evolutionary optimization framework: Comparison of linear vs non-linear isotherm and kinetic parameters. *J. Mol. Liq.* **2020**, *302*, 112526. [CrossRef]
7. Salman, J.M.; Njoku, V.O.; Hameed, B.H. Batch and fixed-bed adsorption of 2,4-dichlorophenoxyacetic acid onto oil palm frond activated carbon. *Chem. Eng. J.* **2011**, *174*, 33–40. [CrossRef]
8. Khan, T.; Sabariah, T.; Abd, B.; Isa, M.H.; Ghanim, A.A.J.; Beddu, S.; Jusoh, H.; Iqbal, M.S.; Ayele, G.T.; Jami, M.S. Modeling of Cu(II) Adsorption from an Aqueous Solution Using an Artificial Neural Network (ANN). *Molecules* **2020**, *25*, 3263. [CrossRef]

9. Wang, Y.; Peng, C.; Padilla-Ortega, E.; Robledo-Cabrera, A.; López-Valdivieso, A. Cr(VI) adsorption on activated carbon: Mechanisms, modeling and limitations in water treatment. *J. Environ. Chem. Eng.* **2020**, *8*, 104031. [CrossRef]
10. Nasrullah, A.; Bhat, A.H.; Naeem, A.; Isa, M.H.; Danish, M. High surface area mesoporous activated carbon-alginate beads for efficient removal of methylene blue. *Int. J. Biol. Macromol.* **2018**, *107*, 1792–1799. [CrossRef]
11. Auta, M.; Hameed, B.H. Preparation of waste tea activated carbon using potassium acetate as an activating agent for adsorption of Acid Blue 25 dye. *Chem. Eng. J.* **2011**, *171*, 502–509. [CrossRef]
12. Kumar, J.A.; Amarnath, D.J.; Sathish, S.; Jabasingh, S.A.; Saravanan, A.; Hemavathy, R.V.; Anand, K.V.; Yaashikaa, P.R. Enhanced PAHs removal using pyrolysis-assisted potassium hydroxide induced palm shell activated carbon: Batch and column investigation. *J. Mol. Liq.* **2019**, *279*, 77–87. [CrossRef]
13. Auta, M.; Hameed, B.H. Optimized waste tea activated carbon for adsorption of Methylene Blue and Acid Blue 29 dyes using response surface methodology. *Chem. Eng. J.* **2011**, *175*, 233–243. [CrossRef]
14. Akar, E.; Altinişik, A.; Seki, Y. Using of activated carbon produced from spent tea leaves for the removal of malachite green from aqueous solution. *Ecol. Eng.* **2013**, *52*, 19–27. [CrossRef]
15. Khan, T.; Mustafa, M.R.U.; Isa, M.H.; Manan, T.S.B.A.; Ho, Y.C.; Lim, J.W.; Yusof, N.Z. Artificial Neural Network (ANN) for Modelling Adsorption of Lead (Pb (II)) from Aqueous Solution. *Water. Air. Soil Pollut.* **2017**, *228*, 1–15. [CrossRef]
16. Tsibranska, I.; Hristova, E. Comparison of different kinetic models for adsorption of heavy metals onto activated carbon from apricot stones. *Bulg. Chem. Commun.* **2011**, *43*, 370–377.
17. Devi, R.; Singh, V.; Kumar, A. COD and BOD reduction from coffee processing wastewater using Avacado peel carbon. *Bioresour. Technol.* **2008**, *99*, 1853–1860. [CrossRef]
18. Van Thuan, T.; Quynh, B.T.P.; Nguyen, T.D.; Ho, V.T.T.; Bach, L.G. Response surface methodology approach for optimization of Cu^{2+}, Ni^{2+} and Pb^{2+} adsorption using KOH-activated carbon from banana peel. *Surf. Interfaces* **2017**, *6*, 209–217. [CrossRef]
19. Lim, W.C.; Srinivasakannan, C.; Al Shoaibi, A. Cleaner production of porous carbon from palm shells through recovery and reuse of phosphoric acid. *J. Clean. Prod.* **2015**, *102*, 501–511. [CrossRef]
20. Ahmed, M.J. Preparation of activated carbons from date (*Phoenix dactylifera* L.) palm stones and application for wastewater treatments: Review. *Process Saf. Environ. Prot.* **2016**, *102*, 168–182. [CrossRef]
21. Ayinla, R.T.; Dennis, J.O.; Zaid, H.M.; Sanusi, Y.K.; Usman, F.; Adebayo, L.L. A review of technical advances of recent palm bio-waste conversion to activated carbon for energy storage. *J. Clean. Prod.* **2019**, *229*, 1427–1442. [CrossRef]
22. Hosseini, S.E.; Wahid, M.A. Utilization of palm solid residue as a source of renewable and sustainable energy in Malaysia. *Renew. Sustain. Energy Rev.* **2014**, *40*, 621–632. [CrossRef]
23. Sidik, S.M.; Jalil, A.A.; Triwahyono, S.; Adam, S.H.; Satar, M.A.H.; Hameed, B.H. Modified oil palm leaves adsorbent with enhanced hydrophobicity for crude oil removal. *Chem. Eng. J.* **2012**, *203*, 9–18. [CrossRef]
24. Ahmad, F.B.; Zhang, Z.; Doherty, W.O.S.; O'Hara, I.M. The outlook of the production of advanced fuels and chemicals from integrated oil palm biomass biorefinery. *Renew. Sustain. Energy Rev.* **2019**, *109*, 386–411. [CrossRef]
25. Elias, N.; Chandren, S.; Attan, N.; Mahat, N.A.; Razak, F.I.A.; Jamalis, J.; Wahab, R.A. Structure and properties of oil palm-based nanocellulose reinforced chitosan nanocomposite for efficient synthesis of butyl butyrate. *Carbohydr. Polym.* **2017**, *176*, 281–292. [CrossRef]
26. Awalludin, M.F.; Sulaiman, O.; Hashim, R.; Nadhari, W.N.A.W. An overview of the oil palm industry in Malaysia and its waste utilization through thermochemical conversion, specifically via liquefaction. *Renew. Sustain. Energy Rev.* **2015**, *50*, 1469–1484. [CrossRef]
27. Tan, I.A.W.; Ahmad, A.L.; Hameed, B.H. Adsorption of basic dye using activated carbon prepared from oil palm shell: Batch and fixed bed studies. *Desalination* **2008**, *225*, 13–28. [CrossRef]
28. Sahu, J.N.; Karri, R.R.; Jayakumar, N.S. Improvement in phenol adsorption capacity on eco-friendly biosorbent derived from waste Palm-oil shells using optimized parametric modelling of isotherms and kinetics by differential evolution. *Ind. Crops Prod.* **2021**, *164*, 113333. [CrossRef]
29. Chew, T.L.; Husni, H. Oil palm frond for the adsorption of Janus Green dye. *Mater. Today Proc.* **2019**, *16*, 1766–1771. [CrossRef]
30. Wafti, N.S.A.; Lau, H.L.N.; Loh, S.K.; Aziz, A.A.; Rahman, Z.A.; Yuen, C. Activated Carbon from oil Palm Biomass as Potential Adsorbent for Palm Oil Mill Effluent Treatment. *J. Oil Palm Res.* **2017**, *29*, 278–290. [CrossRef]
31. Isa, M.H.; Ibrahim, N.; Aziz, H.A.; Adlan, M.N.; Sabiani, N.H.M.; Zinatizadeh, A.A.L.; Kutty, S.R.M. Removal of chromium (VI) from aqueous solution using treated oil palm fibre. *J. Hazard. Mater.* **2008**, *152*, 662–668. [CrossRef] [PubMed]
32. Farma, R.; Lestari, O.; Taer, E.; Apriwandi; Minarni; Awitdrus, A. Removal of Cu, Fe, and Zn from Peat Water by Using Activated Carbon Derived from Oil Palm Leaves. *Adv. Mater. Res.* **2021**, *1162*, 65–73. [CrossRef]
33. Haghbin, M.R.; Niknam Shahrak, M. Process conditions optimization for the fabrication of highly porous activated carbon from date palm bark wastes for removing pollutants from water. *Powder Technol.* **2021**, *377*, 890–899. [CrossRef]
34. Hendges, L.T.; Costa, T.C.; Temochko, B.; Gómez González, S.Y.; Mazur, L.P.; Marinho, B.A.; da Silva, A.; Weschenfelder, S.E.; de Souza, A.A.U.; de Souza, S.M.A.G.U. Adsorption and desorption of water-soluble naphthenic acid in simulated offshore oilfield produced water. *Process Saf. Environ. Prot.* **2021**, *145*, 262–272. [CrossRef]
35. Promraksa, A.; Rakmak, N. Biochar production from palm oil mill residues and application of the biochar to adsorb carbon dioxide. *Heliyon* **2020**, *6*, 1–9. [CrossRef]

36. Shokrollahzadeh, S.; Golmohammad, F.; Naseri, N.; Shokouhi, H.; Arman-Mehr, M. Chemical oxidation for removal of hydrocarbons from gas-field produced water. *Procedia Eng.* **2012**, *42*, 942–947. [CrossRef]
37. Desta, M.B. Mulu Berhe Desta Batch sorption experiments: Langmuir and Freundlich isotherm studies for the adsorption of textile metal ions onto teff straw (Eragrostis tef) agricultural waste. *J. Thermodyn.* **2013**, *2013*, 1–6. [CrossRef]
38. Yurdakal, S.; Garlisi, C.; Özcan, L.; Bellardita, M.; Palmisano, G. Chapter 4—(Photo)catalyst Characterization Techniques: Adsorption Isotherms and BET, SEM, FTIR, UV-Vis, Photoluminescence, and Electrochemical Characterizations. In *Heterogeneous Photocatalysis*; Elsevier: Amsterdam, The Netherlands, 2019. [CrossRef]
39. Khurshid, H.; Mustafa, M.R.U.; Rashid, U.; Isa, M.H.; Chia, H.Y.; Shah, M.M. Adsorptive removal of COD from produced water using tea waste biochar. *Environ. Technol. Innov.* **2021**, *23*, 101563. [CrossRef]
40. Kharrazi, S.M.; Mirghaffari, N.; Dastgerdi, M.M.; Soleimani, M. A novel post-modification of powdered activated carbon prepared from lignocellulosic waste through thermal tension treatment to enhance the porosity and heavy metals adsorption. *Powder Technol.* **2020**, *366*, 358–368. [CrossRef]
41. Dada, A.O.; Olalekan, A.P.; Olatunya, A.M.; Dada, O.J.I.J.C. Langmuir, Freundlich, Temkin and Dubinin–Radushkevich Isotherms Studies of Equilibrium Sorption of Zn^{2+} Unto Phosphoric Acid Modified Rice Husk. *IOSR J. Appl. Chem.* **2012**, *3*, 38–45. [CrossRef]
42. Halim, A.A.; Aziz, H.A.; Johari, M.A.M.; Ariffin, K.S. Comparison study of ammonia and COD adsorption on zeolite, activated carbon and composite materials in landfill leachate treatment. *Desalination* **2010**, *262*, 31–35. [CrossRef]
43. Ahmaruzzaman, M.; Gayatri, S.L. Activated tea waste as a potential low-cost adsorbent for the removal of p-nitrophenol from wastewater. *J. Chem. Eng. Data* **2010**, *55*, 4614–4623. [CrossRef]
44. Wan, S.; Hua, Z.; Sun, L.; Bai, X.; Liang, L. Biosorption of nitroimidazole antibiotics onto chemically modified porous biochar prepared by experimental design: Kinetics, thermodynamics, and equilibrium analysis. *Process Saf. Environ. Prot.* **2016**, *104*, 422–435. [CrossRef]
45. Edet, U.A.; Ifelebuegu, A.O. Kinetics, isotherms, and thermodynamic modeling of the adsorption of phosphates from model wastewater using recycled brick waste. *Processes* **2020**, *8*, 665. [CrossRef]
46. Kohler, T.; Hinze, M.; Müller, K.; Schwieger, W. Temperature independent description of water adsorption on zeotypes showing a type V adsorption isotherm. *Energy* **2017**, *135*, 227–236. [CrossRef]
47. Valderrama, C.; Gamisans, X.; de las Heras, X.; Farrán, A.; Cortina, J.L. Sorption kinetics of polycyclic aromatic hydrocarbons removal using granular activated carbon: Intraparticle diffusion coefficients. *J. Hazard. Mater.* **2008**, *157*, 386–396. [CrossRef]
48. Wang, J.; Guo, X. Adsorption kinetic models: Physical meanings, applications, and solving methods. *J. Hazard. Mater.* **2020**, *390*, 122156. [CrossRef]
49. Van Tran, T.; Bui, Q.T.P.; Nguyen, T.D.; Thanh Ho, V.T.; Bach, L.G. Application of response surface methodology to optimize the fabrication of $ZnCl_2$-activated carbon from sugarcane bagasse for the removal of Cu^{2+}. *Water Sci. Technol.* **2017**, *75*, 2047–2055. [CrossRef]
50. Soysa, R.; Choi, Y.S.; Kim, S.J.; Choi, S.K. Fast pyrolysis characteristics and kinetic study of Ceylon tea waste. *Int. J. Hydrog. Energy* **2016**, *41*, 16436–16443. [CrossRef]
51. Nasri, N.S.; Hamza, U.D.; Ismail, S.N.; Ahmed, M.M.; Mohsin, R. Assessment of porous carbons derived from sustainable palm solid waste for carbon dioxide capture. *J. Clean. Prod.* **2014**, *71*, 148–157. [CrossRef]
52. Deng, S.; Yu, G.; Chen, Z.; Wu, D.; Xia, F.; Jiang, N. Characterization of suspended solids in produced water in Daqing oilfield. *Colloids Surf. A Physicochem. Eng. Asp.* **2009**, *332*, 63–69. [CrossRef]
53. Zbair, M.; Ait Ahsaine, H.; Anfar, Z. Porous carbon by microwave assisted pyrolysis: An effective and low-cost adsorbent for sulfamethoxazole adsorption and optimization using response surface methodology. *J. Clean. Prod.* **2018**, *202*, 571–581. [CrossRef]
54. Ghorbani, F.; Kamari, S.; Zamani, S.; Akbari, S.; Salehi, M. Optimization and modeling of aqueous Cr(VI) adsorption onto activated carbon prepared from sugar beet bagasse agricultural waste by application of response surface methodology. *Surf. Interfaces* **2020**, *18*, 100444. [CrossRef]
55. Eid, M.E.-S. Synthesis of Polyethylenimine-magnetic amorphous carbon nano composite as a novel adsorbent for Hg(II) from aqueous solutions. *Aust. J. Basic Appl. Sci.* **2016**, *10*, 323–335.
56. Soliman, A.M.; Elwy, H.M.; Thiemann, T.; Majedi, Y.; Labata, F.T.; Al-Rawashdeh, N.A.F. Removal of Pb(II) ions from aqueous solutions by sulphuric acid-treated palm tree leaves. *J. Taiwan Inst. Chem. Eng.* **2016**, *58*, 264–273. [CrossRef]
57. Enaime, G.; Baçaoui, A.; Yaacoubi, A.; Lübken, M. Biochar for wastewater treatment-conversion technologies and applications. *Appl. Sci.* **2020**, *10*, 3493. [CrossRef]
58. Mahmoud, A.S.; Farag, R.S.; Elshfai, M.M. Reduction of organic matter from municipal wastewater at low cost using green synthesis nano iron extracted from black tea: Artificial intelligence with regression analysis. *Egypt. J. Pet.* **2020**, *29*, 9–20. [CrossRef]
59. Santoso, E.; Ediati, R.; Kusumawati, Y.; Bahruji, H.; Sulistiono, D.O.; Prasetyoko, D. Review on recent advances of carbon based adsorbent for methylene blue removal from waste water. *Mater. Today Chem.* **2020**, *16*, 100233. [CrossRef]
60. Chaouki, Z.; Hadri, M.; Nawdali, M.; Benzina, M.; Zaitan, H. Treatment of a landfill leachate from Casablanca city by a coagulation-flocculation and adsorption process using a palm bark powder (PBP). *Sci. Afr.* **2021**, *12*, e00721. [CrossRef]
61. Pinto Brito, M.J.; Flores Santos, M.P.; de Souza Júnior, E.C.; Santos, L.S.; Ferreira Bonomo, R.C.; Da Costa Ilhéu Fontan, R.; Veloso, C.M. Development of activated carbon from pupunha palm heart sheaths: Effect of synthesis conditions and its application in lipase immobilization. *J. Environ. Chem. Eng.* **2020**, *8*, 1–11. [CrossRef]

62. Cheraghipour, E.; Pakshir, M. Process optimization and modeling of Pb(II) ions adsorption on chitosan-conjugated magnetite nano-biocomposite using response surface methodology. *Chemosphere* **2020**, *260*, 127560. [CrossRef]
63. Martini, S.; Afroze, S.; Ahmad Roni, K. Modified eucalyptus bark as a sorbent for simultaneous removal of COD, oil, and Cr(III) from industrial wastewater. *Alex. Eng. J.* **2020**, *59*, 1637–1648. [CrossRef]
64. Halim, A.A.; Aziz, H.A.; Johari, M.A.M.; Ariffin, K.S.; Adlan, M.N. Ammoniacal nitrogen and COD removal from semi-aerobic landfill leachate using a composite adsorbent: Fixed bed column adsorption performance. *J. Hazard. Mater.* **2010**, *175*, 960–964. [CrossRef]
65. Azoulay, K.; Bencheikh, I.; Moufti, A.; Dahchour, A.; Mabrouki, J.; El Hajjaji, S. Comparative study between static and dynamic adsorption efficiency of dyes by the mixture of palm waste using the central composite design. *Chem. Data Collect.* **2020**, *27*, 1–19. [CrossRef]
66. El-Sayed, M.; Nada, A.A. Polyethylenimine—Functionalized amorphous carbon fabricated from oil palm leaves as a novel adsorbent for Cr(VI) and Pb(II) from aqueous solution. *J. Water Process Eng.* **2017**, *16*, 296–308. [CrossRef]
67. Isam, M.; Baloo, L.; Sapari, N.; Nordin, I.; Yavari, S.; Al-Madhoun, W. Removal of Lead using Activated Carbon Derived from Red Algae (Gracilaria Changii). *MATEC Web Conf.* **2018**, *203*, 1–10. [CrossRef]
68. Hao, Z.; Wang, C.; Yan, Z.; Jiang, H.; Xu, H. Magnetic particles modification of coconut shell-derived activated carbon and biochar for effective removal of phenol from water. *Chemosphere* **2018**, *211*, 962–969. [CrossRef]
69. Bagheri, R.; Ghaedi, M.; Asfaram, A.; Alipanahpour Dil, E.; Javadian, H. RSM-CCD design of malachite green adsorption onto activated carbon with multimodal pore size distribution prepared from Amygdalus scoparia: Kinetic and isotherm studies. *Polyhedron* **2019**, *171*, 464–472. [CrossRef]
70. Yi, Y.; Huang, Z.; Lu, B.; Xian, J.; Tsang, E.P.; Cheng, W.; Fang, J.; Fang, Z. Magnetic biochar for environmental remediation: A review. *Bioresour. Technol.* **2019**, *298*, 122468. [CrossRef]
71. Pan, Y.; Zhu, Y.; Xu, Z.; Lu, R.; Zhang, Z.; Liang, M.; Liu, H. RETRACTED ARTICLE: Adsorption removal of COD from wastewater by the activated carbons prepared from sugarcane bagasse. In Proceedings of the 5th International Conference on Bioinformatics and Biomedical Engineering, iCBBE, Wuhan, China, 10–12 May 2011. [CrossRef]
72. Gallo-Cordova, A.; Silva-Gordillo, M.D.M.; Muñoz, G.A.; Arboleda-Faini, X.; Almeida Streitwieser, D. Comparison of the adsorption capacity of organic compounds present in produced water with commercially obtained walnut shell and residual biomass. *J. Environ. Chem. Eng.* **2017**, *5*, 4041–4050. [CrossRef]

Article

Marble Dust Effect on the Air Quality: An Environmental Assessment Approach

Qaiser Iqbal [1], Muhammad Ali Musarat [2], Najeeb Ullah [1], Wesam Salah Alaloul [2,*], Muhammad Babar Ali Rabbani [3], Wesam Al Madhoun [4] and Shahid Iqbal [1]

1. Department of Civil and Engineering, Sarhad University of Science & Information Technology, Peshawar 25000, Pakistan; qi.civil@suit.edu.pk (Q.I.); najeebullah@uetpeshawar.edu.pk (N.U.); shahid.civil@suit.edu.pk (S.I.)
2. Department of Civil and Environmental Engineering, Universiti Teknologi PETRONAS, Bandar Seri Iskandar 32610, Malaysia; muhammad_19000316@utp.edu.my
3. College of Science and Engineering, Flinders University, Bedford Park, Adelaide, SA 5042, Australia; rabb0019@flinders.edu.au
4. Faculty of Engineering, Gaza University, Gaza 711226, Palestine; w.madhoun@gu.edu.ps
* Correspondence: wesam.alaloul@utp.edu.my

Abstract: All over the world, increasing anthropogenic activities, industrialization, and urbanization have intensified the emissions of various pollutants that cause air pollution. Marble quarries in Pakistan are abundant and there is a plethora of small- and large-scale industries, including mining and marble-based industries. The air pollution caused by the dust generated in the process of crushing and extracting marble can cause serious problems to the general physiological functions of plants and it affects human life as well. Therefore, the objectives of this study were to assess the air quality of areas with marble factories and areas without marble factories, where the concentration of particulate matter in terms of total suspended particles (TSP) was determined. For this purpose, EPAM-5000 equipment was used to measure the particulate levels. Besides this, a spectrophotometer was used to analyze the presence of PM2.5 and PM10 in the chemical composition of marble dust. It was observed that the TSP concentrations in Darmangi and Malagori areas of Peshawar, Pakistan—having marble factories—were 626 µg/m^3 and 5321 µg/m^3 respectively. The ($PM_{2.5}$, PM_{10}) concentration in Darmangi was (189 µg/m^3, 520 µg/m^3) and in Malagori, it was recorded as (195 µg/m^3, 631 µg/m^3), which was significantly higher than the non-marble dust areas and also exceeded WHO recommended standards. It was concluded that the areas with the marble factories were more susceptible to air pollution as the concentration of TSP was significantly higher than the recommended TSP levels. It is recommended that marble factories should be shifted away from residential areas along with strict enforcement. People should be instructed to use protective equipment and waste management should be ensured along with control mechanisms to monitor particulate levels.

Keywords: marble dust; air pollution; health hazards; environmental pollution

Citation: Iqbal, Q.; Musarat, M.A.; Ullah, N.; Alaloul, W.S.; Rabbani, M.B.A.; Al Madhoun, W.; Iqbal, S. Marble Dust Effect on the Air Quality: An Environmental Assessment Approach. *Sustainability* **2022**, *14*, 3831. https://doi.org/10.3390/su14073831

Academic Editor: Genovaitė Liobikienė

Received: 22 February 2022
Accepted: 18 March 2022
Published: 24 March 2022

Publisher's Note: MDPI stays neutral with regard to jurisdictional claims in published maps and institutional affiliations.

Copyright: © 2022 by the authors. Licensee MDPI, Basel, Switzerland. This article is an open access article distributed under the terms and conditions of the Creative Commons Attribution (CC BY) license (https://creativecommons.org/licenses/by/4.0/).

1. Introduction

Due to extensive urbanization, more than 90% of the world's population is at risk of being affected by air pollution [1,2]. Air pollution occurs when gas, dust particles, and smoke mix with the atmosphere in a harmful way that is toxic to every living organism [3]. The combination of suspended organic and inorganic particles emit various precursors—such as nitrogen oxides (NO_x), sulphur dioxide (SO_2), ozone (O_3), carbon monoxide (CO), lead, and volatile organic chemical compounds—which is a serious problem faced by developing as well as developed countries [2,4]. According to the World Health Organization (WHO), it is estimated that the presence of the most dangerous particulate matter (PM)—that is particles having a 2.5-micrometre diameter (PM)$_{2.5}$—in the air causes approximately 800,000 premature deaths each year and ranks it as the 13th leading cause of death in the world [5]. In the absence of strict controlling measures for air pollution, the number of deaths due to air pollution is expected to rise to 6–9 million deaths per year by 2060 [6]. The number of deaths from air pollution along with a comparison to other reasons for deaths is shown in Table 1.

Table 1. Number of deaths due to various reasons in 2015.

S. No	Reason of Deaths	Number of Deaths (Million)	Source
1	Polluted air	6.4	
2	Household air pollution	2.8	[7,8]
3	Ambient air pollution	4.2	
4	Tobacco	7.0	
5	Acute immunity deficiency syndrome (AIDS)	1.2	[9]
6	Tuberculosis	1.1	
7	Malaria	0.7	

The unwanted gift of the industrial revolution, population explosion, and the rapid expansion of metropolitan areas is air pollution, which is a problem being faced globally, especially in developing countries [10]. Air pollution produces smog and acid rain, which depletes the ozone layer of the atmosphere causes global warming [11]. Polluted air is a perpetual threat and its mitigation is a colossal challenge in terms of achieving sustainability [12]. Moreover, air pollutants have harmful effects on plant life—such as stomata movement, foliar geometry, photosynthesis, membrane permeability, and nutrient transport—which leads to plant growth retardation, low yield, and premature senescence in highly susceptible plants [13]. In a study to evaluate the effect of marble dust on plants, the chlorophyll content of different species of trees was measured. It was found that chlorophyll content was reduced significantly in trees near marble industries [14]. Similarly, poor air quality harms human health [15]. Evidence suggests that these suspended particles are generated by the burning of biomass for energy conversion and fossil fuel combustion, which enter the human body and affect the alveoli of the lungs [16]. Other epidemiological studies revealed that poor air quality is a leading factor in the increase in mortality and causes various cardiovascular and respiratory diseases [17]. The assessment of the diseases caused by air pollution illustrates that more than 2 million premature deaths occur each year that could be attributed to the effects of urban outdoor and indoor air pollution [18]. Medical expenditures are increased significantly due to air-pollution related-diseases, as in 2015, the global economy suffered a loss of USD 21 billion due to air-borne diseases [19].

Dust is regarded as an omnipresent air pollutant [20]. Dust can exist in natural and artificial forms. The natural sources of dust are food and chemical industries, animal debris, the earth's surface, and volcanic eruption, while anthropogenic sources are fossil fuel, factories, mining and quarrying, and stone-working [21]. Besides $PM_{2.5}$, PM_{10} is another highly significant air-pollutant that is generated in the form of dust from road construction activities, mining dust, and manufacturing plants [22]. Dust is produced as a result of a variety of processes such as handling and manufacturing of materials, which consists of transferring, dropping, weighing, and conveying [23].

The marble industry is also one form of construction activity, and it consists of processes and operations—namely cutting, buffing, and polishing—which generate a considerable amount of dust particles [24]. The particulate matter produced in the process of crushing and cutting marble used to make statues is usually larger. Large-scale mining processes also produce many particulate emissions [14]. During marble manufacturing, 40% of the marble waste is equal to consists of the manufactured volume, which is generated when the rock debris is dumped in nearby fields, agricultural lands, and river beds, which produces environmental hazards [25]. The resulting dust particles from marble factories have high levels of toxic PM particles and its exposure is a root cause of many fatal respiratory and carcinogenic diseases—such as nasal cancer, bronchitis, asthma, and lung infection—in marble workers [14]. A similar study was conducted in the Hayatabad residential area in Peshawar, Pakistan that performed the air pollution analysis and its air quality assessment. Based on the results, it was found that the air near the residential area had higher TSP, PM2.5, and PM10 levels than the WHO recommended guidelines [26]. Iran, a neighboring country of Pakistan, also faces mining and quarrying health hazards. A methodology was developed to rank the mines according to safety and sustainable production. A "Multi-Attribute Decision Making" could be used to rank the quarrying and mining factories per safety and environmental standards [27]. Since Iran is one of the main producers of marble all over the globe, the mining industry is in abundance and faces health and occupational issues. To study the environmental and safety of the mining industry, a risk breakdown structure was applied in Fars province due to the high production of marble. Based on the methodology of the analytical hierarchical process, it was found that the employer is at higher risk of health followed by the financial risk in the quarrying sector [28].

Urbanization has increased the demand for construction, which has led to a manifold increase in mineral extraction in many countries [29]. Urban air pollution is associated with inflammation, oxidative stress, blood coagulation, and autonomic dysfunction simultaneously in healthy young humans, with sulfate and O_3 as two major traffic-related pollutants contributing to such effects [30]. According to Environmental Protection Agency (EPA), particles can be carried over long distances by wind and then settle on ground or water [31]. The chemical composition of marble dust is shown in Table 2 [32].

Table 2. Chemical composition of marble dust by percentage.

S. No	Chemical Compounds	Percentage of Marble Dust
1	Calcium carbonate	94.30
2	Lime	50.10
3	Alumina	1.38
4	Silica	1.28
5	Magnesia	1.72
6	Iron oxide	0.54
7	Sulphur trioxide	0.21
8	Alkaline	0.29
9	Loss of ignition	0.39

The marble reserves in Pakistan are estimated to be above 297 billion tons [33]. There are more than 100 different types of marbles available [34]. Marble deposits are present in large quantities in the Khyber Pakhtunkhwa region, Baluchistan region, and Azad Kashmir region [35]. Due to such an abundance of natural deposits of marbles, there is an innumerable number of marble factories in Pakistan and currently, there is no law for proper license and dumping of marble waste [36]. People living nearby marble factories are suffering from polluted water [37], kidney stones [38], radioactive diseases [39], occupational health hazards [40,41], sediment deposition in rivers [42], and polluted landfills [36] due to the marble dust. Besides this, the air quality has worsened due to anthropogenic emissions from these factories [43]. All industrial processes ultimately lead to a decline in air quality standards, which poses health risks in most developing countries such as Pakistan [44]. Table 3 shows the list of most polluted countries based on a dataset containing over 80,000 data points [45].

Table 3. World's most polluted countries 2020 (PM$_{2.5}$) [45].

Rank	Country	2020 Emission (µg/m^3)	2019 Emission (µg/m^3)	Population (2020)
1	Bangladesh	77.10	83.30	164,689,383
2	Pakistan	59.00	65.80	220,892,331
3	India	51.90	58.10	1,380,004,385
4	Mongolia	46.60	62.00	3,278,292
5	Afghanistan	46.50	58.80	38,928,341

Currently, Pakistan is one of the most air-polluted countries in the world [45]. The WHO standard of PM$_{2.5}$ and PM$_{10}$ concentrations are 25 µg/m^3 and 50 µg/m^3 respectively [46]. According to various studies, PM$_{10}$ for Peshawar was calculated as 219 µg/m^3 [47], 540 µg/m^3 for PM$_{10}$, and 160 µg/m^3 for PM$_{2.5}$. The high concentration of PM$_{2.5}$ and PM$_{10}$ in the Peshawar region is due to the dust particles that reach from Afghanistan, which adds air pollution to the local region [48]. The prevalence of air pollution levels surpassing the WHO guidelines beyond the acceptable levels of PM concentration accounts for environmental degradation and health deterioration of people.

In Pakistan, a large number of marble processing units dump their marble waste directly into streams, rivers, and fertile lowlands, which cover the soil pores because there is an absence of awareness and no law about the disposal of waste material [49]. Consequently, soil permeability is reduced, which increases the alkalinity of the soil. Since Pakistan is an agricultural country and most people are dependent on agriculture for their livelihood, losing fertile soil would be disastrous for the people and the national economy. The objectives of the study were to environmentally evaluate the air quality and to investigate the concentration of marble dust in the air of the Peshawar area in the Khyber Pakhtunkhwa region of Pakistan. The outcomes of the study would enable the local communities and the government to know the current levels of harmful particulate matter in the air. This study also proposes mitigation strategies for local decision-makers that could improve the air quality to ensure clean air for living organisms.

2. Methodology

Based on the objectives, various areas of Marble Dust (MD) and Non-Marble Dust (NMD) were selected to measure the air quality using PM concentrations. The air quality was investigated in the Peshawar area of Mattani and Jalozai for marble factories for NMD areas and Warsak road, and Alazizi road for MD having marble factories. For this study, the data were collected for 6 months, from March 2021 to September 2021. The reason for selecting this time for analysis was because the wind speed is dominant at this time; hence, adding the wind factor was considered, which is a crucial parameter for counting the particulate matter. During this time, a fresh supply of marble raw material got supplied to the respective factories; thus, most of the marble factories were operational. By selecting this time for analysis, the highest possible concentration of air particles was captured. The weather condition during this time ranged from a sunny day to broken clouds, wind speed from 10 km/h to 30 km/h, and temperature ranging from 29 °C to 42 °C. The black arrow in Figure 1 indicates the analysis location.

Figure 1. Marble factories in the Darmangi region (Google Earth).

At normal capacity, the marble factories operate from 9 a.m. to 7 p.m. with 40–50 workers. However, due to COVID-19, the operation was being performed below the normal capacity with the restriction of 15 workers with variable shifts. Secondly, the power outages in this study area were also the main factor that was taken into consideration. The power outage was scheduled for three times a day by the electric supply company. Considering this factor, the analysis was stopped when the factories cease to operate.

The dust concentration in the air was measured with HAZ-DUST EPAM-5000 as shown in Figure 2. This apparatus has a size selectable impactor for particulate matter, $PM_{1.0}$, and total suspended particles (TSP). In this research, PM_{10}, $PM_{2.5}$, and TSP were used to analyze the suspended marble particles in areas having marble factories and non-marble factories. A detailed overview of this device is shown in Figure 2.

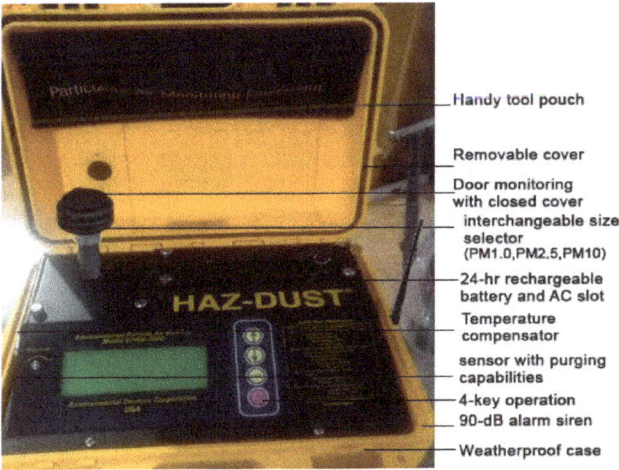

Figure 2. EPAM-5000.

This device works on the principle of light scattering of infrared radiation using the near forward technique to detect the concentration in mg/m^3. The infrared light is measured in a photodetector at a 90° angle. The light is dispersed when the dust is entered into an infrared beam. The amount of aerosol concentration depends on the amount of light detected by the photodetector. As a result, the noise and drift of the light are removed using signal processing, which allows for high stability and accuracy of the baseline results. The schematic principle is shown in Figure 3.

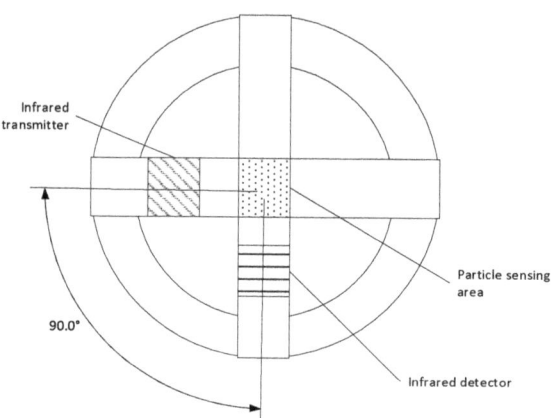

Figure 3. Principle of near forward light scattering.

This device overcomes the limitation of other methods and combines the filter techniques of traditional methods along with real-time monitoring. The advantages of using this unit are that it provides immediate readings, 24 h continuous determination of concentration, and an alarm sound when approaching the hazardous range. Therefore, it is a considerably cheaper and time-saving device as compared to the traditional method of measuring dust particles [50]. EPAM-5000 is calibrated based on the standard protocols of the National Institute for Occupational Safety and Health (NIOSH) for Arizona road dust (ARD) that measures the quality of respiratory air (PM$_{2.5}$) with ±10% accuracy [51]. The readings obtained from this unit are converted using DustComm Pro software that could provide mathematical readings, create graphs, and correct statistical differences between aerosol and calibrated models.

Spectrophotometer

As it uses the combination of concepts of spectrometry and photometry, it is therefore called a spectrophotometer. This instrument measures the optical density of suspension material. Where optical density is defined as the ratio of light-receiving by the material (I_o) to the light transmitted across the material (I_t). Mathematically, it is expressed as I_o/I_t. If the value of this ratio is high, it means that a higher concentration of particles is present in the material and lower values indicate the weak concentration of the particles.

This instrument works on the principle of Beer-Lambert Law, which states that the amount of light absorbed is directly proportional to the concentration and thickness of the solution. The Beer-Lambert law can be expressed as

$$A = \varepsilon bc \qquad (1)$$

where "A" means absorbance, "ε" stands for molar absorptivity, which is the strength of a compound that absorbs the light at a given wavelength, "b" shows the path length, and "c" is the concentration of the sample. Hence, Equation (1) becomes

$$-\log \frac{I_t}{I_o} = \varepsilon bc \tag{2}$$

$$I_t = I_o e^{-\varepsilon bc} \tag{3}$$

The advantage of this technique is that the rate of reaction could be measured directly by measuring the light interacting with the components of the solution rather than stopping the reaction and measuring its rate.

3. Results and Discussion

For investigating air quality, marble dust, and non-marble dust areas were visited and measured the concentration of marble dust in the air by using EPAM-5000. The areas of non-marble dust (NMD) and marble dust (MD) areas are mentioned in Figures 4 and 5 respectively.

Figure 4a,b shows that the area is in its natural soil color because of the absence of any marble factory in the vicinity. However, Figure 5a,b clearly shows that a white residue is left untreated near the marble factories. This means that the lighter marble dust particles such as $PM_{2.5}$ and PM_{10} were transported along with the wind, thereby polluting the air and leaving behind the hardened marble sludge.

The details of the results of different areas with coordinates are mentioned in Table 4. The result shows that areas having marble factories have greater marble dust (MD) levels in the air as compared to non-marble dust (NMD) areas. As per the authors' study, there are 26 marbles factories in the Malagori area, and it is the heart of marble production. While Darmangi area is the neighboring area with adequate plantations and there are less than 10 factories in this area, due to which, the Darmangi area has less concentration than the Malagori area.

Similarly, the $PM_{2.5}$ concentration for marble dust areas (Darmangi and Malagori) are 189 μg/m and 195 μg/m^3, which exceeds the WHO standard limits of 25 μg/m^3. Additionally, the PM_{10} concentration values for the same marble dust areas were calculated as 620 μg/m^3 and 730 μg/m^3 as compared to 50 μg/m^3 recommended values of PM_{10}. While the NMD areas (Mattani and Jalozai) were not quite so affected. Hence, it is evident that the residential areas located near marble factories are at higher risk of exposing people to breathing polluted air. The results are shown in Table 5.

The marble industry is one of the most important industrial sectors that contributes to the socio-economic development of Peshawar residents. This sector adds yields its fair share to the national economy because of the diverse processes of marble factories. Despite having such a crucial role, the dust generated from this sector has had adverse effects on plants, humans, and the environment and poses a potential risk to the living organisms because of polluted water quality and air quality.

The analysis of suspension material of marble dust particles indicates higher absorbance and less transmission of the light, which means that the solution has higher optical density. The lower values of absorbance indicate that there was little concentration of mentioned particles. The concentration of the dust sample is changing over time due to a variety of particulate matters as shown in Table 6.

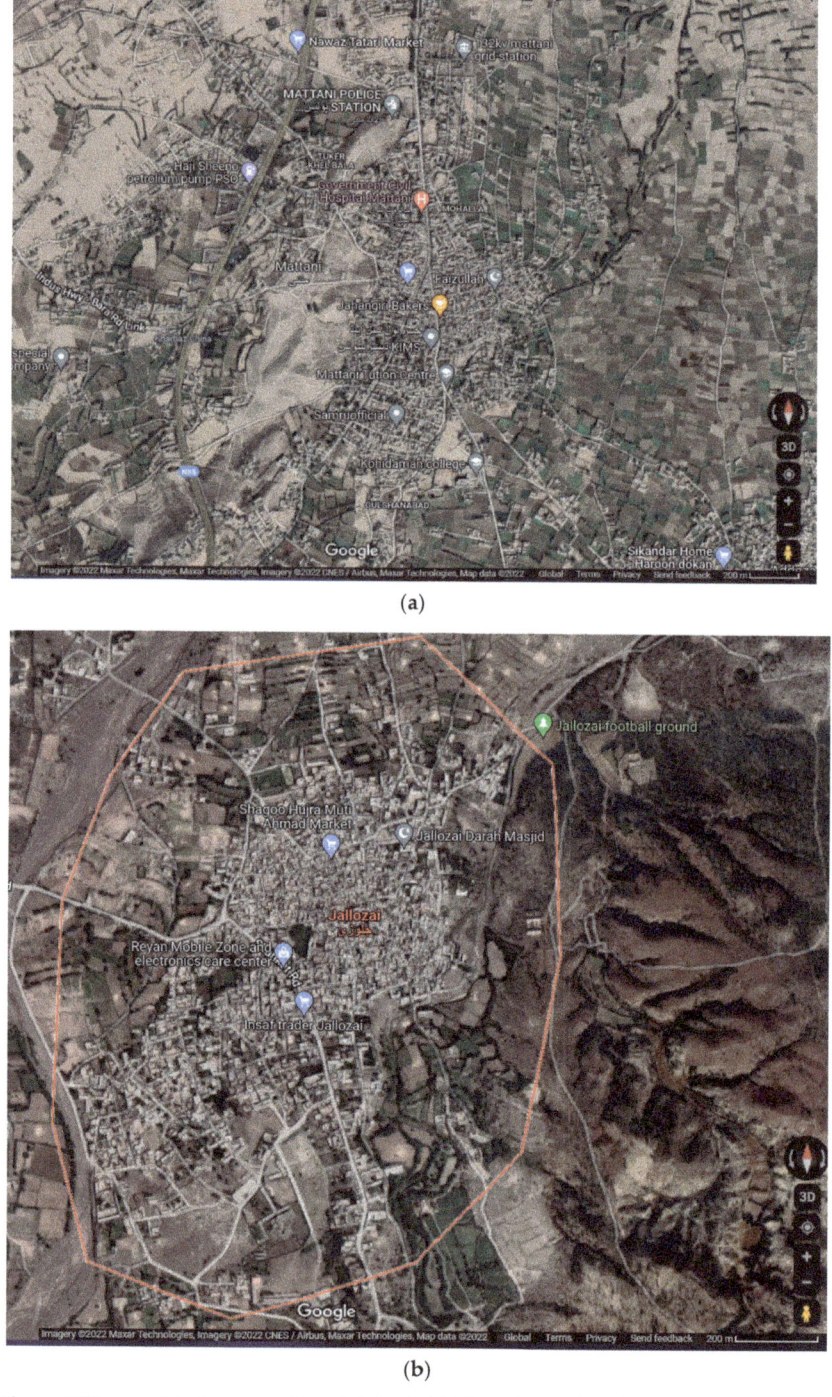

Figure 4. Non-marble dust areas (Google Maps). (a) Mattani area. (b) Jalozai area.

Figure 5. Marble dust area (Google Maps). (**a**) Darmangi Warsak road area. (**b**) Malagori area.

Table 4. Results of marble dust (MD) in air.

S. No	Location	Concentration of TSP ($\mu g/m^3$)	Coordinates	MD/NMD	WHO Guidelines
1	Mattani Kohat Road	2	N: 33.832953° E: 71.565458°	NMD	120 $\mu g/m^3$
2	Jalozai	26	N: 33.914888° E: 71.815810°	NMD	
3	Darmangi Warsak Road	626	N: 34.0484206° E: 71.5205146°	MD	
4	Malagori	5321	N 34.135703° E 71.403719°	MD	

Table 5. Results of $PM_{2.5}$ and PM_{10} in the selected areas of Peshawar.

Description	Mattani $\mu g/m^3$ (NMD)	Jalozai $\mu g/m^3$ (NMD)	Darmangi $\mu g/m^3$ (MD)	Malagori $\mu g/m^3$ (MD)	WHO Guidelines $\mu g/m^3$
Concentration of $PM_{2.5}$ ($\mu g/m^3$)	33	42	189	195	25
Concentration of PM_{10} ($\mu g/m^3$)	109	214	620	730	50

Table 6. Spectrophotometry results.

$PM_{2.5}/PM_{10}$	Absorbance	Transmittance	Wavelength (nm)
NO	0.230	14.2	380
Cu	0.365	7.4	405
Zn	0.415	5.0	430
SO_2	1.130	6.8	530
Mg	1.320	10.7	580
Al	1.590	5.4	630
Fe	1.800	2.7	655
Ca	2.130	19.6	680

The medical records from the last two-year period, indicating 1543 patients with breathing problems, were analyzed. It was found that the highest cases originated from the Malagori area and its neighboring areas. The recorded patient history showed that the breathing problems were not genetic and only male members with ages from 18–57 years of the family were suffering. These people were either worker in the marble factory or were living in the vicinity of these factories. The wages of these factory workers were below USD 100, and they could not afford to pursue other professions and were unable to pay medical bills, which indicates that these people were doing jobs in these factories due to the absence of other skills and were forced to pursue physical jobs.

The water polluted with toxic elements contaminates the pH of water, which in turn affects the turbidity in the water [52]. The turbidity is changed by suspended matter such as silts and organic chemical compounds from these factories. This turbid water enters into the river which makes the water translucent; which consequently blocks the sunlight penetration in bodies of water; hence, the survival of aquatic organisms and algae is compromised [53]. Sodium is another waste generated by marble dust. The excess addition of sodium affects plants and would disrupt the chemical balance in the water, animals, and humans affecting plant life, animal survival, and heart disease among humans respectively [54]. Magnesium is also an important constituent of seafood and vegetables. Magnesium intake is often blocked by the accumulation of marble dust in the soil, which reduced the required magnesium concentration; this lack of magnesium can cause hypomagnesemia, leading to diabetes, low blood pressure, and cardiac arrests [55].

Similarly, the marble dust alters the air quality such as calcium carbonate upon heating releases carbon dioxide from the marble powder, which is released into the atmosphere, thus making the air unfit for the environment. This process is regarded as an important factor in increasing carbon emissions each year [56]. Copper is also released in marble dust and its high concentration in water is dangerous to both humans and crops, which can harm kidneys and can cause cancer in people living and working in marble factories [57]. Zinc is another necessary element, which prevents hearts diseases, acts as an anti-inflammatory agent, and helps in connective tissue formation. Traces of a high amount of zinc were found in marble workers' blood, any amount higher than the necessary intake is harmful [58]. Manganese is also released in dust emanating from steel, marble, and fossil fuel combustion [59]. Higher exposure to manganese especially in marble workers generates strong signals in the human body, triggering liver diseases. Arsenic is a major constituent of dust particles [60]. The emission of $CaCO_3$ produces white dust that reduces visibility and produces asthma problems in people in the vicinity of marble dust [61]. Arsenic pollution is a global dust problem—especially in South Asian countries such as Pakistan, India, and Bangladesh—which can cause hyperkeratosis as well as kidney, liver, cardiovascular, and neurological disorders [62]. Areas with increased concentration of $PM_{2.5}$ also recorded higher rates of casualties due to COVID-19 [63].

This study shows that the areas in the vicinity of the marble factories have reduced air quality due to the presence of a high concentration of particulate matter. However, the absence of marble factories has little to no effect on improving the air quality as the dust from marble factories traveled to NMD areas. To elaborate, $PM_{2.5}$ and PM_{10} concentration levels are higher even in NMD areas due to the abundance of marble activities in the MD areas.

The problem of polluted air is widespread. With industrialization, air pollution has become a global issue. It is estimated that $1 m^3$ of marble, when cut into 2 cm thick slabs, produces 25% marble dust [64]. In the United States of America, there was a 173% increase in gross domestic product, 85% vehicle emissions, energy consumption soared by 19% from 1980 to 2020 along with the addition of 68 million tons of pollutants [65]. In India, marble and mining dust have adverse effects on plants and vegetation near marble industries. Due to the reduced amount of chlorophyll, the trees and plants suffered biochemical, physiological, and morphological changes resulting in 20% reduced growth [66]. The extraction and energy emissions of the marble industry in Italy produce large amounts of marble mining dust during quarrying. The pollution is increased with the number of processes associated with marble production [67]. Turkey's marble production generates dust of 40–60% of the overall manufacturing volume. The mining and quarrying processes involved in marble processing in the Afyon region generate 340,000 tonnes of marble waste [68]. Egypt, being the fifth largest producer of marble, generated 3.5 million tonnes of marble. The Shaq El-Thouban region is suffering from soil alkalinity, airborne diseases, and reduced plant productivity due to the presence of 400 marble factories [69]. The Skikda region in Algeria has cement factories and aggregate manufacturing industries that utilized various processes for marble stone production, which generates harmful particles in the atmosphere. The use of marble in cement preparation is damaging to the air quality; therefore, marble dust recycling is performed at low temperature and reduced humidity [70].

The problem of polluted air can be reduced globally when each country introduces interventions to deal with the pollution of their respective countries. In this regard, sustainable processes, waste treatment, and strict legislation for collective recycling could be utilized. Moreover, awareness campaigns, relocation of marble factories, treatment of wastewater, disposal of effluents away from residential areas, filtering dust and smoke generated, pre-defined schedule of factory operation, using marble dust as admixtures, controlled expansion of factories, action against illegal marble factories, limiting socio-economic activities in a polluted area, and use of personal protective equipment are some of the preventive measures that could prove helpful to make the air, water, and environment suitable for human existence concerning marble quarrying.

4. Conclusions

This study was conducted in the Peshawar region of Pakistan with marble dust (MD) and non-marble dust (NMD) areas to analyze the air quality based on the readings of $PM_{2.5}$, PM_{10}, and TSP. For the determination of air pollution, the areas near marble factories and non-marble factories were compared and it was found that people living near marble factories are more prone to diseases associated with dust inhalation. Based on the results of the air pollution test, the Peshawar area air is unfit to breathe, and it would worsen with time due to the unlicensed and haphazard location of factories because of the absence of regulation. There is a strong need for legislation of establishment of factories in the industrial zone only. Therefore, it is recommended that awareness must be ensured for marble workers and owners about the hazardous results of marble dust. Local communities and stakeholders should be educated about the presence of factories in residential areas. Preventive measures such as using wet processes, dust collection, use of safety gear, avoiding direct skin contact, ventilation systems, using of marble dust as an admixture, and city planning should be implemented.

5. Limitations

The results are limited based on the tenure of the data collected. As monitoring points are random—having diverse localities—it may be possible that the readings of one area differ from the other area. Moreover, the readings can be affected by various factors in settings, such as unpaved roads dust and residential dust, that can interfere with the actual marble dust concentration.

Author Contributions: All authors contributed equally to this study. All authors have read and agreed to the published version of the manuscript.

Funding: This research received no external funding.

Institutional Review Board Statement: Not applicable.

Informed Consent Statement: Not applicable.

Data Availability Statement: Not applicable.

Acknowledgments: The authors would like to appreciate the University Internal Research Funding (URIF) from Universiti Teknologi PETRONAS (UTP), (cost centre #015LB0-051) awarded to Wesam Alaloul for the support provided to this research.

Conflicts of Interest: The authors declare no conflict of interest.

References

1. World Health Organisation (WHO). Ambient (Outdoor) Air Pollution. 2018. Available online: https://www.who.int/news-room/fact-sheets/detail/ambient-(outdoor)-air-quality-and-health (accessed on 27 January 2022).
2. Alaloul, W.S.; Musarat, M.A. Impact of Zero Energy Building: Sustainability Perspective. In *Sustainable Sewage Sludge Management and Resource Efficiency*; InTech: London, UK, 2020.
3. Pénard-Morand, C.; Annesi-Maesano, I. Air pollution: From sources of emissions to health effects. *Breathe* **2004**, *1*, 108–119. [CrossRef]
4. Combes, A.; Franchineau, G. Fine particle environmental pollution and cardiovascular diseases. *Metabolism* **2019**, *100*, 153944. [CrossRef] [PubMed]
5. World Health Organisation (WHO). *The World Health Report 2002: Reducing Risks, Promoting Healthy Life*; World Health Organization: Geneva, Switzerland, 2002.
6. Organisation for Economic Cooperation and Development (OECD). 2016. Available online: https://www.oecd.org/env/air-pollution-to-cause-6-9-million-premature-deaths-and-cost-1-gdp-by-2060.htm (accessed on 8 September 2021).
7. GBD. Risk Factor Collaborators. Global, regional, and national comparative risk assessment of 84 behavioural, environmental and occupational, and metabolic risks or clusters of risks for 195 countries and territories, 1990–2017: A systematic analysis for the Global Burden of Disease Study 2017. *Lancet* **2018**, *392*, 1923–1994. [CrossRef]

8. Prüss-Ustün, A.; Wolf, J.; Corvalán, C.; Neville, T.; Bos, R.; Neira, M. Diseases due to unhealthy environments: An updated estimate of the global burden of disease attributable to environmental determinants of health. *J. Public Health* **2017**, *39*, 464–475. [CrossRef] [PubMed]
9. França, E.B.; Passos, V.M.D.A.; Malta, D.C.; Duncan, B.B.; Ribeiro, A.L.P.; Guimarães, M.D.C.; Abreu, D.M.; Vasconcelos, A.M.N.; Carneiro, M.; Teixeira, R.; et al. Cause-specific mortality for 249 causes in Brazil and states during 1990–2015: A systematic analysis for the global burden of disease study 2015. *Popul. Health Metr.* **2017**, *15*, 39. [CrossRef] [PubMed]
10. Arbex, M.A.; Santos, U.d.P.; Martins, L.C.; Saldiva, P.H.N.; Pereira, L.A.A.; Braga, A.L.F. Air pollution and the respiratory system. *J. Bras. Pneumol.* **2012**, *38*, 643–655. [CrossRef]
11. Choudhary, M.P.; Garg, V. Causes, consequences and control of air pollution. In *All India Seminar on Methodologies for Air Pollution Control*; MNIT: Jaipur, India, 2013.
12. Moraga, R.; Hosseinabad, E.R. A system dynamics approach in air pollution mitigation of metropolitan areas with sustainable development perspective: A case study of Mexico City. *J. Appl. Environ. Biol. Sci.* **2017**, *12*, 164–174.
13. Tripathi, A.K.; Gautam, M.K. Biochemical parameters of plants as indicators of air pollution. *J. Environ. Biol.* **2007**, *28*, 127.
14. Saini, Y.; Bhardwaj, N.; Gautam, R. Effect of marble dust on plants around Vishwakarma Industrial Area (VKIA) in Jaipur, India. *J. Environ. Biol.* **2011**, *32*, 209–212.
15. Quarmby, S.; Santos, G.; Mathias, M. Air Quality Strategies and Technologies: A Rapid Review of the International Evidence. *Sustainability* **2019**, *11*, 2757. [CrossRef]
16. Brook, R.D.; Rajagopalan, S.; Popelll, C.A.; Brook, J.R.; Bhatnagar, A.; Diez-Roux, A.V.; Holguin, F.; Hong, Y.; Luepker, R.V.; Mittleman, M.A.; et al. Particulate matter air pollution and cardiovascular disease: An update to the scientific statement from the American Heart Association. *Circulation* **2010**, *121*, 2331–2378. [CrossRef]
17. Heft-Neal, S.; Burney, J.; Bendavid, E.; Burke, M. Robust relationship between air quality and infant mortality in Africa. *Nature* **2018**, *559*, 254–258. [CrossRef] [PubMed]
18. Jerrett, M. The death toll from air-pollution sources. *Nature* **2015**, *525*, 330–331. [CrossRef] [PubMed]
19. The Vegetarian Resource Group. How Many Youth Are Vegetarian? Available online: http://www.vrg.org/press/youth_poll_20 10.php (accessed on 9 December 2018).
20. Andersen, A.C. In the beginning: The origin of dust. *AIP Conf. Proc.* **2009**, *1094*, 254. [CrossRef]
21. Popescu, F.; Ionel, I. Anthropogenic Air Pollution Sources. In *Air Quality*; InTech: London, UK, 2010; pp. 1–22. [CrossRef]
22. Yan, H.; Ding, G.; Li, H.; Wang, Y.; Zhang, L.; Shen, Q.; Feng, K. Field Evaluation of the Dust Impacts from Construction Sites on Surrounding Areas: A City Case Study in China. *Sustainability* **2019**, *11*, 1906. [CrossRef]
23. Darren, M.; Technikon, M. Hazard Prevention and Control in the Work Environment: Airborne Dust WHO/SDE/OEH/99.14. 2000. Available online: https://apps.who.int/iris/handle/10665/66147 (accessed on 22 September 2021).
24. El-Gammal, M.; Ibrahim, M.; Badr, E.; Asker, S.A.; El-Galad, N.M. Health risk assessment of marble dust at marble workshops. *Nat. Sci.* **2011**, *9*, 144–154.
25. Akbulut, H.; Gürer, C. The environmental effects of waste marble and possibilities of utilization and waste minimization by using in the road layers. In Proceedings of the Fourth National Marble Symposium, Afyonkarahisar, Turkey, 18–19 December 2003; pp. 371–378.
26. Rabbani, M.B.A.; Khan, Q.Z.; Usama, U. Determination of Air Pollutants and Its Modeling: A Machine Learning Approach for Assessment of Air Quality in Industrial Estate, Hayatabad. Presented at the 1st International Conference on Recent Advances in Civil and Earthquake Engineering, Peshawar, Pakistan, 8 October 2021. Available online: https://drive.google.com/file/d/1P7J8 oR1mHm2a0f0J4l0q62dKksrj-fh1/view (accessed on 1 December 2021).
27. Yari, M.; Bagherpour, R.; Almasi, N. An approach to the evaluation and classification of dimensional stone quarries with an emphasis on safety parameters. *Rud. Zb.* **2016**, *31*, 15–26. [CrossRef]
28. Yari, M.; Bagherpour, R.; Khoshouei, M.; Pedram, H. Investigating a comprehensive model for evaluating occupational and environmental risks of dimensional stone mining. *Rud. Zb.* **2020**, *35*, 101–109. [CrossRef]
29. Kumar, N. Effect of different mining dust on the vegetation of district balaghat, MP-A critical review. *Int. J. Sci. Res.* **2015**, *4*, 603–607.
30. Chuang, K.-J.; Chan, C.-C.; Su, T.-C.; Lee, C.-T.; Tang, C.-S. The Effect of Urban Air Pollution on Inflammation, Oxidative Stress, Coagulation, and Autonomic Dysfunction in Young Adults. *Am. J. Respir. Crit. Care Med.* **2007**, *176*, 370–376. [CrossRef]
31. Epa, U.; United States Environmental Protection Agency. Quality Assurance Guidance Document-Model Quality Assurance Project Plan for the PM Ambient Air, Volume 2. 2001. Available online: https://swap.stanford.edu/20131015080356/ (accessed on 30 November 2021).
32. Raghunath, P.; Suguna, K.; Karthick, J.; Sarathkumar, B. Mechanical and durability characteristics of marble-powder-based high-strength concrete. *Sci. Iran.* **2019**, *26*, 3159–3164.
33. Sarkar, R.; Das, S.K.; Mandal, P.K.; Maiti, H.S. Phase and microstructure evolution during hydrothermal solidification of clay–quartz mixture with marble dust source of reactive lime. *J. Eur. Ceram. Soc.* **2006**, *26*, 297–304. [CrossRef]
34. Pre-Feasibility Study—Marble Quarry Project. 2010. Available online: https://www.yumpu.com/en/document/read/11352777 /pre-feasibility-study-marble-quarry-project-sbi-sindh-board-of- (accessed on 22 November 2021).
35. Manan, A.; Iqbal, Y. Phase, Microstructure and Mechanical Properties of Marble in North-Western Part of Pakistan: Preliminary Findings. *J. Pak. Mater. Soc.* **2007**, *1*, 68–71.

36. Fawad, M.; Ullah, F.; Irshad, M.; Shah, W.; Tahir, A.A.; Mehmood, Q.; Ahmed, T. Pollution hotspots and potential impacts on land use in the Mohmand Marble Zone, Pakistan. *Environ. Earth Sci.* **2021**, *80*, 372. [CrossRef]
37. Iqbal, M.; Akbar, F.; Ullah, S.; Anwar, I.; Khan, M.T.; Nawab, A.; Bacha, M.S.; Rashid, W. The effects of marble industries effluents on water quality in Swat, Northern Pakistan. *J. Biodivers. Environ. Sci.* **2018**, *13*, 34–42.
38. Jehangir, K.; Zeshan, A.; Bakht, T.; Faiz, U. Burden of marble factories and health risk assessment of kidney (renal) stones development in district Buner, Khyber Pakhtunkhwa, Pakistan. *Expert Opin. Environ. Biol.* **2015**, *2*, 2.
39. Iqbal, M.; Tufail, M.; Mirza, S.M. Measurement of natural radioactivity in marble found in Pakistan using a NaI (Tl) gamma-ray spectrometer. *J. Environ. Radioact.* **2000**, *51*, 255–265. [CrossRef]
40. Khan, Q.; Maqsood, S.; Khattak, S.B.; Omair, M.; Hussain, A. Evaluation of Activity Hazards in Marble Industry of Pakistan. *Int. J. Eng. Technol.* **2015**, *15*, 73–78.
41. Noreen, U.; Ahmed, Z.; Khalid, A.; Di Serafino, A.; Habiba, U.; Ali, F.; Hussain, M. Water pollution and occupational health hazards caused by the marble industries in district Mardan, Pakistan. *Environ. Technol. Innov.* **2019**, *16*, 100470. [CrossRef]
42. Mulk, S.; Azizullah, A.; Korai, A.L.; Khattak, M.N.K. Impact of marble industry effluents on water and sediment quality of Barandu River in Buner District, Pakistan. *Environ. Monit. Assess.* **2015**, *187*, 8. [CrossRef]
43. Raza, W.; Saeed, S.; Saulat, H.; Gul, H.; Sarfraz, M.; Sonne, C.; Sohn, Z.-H.; Brown, R.J.; Kim, K.-H. A review on the deteriorating situation of smog and its preventive measures in Pakistan. *J. Clean. Prod.* **2021**, *279*, 123676. [CrossRef]
44. Majid, H.; Madl, P.; Alam, K. Ambient air quality with emphasis on roadside junctions in metropolitan cities of Pakistan and its potential health effects. *Health* **2012**, *3*, 79–85.
45. IQAir. World's Most Polluted Countries 2020 (PM2.5). Available online: https://www.iqair.com/world-most-polluted-countries (accessed on 20 September 2021).
46. World Health Organization (WHO). A Prüss-Ustün, J Wolf, C Corvalán, R Bos and M Neira. Preventing Disease through Healthy Environments, A Global Assessment of the Burden of Disease from Environmental Risks. 2018. Available online: https://apps.who.int/iris/rest/bitstreams/908623/retrieve (accessed on 20 December 2021).
47. Ghauri, B.; Lodhi, A.; Mansha, M. Development of baseline (air quality) data in Pakistan. *Environ. Monit. Assess.* **2006**, *127*, 237–252. [CrossRef] [PubMed]
48. Alam, K.; Blaschke, T.; Madl, P.; Mukhtar, A.; Hussain, M.; Trautmann, T.; Rahman, S. Aerosol size distribution and mass concentration measurements in various cities of Pakistan. *J. Environ. Monit.* **2011**, *13*, 1944–1952. [CrossRef] [PubMed]
49. Haider, A.; Amber, A.; Ammara, S.; Mahrukh, K.S.; Aisha, B. Knowledge, perception and attitude of common people towards solid waste management-A case study of Lahore, Pakistan. *Int. Res. J. Environ. Sci.* **2015**, *4*, 100–107.
50. National Institute for Occupational Safety and Health. *The Industrial Environment—Its Evaluation & Control*; US Government Printing Office: Washington, DC, USA, 1973.
51. SKC. Environmental Particulate air Monitor. 1999. Available online: https://www.skcltd.com/images/pdfs/EPAM-5000-manual.pdf (accessed on 23 September 2021).
52. Sabouri, R.; Afkhami, M.; Zarasvandi, A.; Khodadadi, M. Correlation Analysis of Dust Concentration and Water Quality Indicators. *Int. J. Environ. Sci. Dev.* **2011**, *2*, 91–97. [CrossRef]
53. Davies-Colley, R.J.; Smith, D.G. Turbidity suspenied sediment, and water clarity: A review. *Jawra J. Am. Water Resour. Assoc.* **2001**, *37*, 1085–1101. [CrossRef]
54. Kazi, T.G.; Arain, M.B.; Baig, J.A.; Jamali, M.K.; Afridi, H.I.; Jalbani, N.; Sarfraz, R.A.; Shah, A.Q.; Niaz, A. The correlation of arsenic levels in drinking water with the biological samples of skin disorders. *Sci. Total Environ.* **2009**, *407*, 1019–1026. [CrossRef]
55. Smedley, P.L.; Kinniburgh, D.G. A review of the source, behaviour and distribution of arsenic in natural waters. *Appl. Geochem.* **2002**, *17*, 517–568. [CrossRef]
56. Sütçü, M.; Alptekin, H.; Erdogmus, E.; Er, Y.; Gencel, O. Characteristics of fired clay bricks with waste marble powder addition as building materials. *Constr. Build. Mater.* **2015**, *82*, 1–8. [CrossRef]
57. Martin, S.; Griswold, W. Human health effects of heavy metals. *Environ. Sci. Technol. Briefs Citiz.* **2009**, *15*, 1–6.
58. James, W.D.; Elston, D.; Berger, T. *Andrew's Diseases of the Skin E-book: Clinical Dermatology*; Elsevier Health Science: Amsterdam, The Netherlands, 2011; pp. 1–900.
59. Harding, A.K.; Daston, G.P.; Boyd, G.R.; Lucier, G.W.; Safe, S.H.; Stewart, J.; Tillitt, D.E.; Van Der Kraak, G. Endocrine Disrupting Chemicals Research Program of the U.S. Environmental Protection Agency: Summary of a Peer-Review Report. *Environ. Health Perspect.* **2006**, *114*, 1276–1282. [CrossRef]
60. Bhattacharyya, R.; Jana, J.; Nath, B.; Sahu, S.; Chatterjee, D.; Jacks, G. Groundwater As mobilization in the Bengal Delta Plain, the use of ferralite as a possible remedial measure—A case study. *Appl. Geochem.* **2003**, *18*, 1435–1451. [CrossRef]
61. Çelik, M.Y.; Sabah, E. Geological and technical characterisation of Iscehisar (Afyon-Turkey) marble deposits and the impact of marble waste on environmental pollution. *J. Environ. Manag.* **2008**, *87*, 106–116. [CrossRef]
62. Mosley, L.; Singh, S.; Aalbersberg, B. Water quality monitoring in Pacific Island countries. *SOPAC Tech. Rep.* **2005**, *381*, 42.
63. Ali, S.M.; Malik, F.; Anjum, M.S.; Siddiqui, G.F.; Anwar, M.N.; Lam, S.S.; Nizami, A.-S.; Khokhar, M.F. Exploring the linkage between PM2.5 levels and COVID-19 spread and its implications for socio-economic circles. *Environ. Res.* **2021**, *193*, 110421. [CrossRef]
64. Kun, N. Mermer Jeolojisi ve Teknolojisi. *Tezer Printing House.* İzmir **2000**, *1*, 1–149.

65. Environmental Protection Agency. Air Quality—National Summary. 2021. Available online: https://www.epa.gov/air-trends/air-quality-national-summary (accessed on 26 September 2021).
66. Soni, A.; Aseri, G.; Jain, N. Impact of Air Pollution caused by Mining and Marble Dust on Foliar Sensitivity through Biochemical Changes. *Int. J. Eng. Res. Technol.* **2017**, *5*, 1–4.
67. Liguori, V.; Rizzo, G.; Traverso, M. Marble quarrying: An energy and waste intensive activity in the production of building materials. *Wit Trans. Ecol. Environ.* **2008**, *108*, 197–207. [CrossRef]
68. Sabah, E.; Çelik, M. İscehisar (Afyon) Mermer Artıklarının Hayvan Yemi Katkı Maddesi Olarak Kullanılabilirliğinin Araştırılması. In Proceedings of the Türkiye III Mermer Sempozyumu (Mersem '2001) Bildiriler Kitabı, Afyonkarahisar, Turkey, 3 May 2001; pp. 3–5.
69. Abdelkader, H.A.M.; Hussein, M.M.A.; Ye, H. Influence of Waste Marble Dust on the Improvement of Expansive Clay Soils. *Adv. Civ. Eng.* **2021**, *2021*, 3192122. [CrossRef]
70. Seghir, N.T.; Mellas, M.; Sadowski, Ł.; Krolicka, A.; Żak, A.; Ostrowski, K. The Utilization of Waste Marble Dust as a Cement Replacement in Air-Cured Mortar. *Sustainability* **2019**, *11*, 2215. [CrossRef]

Article

Water Level Prediction through Hybrid SARIMA and ANN Models Based on Time Series Analysis: Red Hills Reservoir Case Study

Abdus Samad Azad [1,*], Rajalingam Sokkalingam [1], Hanita Daud [1], Sajal Kumar Adhikary [2], Hifsa Khurshid [3], Siti Nur Athirah Mazlan [1] and Muhammad Babar Ali Rabbani [4]

1. Department of Fundamental and Applied Sciences, Universiti Teknologi PETRONAS, Tronoh 31750, Perak, Malaysia; raja.sokkalingam@utp.edu.my (R.S.); hanita_daud@utp.edu.my (H.D.); siti_19001034@utp.edu.my (S.N.A.M.)
2. Department of Civil Engineering, Khulna University of Engineering & Technology, Khulna 9203, Bangladesh; sajal@ce.kuet.ac.bd
3. Department of Civil Engineering, Universiti Teknologi PETRONAS, Tronoh 31750, Perak, Malaysia; hifsa_18002187@utp.edu.my
4. Department of Civil Engineering, Sarhad University of Science and Information Technology, Peshawar 25000, Pakistan; rabb0019@flinders.edu.au
* Correspondence: Abdus_19000955@utp.edu.my

Abstract: Reservoir water level (RWL) prediction has become a challenging task due to spatio-temporal changes in climatic conditions and complicated physical process. The Red Hills Reservoir (RHR) is an important source of drinking and irrigation water supply in Thiruvallur district, Tamil Nadu, India, also expected to be converted into the other productive services in the future. However, climate change in the region is expected to have consequences over the RHR's future prospects. As a result, accurate and reliable prediction of the RWL is crucial to develop an appropriate water release mechanism of RHR to satisfy the population's water demand. In the current study, time series modelling technique was adopted for the RWL prediction in RHR using Box–Jenkins autoregressive seasonal autoregressive integrated moving average (SARIMA) and artificial neural network (ANN) hybrid models. In this research, the SARIMA model was obtained as SARIMA (0, 0, 1) (0, 3, 2)$_{12}$ but the residual of the SARIMA model could not meet the autocorrelation requirement of the modelling approach. In order to overcome this weakness of the SARIMA model, a new SARIMA–ANN hybrid time series model was developed and demonstrated in this study. The average monthly RWL data from January 2004 to November 2020 was used for developing and testing the models. Several model assessment criteria were used to evaluate the performance of each model. The findings showed that the SARIMA–ANN hybrid model outperformed the remaining models considering all performance criteria for reservoir RWL prediction. Thus, this study conclusively proves that the SARIMA–ANN hybrid model could be a viable option for the accurate prediction of reservoir water level.

Keywords: RWL; time series; RHR; seasonality; prediction; ANN; SARIMA

1. Introduction

Each region has its own set of water quality and quantity concerns, depending on the climatic, geographic, geologic, social, and economic characteristics. The rainfall pattern is likely to shift all over the planet as a result of global warming and climate change. Modelling studies until the year 2050 have anticipated that the world's freshwater distribution is expected to undergo a paradigm shift [1,2]. Therefore, a reliable water management system is necessary, which is a key for the sustainable development of a region or a country.

The management of river water is often critical due to erratic and unexplained flows. Such flows are usually controlled by structural or non-structural measures. Reservoirs are one of the most essential and effective structural measures for regulating both the spatial and temporal distribution of water. They not only offer water supply, hydroelectric energy, and irrigation, but also help to prevent floods and droughts by smoothing out the excessive inflows [1]. However, the efficient operation of reservoirs is undoubtedly very difficult and involves a series of decisions that govern the amount of water that is stored and released over time [2].

Developing and densely populated cities in the Asian region are at high risk of emergency due to uncontrolled river flows and poorly managed reservoir systems. For example, heavy monsoon rains and floods afflicted nearly 40 million people in India, Nepal, and Bangladesh in mid-2017, resulting in over 1200 deaths [3]. Floods are one of India's most serious climate-related disasters, accounting for more than half of all natural disasters since the 1990s [4]. The problem becomes more critical when some regions in the country are hit by both droughts and floods. An example of such a problem is the flooding in cities of South India. Thus, the area has faced numerous challenges in meeting the expanding water demand while dealing with occasional drought and flooding [5].

Chennai is the capital city of Tamil Nadu state in South India. The city has become the third most urbanized city in the country and largely depends on its water resources for the water supply. However, uncontrolled urbanization, increasing population density, and climate change have caused various water resources management problems in the city [5]. The area faced floods in the 2004–2005 period [6] followed by water scarcity and drought during the years from 2006 to 2014 [7]. The 2015 Chennai floods were the worst natural disaster in Tamil Nadu's history [8]. It happened when the Chennai was preparing for another year of drought, the area was hit by its most disastrous floods since 1918 [9]. Thiruvallur, Chennai, Kancheepuram, Villupuram, and Cuddalore districts in Tamil Nadu were under severe floods that were out of control [10]. The damages were projected to be worth up to US $3 billion [11]. Following 2015, flood warnings have been issued over the area in the years 2020 and 2021. The causative factors conclude that higher runoff is expected to occur in the area in the future as well [12].

The Red Hills Reservoir (RHR) is located at the Red Hills Lake of Tamil Nadu state of South India. The reservoir is a vital resource of water supply in the Chennai city and is also expected to be turned into useful resources in the future [13]. However, climate change and current events of droughts and floods seem to create ramifications for the RHR's prospects, which may adversely affect certain aspects of the area and its habitats [5]. Moreover, the main objective of reservoir operation during the dry season is water conservation. A fundamental difficulty in flood control is establishing a trade-off between the different responsibilities of a reservoir. The timing and amount of flood in downstream areas of RHR can be influenced by the reservoir operation. As a result, the availability of reservoirs across the basin is a significant factor to be considered in flood prevention efforts. Flooding can be exacerbated by the faulty reservoir operation. In order to limit the in-evolvement of reservoirs in floods, an accurate and reliable prediction of reservoir inflows is essential in making timely release decisions, especially in the case of RHR, which was constructed for water conservation. Prediction of future reservoir inflows can be valuable in making efficient operating decisions in this phase of natural uncertainty [14]. However, the Central Water Commission (CWC) of India provides inflow information only at 128 out of 5745 reservoirs in the country, with a lead time of 3 days on a trial basis using modelling techniques [4]. Scarce water inflow information of RHR may lead to worse events in future and need to be considered in current time. Therefore, future flow rate fluctuations in the RHR should be evaluated in order to formulate deliberate plans to minimize the repeating of water overflow threats and to avoid losses of lives and economy.

In view of the aforementioned discussion, the development of an accurate and reliable method for the prediction of RWL in RHR is of utmost importance. Hence, the goal of the current study is to develop an implicit system that can effectively predict the RWL in RHR over time. Generally, methods used for water level (WL) prediction problems include data-driven approaches, such as statistical techniques and artificial intelligence (AI) techniques [15]. These techniques include probability characteristics, time series methods, synthetic data generation, multiple regression, pattern detection, and neural network methods [16]. It is commonly acknowledged that time series modelling is a better choice in the areas of prediction problems [17], which describe the pattern or stochastic behavior of a non-linear problem [18,19]. According to the time series modelling, reasonable results have been reported for most areas of the contiguous United States (US) and China. The autoregressive integrated moving average (ARIMA) model is one of the well-known statistical time series models for the prediction of RWL [20,21]. It comes in a variety of forms like AR, MA, or combination of AR and MA, referred to as autoregressive moving average (ARMA) or seasonal autoregressive integrated moving average (SARIMA) [22,23]. It has been found in literature that only a few attempts have been undertaken to predict the WL using the SARIMA model, such as predictions of lake water levels [24] and groundwater levels [25]. Whereas, the SARIMA model has the advantage of requiring few model parameters to describe time series, that show non-stationarity both within and through seasons [26]. This is a significant simplification compared to machine learning (ML) techniques that often require multiple parameters as the input [21]. The current study is focused on the prediction of RWL in RHR considering the past inflows based on the SARIMA-based time series modelling technique. Available literature have also suggested that a hybrid model can take advantage of the strengths of each component of the model to increase modelling precision and adaptability [27]; a hybrid time series modelling technique has been developed and demonstrated in the current study, which combines the SARIMA time series model with the most widely used ML technique, artificial neural network (ANN) model.

In the field of hydrology and water resources, ANN has been mostly used as the ML technique for modelling water flow, analyzing water quality, and predicting water level [28–31]. Ondimu et al. [32] applied ANN model for WL prediction in Lake Naivasha. Rani et al. [33] found the best predicting model for real-time water level of Sukhi Reservoir as feedforward backpropagation ANN (FBPNN). Wan Ishak et al. [34] employed ANN in prediction and control of RWL, and Altunkaynak et al. [35] employed ANN to anticipate WL changes in Lake Van, the largest lake in Turkey. Moreover, a hybrid model of ANN with ARIMA model has also been demonstrated in few studies [36–39]. To the best knowledge of the authors', no study reported the RWL prediction using the time series hybrid modelling technique particularly for RHR in recent times. Therefore, a hybrid time series modelling technique is developed and demonstrated in the current study that combines the SARIMA time series model with the ANN model to describe the linear and non-linear features independently, motivated by the success of the hybrid prediction models. The technique is then employed for the short-term prediction of daily RWL using real datasets from the RHR and to find the peak water level based on time. It is expected that the current study would be supportive to the reservoir management authority in taking timely decisions about the fate of the reservoir and sustainable development.

2. Materials and Methods

In order to forecast RWL, the hybridization of SARIMA and ANN was performed to the dataset. Figure 1 visualizes the overall framework of the study. The approach was carried out in three phases. First, the difference's requirement was identified, a stationarity test was conducted for this reason. By differencing the data, it was made stationary, and the parameters of SARIMA were identified using autocorrelation function (ACF) and partial autocorrelation function (PACF) plots.

Next, the residual of the SARIMA model was determined and the residual was further modelled using the ANN model. Finally, several statistical measures, such as root mean squired error ($RMSE$), mean absolute error (MAE), mean absolute percentage error ($MAPE$), and coefficient of determination (R^2) were used to assess the effectiveness of the developed models.

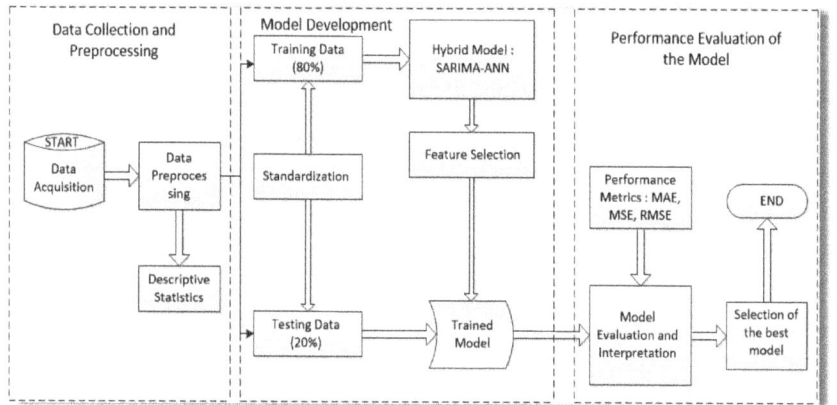

Figure 1. Framework of the study, from data collection, data splitting, and model development to model evaluation and interpretation.

2.1. ACF

The correlation of a time series with its own past and future values is known as autocorrelation. The simple coefficient of the first $N-1$ observation, $t = 1, 2, \ldots, N-1$ $X_t : t = 2, 3, \ldots, N$. The relationship between X_t and X_{t+1} defined as follows:

$$r_1 = \frac{\sum_{t=1}^{N-1}(x_t - X_1)(x_{t+1} - X_2)}{\left[\sum_{t=1}^{N-1}(x_t - X_1)^2\right]\left[\sum_{t=1}^{N-1}(x_t - X_1)^2\right]} \quad (1)$$

$$r_1 = \frac{\sum_{t=1}^{N-1}(x_t - X)(x_{t+1} - X)}{\sum_{t=1}^{N-1}(x_t - X)^2} \quad (2)$$

where X_1 is the first $N-1$ observation's mean. For N substantial large, the variation among the sub-period means X_1 and X_2 may be neglected and r_1 could be estimated by Equation (2):

$$r_k = \frac{\sum_{t=1}^{N-1}(x_t - X)(x_{t+k} - X)}{\sum_{t=1}^{N-1}(x_t - X)^2} \quad (3)$$

2.2. PACF

The PACF defined by the group of partial autocorrelations at various lags k are defined by $(k = 1, 2, 3 \ldots)$. The set of partial autocorrelations at varied lags k defined as follows:

$$r_{kk} = \frac{r_k - \sum_{j=1}^{k-1} r_{k-1,j} r_{k-1}}{1 - \sum_{j=1}^{k-1} r_{k-1,j} r_j} \quad (4)$$

where, $r_{k,j} = r_{k-1,j} - r_{kk} \, r_{k-1}$, $k-1 \, j = 1, 2 \ldots \ldots .k-1$, partially autocorrelations are particularly important for determining the order of an autoregressive model. The PACF of an AR (p) process is zero at lag $p+1$ and greater.

2.3. Study Area

Red Hills Reservoir (RHR) is taken as the study area in this study, which is also known as the Puzhal Lake. The reservoir is located in Chennai City, Red Hills, Thiruvallur district, Tamil Nadu, South India, which is shown in Figure 2. The area is bounded by 13°11′53″ N latitude and 80°11′54″ E longitude. The reservoir is spread over an area about 20.89 km^2 and has a total storage capacity of 3300 million ft^3 (93 million m^3).

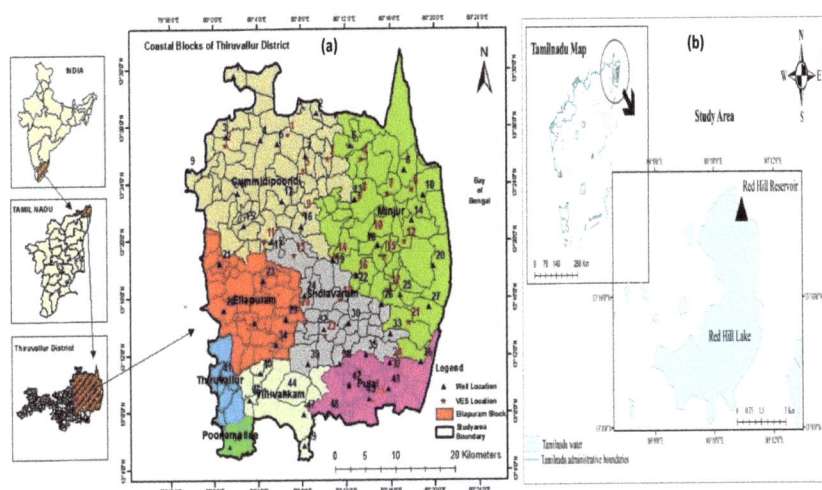

Figure 2. Locational map of the Red Hills Reservoir in the study area. (**a**) topographic map of India [40] and (**b**) topographic map of location.

2.4. Data Collection

The daily RWL data for the RHR were collected from Chennai Water Management for the period from January 2004 to November 2020 [41]. The daily data was converted to average monthly data, which were used for developing and testing the models in the current study. The monthly average RWL for the RHR is shown in Figure 3. As can be seen from the figure, the highest amount of RWL was nearly 32 million ft^3 in January 2011, whereas the lowest amount was found to be 0 in September 2004.

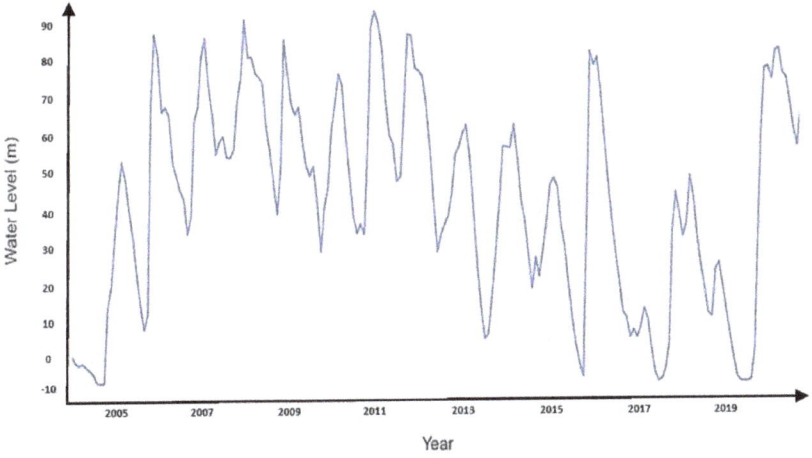

Figure 3. Average monthly RWL of RHR from 2004–2020.

2.5. Seasonal ARIMA (SARIMA) Model

In the 1930s and 1940s, an electrical engineer called Norbert Wiener et al. created the ARIMA idea later named the well-known Box–Jenkins technique. The ARIMA model, also known as (p, d, q) model, is a stochastic model that has been commonly used in hydrological prediction studies [42,43]. The ARIMA model is made up of three components: AR, I, and MA. The AR model denotes the link between current and previous data, the MA denotes the auto correlation frame work of error, and the I denote the series' differencing level. It provides a time series approach towards problems by making a prediction. Peter Whittle proposed the first general version of ARMA in 1951 [44], which may be written as:

$$X_t = c + \varepsilon_t + \sum_{i=1}^{p} \varphi_t X_{t-1} + \sum_{i=1}^{q} \theta_i \varepsilon_{t-i} \tag{5}$$

where ε_t was denoted a white noise and ϕ, θ were denoted the time series coefficients. Equations (6) and (7), which were presented by, show the numerical structure of AR (p) and MA (q) [45]:

AR (p),

$$y_t = c + \beta_1 y_{t-1} + \beta_2 y_{t-2} + \beta_3 y_{t-3} + \cdots + \beta_p y_{t-p} + \varepsilon_t \tag{6}$$

This is an instance of multiple regressions with lagged y_t values as predictors. It's denoted as AR (p):

MA (q)

$$y_t = c + \epsilon_t + \alpha_1 \epsilon_{t-1} + \alpha_2 \epsilon_{t-2} + \alpha_3 \epsilon_{t-3} \ldots + \alpha_q \epsilon_{t-q} \tag{7}$$

ARIMA (p, d, q),

$$y_t = c + \beta_1 y_{t-1} + \beta_2 y_{t-2} + \beta_3 y_{t-3} + \cdots + \beta_p y_{t-p} \epsilon_t + \alpha_1 \epsilon_{t-1} + \alpha_2 \epsilon_{t-2} + \alpha_3 \epsilon_{t-3} \cdots + \alpha_q \epsilon_{t-q} \tag{8}$$

the term β coming from AR and ε the error terms coming from MA model.

The SARIMA was used to eliminate seasonal variance characteristics of data via seasonal differences [46]. They're the same as in the ARIMA model, as follows:

p: Order of trend autoregression;
d: Order of trend difference;
q: Order of trend moving average.

There are four seasonal components that must be adjusted that are not part of ARIMA:

P: Order of Seasonal autoregressive;
D: Order of Seasonal difference;
Q: Order of Seasonal moving average;
m: A single seasonal period's number of time steps.

The general equations for the SARIMA model can be defined by Equations (9)–(13):

$$\phi_p(L)\Phi_p(L^s)(1-L)^d(1-L^s)^D Z_t = \theta_q(L)\Theta_Q(L^s)\varepsilon_t \tag{9}$$

$$\phi_p(L) = 1 - \phi_1 L - \phi_2 L^2 - \cdots - \phi_p L^P \tag{10}$$

$$\theta_q(L) = 1 - \theta_1 L - \theta_2 L^2 - \cdots - \theta_q L^q \tag{11}$$

$$\Phi_P(L^s) = 1 - \Phi_S\left(L^S\right) - \Phi_{2S}\left(L^{2S}\right) - \cdots - \Phi_{PS}\left(L^{PS}\right) \tag{12}$$

$$\Theta_Q\left(L^S\right) = 1 - \Theta_S L^S - \Theta_{2S} L^{2S} - \cdots - \Theta_{QS} L^{QS} \tag{13}$$

where, Z_t stands for the observed value and ε_t stands for the lagged error at time t; L (lag operator) defined by:

$$L^k Z_t = Z_{t-k}; \; \phi_p(p=1,2,\ldots,p), \; \Phi_p(P=1,2,\ldots,P), \; \theta_q(q=1,2,\ldots,q), \; \Theta_q(Q=1,2,\ldots,Q) \tag{14}$$

The orders of autoregressive and moving average were represented by p and q, respectively. P and Q indicate the seasonal autoregressive and seasonal moving average orders, accordingly. S stands for seasonal length, whereas d and D stand for difference order and seasonal difference, respectively. Figure 4 shows the flowchart of SARIMA model.

The mean, variance, and autocorrelation functions of the data were tested for stationarity with respect to time as a prerequisite for using the SARIMA model. ε_t (random error) was also made independent and distributed similarly to a standard zero-mean dispersion. Higher weights were regarded as indicators of a better prediction model when the SARIMA model was tested for weights [47,48]. After the selection of weights, the SARIMA model was modelled in four stages including: (i) stationarity check; (ii) identification and estimation stage; (iii) diagnosis stage; and (iv) prediction stage. In stage one, the time series data were checked for stationarity. The stationarity of data was checked followed by the Augmented Dickey Fuller (ADF) test, which examines a time series for the null hypothesis of the existence of a unit root [49]. ADF's mathematical expression is as follows:

$$\Delta y = \alpha + \beta\, t + \gamma\, y_{t-1} + \delta_1 \Delta y_{t-1} + \cdots + \delta_{p-1} \Delta y_{t-p+1} + \epsilon_t \tag{15}$$

where α is constant, β is time trend coefficient, p lag order, and ϵ_t is the error term. Before executing the test for the null hypothesis $\gamma = 0$, the appropriate lags of order p were selected. If the time series is non-stationary, stationarity can be obtained by regressing or differencing the data until it becomes stationary [50]. Non-seasonal and seasonal differencing are the two types of differencing with order d. Seasonal difference is a technique for removing seasonal components from data and by eliminating trend characteristics from the data, the non-seasonal difference aims to address data instability [51]. This can be achieved by determining the order of the SARIMA model, estimating unknown parameters, accumulating model candidates with a p-value of less than 0.05, and evaluating the goodness of fit on the anticipated errors; then, predicting a value in the future using the data that is available. To establish the ordering of the SARIMA, ACF and PACF charts were required [52].

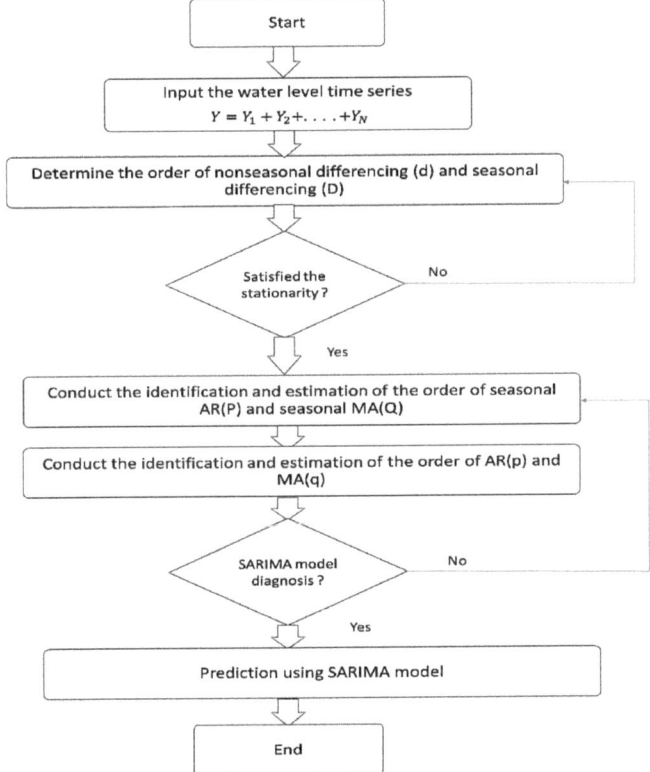

Figure 4. Flowchart of SARIMA model.

2.6. ANN Model

The capacity of an artificial neural network is to represent complicated nonlinear relationships [48,53–57]. One of the most extensive ANN for time series modelling and prediction is the multilayer perceptron, especially individuals with one hidden layer. A network of three layers of functioning is linked by acyclic linkages. The mathematical equation between the output (y_t) and the inputs ($y_{t-i}, \ldots, y_{t-p}$) as follows:

$$y_t = w_0 + \sum_{j=1}^{Q} w_g g \left(w_{0j} + \sum_{i=1}^{p} w_{i,j} y_{t-i} \right) + e_t \qquad (16)$$

where $w_{i,j}$ ($i = 0, 1, 2, \ldots, P, j = 1, 2, \ldots, Q$) and w_j ($j = 0, 1, 2, \ldots, Q$) stands for model parameters, which are also known as connection weights; the number of input nodes is indicated by P; and the number of hidden nodes is represented by Q. For hidden layers, the sigmoid function is frequently employed transfer function, that is:

$$Sig(x) = \frac{1}{1 + \exp(-x)} \qquad (17)$$

$$y_t = f(y_{t-i}, \ldots, y_{t-P}, W) + e_t \qquad (18)$$

As a result, the ANN model of (18) conducts mapping from historical data to projected values y_t, i.e.,:

where $f(.)$ is a function based on the network structure and connection weights and W is a vector containing all parameters. In the output layer, the formulation (18) implies one output node, which is generally employed for a step-ahead prediction. The basic network represented by (18) is remarkably strong in that it can estimate any function when the neurons of hidden nodes (Q) are high enough [58]. In out-of-sample predictions, a basic network layout with a modest number of hidden nodes frequently works effectively. This might be related to the over-fitting phenomenon that occurs frequently in neural network models [59].

As Q is data dependent, there is no methodical procedure for determining this parameter. The selection of the number of lagged observations, P, and the dimensionality of the input vector is another essential part in ANN modelling of a time series [59], in addition to determining the adequate number of hidden nodes. Because it determines the (nonlinear) autocorrelation frameworks of the time series, it is likely the most critical parameter to estimate in an ANN model. There is, nevertheless, no hypothesis that can be utilized to assist in P selection. As a result, studies are frequently undertaken to find a suitable P and Q.

2.7. Hybrid SARIMA-ANN Model

The SARIMA models' approximation of complicated nonlinear issues may not be acceptable. The use of artificial neural networks to represent linear issues has generated mixed results. Denton et al. [60], for example, demonstrated that when the data contain outliers or multicollinearity, neural networks outperformed linear regression algorithms considerably. The effectiveness of ANNs for linear regression issues is similarly influenced by sample size and noise level, according to Markham et al. [61]. As a consequence, employing ANN is unwise since it is unfeasible to adequately comprehend the characteristics of data in a real situation. A hybrid technique which merges the linear and nonlinear skills might be a useful strategy in practice. In the first phase, the SARIMA model is employed to extract the linear component of the time series. The residuals of SARIMA and the lagged data are then employed as input for statistical ML techniques throughout the second step. Lastly, predictions were estimated using best suited model in the third step. The following sections go into the specifics of these steps:

$$y_t = L_t + N_t \qquad (19)$$

where L_t represents the linear element and N_t indicates the nonlinear component, these two factors must be calculated based on the data. At first, the SARIMA model was employed as the linear module, and then the linear model's residual was determined from the model. Let e_t stand for the linear model's residual at time t, then:

$$e_t = y_t + \hat{L}_t \qquad (20)$$

where $\hat{L}_t$ is the calculated relationship's prediction value for time t. The diagnosis of the sufficiency of linear models relies heavily on residuals. If there are still linear correlation patterns in the residuals, a linear model is insufficient. Residual analysis, on the other hand, is unable to find any nonlinear correlations. In reality, no universal diagnostic statistics for nonlinear auto correlation connections exist at this time. As a result, if the model passes diagnostic testing, it might still be insufficient since nonlinear interactions have not been adequately represented. The SARIMA's restriction will be shown by any major nonlinear pattern in the residuals. Modeling residuals with ANNs may be used to investigate nonlinear linkages. For residuals, an ANN model will be used as follows:

$$e_t = f(e_{t-1}, e_{t-2}, \ldots, e_{t-n}) + \varepsilon_t \qquad (21)$$

where f is non-linear function and ε_t is the random error. It's worth mentioning that if model f isn't adequate, the error term isn't always random. As a consequence, it's crucial to have the right model. The prediction from (18) will be denoted as $\hat{N}_t$, and the combined prediction will be:

$$\hat{y}_t = \hat{L}_t + \hat{N}_t \qquad (22)$$

In summary, the suggested hybrid technique comprises of two parts. During the first phase, the SARIMA was utilized to investigate the linear component of the problem. In the second phase, the residuals from the SARIMA model were modelled by ANN. The residuals of the linear model will contain information on the nonlinearity because the SARIMA model could not account for data's nonlinearity. The findings of the neural network may be utilized to anticipate the SARIMA model's error terms. In identifying diverse patterns, the hybrid model uses the unique features and strengths of both the SARIMA and ANN models. This might be beneficial to analyze linear and nonlinear trends independently employing various techniques and then integrate the predictions to enhance overall modelling and prediction performance.

Figure 5 shows the SARIMA and ANN hybrid model in which the SARIMA model is combined with the ANN model. As described earlier, in developing SARIMA and ANN models, subjective interpretation of the order and model appropriateness is often needed. In the hybrid technique, it is conceivable that suboptimal models will be utilized. Box–Jenkins technique, for example, relies on low order auto correlation. Even if substantial auto correlations of higher order exist, the model was regarded acceptable if low order auto correlations were insignificant. The hybrid model's usefulness may not be blemished by this suboptimality. In 1989, Granger pointed out that the component model in a hybrid model should be suboptimal in order for the hybrid model to generate the enhanced prediction [62,63].

2.8. Performance Evaluation of the Models

The outcomes of the SARIMA, ANN, and SARIMA–ANN hybrid models were monthly RWL of Red Hills Reservoir for the January 2004–November 2020 period. The predicted values obtained by the models were compared with the actual dataset. Several statistical performance indicators, including $RMSE$, MAE, $MAPE$, and R^2 (Equations (23)–(26)) were used to evaluate the performance of each model. A lower value of MAE, $RMSE$, and $MAPE$ indicate a good correlation between the observed and predicted datasets. The value of R^2 closer to 1 demonstrates a good correlation between the observed and predicted data sets:

$$MAE = \frac{1}{n}\sum_{i=1}^{n}|O_i - P_i| \qquad (23)$$

$$RMSE = \sqrt{\frac{\sum_{i=1}^{n}(O_i - P_i)^2}{n}} \qquad (24)$$

$$MAPE = \frac{1}{n}\sum_{t=1}^{n}\left|\frac{(O_i - P_i)}{O_i}\right| \times 100 \qquad (25)$$

$$R^2 = 1 - \frac{\sum_{i=1}^{n}(O_i - P_i)^2}{\sum_{i=1}^{n}(O_i - \hat{P}_i)^2} \qquad (26)$$

Here, O_i represents the observations, P_i represents the predictions at each time step, and n represents total time step numbers.

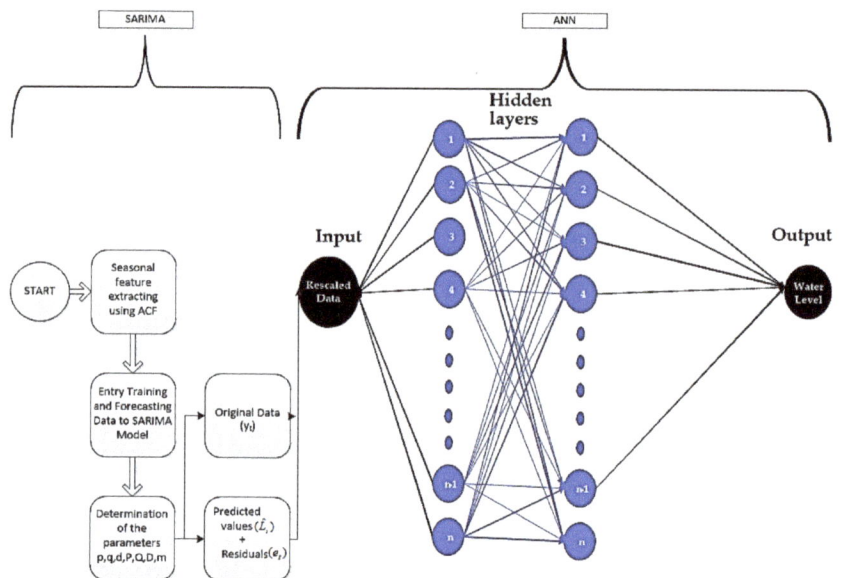

Figure 5. Flowchart of the SARIMA-ANN Hybrid model.

3. Results

The entire obtained data were separated into two portions for training and testing of the model in order to assess and compare the adopted modelling techniques. The training datasets were those from January 2004 to March 2017, accounting for 80% of the dataset, while the remaining dataset (20%) were used for model testing purpose.

To anticipate the SARIMA model, firstly we had to analyze the fluctuation of the RWL data based on Figure 2. The approach of the SARIMA model was carried out in three phases. First, the difference's requirement was identified, a stationarity test was accomplished for this reason. The samples were stationary adjusted through differencing, and then SARIMA p, d, q, m and P, D, Q parameters were identified using ACF plots and PACF.

Step 1: Stationarity test of the Data

The ADF test indicates that the null hypothesis that the dataset has a unit root (non-stationary) at the 5% significance level may be rejected. According to the ACF graph (Figure 6), the seasonally differenced trend shows significant spikes in negative values at the 1st lag, and ACF shuts out after that for the non-seasonal element.

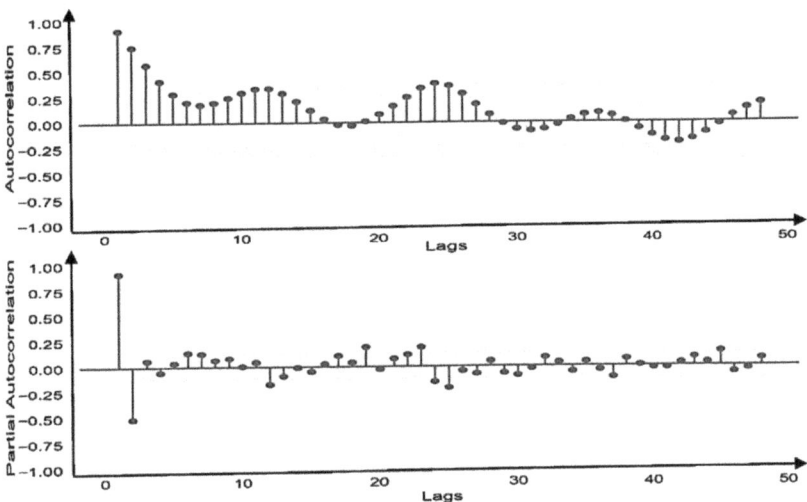

Figure 6. Autocorrelation function (ACF) and partial autocorrelation functions (PACF) for RWL Dataset. The light blue bands behind the dots on each plot denote the corresponding 95% confidence interval.

Due to seasonal influences, the ACF exhibits strong spikes at numerous lags that demonstrate a periodic order across 12 months, non-seasonal differencing is thus unnecessary, whereas seasonal differencing is essential due to seasonal stationarity.

Furthermore, substantial spikes were noticed after first order differencing at intervals of every 12 months (12th, 24th, 36th lags ...) on the ACF plot in Figure 7. As a result, to reduce seasonality, a seasonal differencing technique continued till third order differencing. A chart of ACF and PACF with seasonal differences is shown in Figure 8 after third order seasonal differencing.

Step 2: Model Identification

This stage is to estimate the suitable values of p, d, and q employing correlogram and partial correlogram and ADF Test. The preliminary model's order was determined using the ACF and PACF. The ACF exhibits strong spikes at numerous lags, which demonstrate a periodic sequence over 12 months due to seasonal impacts, as seen by the correlogram. At many lags, the PACF shows substantial increases; therefore, the model might be a SARIMA model.

The observed RWL samples were subjected to seasonal differencing $(D = 3)$ in order to create a time series that was seasonally stationary. For future exploration, SARIMA $(p, 0, q)$ $(P, 3, Q)12$ are recommended. Initial, parameters of p, q, P and Q were identified by ACF and PACF plots (Figure 5).

The seasonal component of ACF displays a positive substantial spike at the 12th and 24th lags. As a result, for model identifier, two seasonal (SMA) and one non-seasonal MA values are recommended. In the same way, for PACF, there were no seasonal or non-seasonal significant spikes detected. Hence, zero AR for non-seasonal and zero seasonal (SAR) are recommended for inclusion in the SARIMA model. As a consequence, SARIMA $(0, 0, 1)$ $(0, 3, 2)_{12}$ was identified for further evaluation.

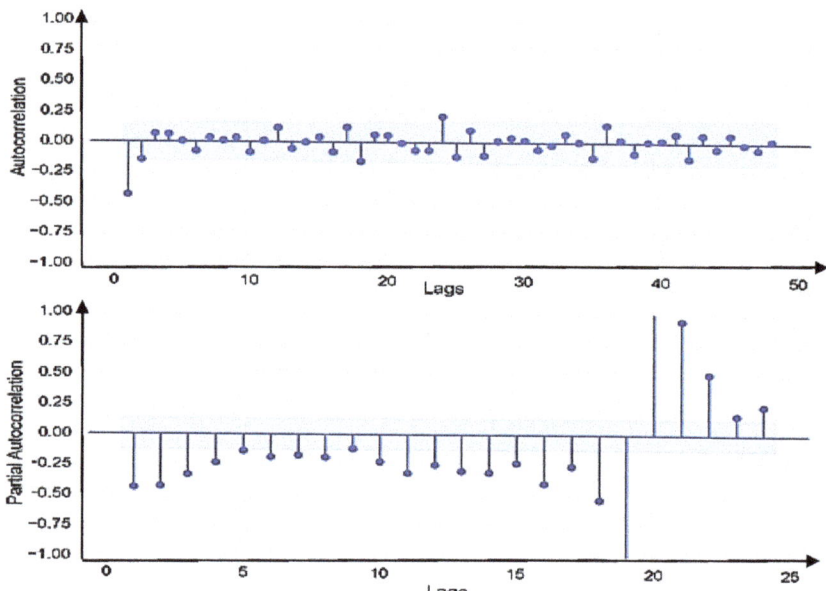

Figure 7. Autocorrelation function (ACF) and partial autocorrelation functions (PACF) after first seasonal differencing for RWL Dataset. The light blue bands behind the dots on each plot denote the corresponding 95% confidence interval.

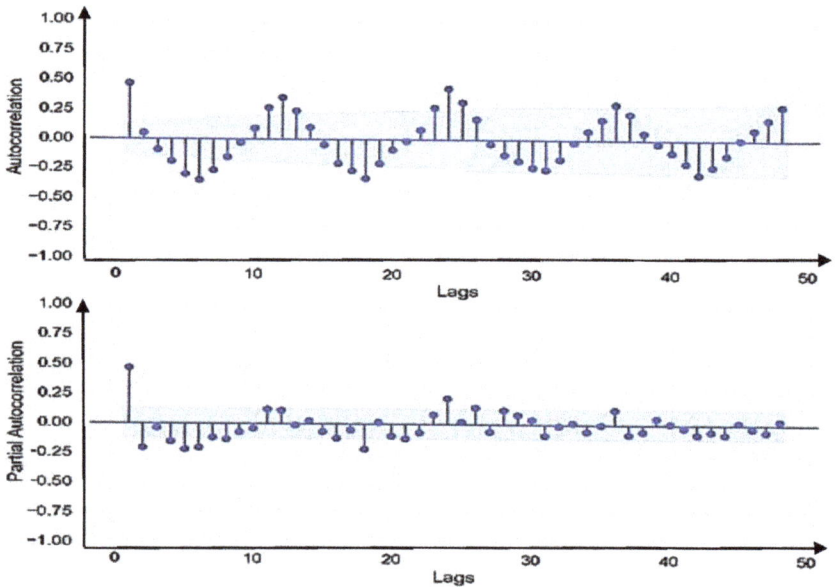

Figure 8. ACF and PACF plots after third seasonal differencing for RWL Dataset. The light blue bands behind the dots on each plot denote the corresponding 95% confidence interval.

Step 3: Parameter Estimation
The parameters of the AR and MA were estimated in this stage.

Here the parameter estimates in Table 1 and the performance values are shown in Table 2 for SARIMA model.

Table 1. Parameter estimates for SARIMA (0,0,1) (0,3,2) model.

Parameter	θ_1	Θ_1	Θ_2
Value	0.7993	−1.5737	0.5760

Explanations: θ_1 = MA parameter of non-seasonal components, Θ_1, Θ_2 = MA parameters of seasonal components.

Table 2. Performance values of selected models.

Akaike information criterion	2010.938
Bayesian information criterion	2022.283
Hannan-Quinn criterion	2015.547
Ljung-Box	27.74
Heteroskedasticity	2.00

Step 4: Diagnostic Checking

At this stage, residual's diagnosis, standard residual, histogram plus estimated density, normal $Q-Q$, and correlogram were checked to analyze the model.

Figure 9a shows that the residual errors seem to fluctuate around a mean of zero and have a uniform variance. Figure 9b illustrates the density plot. It suggests a normal distribution with a mean zero. Figure 9c demonstrates that all the dots fall closely with the red line. Any significant deviations would imply that the distribution is skewed. Figure 9d reveals that the residual errors are not autocorrelated.

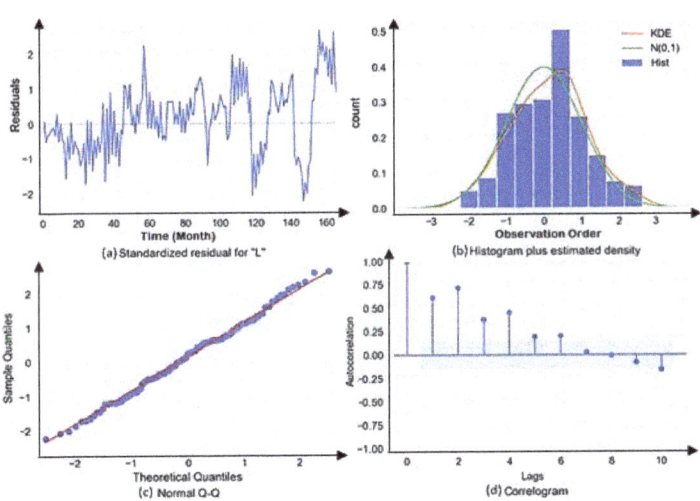

Figure 9. Residual's diagnosis (**a**) Standard residual for "L" (**b**) Histogram plus estimated density (**c**) Normal Q-Q (**d**) Correlogram.

Figure 10 shows the ACF and PACF plots of residuals for RWL dataset. The ACF and PACF of the residual is showing inadequate results and the presence of autocorrelation in residuals may be determined employing the Durbin Watson (DW) test. The DW value should be in the range of 1.5 and 2.5 [64,65]; here, the value is 0.99, which indicates that the SARIMA (0, 0, 1) (0, 3, 2)12 model is not well fitted for prediction. The alternative which used to resolve the problem is building a residual model of SARIMA using ANN which is no regression assumption.

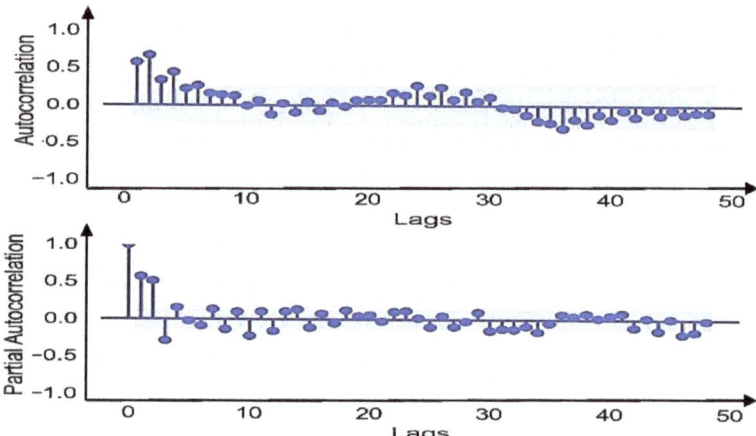

Figure 10. ACF and PACF plots of residuals for RWL Dataset. The light blue bands behind the dots on each plot denote the corresponding 95% confidence interval.

The information about the neural network architecture shows that network has an input layer, two hidden layers with 512 and 4 nodes, and an output layer with one output node. For the best network structure, an ANN should be used with the appropriate number of hidden layers and neurons in each hidden layer. The enumeration approach is based on least MSE used in the ANN modelling to discover the best number of layers and associated neurons in each hidden layer. All of the sample data have to transform into a value between 0 and 1 because of the activation function which is used in artificial neural network is sigmoid function. The error is the sum-of-squares error because identity and activation function are applied to the output layer. Initialization, feed-forward, error assessment, propagation, and adjustment are the learning methods of the artificial neural network. An optimised network of topology 2-512-4-1 was determined to be superior to the other studied network topologies based on MSE criteria. The training cycle was set at 500 epochs, while bath size and validation split are 4 and 0.2, respectively. A neural network is typically initialized using random weights during the initialization process [66].

Figure 11 reveals the SARIMA residuals plot of RWL dataset employed to test the existence of nonlinearity. The best selected SARIMA, ANN, and SARIMA–ANN models were compared based on MAE, $MAPE$, $RMSE$, and R^2 values using Equations (23)–(26). The comparison results are given in Table 3. The results show that the SARIMA–ANN model performed better than single SARIMA and ANN models in prediction of data, with an R^2 value of 0.84, MAE value of 328.69, $MAPE$ value of 32,868.51, MSE value of 174,043.217, and $RMSE$ value of 417.185. Furthermore, the RWL number's projected value is almost identical to its actual value.

Table 3. Evaluation results for SARIMA, ANN and SARIMA-ANN models.

Model	MAE	$MAPE$	$RMSE$	R^2
SARIMA	798.10	79,810.15	891.994	0.30
ANN	660.32	66,032.258	806.062	0.51
SARIMA-ANN	343.23	34,323.06	430.728	0.84

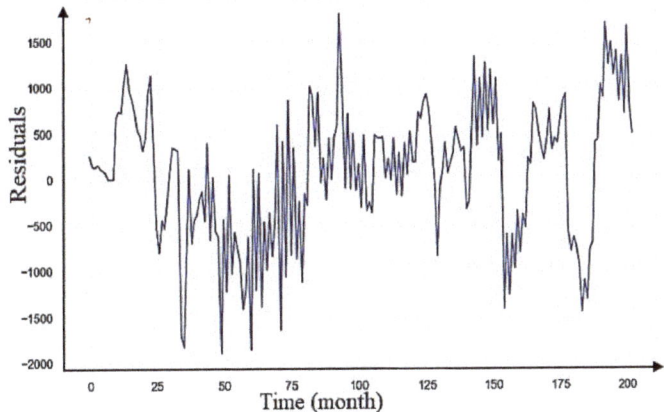

Figure 11. Residuals of SARIMA model.

Figures 12–14 shows the results for predicted RWL using SARIMA, ANN, and hybrid SARIMA–ANN models, respectively. The graphs shown that while applying ANN alone in the test dataset, the performance is worse comparing to the SARIMA and SARIMA–ANN models. It can be observed that the predicted monthly RWL obtained from the fitted SARIMA–ANN model is matching closely with the pattern of the curve of actual RWL.

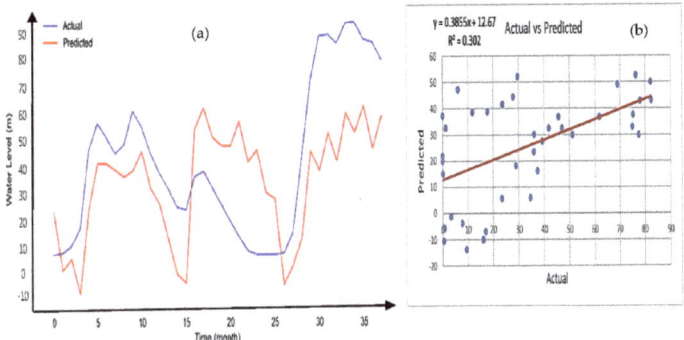

Figure 12. (**a**) Actual and Prediction plot using SARIMA Model. (**b**) Model prediction versus actual correlation.

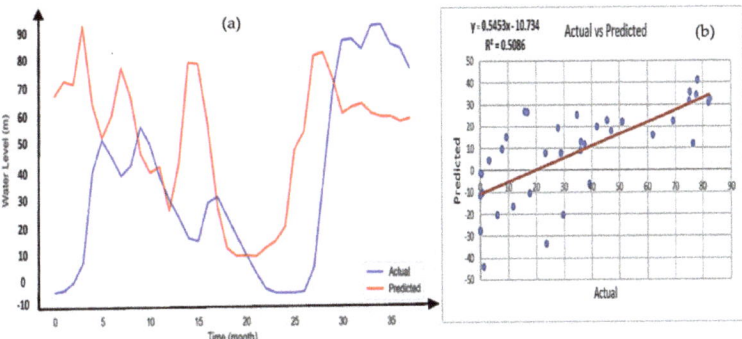

Figure 13. (**a**) Actual and Prediction plot using ANN Model. (**b**) Model prediction versus actual correlation.

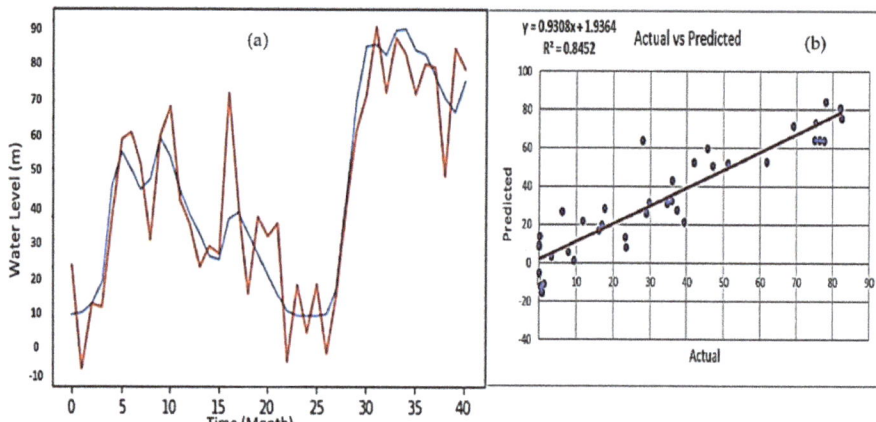

Figure 14. (**a**) Actual and Prediction plot using SARIMA-ANN Model. (**b**) Model prediction versus actual correlation.

4. Discussions

When it comes to river systems, India is a wealthy country. The Aravalli, Ganges, Brahmaputra, and Indus are four significant river systems in India, each with a substantial catchment area and drainage density. All of these river systems have several tributaries that run the length and width of India, making it more vulnerable to flooding [67]. Consequently, during the years 1987 and 1993, the cycle of floods followed by severe drought and its impact on water shortage in Chennai was at its peak. A severe drought struck Chennai City in 1986. Only 40% of the rainfall was reported and 17% of the total available water in the city's three lakes was used. Legislation to regulate the exploitation of groundwater was required. The groundwater level in Chennai was roughly 8 m deep in 1988, but it rose to 4.08 m in 2007.

From 1988 through 1996, there was increasing commercial exploitation. Proactive detection of water intensity in a given location is always useful for planning with scarce resources and implementing effective intervention techniques. A high-quality water level prediction is required to optimize the net benefits of water management. This surface water is critical to the region's socioeconomic development and expansion. Water infrastructure development, floods, and droughts all have an impact on industrial operations, necessitating smart resource management. Precision water level prediction not only decreases the danger of mis-operation and the likelihood of damage, but it also boosts earnings. There is a lack of statistical computational analysis that might use prediction models to anticipate water increase in Chennai. In such cases, advanced computational models, such as SARIMA and ANN were chosen with the goal of predicting water level and, as a result, filling the gap in the specific region.

The selected models for the water level prediction in this study could be used confidently to estimate the water level in the Red Hills Reservoir. The results obtained in the present study are in validation of other previous studies that were performed for the prediction of water level in India using ANN, such as M.Y.A. Khan et al. [68], that has put the application of ANN into use for the prediction of water level for the Ramganga river, which suggests the prediction accuracy of 93.42%. Additionally, machine learning techniques, such as wavelet and support vector machine implementation has also been performed by Yadav et al. [69] to predict the daily water level of Loktak lake (India). The prediction accuracy for accumulated data in this study was found to be higher than the original data series based on the performance evaluation using *RMSE*, because it accounted for the past data for analysis that increased the prediction efficiency. Moreover, a hybrid system of ANN and machine learning was used to forecast the water level in Jhelum river at Sangam station of Kashmir valley in India to improve the early warning system for flood prevention. It was found that the condition of accuracy depends on previous data and forecasted values of temperature and precipitation [70].

Based on the application of previous and present studies, the conditions for the application of predicting the water level include the applicability of the past data collected, data collection techniques, and a hybrid prediction algorithm rather than standalone method. The accuracy of prediction of water level depends on the forecasted values of precipitation, weather conditions, and location. Therefore, the short-term prediction results are more promising than the long-term prediction.

In other words, monthly water level data may be examined without any adjustments to meteorological data, such as temperature, precipitation, wind speed, humidity, sun hours and UV index, and amount of the location, which would be the reason for its global generalization, because multivariate data is not required to assess the prediction of water level data. Apart from the findings, it will be impressive to observe if the hybrid strategy is useful in crucial conditions, such as when the water level rises dramatically (suggesting a peak in water level). The visual comparisons of all approaches for water level prediction are shown in Figures 12–14. The Figures show that the SARIMA and ANN hybrid model could represent the trend of the actual data fairly well, despite the fact that single techniques do not function effectively in any of these circumstances. It is also worth mentioning that the results are solely applicable to the examined region owing to the statistical application of the analysis. As a result, due to the changing nature of statistical data, any model that successfully predicts reservoir water level data for this research may not be useful for other areas. To put it another way, the SARIMA and ANN hybrid models utilized in this work for reservoir water level prediction in various areas may differ due to regional differences. The type of data, such as seasonality, residuals, autocorrelation, and the data's explanatory power, would have a significant impact on the prediction utilizing SARIMA and ANN for any region.

5. Conclusions

The probabilistic aspect of RWL prediction is investigated in this study using a hybrid model, SARIMA and ANN model for the Red Hills Reservoir (RHR). Time series data originating from various applications, in general, comprise both linear and nonlinear variations. Linear SARIMA and nonlinear ANN models cannot individually take out consequence data accurately. Hybrid models, which combine the strengths of SARIMA and ANN models, are better than individual types of models as they are capable of exploiting the advantages of both types of models simultaneously. The results show that the hybrid SARIMA–ANN model performed better than the SARIMA and ANN model for RHR in the RWL prediction. The prediction from the hybrid model is obtained by adding the predicting values from the two individual models. Thus, the hybrid model proposed in the present paper becomes a simple and accurate prediction model in many applications.

6. Limitations and Future Scope

A mix of linear and nonlinear models were utilized in this research. In their respective linear or nonlinear domains, both SARIMA and ANN models have great success in the particular area. Neither of them, however, are universal paradigms that can be used in all situations. It is possible that the SARIMA models' approximation of complicated nonlinear issues is insufficient. ANNs, on the other hand, have had inconsistent success when used to simulate linear issues. To improve model accuracy, most external parameters, including meteorological data, such as temperature, precipitation, wind speed, humidity, sun hours, and UV index should be added with the passage of time. The results of this study could be a good reference for the facilitators and decision-making stakeholders by performing predictions in a quick and easy way by incorporating the methodology adopted in this study. Because, unlike other techniques for prediction, this approach does not require a colossal set of data that is collected through epidemiological retrospective information.

For future studies, meteorological elements should be used as independent variables to increase the accuracy of anticipated findings. This research might be used for such locations where natural disasters cause the majority of water shortages.

Author Contributions: Conceptualization, A.S.A. and R.S.; methodology, A.S.A. and R.S.; software, A.S.A.; validation, R.S. and H.D.; formal analysis, A.S.A., S.K.A. and H.K.; investigation, A.S.A.; resources, R.S. and H.D.; data curation, A.S.A. and M.B.A.R.; writing—original draft preparation, A.S.A.; writing—review and editing, R.S., H.D., S.K.A. and H.K.; visualization, R.S. and H.D., S.K.A.; supervision, R.S.; project administration, R.S.; funding acquisition, S.N.A.M. All authors have read and agreed to the published version of the manuscript.

Funding: This research was funded by YUTP research project (Cost center: 015LC0-114).

Institutional Review Board Statement: Not applicable.

Informed Consent Statement: Not applicable.

Data Availability Statement: Not applicable.

Acknowledgments: The financial support provided by Universiti Teknologi PETRONAS under YUTP research project (Cost center: 015LC0-114) is highly appreciated. The authors also would like to thank the Fundamental and Applied Science Department and Centre of Graduate Studies of Universiti Teknologi PETRONAS for their support and funding under graduate assistantship scheme (GA).

Conflicts of Interest: The authors declare no conflict of interest.

References

1. Castillo-Botón, C.; Casillas-Pérez, D.; Casanova-Mateo, C.; Moreno-Saavedra, L.M.; Morales-Díaz, B.; Sanz-Justo, J.; Gutiérrez, P.A.; Salcedo-Sanz, S. Analysis and prediction of dammed water level in a hydropower reservoir using machine learning and persistence-based techniques. *Water* **2020**, *12*, 1528. [CrossRef]
2. Paul, N.; Elango, L. Predicting future water supply-demand gap with a new reservoir, desalination plant and waste water reuse by water evaluation and planning model for Chennai megacity, India. *Groundw. Sustain. Dev.* **2018**, *7*, 8–19. [CrossRef]
3. Young, C.C.; Liu, W.C.; Hsieh, W.L. Predicting the Water Level Fluctuation in an Alpine Lake Using Physically Based, Artificial Neural Network, and Time Series Forecasting Models. *Math. Probl. Eng.* **2015**, *2015*, 708204. [CrossRef]
4. Krishna, R.N.; Ronan, K.; Spencer, C.; Alisic, E. The lived experience of disadvantaged communities affected by the 2015 South Indian floods: Implications for disaster risk reduction dialogue. *Int. J. Disaster Risk Reduct.* **2021**, *54*, 102046. [CrossRef]
5. Nanditha, J.S.; Mishra, V. On the need of ensemble flood forecast in India. *Water Secur.* **2021**, *12*, 100086. [CrossRef]
6. Venkatesan, D. Impact of Water-Level Fluctuation in Redhills Reservoir on Population Dynamics of Chennai City. *Int. J. Eng. Appl. Sci. Technol.* **2019**, *4*, 99–104. [CrossRef]
7. Jameson, S.; Baud, I. Varieties of knowledge for assembling an urban flood management governance configuration in Chennai, India. *Habitat Int.* **2016**, *54*, 112–123. [CrossRef]
8. Selvaraj, K.; Pandiyan, J.; Yoganandan, V.; Agoramoorthy, G. India contemplates climate change concerns after floods ravaged the coastal city of Chennai. *Ocean Coast. Manag.* **2016**, *129*, 10–14. [CrossRef]
9. Mohan, P.R.; Srinivas, C.V.; Yesubabu, V.; Baskaran, R.; Venkatraman, B. Simulation of a heavy rainfall event over Chennai in Southeast India using WRF: Sensitivity to microphysics parameterization. *Atmos. Res.* **2018**, *210*, 83–99. [CrossRef]
10. Bhuvana, N.; Arul Aram, I. Facebook and Whatsapp as disaster management tools during the Chennai (India) floods of 2015. *Int. J. Disaster Risk Reduct.* **2019**, *39*, 101135. [CrossRef]
11. Veerasingam, S.; Mugilarasan, M.; Venkatachalapathy, R.; Vethamony, P. Influence of 2015 flood on the distribution and occurrence of microplastic pellets along the Chennai coast, India. *Mar. Pollut. Bull.* **2016**, *109*, 196–204. [CrossRef]
12. Lakshmi, D.D.; Satyanarayana, A.N.V. Influence of atmospheric rivers in the occurrence of devastating flood associated with extreme precipitation events over Chennai using different reanalysis data sets. *Atmos. Res.* **2019**, *215*, 12–36. [CrossRef]
13. Correspondent, S. Shutters of Chembarambakkam, Red Hills Reservoirs Opened Again. Available online: https://www.thehindu.com/news/cities/chennai/as-rain-continues-in-chennai-shutters-of-chembarambakkam-red-hills-reservoirs-to-be-opened-again/article33499670.ece (accessed on 24 November 2021).
14. Murugesan, A.; Bavana, N.; Vijayakumar, C.; Vignesha, D.T. Drinking Water Supply And Demand Management In Chennai City-A Literature Survey. *IJISET International J. Innov. Sci. Eng. Technol.* **2015**, *2*, 715–728.
15. Chang, F.J.; Chang, Y.T. Adaptive neuro-fuzzy inference system for prediction of water level in reservoir. *Adv. Water Resour.* **2006**, *29*, 1–10. [CrossRef]
16. Zhang, X.; Liu, P.; Zhao, Y.; Deng, C.; Li, Z.; Xiong, M. Error correction-based forecasting of reservoir water levels: Improving accuracy over multiple lead times. *Environ. Model. Softw.* **2018**, *104*, 27–39. [CrossRef]

17. Bourdeau, M.; Zhai, X.Q.; Nefzaoui, E.; Guo, X.; Chatellier, P. Modeling and forecasting building energy consumption: A review of data-driven techniques. *Sustain. Cities Soc.* **2019**, *48*, 101533. [CrossRef]
18. Karunasinghe, D.S.K.; Liong, S.Y. Chaotic time series prediction with a global model: Artificial neural network. *J. Hydrol.* **2006**, *323*, 92–105. [CrossRef]
19. Musarat, M.A.; Alaloul, W.S.; Rabbani, M.B.A.; Ali, M.; Altaf, M.; Fediuk, R.; Vatin, N.; Klyuev, S.; Bukhari, H.; Sadiq, A.; et al. Kabul river flow prediction using automated ARIMA forecasting: A machine learning approach. *Sustainability* **2021**, *13*, 10720. [CrossRef]
20. Islam, M.N.; Sivakumar, B. Characterization and prediction of runoff dynamics: A nonlinear dynamical view. *Adv. Water Resour.* **2002**, *25*, 179–190. [CrossRef]
21. Yu, Z.; Lei, G.; Jiang, Z.; Liu, F. ARIMA modelling and forecasting of water level in the middle reach of the Yangtze River. In Proceedings of the 2017 4th International Conference on Transportation Information and Safety (ICTIS), Banff, AB, Canada, 8–10 August 2017; pp. 172–177. [CrossRef]
22. Viccione, G.; Guarnaccia, C.; Mancini, S.; Quartieri, J. On the use of ARIMA models for short-term water tank levels forecasting. *Water Sci. Technol. Water Supply* **2020**, *20*, 787–799. [CrossRef]
23. Ghimire, B.N. Application of ARIMA Model for River Discharges Analysis. *J. Nepal Phys. Soc.* **2017**, *4*, 27. [CrossRef]
24. Birylo, M.; Rzepecka, Z.; Kuczynska-Siehien, J.; Nastula, J. Analysis of water budget prediction accuracy using ARIMA models. *Water Sci. Technol. Water Supply* **2018**, *18*, 819–830. [CrossRef]
25. Aytek, A.; Kisi, O.; Guven, A. A genetic programming technique for lake level modeling. *Hydrol. Res.* **2014**, *45*, 529–539. [CrossRef]
26. Mirzavand, M.; Ghazavi, R. A Stochastic Modelling Technique for Groundwater Level Forecasting in an Arid Environment Using Time Series Methods. *Water Resour. Manag.* **2015**, *29*, 1315–1328. [CrossRef]
27. Fang, T.; Lahdelma, R. Evaluation of a multiple linear regression model and SARIMA model in forecasting heat demand for district heating system. *Appl. Energy* **2016**, *179*, 544–552. [CrossRef]
28. Zhang, P.G. Time series forecasting using a hybrid ARIMA and neural network model. *Neurocomputing* **2003**, *50*, 159–175. [CrossRef]
29. Seo, Y.; Kim, S.; Kisi, O.; Singh, V.P. Daily water level forecasting using wavelet decomposition and artificial intelligence techniques. *J. Hydrol.* **2015**, *520*, 224–243. [CrossRef]
30. Kasiviswanathan, K.S.; He, J.; Sudheer, K.P.; Tay, J.H. Potential application of wavelet neural network ensemble to forecast streamflow for flood management. *J. Hydrol.* **2016**, *536*, 161–173. [CrossRef]
31. Nouri, H.; Ildoromi, A.; Sepehri, M.; Artimani, M. Comparing Three Main Methods of Artificial Intelligence in Flood Estimation in Yalphan Catchment. *Geogr. Environ. Plan.* **2019**, *29*, 35–50.
32. Adhikary, S.K.; Muttil, N.; Yilmaz, A.G. Improving streamflow forecast using optimal rain gauge network-based input to artificial neural network models. *Hydrol. Res.* **2018**, *49*, 1559–1577. [CrossRef]
33. Ondimu, S.; Murase, H. Reservoir Level Forecasting using Neural Networks: Lake Naivasha. *Biosyst. Eng.* **2007**, *96*, 135–138. [CrossRef]
34. Rani, S.; Parekh, F. Application of Artificial Neural Network (ANN) for Reservoir Water Level Forecasting. *Int. J. Sci. Res.* **2014**, *3*, 1077–1082.
35. Lukman, Q.A.; Ruslan, F.A.; Adnan, R. 5 Hours ahead of time flood water level prediction modelling using NNARX technique: Case study terengganu. In Proceedings of the 2016 7th IEEE Control and System Graduate Research Colloquium (ICSGRC), Shah Alam, Malaysia, 8 August 2016; pp. 104–108. [CrossRef]
36. Altunkaynak, A. Forecasting surface water level fluctuations of lake van by artificial neural networks. *Water Resour. Manag.* **2007**, *21*, 399–408. [CrossRef]
37. Xu, G.; Cheng, Y.; Liu, F.; Ping, P.; Sun, J. A water level prediction model based on ARIMA-RNN. In Proceedings of the 2019 IEEE Fifth International Conference on Big Data Computing Service and Applications (BigDataService), Newark, CA, USA, 4–9 April 2019; pp. 221–226. [CrossRef]
38. Khandelwal, I.; Adhikari, R.; Verma, G. Time series forecasting using hybrid arima and ann models based on DWT Decomposition. *Procedia Comput. Sci.* **2015**, *48*, 173–179. [CrossRef]
39. Phan, T.T.H.; Nguyen, X.H. Combining statistical machine learning models with ARIMA for water level forecasting: The case of the Red river. *Adv. Water Resour.* **2020**, *142*, 103656. [CrossRef]
40. Lola, M.S.; Zainuddin, N.H.; Tajuddin, M.; Abdullah, M.T.; Ponniah, V. Improving the Performance of Ann-Arima Models for Predicting. Mproving Perform. Ann-Arima Model. Predict. *Water Qual. Offshore* **2018**, *13*, 27–37.
41. Vinodh, S.S.K.; Babu, G.J.; Arulprakasam, B.G.V. Integrated seawater intrusion study of coastal region of Thiruvallur district, Tamil Nadu, South India. *Appl. Water Sci.* **2019**, *9*, 124. [CrossRef]
42. Kumar, S. Chennai Water Management Water Resources Availability Data for Chennai. Available online: https://www.kaggle.com/sudalairajkumar/chennai-water-management (accessed on 6 December 2021).
43. Choubin, B.; Malekian, A. Combined gamma and M-test-based ANN and ARIMA models for groundwater fluctuation forecasting in semiarid regions. *Environ. Earth Sci.* **2017**, *76*, 538. [CrossRef]
44. Mullainathan, S.; Spiess, J. Machine learning: An applied econometric approach. *J. Econ. Perspect.* **2017**, *31*, 87–106. [CrossRef]
45. Gurland, J.; Whittle, P. Hypothesis Testing in Time Series Analysis. *J. Am. Stat. Assoc.* **2015**, *49*, 197–200. [CrossRef]

46. Jeong, K.; Koo, C.; Hong, T. An estimation model for determining the annual energy cost budget in educational facilities using SARIMA (seasonal autoregressive integrated moving average) and ANN (artificial neural network). *Energy* **2014**, *71*, 71–79. [CrossRef]
47. George, E.P.B.; Gwilym, M.J. Time Series Analysis Forecasting and Control. *J. Am. Stat. Assoc.* **2014**, *68*, 493–494.
48. Kandananond, K. A comparison of various forecasting methods for autocorrelated time series. *Int. J. Eng. Bus. Manag.* **2012**, *4*, 4. [CrossRef]
49. Wang, P.; Xu, L.; Zhou, S.M.; Fan, Z.; Li, Y.; Feng, S. A novel Bayesian learning method for information aggregation in modular neural networks. *Expert Syst. Appl.* **2010**, *37*, 1071–1074. [CrossRef]
50. Taylor, P.; Dickey, D.A.; Fuller, W.A.; Dickey, D.A.; Fuller, W.A. Journal of the American Statistical Association Distribution of the Estimators for Autoregressive Time Series with a Unit Root Distribution of the Estimators for Autoregressive Time Series with a Unit Root. *J. Am. Stat. Assoc.* **1979**, *74*, 427–431. [CrossRef]
51. Wayne, A.; Fuller, J.T. *Introduction to Statistical Time Series*, 2nd ed.; A Wiley-Interscience Publication: Hoboken, NJ, USA, 1978; pp. 308–311.
52. Guresen, E.; Kayakutlu, G.; Daim, T.U. Using artificial neural network models in stock market index prediction. *Expert Syst. Appl.* **2011**, *38*, 10389–10397. [CrossRef]
53. Chuang, A.; Wei, W.W.S. Time Series Analysis: Univariate and Multivariate Methods. *Technometrics* **1991**, *33*, 108. [CrossRef]
54. Li, H.X.; Da, X.L. A neural network representation of linear programming. *Eur. J. Oper. Res.* **2000**, *124*, 224–234. [CrossRef]
55. Li, H.X.; Li, L.X. Representing diverse mathematical problems using neural networks in hybrid intelligent systems Hong. *Expert Syst. Appl.* **1999**, *16*, 271–281.
56. Li, L.; Ge, R.L.; Zhou, S.M.; Valerdi, R. Guest editorial integrated healthcare information systems. *IEEE Trans. Inf. Technol. Biomed.* **2012**, *16*, 515–517. [CrossRef]
57. Zhou, S.M.; Xu, L.D. A new type of recurrent fuzzy neural network for modeling dynamic systems. *Knowledge-Based Syst.* **2001**, *14*, 243–251. [CrossRef]
58. Yin, Y.H.; Fan, Y.J.; Xu, L.D. EMG and EPP-integrated human-machine interface between the paralyzed and rehabilitation exoskeleton. *IEEE Trans. Inf. Technol. Biomed.* **2012**, *16*, 542–549. [CrossRef] [PubMed]
59. Zhang, G.; Eddy Patuwo, B.; Hu, M.Y. Forecasting with artificial neural networks: The state of the art. *Int. J. Forecast.* **1998**, *14*, 35–62. [CrossRef]
60. Morgan, N.; Bourlad, H. Generalization and Parameter Estimation in Feedforward Nets: Some Experiments. *Adv. Neural Inf. Process. Syst.* **1990**, *2*, 630–637.
61. Denton, J.W. How good are neural networks for causal forecasting. *J. Bus. Forecast.* **1995**, *14*, 17.
62. Markham, I.S.; Rakes, T.R. The effect of sample size and variability of data on the comparative performance of artificial neural networks and regression. *Comput. Oper. Res.* **1998**, *25*, 251–263. [CrossRef]
63. Granger, C.W.J. Combining Forecasts-Twenty Years Later. *J. Forecast.* **1989**, *8*, 167–173. [CrossRef]
64. Perrone, M.P.; Cooper, L.N. *When Networks Disagree: Ensemble Methods for Hybrid Neural Networks*; Chapman and Hall: London, UK, 1995; pp. 342–358. [CrossRef]
65. Brooks, C. *Introductory Econometrics for Finance*, 2nd ed.; Cambridge University Press: Cambridge, UK, 2008; Volume 148.
66. Durbin-Watson and Interactions for Regression in SPSS Investigating Outliers and Influential Observations. Available online: https://www.sheffield.ac.uk/polopoly_fs/1.531431!/file/MASHRegression_Further_SPSS.pdf (accessed on 2 January 2021).
67. Yollanda, M.; Devianto, D. Hybrid Model of Seasonal ARIMA-ANN to Forecast Tourist Arrivals through Minangkabau International Airport. In Proceedings of the 1st International Conference on Statistics and Analytics, ICSA 2019, Bogor, Indonesia, 2–3 August 2019. [CrossRef]
68. Tatipamul, R. Application of geospatial technology in environmental and disaster management Application of Geospatial Technology In Environmental Hazards And Disaster Managment. *Aayushi Int. Interdiscip. Res. J.* **2019**, *1*, 143–145.
69. Khan, M.Y.A.; Hasan, F.; Panwar, S.; Chakrapani, G.J. Neural network model for discharge and water-level prediction for Ramganga River catchment of Ganga Basin, India. *Hydrol. Sci. J.* **2016**, *61*, 2084–2095. [CrossRef]
70. Yadav, B.; Eliza, K. A hybrid wavelet-support vector machine model for prediction of Lake water level fluctuations using hydro-meteorological data. *Meas. J. Int. Meas. Confed.* **2017**, *103*, 294–301. [CrossRef]

Article

Life Cycle Impact Assessment of Recycled Aggregate Concrete, Geopolymer Concrete, and Recycled Aggregate-Based Geopolymer Concrete

Lahiba Imtiaz [1], Sardar Kashif-ur-Rehman [1], Wesam Salah Alaloul [2,*], Kashif Nazir [3], Muhammad Faisal Javed [1,*], Fahid Aslam [4] and Muhammad Ali Musarat [2]

1. Department of Civil Engineering, Abbottabad Campus, COMSATS University Islamabad, Abbottabad 22060, Pakistan; laibaimtiaz01@gmail.com (L.I.); skashif@cuiatd.edu.pk (S.K.-u.-R.)
2. Department of Civil and Environmental Engineering, Universiti Teknologi PETRONAS, Bandar Seri Iskandar 32610, Perak, Malaysia; muhammad_19000316@utp.edu.my
3. Department of Civil Engineering, School of Engineering, Nazabayev University, Astana 010000, Kazakhstan; kashif.nazir@nu.edu.kz
4. Department of Civil Engineering, College of Engineering in Al-Kharj, Prince Sattam Bin Abdulaziz University, Al Kharj 11942, Saudi Arabia; f.aslam@psau.edu.sa
* Correspondence: wesam.alaloul@utp.edu.my (W.S.A.); arbab_faisal@yahoo.com (M.F.J.)

Citation: Imtiaz, L.; Kashif-ur-Rehman, S.; Alaloul, W.S.; Nazir, K.; Javed, M.F.; Aslam, F.; Musarat, M.A. Life Cycle Impact Assessment of Recycled Aggregate Concrete, Geopolymer Concrete, and Recycled Aggregate-Based Geopolymer Concrete. *Sustainability* **2021**, *13*, 13515. https://doi.org/10.3390/su132413515

Academic Editor: Constantin Chalioris

Received: 27 October 2021
Accepted: 30 November 2021
Published: 7 December 2021

Publisher's Note: MDPI stays neutral with regard to jurisdictional claims in published maps and institutional affiliations.

Copyright: © 2021 by the authors. Licensee MDPI, Basel, Switzerland. This article is an open access article distributed under the terms and conditions of the Creative Commons Attribution (CC BY) license (https://creativecommons.org/licenses/by/4.0/).

Abstract: This study presents a life cycle impact assessment of OPC concrete, recycled aggregate concrete, geopolymer concrete, and recycled aggregate-based geopolymer concrete by using the midpoint approach of the CML 2001 impact-assessment method. The life cycle impact assessment was carried out using OpenLCA software with nine different impact categories, such as global warming potential, acidification potential, eutrophication potential, ozone depletion potential, photochemical oxidant formation, human toxicity, marine aquatic ecotoxicity, and freshwater and terrestrial aquatic ecotoxicity potential. Subsequently, a contribution analysis was conducted for all nine impact categories. The analysis showed that using geopolymer concrete in place of OPC concrete can reduce global warming potential by up to 53.7%. Further, the use of geopolymer concrete represents the reduction of acidification potential and photochemical oxidant formation in the impact categories, along with climate change. However, the potential impacts of marine aquatic ecotoxicity, freshwater aquatic ecotoxicity, human toxicity, eutrophication potential, ozone depletion potential, and terrestrial aquatic ecotoxicity potential were increased using geopolymer concrete. The increase in these impacts was due to the presence of alkaline activators such as sodium hydroxide and sodium silicate. The use of recycled aggregates in both OPC concrete and geopolymer concrete reduces all the environmental impacts.

Keywords: life cycle; impact assessment; recycled material; geopolymer concrete; sustainability

1. Introduction

Concrete is the most widely used construction material and the second most-consumed substance on earth, after water [1]. Ordinary portland cement (OPC) used in concrete production has detrimental effects on the environment due to the release of a high amount of greenhouse gas, especially CO_2. One ton of CO_2 is released by the production of one ton of OPC [1,2]. According to research conducted in Hawaii in the year 2020, it was reported that the amount of CO_2 in the atmosphere reached the highest level of 417.1 ppm [3]. Moreover, there exist other environmental issues, such as the dumping of construction and demolition wastes. Hence, it is crucial to develop environmentally sustainable solutions in the construction industry.

Researchers developed the idea of introducing alternative binders to OPC that not only reduce CO_2 production but also resolve the disposal problems. One such alternative is a geopolymer concrete (GPC) technology that promotes sustainable development. Its efficient performance depends on both the composition of the source material and the activator used [4]. Moreover, the use of recycled aggregates (RA) in GPC provides both eco-friendly and economical solutions by addressing the issue of dumping demolition wastes.

Peem et al. [5] investigated the influence of RA on high-calcium fly ash (FA)-based GPC at different molarities. It was found that the RA can be used in FA-based GPC with an early age strength of about 30.6–38.4 MPa, slightly lower than normal aggregate FA-based GPC. Likewise, Xie et al. [6] studied the combined effect of FA and ground granulated blast-furnace slag (GGBFS) on recycled aggregate geopolymer concrete (RAGC), and reported that RAGC with a 50% FA and 50% GGBFS binder content exhibits a superior synergetic effect on mechanical and workability properties. Further, it was found that the use of metakaolin fly ash (MK-FA)-based binders in RAGC resulted in better mechanical and durability properties [7]. Hence, the use of MK-FA-based GPC with 100% recyclable coarse aggregates provides an environmentally sustainable solution. While assessing the environmentally sustainable performance of GPC compared to OPC concrete, there are a lot of techniques and procedures used. One such assessment technique used is the life cycle assessment (LCA).

LCA is a 'cradle-to-grave' or 'cradle-to-gate' assessment technique used to evaluate the environmental impacts from raw material extraction to the demolition application stage [8,9]. This tool plays an important role in the environmental management of a given product system that further involves an environmental comparison of different prototypes. It is a policy or program applied in GPC technology to justify that GPC has less potential to degrade the environment, compared to OPC concrete.

Different studies have been conducted to assess the global warming potential (GWP) and environmental impact assessment of GPC [4,8–13]. Daniel et al. [9] analyzed the life cycle inventory of GPC and OPC concrete from lab to industrial scales, based on the source of a sodium hydroxide (NaOH) activator. It was found that GPC exhibits 64% less GWP than OPC concrete if the source of NaOH is local solar salt. Rishabh et al. [10] investigated the environmental impact assessment of SF-FA-based GPC activated with both NaOH and sodium silicate (Na_2SiO_3) separately. It was concluded that OPC has a greater GWP than GPC, and further, that SF-FA-based GPC activated with NaOH has a lesser environmental potential when compared to GPC activated with Na_2SiO_3.

Many researchers have investigated the LCA of GPC [7,9–11,14,15], but as per the authors' knowledge, no systematic and detailed study has been devised to study the LCA of RAGC along with its comparison with mixtures of GPC, RAC, and OPC concrete. For example, how does RA impact the LCA of GPC and OPC concrete? What is the GWP of RAGC? How do RAGC and other mixtures impact the other environmental factors, such as GWP, acidification potential (ADP), photochemical oxidants formation (POF), and ozone depletion? These opacities still need to be answered. Hence, endorsing that idea, this study intends to investigate the environmental impact assessment of RAGC using the LCA approach.

2. Materials and Methods

The LCA methodology for all four mixes is performed in four steps, as per ISO 14040 and 14044 [16]. The first step is to define the goal and scope of the research, while the second and third steps are to conduct inventory analyses and life cycle impact assessments (LCIA), respectively. The last step is to conduct an interpretation based on inventory and impact-assessment analysis. Lastly, the methodology adopted to conduct the life cycle inventory is presented in the form of a flow chart.

2.1. Goal and Scope

In the present research, the goal of the LCA is to find out the impact of the inclusion of RA in both concrete and GPC on the environment, and to compare the environmental impacts of four mixes, i.e., the OPC concrete, RAC, GPC, and RAGC. The scope of the LCA begins with the extraction of the natural resources, including aggregates, the raw material for cement, and alkali activators, and ends with the GPC production with processed RA. The raw material from natural reserves is utilized in producing OPC concrete, which, after demolition, can be utilized as RA for the GPC mix.

For the RAGC mix, the FA was considered as an aluminosilicate source while sodium hydroxide and sodium silicate were used as an activator. The production of silicates and hydroxide from the raw material to the end product was considered in conducting the life cycle inventory analysis. After defining the goal and scope, the functional unit was set as 1 m^3 of GPC, RAGC, and RAC of a specific strength and compared with OPC concrete. The strength conditions considered in this research study varied from 25–30 MPa for all four types of mixtures. However, the system boundaries specified in this research study started from the collection of their ingredients to their production, as presented in Figure 1.

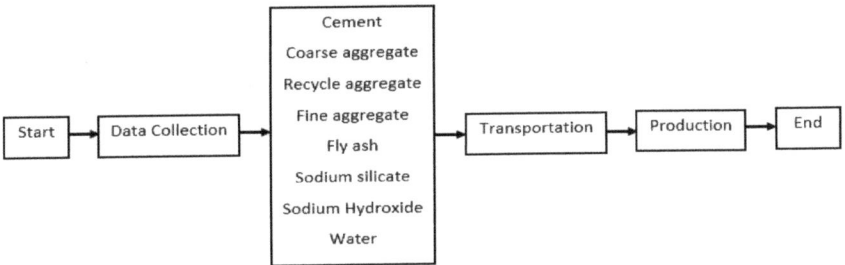

Figure 1. System boundaries of concrete.

Further, the mix design data of RAC, GPC, RAGC, and OPC concrete were taken from the literature [16,17]. Further, the emissions data of the respective activities in the production of all concrete mixtures were taken on the basis of the geography of Pakistan. However, the missing data were taken from the ecoinvent database source [18]. In addition, it was assumed that the production and transportation conditions of all concrete mixtures are the same as that of the conditions that exist in the respective locations from where the emissions data were collected. The output emissions depend only on the consumption of energy in the production processes of all the concrete mixtures.

2.2. Study Area

The LCA methodology for four different concrete mixtures was applied based on the inventory analysis applied in the city of Abbottabad, Pakistan. The location of COMSATS University Islamabad, Abbottabad Campus (CUI, atd), was assumed as the production location for all the four mixtures. Furthermore, the collection of cement, coarse aggregate, and fine aggregate were considered at the location of the Bestway Cement factory in Haripur, the Choona Crushing plant in Abbottabad, and Thore in Muzaffarabad, respectively.

2.3. Mix Design adaptation

The mix design of normal concrete and the RAC mixture was taken from the study conducted in Pakistan [16]. The natural aggregate was fully replaced (i.e., 100% replacement) with RA in this study. Both types of concrete have 28 days' compressive strengths of 28 MPa. However, the mix design procedure for the GPC and RAGC mixtures was taken from [17], with the compressive strength of 30 MPa and 27 MPa, respectively. A total of 40% of the normal coarse aggregate was replaced with RA in the RAGC mixture. The mix design for all four mixtures is presented in Table 1.

Table 1. Mix design of four concrete mixtures.

Ingredients	OPC Concrete (kg/m^3) [16]	RAC (kg/m^3) [16]	GPC (kg/m^3) [17]	RAGC (kg/m^3) [17]
Cement	415	415	-	-
Fly ash	-	-	408	408
Fine aggregate	620	620	554	554
Coarse aggregate	1040	-	1243	746
Recycled aggregate	-	1040	-	497
Water	185	185	20	20
Sodium Hydroxide	-	-	41	41
Sodium Silicate	-	-	103	103
Superplasticizer SP (% of cement)	0.5	1.5	-	-

2.4. Inventory Analysis

The life cycle inventory is the next step, after defining the goal and scope. Mostly, the data of concrete production were based on a questionnaire survey of local producers and suppliers in the city of Abbottabad, Pakistan. However, missing data were taken from the literature and the ecoinvent database, version 3.7.1. The cutoff classification method was considered the system model in the ecoinvent database [18]. The cutoff approach was based on the assumption that the primary producer of any material is allocated to a primary consumer and has no impact or credit on its recycled material.

In the present research, the data of emissions of different production processes for the geography of Pakistan are generated on the basis of an emission/energy ratio method. The inventory data of emissions and energy for all the ingredients of OPC concrete and GPC are taken from the literature [18–21]. In the next step, the ratio of emission/energy (kg/MJ) is calculated for each ingredient in the respective technical paper. After taking the average of the emission/energy ratio of each ingredient, it was then multiplied by the energy produced (in MJ) by every ingredient, with respect to the location in Pakistan. Moreover, the flow of taking inventory data was presented in the form of a flow chart (Figure 2).

2.5. Questionnaire Survey

The data for calculating the total energy produced by each ingredient, i.e., fine aggregate, coarse aggregate, RA, and cement, are based on the questionnaire survey in the region of Abbottabad city. The data for the cement are taken from the Bestway Cement industry in the Hattar Industrial Estate, KPK. Further, the data for the coarse aggregate are taken from the Choona crushing plant in Abbottabad (Figure 3). The fine aggregate is taken from the Neelum riverbed at the Thore Site in Muzaffarabad (Figure 4). The energy data of the sand are based on the excavator method, adopted at the location of Thore, Muzaffarabad, using a medium-load truck for transportation to the final location. Moreover, the total energy considered is the summation of both manufacturing energy and transportation energy, with respect to the selected reference point.

For determining the transportation energy of the cement and the aggregates, the material suppliers near COMSAT University Islamabad, Abbottabad Campus, were assumed to be the final location where the all ingredients of concrete are supplied. The calculation of transportation distances relied on the Google Maps application. In addition, the transportation of cement, coarse aggregate, and RA to the destination used the heavily loaded truck. In case of RA, the site near Dhamtor, Abbottabad was considered as the dumping site for construction and demolition waste. For the RAC, the natural raw material and mining activity for aggregate production was set to zero. Similarly, the raw material was taken as zero for FA aluminosilicate, as it is a by-product of the coal industry. The cutoff allocation procedure was assumed, which showed that FA is a by-product and has no impact on life cycle inventory emissions. However, the raw materials inventory for the silicate production was taken from the literature. Additionally, the inventory for sodium hydroxide production was based on the ecoinvent database. The life cycle inventory, based on the questionnaire survey, is presented in Table 2.

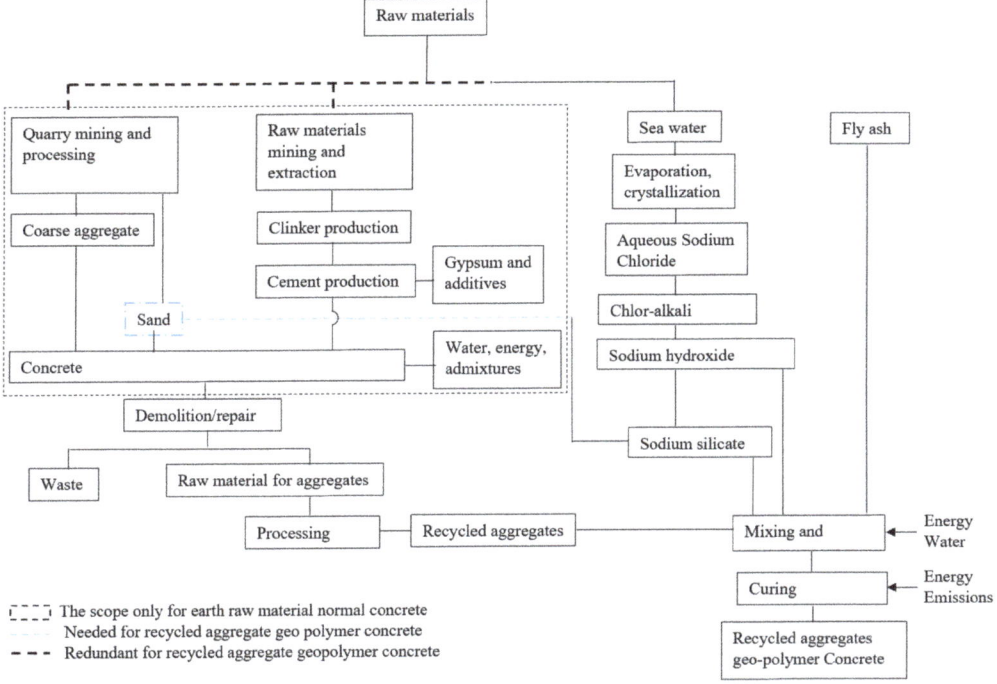

Figure 2. Life cycle inventory flow of OPC concrete and RAGC.

Figure 3. Mining and Crushing at the Choona Site, Abbottabad.

Figure 4. Sand Extraction near the Neelam Riverbed, Muzaffarabad.

Table 2. Life Cycle Inventory for ingredient of concrete.

Ingredient	Cement	Coarse Aggregate	Fine Aggregate	Recycled Aggregate
Total Energy (MJ/kg)	2.973	0.0154	0.0136	0.00833
Emissions (kg)				
CO_2	0.614	0.00173	0.00095	0.00124
SO_2	0.0014	6.976×10^{-6}	1.99×10^{-6}	2.091×10^{-6}
CO	0.0026	0.001437	3.46×10^{-6}	2.394×10^{-6}
NO_x	0.00141	1.128×10^{-5}	7.25×10^{-6}	8.202×10^{-6}
PM < 10	0.000267	1.281×10^{-5}	1.1×10^{-5}	7.097×10^{-6}
NMVOC	0.000161	6.455×10^{-7}	6.4×10^{-10}	4.320×10^{-7}
NH_3	1.893×10^{-5}	-	3.37×10^{-9}	-
N_2O	1.357×10^{-6}	2.813×10^{-8}	3.29×10^{-7}	1.535×10^{-8}
CH_4	0.000655	6.979×10^{-7}	1.88×10^{-8}	3.629×10^{-7}

2.6. Life Cycle Impact Assessment (LCIA)

The impact was analyzed by using OpenLCA software with a mid-point approach, called CML 2001 baseline (Centrum voor Milieukunde Leiden). There were a number of impact categories that were analyzed by the CML approach for the ecoinvent dataset. However, in the present research work, nine impact categories were analyzed, i.e., GWP, ADP, photochemical oxidants formation (POF), ozone depletion, human toxicity, marine aquatic ecotoxicity, freshwater aquatic ecotoxicity, and eutrophication potential. The above-mentioned impact categories were analyzed and compared to four types of mixes, i.e., concrete mix, RAC, GPC, and RAGC. The category indicators can be expressed in the form of equations, as presented below:

$$GWP = \sum Load\ (i) \times GWP\ (i)$$

$$ODP = \sum Load\,(i) \times ODP\,(i)$$
$$ADP = \sum Load\,(i) \times ADP\,(i)$$
$$POF = \sum Load\,(i) \times POF\,(i)$$
$$HTP = \sum Load\,(i) \times HTP\,(i)$$
$$EP = \sum Load\,(i) \times EP\,(i)$$

where,
Load (i) is the environmental load of the respective inventory item (i);
GWP (i), ODP (i), ADP (i), POF (i), HTP (i), and EP (i) are the characterization factors for the GWP, ODP, ADP, POF, HTP, and EP inventory items (i), respectively.

3. Results and Discussion

In this section, the environmental impacts and process contributions of four different concrete mixtures are analyzed and compared using a mid-point approach, called CML 2001. In the first section, the life cycle inventory results for the ingredients of concrete are reported. In the next section, the numbers of impact categories are analyzed for four types of mixes, i.e., concrete mix, RAC, GPC, and RAGC. At last, the contribution analyses of all four concrete mixtures are presented.

3.1. Life Cycle Inventory Results

Based on the questionnaire survey and the emission/energy procedure, it was concluded that the total energy (the sum of electric, coal, and transportation energy) required for one kilogram of cement is 2.973 MJ/kg. However, the total energy required for the coarse aggregate (the sum of mining, crushing, and transportation energy) and RA (the sum of crushing and transportation energy) are 0.0154 MJ/kg and 0.00834 MJ/kg, respectively. The total energy required for sand production and transportation is 0.0136 MJ/kg. The energy data for all ingredients, along with the transportation energy, are given in Table 3.

Table 3. Energy production by all ingredients through questionnaire survey.

Ingredients	Production Energy (MJ/kg)	Transportation Energy (MJ/kg)
Cement	2.918	0.055
Fine Aggregate	0.00565	0.00795
Coarse Aggregate	0.00873	0.00630
Recycled Aggregate	0.00524	0.00309

3.2. Environmental Impact Analysis of Four Mixes

In this study, the environmental impacts were analyzed for the comparison of normal concrete and GPC along with their RAC. From the Open LCA software, the impacts were analyzed that represented that the inclusion of an alternative binder or RA could help to reduce the certain environmental impacts were analyzed. The most concerning impact category in the construction industry is the GWP that results from CO_2 production and the emissions of GHGs [14]. The GWP-100a (100-year global warming potential) of OPC concrete, RAC, GPC, and RAGC are compared and presented in Figure 5. It is shown that OPC concrete has the highest GWP when compared to the other three mixes. The GWP follows a decreasing pattern from normal concrete, > RAC > GPC > RAGC, as shown in the respective figure. This pattern provides the idea that the mixes containing higher contents of cement have higher GWPs, when compared to the others. However, with the inclusion of RA in the mix, the net impact of global warming is reduced [17,22–26].

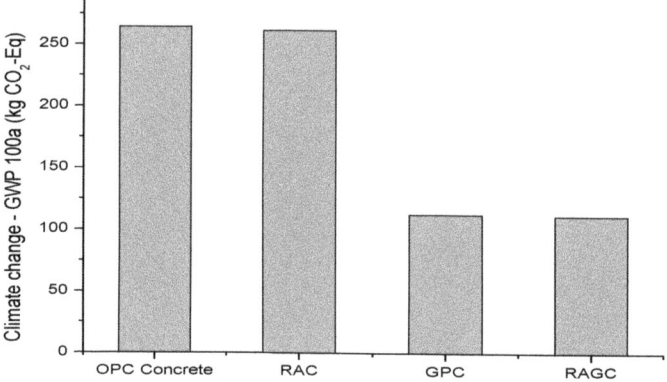

Figure 5. Climate change GWP of four mixes.

In addition, ADP follows the same pattern as GWP for the four types of concrete mixtures. The normal concrete has the highest impact on acidification due to high emissions of air pollutants, such as NOx, SO_2, NH_3, etc., during cement production. From Figure 6, it is concluded that RAC and RAGC exhibit lower ADP when compared to normal concrete and GPC, respectively, due to the recycling of coarse aggregates. The recycling of coarse aggregates requires lower energy than the normal aggregate, due to the elimination of mining energy and the reduction in transportation energy. It is reported that the net environmental impacts of RAC are also influenced by the transportation distance [20,21,27,28]. The environmental impact of RAC has a lower influence if the transportation distance is less than 20 km for the considered natural aggregate when it is compared [26,27].

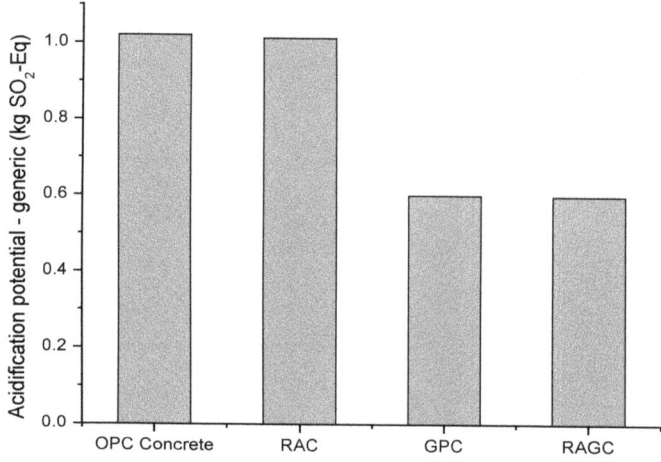

Figure 6. Acidification potential for four mixes.

Additionally, the ozone depletion potential of the four types of mixes is presented in Figure 7, which clearly shows that the production of concrete and RAC mixtures has no direct impact on ozone depletion for a specified functional unit of 1 m^3 of concrete samples. Conversely, both type of geopolymer mixes, i.e., GPC and RAGC, have a significant impact on ozone depletion. This impact is due to the presence of a sodium hydroxide activator in FA-based GPC mixtures. The production of sodium hydroxide through the process of chlor-alkali electrolysis, using a membrane cell, emits some amount of tetra-chloro methane in the atmosphere, which could impact the ozone layer.

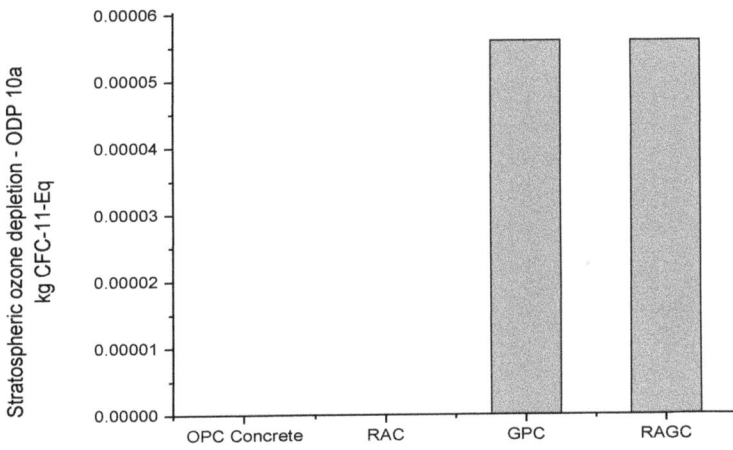

Figure 7. Ozone depletion potential of four mixes.

The impact of different environmental pollutants on air pollution, specifically ozone depletion, is a very complex process. The characteristics of environmental pollutants depend on their nature—whether either is a primary or secondary pollutant. The direct emissions of gases, fumes, and smoke from the exhausts of vehicles and combustion factories, along with the burning of fossil fuels, are the causes of the primary pollutants. The primary pollutants, such as particulates, hydrocarbon, nitrogen oxides, and carbon monoxide, etc., when coming in contact with other pollutants such as VOC or compounds of ammonia (coming from other developmental activities), form the secondary pollutants. Their chemical reactions in the atmosphere increase the impact on urban air quality by acid deposition and the formation of ground-level ozone (bad ozone or tropospheric ozone). However, the presence of chemicals, such as manufactured halocarbon refrigerants, propellants, solvents, and foam-blowing agents (CFCs, HCFCs, and halons), promotes the depletion of the ozone hole (beneficial ozone or stratospheric ozone). The emissions resulting from the production of concrete influence the presence of photochemical oxidants that affect the tropospheric ozone.

However, the photochemical oxidation of four concrete mixes is represented in Figure 8, which shows that the concrete mixture has the highest ability to produce photochemical oxidants in the atmosphere. These oxidants are produced from the reaction of primary air pollutants such as NO_x, SO_x, and hydrocarbons under the action of sunlight [29]. The decreasing pattern of this impact category starts from concrete to the RAGC mixture, i.e., normal concrete > GPC > RAGC > RAC. The production of elementary environmental pollutants during cement production and transportation is responsible for the highest photochemical oxidation when compared to the other mixtures. The production of photochemical oxidants adversely influences the atmosphere by the incorporation of unwanted ozone molecules in the troposphere and, thus, causes smog, along with other environmental effects.

The impact category, namely, the ETP of the four concrete mixtures, is shown in Figure 9, which concludes that GPC has the highest ETP (0.1148 kg PO_4-Eq/m^3 of GPC) when compared to the other mixtures. This is due to the presence of hydroxide and silicate sources in GPC [15]. On the other hand, the RAC and RAGC represent a slight decrease of impact categories, when compared to OPC concrete and GPC, respectively. The use of RA is responsible for less NO_x, SO_2, and ammonia emissions, when compared to normal aggregate production.

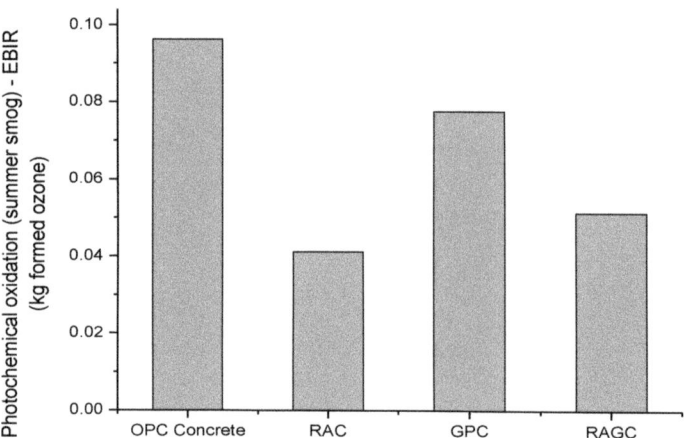

Figure 8. Photochemical oxidation of four mixes.

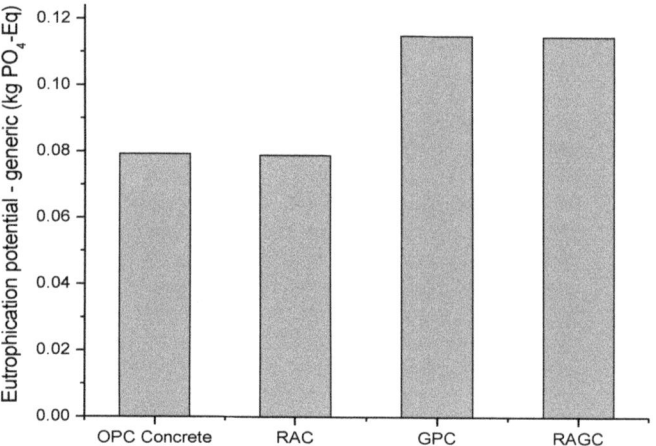

Figure 9. Eutrophication potential of four mixes.

However, HTP describes the potential damage of the chemical unit that is released in the atmosphere. Its potentiality depends on both the inherent toxicity of the chemical and its potential dose. From Figure 10, it is represented that GPC has a higher impact on human toxicity when compared to the other three mixes. The pattern of HTP for all four mixtures represents that the GPC binder with natural or RA shows a higher potency due to presence of alkaline activators, especially a sodium silicate source [10,12,30]. Similarly, the impact category, called marine aquatic ecotoxicity potential (MAETP), is shown in Figure 11. The MAETP of OPC concrete, RAC, GPC, and RAGC are 4.57×10^{-5}, 4.47×10^{-5}, 136.45, and 136.45 kg of 1.4 DCB-Eq/m^3 of mixture, respectively. The values predict that the OPC and RAC concrete has minute impact on aquatic ecotoxicity due to the absence of an alkaline activator, as in the case of the geopolymer mixtures. The aquatic ecotoxicity can be hindered by using sustainable production sources of alkaline activators.

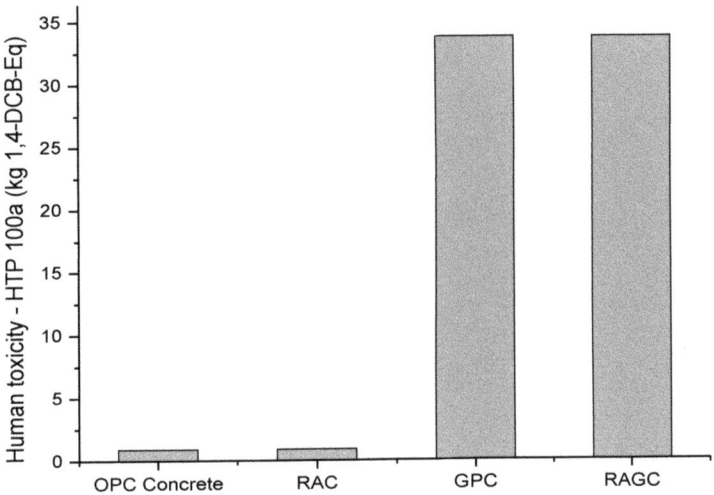

Figure 10. HTP of OPC concrete, RAC, GPC, and RAGC.

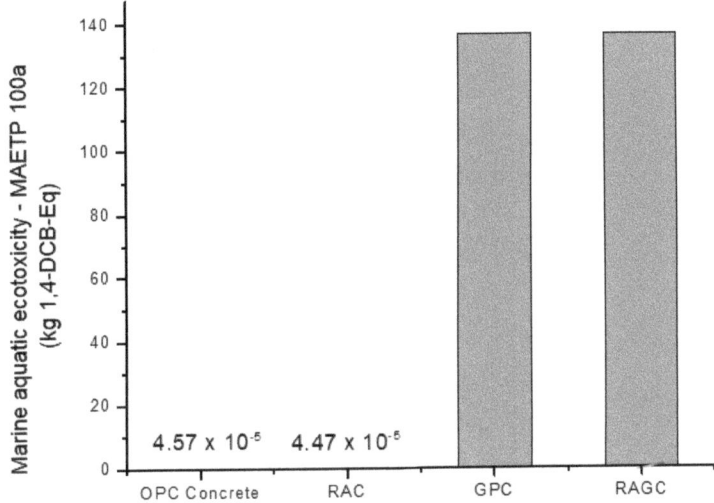

Figure 11. MAETP of OPC concrete, RAC, GPC, and RAGC.

The last two impact categories considered in this research study are freshwater aquatic ecotoxicity potential (FAETP) and terrestrial aquatic ecotoxicity potential (TAETP), as shown in Figures 12 and 13, respectively. From Figure 12, it is clearly seen that the GPC mixture has a higher FAETP value when compared to the other mixtures. The decreasing pattern of both impact categories, i.e., GPC > RAGC > OPC concrete > RAC, depicts that the presence of a silicate and hydroxide source in GPC mixtures is responsible for a higher ecotoxicity impact [10,13,30]. However, the terrestrial aquatic ecotoxicity of both GPC mixtures shows the same value of 0.0107 kg 1,4 DCB-Eq, as presented in Figure 13. This depicts that the impact category is only affected by the presence of a sodium silicate and sodium hydroxide source in both mixtures.

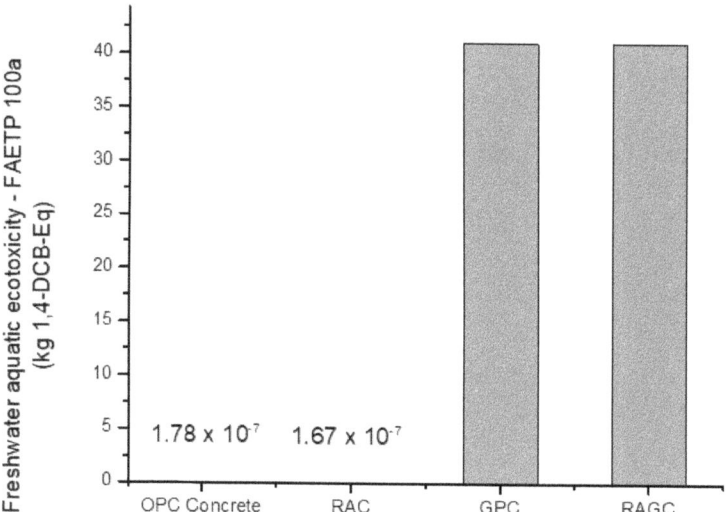

Figure 12. FAETP of OPC concrete, RAC, GPC, and RAGC.

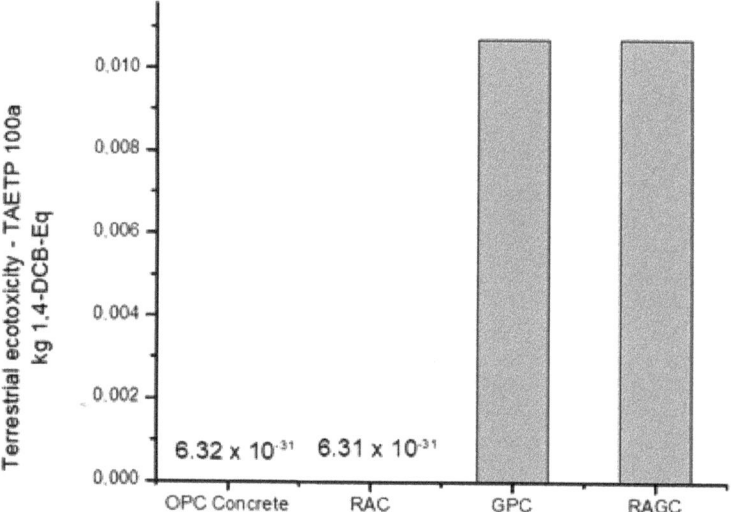

Figure 13. TAETP of OPC concrete, RAC, GPC, and RAGC.

In the present impact assessment analysis, it is concluded that the OPC concrete has potentially higher impacts than GPC and recycled mixtures in the impact categories GWP, ADP, ETP, and POF. The use of GPC can reduce GWP significantly—up to 57.34%—when compared to normal concrete. However, other impact categories, such as FAETP, MAETP, stratospheric ozone depletion, HTP, and TAETP, show a greater impact of GPC than normal concrete. This is due to the presence of alkaline activators, such as a silicate source, in the GPC [10,12,30]. Moreover, the recycling of coarse aggregates in both concrete and GPC mixtures can reduce the overall environmental impacts. The values of the potential impact categories of the four mixtures by the CML baseline method are presented in Table 4.

Table 4. Impact categories by CML baseline method.

Indicator	OPC Concrete	RAC	GPC	RAGC	Units
Acidification Potential—Generic	1.01904	1.01165	0.60119	0.59769	kg SO_2-Eq
Climate Change—GWP	264.181	261.315	112.743	111.377	kg CO_2-Eq
Eutrophication Potential	0.07922	0.0788	0.11483	0.11463	kg PO_4-Eq
Freshwater Ecotoxicity	1.78×10^{-7}	1.677×10^{-7}	40.940	40.940	kg 1,4-DCB-Eq
Human toxicity	0.8952	0.8860	33.70	33.68249	kg 1,4-DCB-Eq
Marine aquatic Ecotoxicity	4.575×10^{-5}	4.475×10^{-5}	136.45	136.45	kg 1,4-DCB-Eq
Photochemical Oxidation	0.0963	0.0411	0.0777	0.0513	kg ozone formed
Stratospheric ozone depletion	0	0	5.59×10^{-5}	5.59×10^{-5}	kg CFC-11-Eq
Terrestrial Ecotoxicity	6.32×10^{-31}	6.30×10^{-31}	0.0107	0.0107	kg 1,4-DCB-Eq

However, the nine considered environmental indicators in this research work were scaled while keeping the potential environmental damage to the surrounding atmosphere in view. The GWP is ranked highest, followed by ODP, POF, HTP, ADP, EP, FAETP, MAETP, and TAETP. From the weighted average of all the indicators from all the mixtures, it is concluded that the RAGC mixture is more sustainable for the environment, followed by GPC, RAC, and OPC concrete mixtures. The ranking of all the mixtures regarding their environmentally sustainable performance is given in Table 5. This ranking will provide an idea to civil society about which concrete mixture efficiently provides for structural needs and offers sustainable solutions to the environment. Depending on the strength requirement, the audience can select the required aluminosilicate and activator source along with the choice of selection of recycled aggregate or natural aggregate. In the present research work, the RAGC is the best-optimized mixture for meeting the structural needs and for hastening sustainable developments.

Table 5. Ranking of mixtures on the basis of environmentally sustainable performance.

Ranking	Mixture
1st	RAGC
2nd	GPC
3rd	RAC
4th	OPC

3.3. Contribution Analysis

A contribution analysis for the four selected mixtures was performed to check the contribution of the selected processes to the chosen LCIA impact category. The contribution of coarse aggregate, fine aggregate, cement, and the mixing process was checked in the analysis of OPC concrete, while the contribution of RA, fine aggregate, cement, and the mixing process was checked in the RAC analysis. Figures 14 and 15 show that the cement had the highest negative impacts on the chosen environmental categories [10,11,31]. In the case of OPC concrete, cement had the highest impact, followed by coarse aggregate and fine aggregate. The categories GWP, ADP, HTP, and EP are mostly affected by cement because of higher CO_2, SO_x, and NO_x emissions created during its manufacturing and transportation. However, coarse aggregate and cement contribute 57.4% and 41.5% to the POF, respectively. This is due to the presence of both mining and crushing activities that lead to more emissions of particulate matter (PM), volatile organic compounds, SO_2, and NO_x [10].

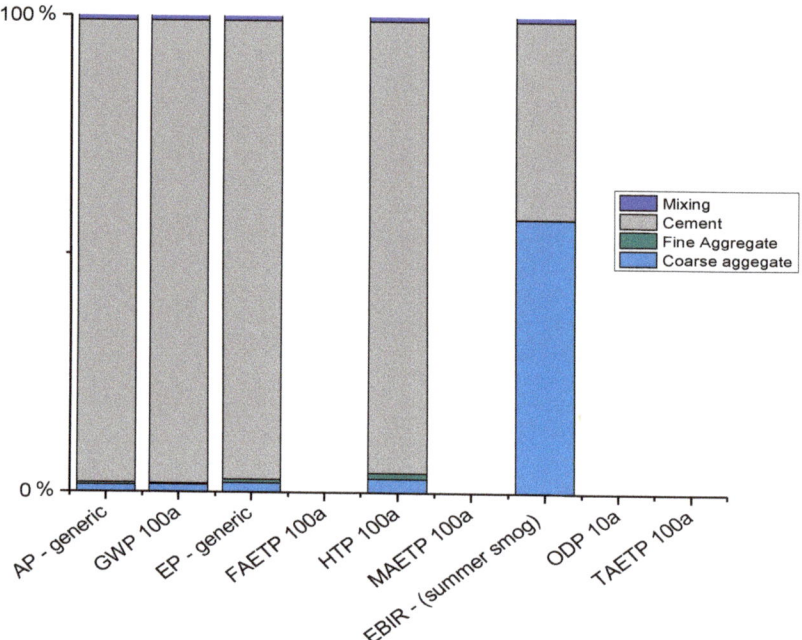

Figure 14. Contribution Analysis for OPC concrete.

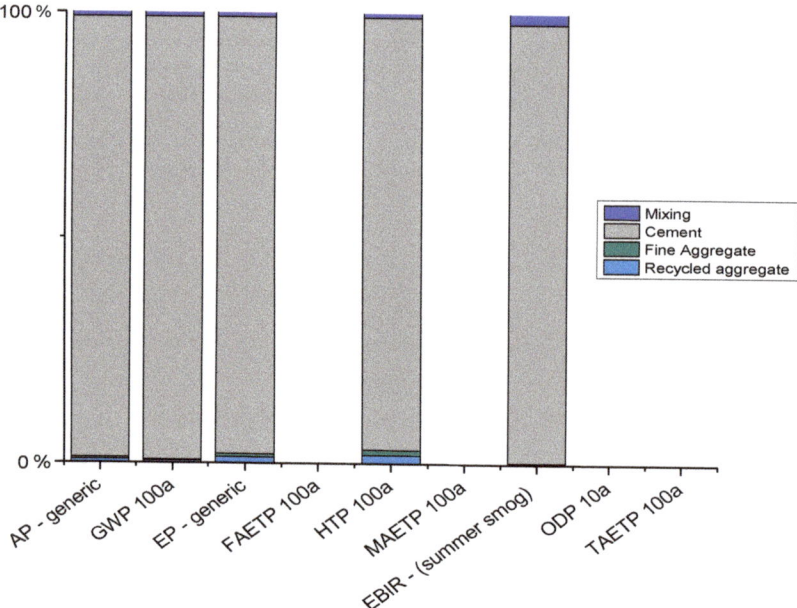

Figure 15. Contribution Analysis for RAC.

Moreover, it is clear from Figure 15 that replacing the coarse aggregate with RA can reduce all environmental impacts. All the impact categories are mostly affected by the use of OPC cement. The lesser contribution of RA to LCIA categories is due to the elimination of mining activity and a lesser transportation distance [21,23,32]. The fine aggregate has the lowest contribution due to its source from the riverbed. Its impact on the LCIA categories mostly depends on the transportation activity. Furthermore, the production of OPC concrete and the RAC mixture shows the lowest impact on the impact categories FAETP, MAETP, and TAETP. This is due to presence of less water emissions, due to its ingredients' activities and production.

The contribution analysis for the GPC and RAGC mixtures is presented in Figures 16 and 17, respectively. For the GPC, the contribution of coarse aggregate, fine aggregate, sodium hydroxide, sodium silicate, and mixing to all LCIA categories is checked. These contributions are checked to predict and verify which ingredient impacts and contributes to the four different concrete mixtures. It is noticed, from Figure 16, that the presence of activators has a higher contribution than the aggregates. The contribution of silicate and hydroxide sources is because of the presence of separate manufacturing processes. Each activator requires considerable chemicals and products for their manufacturing, which leads to higher GHG emissions, along with emissions of certain elements and the addition of chemicals to water systems [9,33]. From Figure 17, it is represented that sodium hydroxide had the highest contribution to all LCIA categories, followed by sodium silicate and coarse aggregate. This contribution depends on the inventory data of the hydroxide and silicate source. The inventory data for sodium hydroxide are based on the chlor-alkali electrolysis method through a membrane cell. In addition, the contribution of coarse aggregate to the impact category of POF—summer smog is higher in both the GPC and RAGC mixtures. This contribution is due to the production of oxides of nitrogen and NMVOC in coarse aggregate manufacturing.

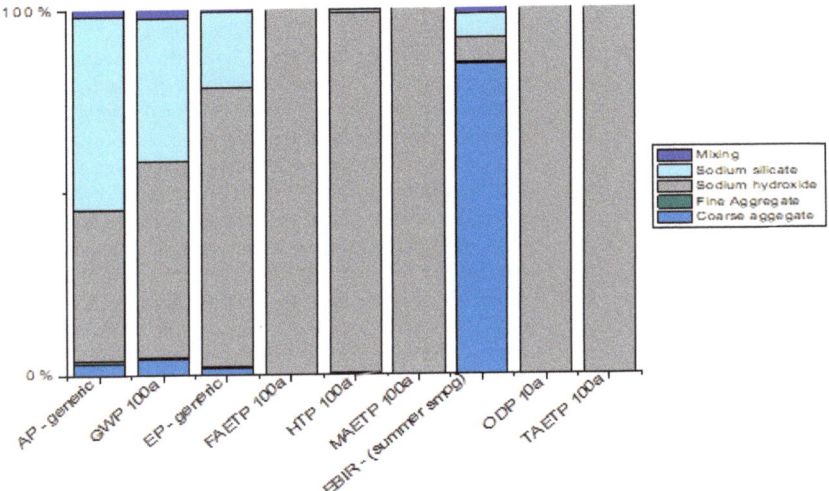

Figure 16. Contribution Analysis for GPC.

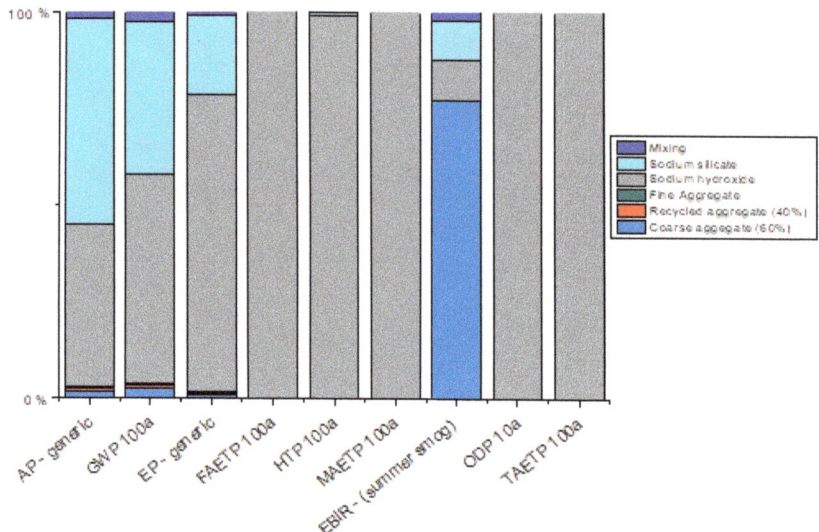

Figure 17. Contribution Analysis for RAGC.

4. Conclusions and Recommendations

Based on the LCA analysis of OPC concrete, the RAC, and the RAGC mixtures, the following conclusions can be drawn:

- A questionnaire survey was conducted to calculate the production and transportation energy of all products of concrete mixtures per kilogram. It is reported from the survey that the total energy (the sum of electric, coal, and transportation energy) required for one kilogram of cement is 2.973 MJ. However, the total energy required for coarse aggregate (the sum of mining, crushing, and transportation energy) and RA (the sum of crushing and transportation energy) are 0.0154 MJ and 0.00834 MJ, respectively. The total energy required for sand production and transportation is 0.0136 MJ;
- A LCA analysis was conducted using OpenLCA software with the aid of the CML 2001 baseline method. Nine different impact categories were analyzed and compared for each mixture in order to evaluate the best mixture for the environment. On the basis of the LCA analysis, it is concluded that OPC in concrete mixtures is the major contributor to the production of negative environmental impacts;
- The use of aggregates also contributes to different environmental impacts, such as GWP, EP, ADP, and POF. The use of RA in both RAC and RAGC mixtures help to reduce the overall environmental impacts. However, the use of RA in concrete mixtures depends on the transportation distances;
- The inclusion of FA as an aluminosilicate in GPC concrete reduces some of the environmental impacts, such as GWP, ADP, and photochemical oxidation. However, the use of alkaline activators, such as sodium silicate and sodium hydroxide, is a major contributor to other environmental impacts, such as FAETP, TAETP, MAETP, ODP, and HTP. Hence, it is important to select the suitable and sustainable manufacturing method for alkaline activators;
- The use of GPC and RAGC mixtures is a more suitable option for reducing the GWP produced due to normal concrete cement. It is concluded that use of GPC lowers the GWP impact up to 57.34%, when compared to OPC concrete. However, categories other than GWP are affected by use of GPC mixtures;
- The use of an alkaline activator is a major contributor to environmental impacts, both in the case of GPC and RAGC. Hence, it is important to select the appropriate source of alkaline activators to be used in the GPC mixture. If the sodium hydroxide is taken

from the seabed, or the sodium silicate is taken from any sustainable source, then the overall environmental impacts of GPC mixtures can be reduced.

However, this study recommends the following future research:

- Research on generating and collecting the LCI data for condition of Pakistan
- The comparison of GPC and RAGC concrete at different percentages of RA
- The comparison of concrete mixtures by using different impact assessment methods
- The investigation of LCA of GPC mixtures by using different manufacturing processes for the alkaline activators
- The investigation of different transportation scenarios while comparing different mixtures of concrete
- The use of normalization and weighting set analysis after the comparison of different concrete mixtures

Author Contributions: L.I., S.K.-u.-R., W.S.A., K.N., M.F.J., F.A. and M.A.M. contributed equally to this research work. All authors have read and agreed to the published version of the manuscript.

Funding: This research received no external funding.

Institutional Review Board Statement: Not applicable.

Informed Consent Statement: Not applicable.

Data Availability Statement: All data is available within the manuscript.

Acknowledgments: The authors would like to appreciate the Shale Gas Research Group (SGRG) in UTP and Shale PRF project (cost center # 0153AB-A33) awarded to E. Padmanabhan for the support.

Conflicts of Interest: The authors declare no conflict of interest.

Abbreviations

ADP	Acidification Potential
CFC	Chlorofloro Carbons
CML	Centrum voor Milieukunde Leiden
CO	Carbon Monoxide
CO_2	Carbon Dioxide
EPD	Environmental Product Declaration
ER	Environmental Reports
ETP	Eutrophication Potential
EBIR	Equal Benefit Incremental Reactivity
FA	Fly Ash
FAETP	Freshwater Aquatic Ecotoxicity Potential
GGBFS	Ground Granulated Blast Furnace Slag
GHG's	Greenhouse Gases
GPC	Geopolymer Concrete
GWP	Global Warming Potential
HTP	Human Toxicity Potential
LCA	Life Cycle Assessment
LCIA	Life Cycle Impact Assessment
MAETP	Marine Aquatic Ecotoxicity Potential
MK	Metakaolin
MIR	Maximum Incremental Reactivity
NO_x	Nitrogen Oxides
NMVOC	Volatile Organic Compounds
PM	Particulate Matter
MOIR	Maximum Ozone Incremental Reactivity
OPC	Ordinary Portland cement

POCP	Photochemical Ozone Creation Potential
RA	Recycled Aggregate
POF	Photochemical Oxidant Formation
RAC	Recycled Aggregate Concrete
SF	Silica Fume
SO_x	Sulphur Oxides
RAGC	Recycled Aggregate Geopolymer Concrete
SETAC	Society of Environmental Toxicology and Chemistry
GWP-100a	100-year Global warming potential
TAETP	Terrestrial Aquatic Ecotoxicity Potential
UNEP	United Nations Environmental Program
ISO	International Organization for Standardization
TRACI	Tool for Reduction and Assessment of Chemical and other Environmental impacts

References

1. Aprianti, E.; Shafigh, P.; Bahri, S.; Farahani, J.N. Supplementary cementitious materials origin from agricultural wastes–A review. *Constr. Build. Mater.* **2015**, *74*, 176–187. [CrossRef]
2. Oss, H.G.; Padovani, A.C. Cement Manufacture and the Environment Part II: Environmental Challenges and Opportunities. *J. Ind. Ecol.* **2003**, *7*, 93–126. [CrossRef]
3. WORLD, H. Rise of Carbon Dioxide in the Atmosphere Continues Unabated. 2020. Available online: https://earthsky.org/earth/rising-carbon-dioxide-co2-record-high-june2020/ (accessed on 25 November 2021).
4. Li, N.; Shi, C.; Zhang, Z.; Wang, H.; Liu, Y. A review on mixture design methods for geopolymer concrete. *Compos. Part B Eng.* **2019**, *178*, 107490. [CrossRef]
5. Nuaklong, P.; Sata, V.; Chindaprasirt, P. Influence of recycled aggregate on fly ash geopolymer concrete properties. *J. Clean. Prod.* **2016**, *112*, 2300–2307. [CrossRef]
6. Xie, J.; Wang, J.; Rao, R.; Wang, C.; Fang, C. Effects of combined usage of GGBS and fly ash on workability and mechanical properties of alkali activated geopolymer concrete with recycled aggregate. *Compos. Part B Eng.* **2019**, *164*, 179–190. [CrossRef]
7. Nuaklong, P.; Sata, V.; Chindaprasirt, P. Properties of metakaolin-high calcium fly ash geopolymer concrete containing recycled aggregate from crushed concrete specimens. *Constr. Build. Mater.* **2018**, *161*, 365–373. [CrossRef]
8. E McGrath, T.; Cox, S.; Soutsos, M.; Kong, D.; Mee, L.P.; Alengaram, J.U.J. *Life Cycle Assessment of Geopolymer Concrete: A Malaysian Context*; IOP Conference Series: Materials Science and Engineering; IOP Publishing: Kuala lumpur, Malaysia, 2018.
9. Salas, D.A.; Ramirez, A.D.; Ulloa, N.; Baykara, H.; Boero, A. Life cycle assessment of geopolymer concrete. *Constr. Build. Mater.* **2018**, *190*, 170–177. [CrossRef]
10. Bajpai, R.; Choudhary, K.; Srivastava, A.; Sangwan, K.S.; Singh, M. Environmental impact assessment of fly ash and silica fume based geopolymer concrete. *J. Clean. Prod.* **2020**, *254*, 120147. [CrossRef]
11. Habert, G.; d'Espinose de Lacaillerie, J.B.; Roussel, N. An environmental evaluation of geopolymer based concrete production: Reviewing current research trends. *J. Clean. Prod.* **2011**, *19*, 1229–1238. [CrossRef]
12. Zain, H.; Abdullah, M.M.A.B.; Hussin, K.; Ariffin, N.; Bayuaji, R. Review on Various Types of Geopolymer Materials with the Environmental Impact Assessment. In *MATEC Web of Conferences*; EDP Sciences: Perlis, Malaysia, 2017.
13. Habert, G.; Lacaillerie, J.D.D.; Lanta, E.; Roussel, N. Environmental evaluation for cement substitution with geopolymers. In Proceedings of the 2nd International Conference on Sustainable Construction Materials and Technologies, Ancone, Italy, 28–30 June 2010.
14. Abbas, R.; Khereby, M.A.; Ghorab, H.Y.; Elkhoshkhany, N. Preparation of geopolymer concrete using Egyptian kaolin clay and the study of its environmental effects and economic cost. *Clean Technol. Environ. Policy* **2020**, *22*, 669–687. [CrossRef]
15. Pozzo, A.D.; Carabba, L.; Bignozzi, M.C.; Tugnoli, A. Life cycle assessment of a geopolymer mixture for fireproofing applications. *Int. J. Life Cycle Assess.* **2019**, *24*, 1743–1757. [CrossRef]
16. Finkbeiner, M.; Inaba, A.; Tan, R.; Christiansen, K.; Klüppel, H.-J. The New International Standards for Life Cycle Assessment: ISO 14040 and ISO 14044. *Int. J. Life Cycle Assess.* **2006**, *11*, 80–85. [CrossRef]
17. Rashid, K.; Rehman, M.U.; de Brito, J.; Ghafoor, H. Multi-criteria optimization of recycled aggregate concrete mixes. *J. Clean. Prod.* **2020**, *276*, 124316. [CrossRef]
18. Wernet, G.; Bauer, C.; Steubing, B.; Reinhard, J.; Moreno-Ruiz, E.; Weidema, B. The ecoinvent database version 3 (part I): Overview and methodology. *Int. J. Life Cycle Assess.* **2016**, *21*, 1218–1230. [CrossRef]
19. Galvin, B.; Lloyd, N. Fly ash based geopolymer concrete with recycled concrete aggregate. In Proceedings of the CONCRETE 2011 Conference, Perth, Australia, 12 October 2011; The Concrete Institute of Australia: Sydney, Australia.
20. Mohammadi, J.; South, W. Life cycle assessment (LCA) of benchmark concrete products in Australia. *Int. J. Life Cycle Assess.* **2017**, *22*, 1588–1608. [CrossRef]
21. Frischknecht, R. LCI modelling approaches applied on recycling of materials in view of environmental sustainability, risk perception and eco-efficiency. *Int. J. Life Cycle Assess.* **2010**, *15*, 666–671. [CrossRef]

22. Ding, T.; Xiao, J.; Tam, V.W.Y. A closed-loop life cycle assessment of recycled aggregate concrete utilization in China. *Waste Manag.* **2016**, *56*, 367–375. [CrossRef]
23. Guo, Z.; Tu, A.; Chen, C.; Lehman, D.E. Mechanical properties, durability, and life-cycle assessment of concrete building blocks incorporating recycled concrete aggregates. *J. Clean. Prod.* **2018**, *199*, 136–149. [CrossRef]
24. Marinkovic, S.; Ignjatović, I.; Radonjanin, V. Life-cycle assessment (LCA) of concrete with recycled aggregates (RAs). In *Handbook of Recycled Concrete and Demolition Waste*; Elsevier: Amsterdam, The Netherlands, 2013; pp. 569–604.
25. Serres, N.; Braymand, S.; Feugeas, F. Environmental evaluation of concrete made from recycled concrete aggregate implementing life cycle assessment. *J. Build. Eng.* **2016**, *5*, 24–33. [CrossRef]
26. Knoeri, C.; Sanyé-Mengual, E.; Althaus, H.-J. Comparative LCA of recycled and conventional concrete for structural applications. *Int. J. Life Cycle Assess.* **2013**, *18*, 909–918. [CrossRef]
27. Shi, X.; Mukhopadhyay, A.; Zollinger, D.; Grasley, Z. Economic input-output life cycle assessment of concrete pavement containing recycled concrete aggregate. *J. Clean. Prod.* **2019**, *225*, 414–425. [CrossRef]
28. Colangelo, F.; Navarro, T.G.; Farina, I.; Petrillo, A. Comparative LCA of concrete with recycled aggregates: A circular economy mindset in Europe. *Int. J. Life Cycle Assess.* **2020**, *25*, 1790–1804. [CrossRef]
29. Marinković, S.; Radonjanin, V.; Malešev, M.; Ignjatović, I. Comparative environmental assessment of natural and recycled aggregate concrete. *Waste Manag.* **2010**, *30*, 2255–2264. [CrossRef]
30. Shan, X.; Zhou, J.; Chang, V.W.-C.; Yang, E.-H. Life cycle assessment of adoption of local recycled aggregates and green concrete in Singapore perspective. *J. Clean. Prod.* **2017**, *164*, 918–926. [CrossRef]
31. Kim, T.; Tae, S.; Chae, C.U. Analysis of Environmental Impact for Concrete Using LCA by Varying the Recycling Components, the Compressive Strength and the Admixture Material Mixing. *Sustainability* **2016**, *8*, 389. [CrossRef]
32. Asadollahfardi, G.; Katebi, A.; Taherian, P.; Panahandeh, A. Environmental life cycle assessment of concrete with different mixed designs. *Int. J. Constr. Manag.* **2021**, *21*, 665–676. [CrossRef]
33. Turner, L.K.; Collins, F.G. Carbon dioxide equivalent (CO_2-e) emissions: A comparison between geopolymer and OPC cement concrete. *Constr. Build. Mater.* **2013**, *43*, 125–130. [CrossRef]

Article

Exploring Perceptions of the Adoption of Prefabricated Construction Technology in Pakistan Using the Technology Acceptance Model

Muhammad Hamza [1], Rai Waqas Azfar [1,*], Khwaja Mateen Mazher [2], Basel Sultan [3], Ahsen Maqsoom [4], Shabir Hussain Khahro [3,*] and Zubair Ahmed Memon [3]

1. Department of Construction Engineering and Management, NUST College of Civil Engineering, National University of Sciences and Technology, Risalpur 23200, Pakistan
2. Department of Construction Engineering and Management, King Fahd University of Petroleum and Minerals, Dhahran 31261, Saudi Arabia
3. Department of Engineering Management, College of Engineering, Prince Sultan University, Riyadh 11586, Saudi Arabia
4. Department of Civil Engineering, COMSATS University Islamabad, Wah Campus, Rawalpindi 47040, Pakistan
* Correspondence: waqas.azfar@mce.nust.edu.pk (R.W.A.); shkhahro@psu.edu.sa (S.H.K.)

Abstract: Prefabricated construction is being pursued globally as a critically important sustainable construction technology. Prefabricated construction technology (PCT) provides opportunities to effectively manage construction waste and offers venues to address the poor productivity and lackluster performance of construction projects, which are often expected to miss their budget and schedule constraints. Despite the significant benefits inherent in the adoption of PCT, research has shown an unimpressive exploitation of this technology in the building sector. A modified version of the popular technology acceptance model (TAM) was used to understand Pakistan's building construction industry stakeholder's acceptance of PCT and the factors that influence its usage. Data were collected from 250 building construction experts in the industry to test the hypotheses derived from the proposed model. Data analysis using covariance-based structural equation modeling revealed that construction industry stakeholders' perceptions of perceived ease-of-use, perceived usefulness, trust, and satisfaction all strongly influenced PCT acceptance behavior. Moreover, results also confirmed the total direct and indirect effects of the perceived usefulness and perceived ease-of-use of behavioral intention toward using PCT, with trust and user satisfaction as mediators. The results of this research are expected to serve as a guide for the construction industry stakeholders to effectively plan, strategize, encourage, and increase the adoption of PCT to achieve sustainable construction outcomes in the building construction sector.

Keywords: prefabricated construction; structural equation modeling; mediation analysis; trust; satisfaction

1. Introduction

The construction industry contributes significantly to national and global economies. Construction is one of Pakistan's most neglected industries. Comparing the construction industry to other industries, such as manufacturing, the construction industry is considered backward [1]. The industry lacks regulations, standards, mechanization, advanced technology, and a waste management plan [2]. Construction processes have always been criticized for project time overruns, low productivity, a lack of security, and waste generation [3]. There are still issues with traditional onsite construction practices, such as a lack of regulations and high construction waste generation [4]. Therefore, there is a need to change the construction industry environment through technological advancements for sustainable growth in the construction industry. To improve overall quality and efficiency, the process of construction must be improved based on sustainability parameters [5]. The key is to innovate and remove the many obstacles that prevent the sector from creating a sustainably built environment. Increasing construction demand, rising costs, and critical environmental issues have prompted a worldwide search for innovative sustainable alternatives, such as prefabricated construction. Prefabricated construction has been identified as a vital means of overcoming these issues and ensuring sustainable development and green buildings in the construction industry. In today's world, prefabricated buildings are associated with modern and innovative ecological qualities. Environmentally friendly building materials are simple to use in prefabricated buildings. The customization of prefabricated buildings based on location, layout, and materials used is possible. All of this gives the end user more options and flexibility [6]. Thus, it is critical to assess the current landscape to properly discuss the benefits of prefabrication and other construction methods. Therefore, it is indispensable to study advanced construction techniques.

Previously, there are a number of studies that have been published relating to the acceptance of prefabricated construction technology. Lee and Kim (2018) highlighted the variables that are driving the construction industry to adapt modular construction and they give some recommendations for future vertical extension [7]. Karthik et al. (2020) investigated the benefits and limitations of modular construction and compared benefits with conventional construction [8]. Similarly, Seo and Lin (2020) conducted a case study to examine the environmental implications of prefabricated construction and its characteristics, and also shed light on some advantages of prefabricated construction [9]. The current status of prefabrication adoption in small-scale construction projects was evaluated by Khahro et al. (2019) [10]. Similarly, Adindu et al. (2020) experimentally explored the understanding, adoption, possibilities, and difficulties of applying the prefabricated construction technology method to construction infrastructure projects [11]. Attempts were made by El-Abidi et al. (2019) to investigate the current state of prefabricated construction systems and the potential of these systems to satisfy the housing needs of the people of Libya [12]. Wu et al. (2019) examined the impact of technology promotion and cleaner production on the use of prefabricated construction technology in China to improve its application [13].

In prefabricated construction technology related previous studies, various research gaps were found; first, prefabricated construction technology acceptance has not been studied quantitatively [14]. Second, no study investigated the effect of trust and satisfaction as mediators in technology acceptance literature related to the construction industry [15]. Third, no study investigated the technology acceptance model (TAM) to establish a theoretical framework to account for prefabricated construction technology acceptance in the construction industry. To fill these research gaps, the present study proposed a prefabricated construction technology-based technology acceptance model for construction stakeholders that integrates the technology acceptance model with trust and user satisfaction as mediators. Therefore, developing and presenting a proposed model for prefabricated construction technology adoption based on the technology acceptance model is the prime objective of this research. The technology acceptance model was selected because it has been widely used to explain technology acceptance and human behavior in different research areas; for example, BIM and AR integration in the construction industry [16], and construction safety [17]. Relationships between the perceived usefulness (PU) and perceived ease-of-use (PEOU) of prefabricated construction technology are explored and developed in the presence of certain influencing factors. After reviewing the literature on these technology acceptance model variables and external variables, stakeholders' behavioral intention toward prefabricated construction technology adoption is being studied using an extended technology acceptance model. An analysis of the hypothesized relationships is conducted, with conclusions drawn.

The theoretical and empirical contributions of this study are unique and essential. To explain behavioral intentions, the technology acceptance model uses only two variables, even though it is used as a unique and robust concept by other researchers [18]. As a basic research approach, the technology acceptance model does not inform us what variables might impact customers' intentions to use prefabricated construction technology. Behavioral intentions [19,20] and performance have been considered to be influenced by trust [21]. Attitudes of people in the tourism sector are strongly influenced by trust, according to research by Kaushik et al. (2015) [20]. Similarly, there has been a considerable influence of customer satisfaction on behavioral intention, as argued by Durdyev et al. (2018) [22]. The combination of satisfaction, and trust (technology acceptance model) is good in this regard. Because of this, the primary goal of this study is to examine the elements that influence people's intentions to use prefabricated construction technology. This new technology acceptance model based integrated model contends that trust and satisfaction mediate the relations between perceived ease-of-use, perceived usefulness, and behavioral intention to adopt prefabricated construction technology. To author's knowledge, this study is one of the first to use trust and satisfaction to explore perceived usefulness and perceived ease-of-use in behavioral intention related to the construction industry. Furthermore, this research findings provide valid evidence in favor of utilizing the technology acceptance model in conjunction with additional constructs (external variables) to anticipate prefabricated construction technology acceptance.

The paper is organized as follows. According to previous studies, prefabricated construction technology adoption in construction has been substantiated. Second, a detailed set of hypotheses is presented based on a literature assessment of each measured item in the prefabricated construction technology acceptance model. Third, the survey's methodology and findings are discussed. Finally, recommendations for further research and the implications are discussed. The research model was tested using data from a sample of construction industry stakeholders (consultants, contractors, architects/engineers, etc.), who were familiar with prefabrication. To generalize the outcomes, participants were selected from different construction project sites. Structural equation modeling (SEM) using AMOS software was employed for hypothesis testing. Based on Anderson and Gerbing's (1988), research, a two-phased strategy was employed [23]. First, a measurement model was estimated using exploratory (EFA) and confirmatory factor analysis (CFA) to assess overall model fit, validity, and reliability. Second, the structural model was used to test the hypotheses.

2. Literature Review and Model Development

2.1. Literature Review

2.1.1. Technology Acceptance-Related Theories

User perceptions of technology are key factors in adopting, accepting, and using new technologies, and targeting these perceptions can help avoid resistance and increase the chances of success [12]. Many studies have described the role of user perceptions in accepting new technology in terms of the theory of reasoned action [24], such as the technology acceptance model (TAM) [25], the technology acceptance model 2 (TAM2) [26], and the unified theory of acceptance and use of technology (UTAUT) [27].

The theory of reasoned action was created by Fishbein and Ajzen (1975) [24]. It was the basis for most later theories, such as TAM, TAM2, TAM3, and UTAUT. It clarifies human behavior for technology adoption from a social psychology standpoint. The theory asserts that behavioral intention is affected by two main constructs: subjective norms and attitude toward behavior. The technology acceptance model proposed by Davis was modified from the TRA of Fishbein and Ajzen (1975) [24] and is one of the most widely used research models to predict the acceptance and use of information technology systems. The technology acceptance model was developed to overcome the various weaknesses of the TRA. According to F. Davis (1986), the deletion of subjective norms was justifiable because participants did not have enough information regarding the social influence at the acceptance testing stage [28].

The technology acceptance model is an extension of the TRA as a general psychological theory for individual behavior prediction in information systems [25]. Both perceived usefulness (PU) and perceived ease-of-use (PEOU) were used as technology acceptance model external constructs. A learner's perception of the usefulness of technology is measured by perceived usefulness. Perceived ease-of-use refers to the assumption that learning using technology requires no intellectual effort. The technology acceptance model is widely used in information systems, electronics, and construction to describe technology acceptance.

Venkatesh and Davis (2000) developed TAM2 to overcome the drawbacks of the technology acceptance model and enhance the model's explanatory power (R^2) [26]. TAM2 has the primary determinants of the original TAM, namely perceived usefulness and PEU. It also considers social impact, including subjective norms, images, and the cognitive instrumental processes, which include output quality, job relevance, and result demonstrability. TAM2 and TAM are widely utilized for explaining an individual's adoption and technology acceptance in various settings and contexts [29].

The unified theory of acceptance and use of technology (UTAUT) was constructed by Venkatesh et al. (2003) to address the various weaknesses of previous theories [27]. The UTAUT integrates eight of the most well-known previous theories and includes four determinant constructs: social influence, facilitating condition, effort expectancy, and performance expectancy, all of which affect BI [27].

TAM3 reveals the moderating effect of experience, which Venkatesh (2000) [30] and Venkatesh and Davis (2000) [26] did not empirically test. The relationships between (a) PEOU and PU; (b) computer anxiety and PEOU; and (c) PEOU and behavioral intention [31] are moderated by experience. Through the determinants of PU and PEOU, TAM3 has made significant theoretical contributions. In TAM3, there are complementary elements of context, content, process, and individual differences [31].

2.1.2. Prefabricated Construction

Structures that are built onsite using prefabricated components are called prefabricated construction. It is the method of construction in which construction is performed with the help of separate components of structures, e.g., walls or roofs already being constructed in an established offsite factory-based environment before their fabrication at the construction site [32]. PCT modules come in many shapes and sizes for usage in the construction industry. Prefabricated construction is considered a technology in this study, which includes all of the prefabricated components such as stairs, façades, slabs, air-conditioning panels, and balconies. Prefabrication can be divided into three categories: semi-prefabrication, comprehensive prefabrication, and volumetric modular building [33]. On the contrary, modular construction includes transforming a structure into a panel or volumetric-style unit [34]. Modular construction is the practice of construction in which a whole part of a building or structure, such as a room of a building or a whole house, is transported to the construction site for assembling, after finalizing its construction at an offsite facility.

The topic of green construction is primarily focused on the practice of prefabrication [35]. The implementation of prefabricated assembly techniques can significantly reduce the ecological footprint of the construction process while also optimizing the allocation of project resources. Prefabricated buildings have a low environmental impact due to their effective conservation of project resources and significant potential for future growth. This has the potential to facilitate the progression of society toward a more sustainable mode of development [36]. Prefabricated buildings have been advocated as a sustainable development strategy in the construction sector due to the traditional construction method's lack of suitability for cleaner production [37,38]. Ai et al. (2023) investigated the limits of government regulation of prefabricated buildings and concluded that such structures had the potential to not only exceed previous growth projections for developers but also to fully embody the idea of green development and significantly impact China's future sustainable development [36]. Prefabricated buildings can boost the sustainability performance of construction initiatives, which in turn encourages the sustainable development of society [39]. Research conducted by Rahardjo and Dinariana (2016) using Bantul and Bandung in Badan city as case studies demonstrated that precast systems were a form of green building since they conserved wood, lowered construction costs, and safeguarded the environment [40].

PCT can help the construction industry in achieving lower costs [41], better HSE [10], improved productivity [41], efficiencies of material and labor resources [41], sustainability [41], quality [10], and a reduction in construction duration [41]. The advantages of PCT in the construction industry show that fully automated production, modernization control, and industrialized production are achievable goals. PCT not only improves quantity, cost, schedule, and material utilization in construction projects but also helps to achieve high mechanization and increased work efficiency. Despite these benefits, the use of PCT in the construction industry has been slow because of higher initial costs [13,42], dominated project processes [43], inadequate policies and regulations [43], a lack of knowledge and expertise [44], a lack of social climate and acceptance [45], and ineffective logistics ([39] vand a limited availability of design options/complicated designing [46]).

Buildings in the housing sector are responsible for most of the new construction. Data on the attitudes of Australian builders concerning prefabrication is provided by Steinhardt and Manley (2016) using the theory of planned behavior (TPB) and the technology acceptance model (TAM), resulting in a clarification of beliefs that can guide efforts to enhance the market share of prefabrication [14]. However, despite a challenging stakeholder network and an industrial setting, their views on prefabrication for Australian housing are positive. Due to a lack of industry infrastructure, prefabrication adoption has been slow and is almost entirely unsupportive. The study on offsite technologies in housing by Nanyam et al. (2017) defines a holistic selection framework with a set of offsite-specific attributes along with a set of standard attributes that are necessary and favored for the acceptance of offsite technologies for affordable housing [47]. They also tested and validated the framework in an offsite case study.

PCT is still in its infancy in China, but it will undoubtedly be the future of Chinese construction industrialization. Jiang et al. (2020) focus on the interrelationships of factors affecting PCT promotion [48]. The overall relationship of each factor was quantitatively modeled (SEM). The results show that the policy factor dominates, followed by the management and market factors. In another study on rural residential buildings, Zhou et al. (2019) aim to design a model for determining the suitable strategy for prefabrication implementation [49]. Similarly, Imran et al. (2019) studied the level of adoption of prefabrication in the construction industry of Pakistan [1]. Buildings, roads, and bridges were selected to form a literature review and questionnaire survey.

In the PCT literature, many previous research efforts have tried to investigate user acceptance, establishing frameworks and models for the development of PCT technologies. However, they lack the consideration of the context of establishing a theoretical framework that determines the extent of end-user acceptance. Therefore, using the technology acceptance model may fill this gap by explaining the variance between factors and discussing the significance of external factors to predict and explain the adoption of PCT in developing countries' construction industry. Additionally, based on this model, construction professionals' acceptance of PCT will be predicted and explained in terms of perceived usefulness, perceived ease-of-use, and related variables. Therefore, there is a need to identify additional factors (if any) related to prefabricated construction and its validation by the technology acceptance model in the current scenario.

Nonetheless, there is a scarcity of research in Pakistan on PCT acceptance models based on the perspectives of construction industry stakeholders; as a result, the mechanisms for achieving or accepting PCT have yet to be defined.

2.2. Model Development

2.2.1. Overview of Proposed Model

Despite widespread agreement on PCT's potential applicability and advantages, it remains unclear how PCT could be employed and what its advantages are. As a result, construction research and practice continue to focus on how people perceive PCT acceptance. Therefore, the goal of the study is to understand how PCT is accepted based on empirically validated and proven research models, such as technology acceptance model-related concepts [26,31,50,51].

There is a theoretical basis for each component in the proposed model, as well as additional factors based on previous research on PCT use. A research model for PCT acceptance is provided based on the above concepts. The model comprises (1) project resources, site management, project coordination, and technological factors as an external variable for PCT acceptance, and (2) technology acceptance model-related factors (PEOU, PU) and user trust and satisfaction as mediation factors for the intention to adopt PCT.

2.2.2. External Variables for Prefabricated Construction Acceptance

Extending the standard technology acceptance model, this research proposes that the influence of external variables (e.g., project resources, site management, project coordination, and technological factors) on the intention to use are mediated by user trust, satisfaction, perceived usefulness, and perceived ease-of-use. As a result of selecting external variables, theory development and technological adoption are both enhanced. The existence of external variables directs the steps required to influence increased use by providing a better understanding of what drives perceived usefulness and perceived ease-of-use. The most important variables influencing the adoption of PCT in construction firms are the external variables.

A total of 54 critical factors for PCT adoption were considered from previous studies for this study. The factors were classified into four categories: project resources, site management, project coordination, and technological factors.

In an attempt to develop a technology acceptance model for PCT adoption, this research identified the external factors by exploring the constructs that can influence PCT adoption by diverse stakeholders. As previously stated, this research examines the details of enablers and drivers in the literature. Even though numerous studies have looked at what makes PCT usable, these earlier studies have limitations since they do not account for the potential effects of factors related to the execution stage and may, therefore, only partially explain the reason why actual technology use is insufficient. One of the most common obstacles to PCT cited is the execution stage, policies, etc. Since the successful completion of a building project is a crucial component, these particular types of construction present unique challenges. As a result, this research focuses on identifying external factors that could affect perceived usefulness and perceived ease-of-use from an execution perspective. It could be categorized into the content of project resources, site management, project coordination, and technological features.

Project resources are related to low-cost and sustainable approaches undertaken in prefabricated building construction. Building construction performance depends on the performance of parties and resource availability. Consequently, the optimized use of resources and materials seems to be another indicator for the prolonged advancement of project sustainability performance. An organization's willingness to embrace the notion of a sustainable economy and take constructive efforts toward sustainable development can be shown in its propensity to organize its activities and resource use strategy to respect the rights of subsequent generations to environmental resources. So, to comprehend the mechanism of PCT adoption, this research includes project resources.

Site management is related to time, quality, safety, and logistics/site operations. Creating a strong organizational culture is a powerful tool to influence employees' behavior and improve their performance. Successful site management includes time management, improved quality of construction, safer construction, and better site operations in a factory-controlled environment. The site management factors that are involved and the implementation of the variables in those factors that lead to the improvement of the prefabricated construction building performance results in the inclusion of this construct in our model.

Project coordination is related to the coordination of staff and tasks and the simplification of activities. Collaboration has been a facilitator of PCT, helping to change problem-addressing behavior, and as a crucial component in PCT practitioners' relationships. Simply moving the building process inside a factory was noted for the benefit of a central coordination point for organizing staff across multiple projects. The construction sector is said to be highly disorganized, dependent on cooperation, and reliant on communication. The success of a project depends on timely, precise communication among all involved stakeholders. All of these promote collaboration in PCT procurement. In this respect, the concept of project coordination was included in the proposed research model.

Technological features are related to technology and innovation, industry/market culture and knowledge, improved productivity, and the efficiency of materials and labor. The availability of locally manufactured plants and equipment, skilled personnel resources, the breadth and depth of local material resources, and the depth of use of such local construction resources are all indicators of suitable construction technology, resulting in alleviating the industry's performance. Similarly, the importance and role of innovation in construction and its future are also very much evident in the literature. Based on this, the construct of technological features was included in the research model. These factors were employed to establish hypotheses to comprehend the psychological mechanism of PCT acceptance. This led to the creation of an extended technology acceptance model that accounts for external factors.

2.2.3. Mediating Variables

In construction management research, the role of mediating factors and their mechanisms remains relatively understudied. Little is known about how perceived usefulness and perceived ease-of-use lead to user trust and satisfaction and how they interact to facilitate PCT adoption. That is why it is being argued that mediating factors are important to help us understand the prefabricated building processes that influence PCT adoption.

When discussing technology, "trust" is synonymous with confidence in the system's ability to function as intended. To be more precise, it is the belief that a piece of technology will aid one in accomplishing a task because of its usefulness, dependability, and functionality. In the early stages of adopting new technology, this can be a decisive component in overcoming the risk and skepticism that users may feel. The incorporation of trust in the technology acceptance model revealed its importance in predicting customers' intention to use new technologies in many studies, which why it has been in the research model as a mediator that is considered along with satisfaction.

Customer satisfaction is defined as "the degree to which a customer is pleased or dissatisfied with the performance or outcome of a product as compared to the customer's expectations." Satisfaction is considered an important variable due to its high effects on customers' future behavior and attitudes about certain products or services. The level of client satisfaction an organization achieves is inextricably linked to the quality, pricing, timeliness, and accessibility of the items it offers. Therefore, to examine the influence of satisfaction of users in the context of prefabricated building construction, it has been incorporated in the proposed model as a mediator.

According to the concept of the technology acceptance model, it is, therefore, argued that satisfaction and trust mediate the effect of perceived ease-of-use and perceived usefulness on behavioral intention. Perceived ease-of-use could elicit behavioral intention along the first mediating path (through trust). We argue that clients who consider the PCT technologies as beneficial and simple to use and have high trust have a higher intention to employ the technologies. The second mediating path emphasizes satisfaction as a vital foundation to enhance behavioral intention. It is anticipated that people with a high level of satisfaction with a certain technological product are more inclined to adopt it.

According to the literature review, the success of a construction project can be evaluated by a set of criteria specified by various scholars. According to Sue and Ritter (2012), when the terminology and terms used in the survey are inaccurate, the survey's validity is compromised [52]. As a result, the content of the designed survey was evaluated before the study's final version. Experts are asked to assess whether the constructed survey measures the necessary content, which is known as face validity [53,54]. This will help to elicit recommendations from reviewers based on prior knowledge and expertise [55]. In this research, the developed questionnaire was tested in collaboration with three academic experts. The face validity test was successful, and several different versions of the questionnaire were created before the final one. Several grammatical issues were raised. Factors were reduced to 36. Various scale items were replaced and rephrased. An overall positive response was received with some remarks on the questionnaire layout design and question wording. The questionnaire was modified because of these suggestions.

As a result, before the data collection step, the questionnaire items were clear and understandable. As stated in Table 1, the hypotheses are scientifically formulated based on some rationale and supporting research. The relevant literature is the outcome of studies that are related to every research variable.

Table 1. Literature review for the research hypotheses.

No.	Hypotheses	Description of Category	Relevant Literature
H1	Project resources affect perceived usefulness positively.	External factors	[37,56]
H2	Site management affects perceived usefulness positively.		[42,57]
H3	Project coordination and collaboration affect perceived ease-of-use positively.		[14,58]
H4	Technological features affect perceived ease-of-use positively.		[57,59]
H5	Perceived ease-of-use affects perceived usefulness positively.	Technology acceptance model factors	[50]
H6	Perceived usefulness affects behavioral intention positively.		[50]
H7	Perceived usefulness affects user trust positively.		[60]
H8	Perceived usefulness affects user satisfaction positively.		[61,62]
H9	Perceived ease-of-use affects user trust positively.		[60,63]
H10	Perceived ease-of-use affects user satisfaction positively.		[64,65]
H11	Perceived ease-of-use affects behavioral intention positively.		[50]
H12	User trust affects behavioral intention positively.	Mediating factors	[66]
H13	User satisfaction affects behavioral intention positively.		[22,67]
H14a	Perceived usefulness mediates the relationship between PEOU and the intention to use PCT.		[68]
H14b	Perceived usefulness mediates the relationship between PEOU and user satisfaction.		Proposed by the authors
H14c	Perceived usefulness mediates the relationship between PEOU and user trust.		Proposed by the authors
H15a	User trust mediates the relationship between PEOU and the intention to use PCT.		[69]
H15b	User trust mediates the relationship between PU and the intention to use PCT.		[69]
H16a	User satisfaction mediates the relationship between PU and the intention to use PCT.		[70,71]
H16b	User satisfaction mediates the relationship between PEOU and the intention to use PCT.		[70]

2.2.4. Proposed Model

In this paper, a research model for empirical analysis of the intention to accept PCT is proposed, based on the technology acceptance model's previous literature review (Figure 1). Project resources, site management, project coordination, technological factors, perceived ease-of-use, perceived usefulness, and behavioral intention to accept PCT are among the 36 observed indicators in the proposed model, and 9 latent constructs are described here (assessment items and factors).

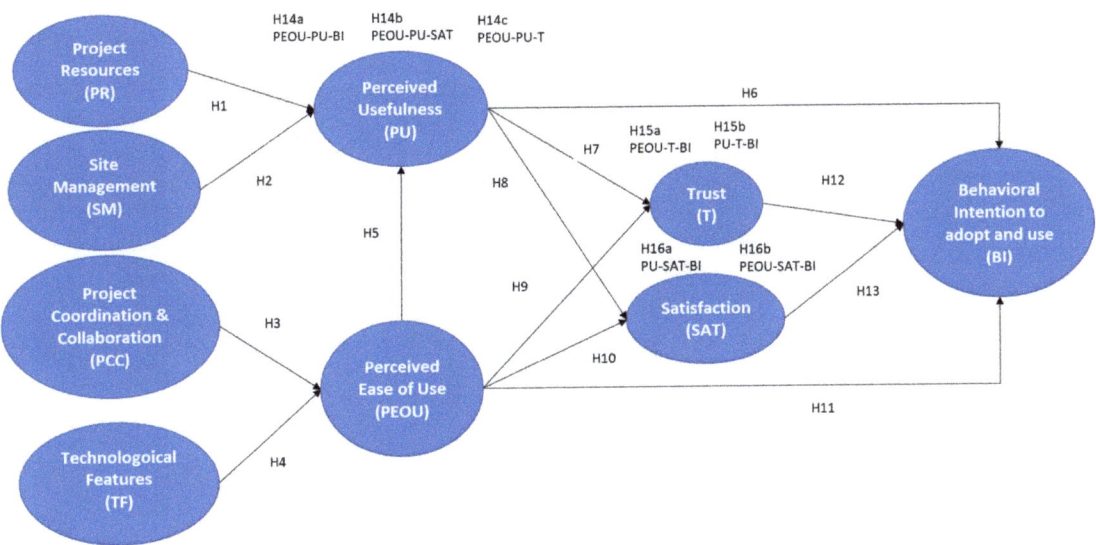

Figure 1. Hypothesized model.

Based on the proposed research model, hypotheses are formulated (Table 2). The basis for the developed hypotheses was described in the previous section. Hypotheses were tested using AMOS 23.0 and SEM. The measurement model was first estimated to test the overall fit of the model, as well as its validity and reliability. Second, using the structural model, the hypotheses were tested between constructs.

Table 2. Constructs with measurement items.

Construct	Definition	Items	Measured Variables
Project Resources (PR)	Project resources are related to low-cost and sustainable approaches undertaken in prefabricated building construction.	PR1	I find prefabricated construction technologies affordable/low costly.
		PR2	I think using prefabricated construction yields savings due to on-site labor reduction.
		PR3	I find reduction in construction waste while adopting and using prefabricated construction.
		PR4	Using prefabricated construction reduces energy consumption.
Site Management (SM)	Site management is related to time, quality, safety, and logistics/site operations.	SM1	I find more control of construction quality while using prefabricated construction.
		SM2	I find it feasible to adopt and use prefabricated construction as it enhances easy transport of heavy components (size restrictions) and increases site accessibility.
		SM3	Using prefabricated construction enhances speed of construction.
		SM4	I find reduction in elevated work and dangerous activities by adopting and using prefabricated construction.

Table 2. Constructs with measurement items.

Construct	Definition	Items	Measured Variables
Project Coordination and Collaboration (PCC)	Project coordination is related to the coordination of staff and tasks and the simplification of activities.	PCC1	Using prefabricated construction enhances logistics processes coordination/Supply chain coordination.
		PCC2	It is desirable to use prefabricated construction involving information transparency.
		PCC3	I find prefabricated construction to enhance simplification of the construction process.
		PCC4	Given the ease of handling, I intend to use prefabricated construction.
Technological Features (TF)	Technological features are related to technology and innovation, industry/market culture and knowledge, improved productivity, the efficiency of materials, and labor.	TF1	Given the industry knowledge, experience, awareness, availability modular producers and suppliers, I expect that I would use it.
		TF2	Using prefabricated construction enhances reduction of material use.
		TF3	I think adopting and using prefabricated construction to be feasible due to improved flexibility and adaptability.
		TF4:	It is desirable to use prefabricated construction due to work on-site continues irrespective of whether the product is being made elsewhere.
Trust (T)	Trust is the belief that a piece of technology will aid one in accomplishing a task because of its usefulness, dependability, and functionality.	T1	Based on my experience with the prefabricated construction technologies will have integrity.
		T2	Centered on my experience with the prefabricated construction will be reliable.
		T3	Overall, prefabricated construction techniques will be trustworthy.
		T4	Based on my experience with prefabricated construction, I believe that this technology will provide good services.
Satisfaction (SAT)	Customer satisfaction is defined as "the degree to which a customer is pleased or dissatisfied with the performance or outcome of a product as compared to the customer's expectations".	SAT1	I am satisfied with the performance of prefabricated construction technologies.
		SAT2	I am pleased with the experience of using the prefabricated construction technologies.
		SAT3	I am satisfied with the prefabricated construction efficiency and effectiveness.
		SAT4	Overall, I am satisfied with the prefabricated construction.
Perceived Usefulness (PU)	Perceived usefulness is defined as the extent to which stakeholders believe that utilizing PCT is useful.	PU1	Using prefabricated construction would improve construction industry performance.
		PU2	Using prefabricated construction would increase construction productivity.
		PU3	Using prefabricated construction would enhance the effectiveness of the construction industry.
		PU4	I would find prefabricated construction useful in the construction industry.
Perceived Ease-of-Use (PEOU)	Perceived Ease-of-use is defined as the extent to which construction industry stakeholders perceive that using PCT does not involve substantial effort.	PEOU1	Learning to operate prefabrication techniques would be easy for me.
		PEOU2	I would find it easy to get prefabricated construction to do what I want it to do.
		PEOU3	It would be easy for me to become skillful at using prefabricated construction.
		PEOU4	I would find prefabricated construction easy to use.

Table 2. Constructs with measurement items.

Construct	Definition	Items	Measured Variables
Behavioral Intention (BI)	Behavioral intention is the probability or a measure of the strength of one's intention to perform a specific behavior toward using technology and, therefore, it determines technology acceptance.	BI1	I support the adoption and use of prefabricated construction in construction industry projects.
		BI2	Intend to increase use of prefabricated construction to perform construction activities in the future.
		BI3	Given that I had access to prefabricated construction (technology, materials and equipment's), I predict that I would use it in construction industry projects.
		BI4:	I will recommend others to use prefabricated construction in performing construction activities.

The constructs, definitions, and measurement items, along with the respective codes used in this research, are shown in following Table 2.

3. Research Methodology

The relationships between the research model's constructs were investigated using a quantitative cross-sectional design. Nine constructs were derived from previously validated instruments used in similar circumstances to produce research instruments. There are references in Appendix A to prior studies from which the items in the questionnaire for these constructs were obtained. There were multiple items for each construct, measured on a five-point Likert scale (1–5), i.e., (1) strongly disagree to (5) strongly agree.

The pilot study is used to remedy any lack of required quality, assure the clarity of the questionnaire items, and eliminate phrasing errors, according to [54]. Few researchers used small sample sizes in their studies [72,73]. Before collecting full-scale data, a 30-person pilot study was conducted to test instrument reliability and validity [74]. The questionnaire was sent to a random sample of construction industry stakeholders (engineers, contractors, consultants) in Pakistan. The value of Cronbach's alpha calculated for the pilot study sample considering the entire questionnaire (36 items) was 0.890. The pilot study gave acceptable results for the measurement items through Cronbach's alpha test.

3.1. Sampling and Procedure

The population of interest for this study was stakeholders of all construction companies in Pakistan. This research study's sampling frame consisted of only major cities in Pakistan due to time and cost constraints. Companies from both the public and private sectors were involved. Researchers, project managers, contractors, and employees make up the unit of analysis. A sample of at least 200 participants was necessary for the data analysis technique and the analysis of moment structures (AMOS) [75,76]. A total of 350 questionnaires were distributed. The sample was recruited from different construction companies. This sample was selected to meet specific criteria: participants who were 18+ years old, owned or used offsite construction techniques and other prefab-related technologies, and involved in the construction industry.

The current study utilized the Statistical Package for the Social Sciences (SPSS) and analysis of moment structures (AMOS) software to perform statistical analyses. The statistical software package SPSS was utilized for data coding, data cleaning, the verification of assumptions, and conducting exploratory factor analysis. In contrast, AMOS was used to evaluate a measurement model's validity, reliability, discriminant validity, and goodness-of-fit indices. The structural model in AMOS was utilized to conduct hypothesis testing. Finally, mediation analysis was carried out using PROCESS Macro v4.0.

3.2. Common Method Bias

Harman's single-factor test is a prevalent technique used in academic literature that serves as a simple and frequently utilized approach to identify and mitigate common method bias (CMB). CMB is a plausible origin of measurement error that has the potential to distort the associations between variables and compromise the credibility of research outcomes [77]. The present study involved the implementation of Harman's single factor test utilizing an un-rotated single factor constraint within the SPSS v21 software. The initial component extracted was found to explain a variance of less than 22.472%. The study's results indicate that the presence of common method bias was not a significant issue.

Common method bias (CMB) [77] was tested on the measurement items used in this study, which represents the variance attributed to the method of measurement rather than the variance explained by the constructs in this study. Harman's single-factor test [77,78], which uses an unrotated factor analysis using a single component, is a standard method to detect CMB. The test revealed that the first factor explained only 22.47% (below 50%) [78] of the variance in the data, indicating the lack of dominance of a single factor, and CMB was not a substantial issue in this study. The research method consisted of two phases. The measurement model was first validated. A structural model and path analysis were used to analyze the relationship between the constructs for hypothesis testing. IBM SPSS and AMOS were used to analyze data.

4. Results and Discussion

The results of the analysis are revealed in this section. The assumption of linearity and normality of constructs was made before the statistical analysis of the obtained data. The scale's reliability and validity were tested using a dataset of $n = 250$. For the hypotheses that were developed for this study, the following are the measurements, structural models, and path analysis results.

4.1. Descriptive Statistics

The questionnaires were distributed via databases of organizations throughout Pakistan that are related to the construction industry. Out of 300 distributed questionnaires, 260 responses (representing 86% of the total distributed) were valid, and 3% of the survey responses were invalid or missing, resulting in 250 remaining responses. Most participants ($n = 102$) were architects/engineers, and these constituted about 41% of the sample. A total of 57 respondents were project/construction managers, 22 respondents were consultants, 40 were contractors, and 29 respondents were academics/ researchers. The percentages of the sample size constitute about 22.8%, 8.8%, 16%, and 11% of the sample, respectively. A 5-point Likert scale ranging from "strongly disagree" to "strongly agree" was used for each question.

4.2. Reliability of Constructs

Cronbach's reliability test was utilized to determine the robustness of the measurement model used for the final SEM evaluation. For Cronbach's alpha, a cutoff value of 0.7 was used to indicate acceptable levels of initial consistency [75]. There was good reliability (0.890) for each of the items in the final SEM that measured all latent variables.

4.3. Exploratory Factor Analysis

As far as current knowledge permits, this study represents an initial attempt to investigate the factor structure underlying the adoption of PCT. Thus, exploratory factor analysis was used because the author's could not make a confident estimate of the number of factors contained in this measure. To ascertain the dimensions that underlie the various variables incorporated in this investigation, we employ the maximum likelihood approach, with the Promax and Kaiser normalization rotation methods, for optimal results. [79]. The cutoff criterion that is acknowledged in the literature for its practical consequences is maintaining items with factor loading minimums that are less than 0.35 [79].

To determine whether the data on the respondents is suitable for factor analysis before the factors are extracted, many tests need to be performed [80]. Two statistical tests commonly used in research are the Kaiser–Meyer–Olkin (KMO) measure of sampling adequacy and Bartlett's test of sphericity. In factor analysis, the KMO index is utilized with a threshold of 0.5, and its values fall within the range of 0 to 1. The application of factor analysis is deemed suitable solely when the statistical significance of Bartlett's test of sphericity is established ($p < 0.05$).

4.4. Measurement Model Assessment

Purification of items is vital. Thus, EFA was used to assess the observed variables' transparency. The 36-item survey was subjected to exploratory factor analysis (EFA) to see if the underlying structure in the data was present. Initially, EFA was used to assess the validity of the measurement model using maximum likelihood in SPSS 21. Using the Kaiser–Meyer–Olkin (KMO) sample adequacy method, its value is 0.830, and any value above 0.5 is considered acceptable [81]. Using nine constructs, 60.97 percent of the total variance is explained, which is quite high. In these tests, all nine constructs and responses were sufficient for factor analysis. EFA uses the maximum likelihood method, with Promax rotation and Kaiser normalization for better results [79]. After EFA, a confirmatory factor analysis (CFA) was performed by SEM using AMOS to verify recognized factors.

The measurement model was evaluated using CFA, and validation of the measurement model was performed with discriminant and convergent validity and reliability [75]. Construct validity was established by following the criteria specified by [82] and, which states the following: When performing a CFA, convergent and discriminant validity must be established. Testing a causal model is useless if factors do not show adequate validity and reliability. Measures of validity and reliability include composite reliability (CR), average variance extracted (AVE), maximum shared variance (MSV), and average shared variance (ASV). The thresholds [75] for these values are as follows: reliability (CR > 0.7), convergent validity (AVE > 0.5), discriminant validity (MSV < AVE), and the square root of AVE greater than inter construct correlations. AVE > 0.50 indicates adequate convergent validity. All AVE values in Table 3 are above this threshold, indicating convergent validity. Construct reliability (CR) > 0.70 indicates internal consistency or adequate convergence. As shown in Table 3, all CR values are above this threshold, indicating internal consistency.

Table 3. Validity and reliability.

	PR	PU	SAT	PEOU	T	PCC	TF	SM	BI
Cronbach's Alpha	0.732	0.912	0.858	0.877	0.837	0.851	0.85	0.783	0.87
CR	0.755	0.861	0.812	0.854	0.84	0.853	0.852	0.757	0.821
AVE	0.515	0.756	0.59	0.661	0.569	0.593	0.59	0.51	0.606
MSV	0.088	0.497	0.57	0.364	0.28	0.021	0.088	0.063	0.57
MaxR (H)	0.83	0.863	0.813	0.862	0.852	0.861	0.857	0.765	0.825
PR	**0.718**	0.275	0.049	0.171	0.091	0.056	0.297	0.209	0.109
PU		**0.87**	0.622	0.469	0.344	−0.028	0.067	0.218	0.705
SAT			**0.768**	0.508	0.382	0.087	0.003	0.251	0.755
PEOU				**0.813**	0.353	0.144	0.285	0.081	0.603
T					**0.754**	0.078	0.108	0.11	0.529
PCC						**0.77**	0.052	−0.057	0.114
TF							**0.768**	−0.066	0.142
SM								**0.714**	0.146
BI									**0.778**

Note: Values in bold are the square root of the AVE for each construct.

The utilization of SPSS was initially employed to assess the reliability and validity of the data. As shown in Table 3, Cronbach's alpha values for the variables ranged from 0.732 to 0.912, while the CR values ranged from 0.755 to 0.861. These values surpass the standard threshold of 0.7, signifying the credibility and reliability of the scale data. Additionally, the measured items displayed strong internal consistency and overall reliability.

Construct validity was evaluated via convergent and discriminant validity. Convergent validity was established by making sure the average variance extracted (AVE) values of the variables exceeded 0.5, and the composite reliability (CR) values were above 0.7, as illustrated in Table 3. These findings show acceptable convergence validity. By confirming that the square root of the AVE values for each variable was greater than the correlation coefficients between the variables, discriminant validity was established. Table 3 shows that for all variables, the square root of the AVE values was greater than the correlation coefficients, demonstrating good discriminant validity.

The measurement model indicates covariance between latent variables and standardized weights or indicator loadings for the variables. Two variables from PU, and one each from SAT, PEOU, SM, PR, and BI were removed to improve model fit. The values of the standard error of the coefficient imply more reliable predictions and smaller confidence intervals [82]. As a result, there is no need to reduce the variables, and they can be used as predictors of their respective latent variables. In this study, the measurement model has a high level of reliability, convergent validity, and discriminant validity.

Table 4 provides a mix of absolute and incremental measurement model fit indices commonly reported in SEM-related literature. The results show that the model fit is good, as none of the fit indices are outside the acceptable range. In the absence of any issues with fit indices, no further model modifications were made. The chi-square and values of CFI and SRMR indicate excellent model fit [82,83].

Table 4. Model fit indices of the measurement model (adopted from Hanif et al. (2018) [84]).

Model Fit Indices	Reference Range	Measurement Model Estimate	Structural Model Estimate
CMIN		488.233	452.133
DF		338	328
CMIN/DF	Between 1 and 3	1.444	1.378
CFI	>0.95	0.953	0.959
SRMR	<0.08	0.049	0.058
RMSEA	<0.06	0.042	0.039
PCLOSE	>0.05	0.943	0.985

The study determined the fit indices of the proposed model, which yielded root mean square error of approximation (RMSEA) values of 0.042 and 0.039, as well as a comparative fit index (CFI) of 0.953 and 0.959. The presence of these indices indicates an adequate level of fitness. Table 4 shows the goodness-of-fit indices, which serve as an indicator of the degree to which the data align with the model. The SEM path analysis did not violate the thresholds of the fit indices, as reported in previous studies.

4.5. Structural Model Assessment

This study conducted a multicollinearity test to see if there was a problem with multicollinearity. The value of VIF for multicollinearity assessment should be around 1.0 and less than 3.0. There was no evidence of multicollinearity because all VIF values were less than 3.0 and around 1.0.

Because relations are allocated between constructs based on studies, the structural model attempted to identify dependencies between model constructs. The structural model was evaluated using a two-step approach, as suggested by Cheng, (2001) [85]. The structural model's goodness-of-fit indices (GOF) are evaluated first, and then standardized parameter estimates are utilized to support causal relationships and testing hypotheses. The first step is to evaluate and test the overall model, GOF, using the same criteria as the measurement model. It is preferable to have a structural model GOF nearer to the measurement model's fit values. The hypothesized structural model is presented in Table 4.

The standardized coefficients and hypothesis testing results are in Table 5. AMOS results show a strong significant effect for all factors except project coordination and collaboration (PCC), which is not significant (H3). It appears that the hypothesized model is adequately fitted, as the GOF statistics are within acceptable parameters. Considering these results (Table 5), Figure 2 summarizes the proposed model.

Table 5. Parameter estimates and results of the hypotheses. (*** Indicating significance at $p < 0.001$).

			Estimate	S.E.	C.R.	p
PEOU	<—	PCC	0.182	0.096	1.891	0.059
PEOU	<—	TF	0.342	0.089	3.842	***
PU	<—	SM	0.306	0.133	2.306	0.021
PU	<—	PR	0.206	0.099	2.090	0.037
PU	<—	PEOU	0.386	0.067	5.723	***
SAT	<—	PU	0.480	0.078	6.196	***
T	<—	PU	0.219	0.080	2.726	0.006
SAT	<—	PEOU	0.229	0.065	3.553	***
T	<—	PEOU	0.195	0.071	2.757	0.006
BI	<—	PU	0.284	0.075	3.785	***
BI	<—	SAT	0.406	0.090	4.515	***
BI	<—	T	0.221	0.058	3.814	***
BI	<—	PEOU	0.171	0.057	3.015	0.003

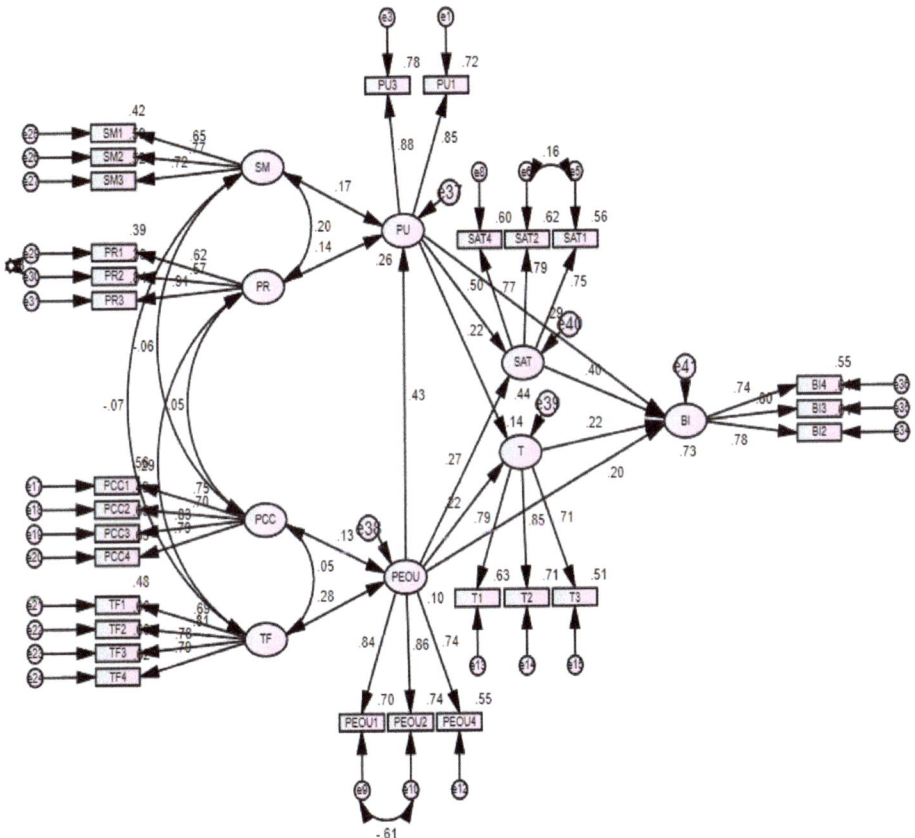

Figure 2. Hypothesized model (standardized).

The summary of the hypotheses testing is presented in Table 5, which shows that twelve hypotheses are accepted and only one is rejected. The output of the structural equation model revealed that the hypothesized constructs, including PR, SM, TF, PU, PEOU, T, and SAT, have a significant positive effect ($p < 0.05$). PU was significantly influenced by two constructs, i.e., PR ($p = 0.037$) and SM ($p = 0.021$). Thus, H1 and H2 are supported. Similarly, PCC had a significant influence on PEOU ($p = 0.059$) but not TF ($p < 0.001$). Therefore, H4 was supported, However, PCC (H3) tends to be not significant and rejected as a result of this analysis. Trust was associated with PU ($p = 0.006$) and PEOU ($p = 0.006$), and both of these were also predictors of satisfaction, with statistics supporting their significance ($p < 0.001$). Thus H7, H8, H9, and H10 were supported. PU ($p < 0.001$), PEOU ($p = 0.003$), T ($p < 0.001$), and SAT ($p < 0.001$) were four predictors of BI, supporting H6, H11, H12, and H13, respectively.

4.6. Mediation Analysis

Recommendations described by Preacher et al. (2007) were followed to test the hypothesized meditation model, including a bootstrapping procedure [86]. PROCESS macro developed by Hayes (2017) was used, which was based on the analytical conceptualization of Preacher et al. (2007) [86,87]. To test the hypotheses, the first hierarchical multiple models (mediated relationships) were developed, as suggested by hypotheses H14a,b,c-H15a,b-H16a,b, and then in SPSS, the independent variables, mediators, and dependent variables were analyzed in separate steps.

Table 6 shows the total, direct, and indirect effects of the perceived ease-of-use (PEOU) on behavioral intention (BI) toward using PCT, with perceived usefulness (PU) as a mediator. PEOU had a significant effect on BI = 0.5535, $p = 0.000$, according to the mediation results. PEOU had a significant and positive impact on BI $\beta = 0.2960$, $p = 0.000$, revealing that the increase of one unit in PEOU will result in a change of 0.2960 units in BI. The findings also reveal that the indirect effect of PEOU on BI across PU is also substantial $\beta = 0.2576$ (0.1860, 0.3380). This shows that PU acts as a mediator between PEOU and BI. The results of this study strongly confirm hypothesis H14a, which states that PU partially mediates PEOU and BI.

Table 6. Mediation analysis.

Mediation Path	Total Effect (β)	Direct Effect (β)	Indirect Effect (β)	S.E	Indirect Effect and 95% Confidence Interval	
					LL	UL
PEOU-PU-BI	0.5535	0.2960	0.2576	0.0391	0.1860	0.3380
PEOU-SAT-BI	0.5535	0.2289	0.3246	0.0404	0.2483	0.4062
PEOU-T-BI	0.5535	0.4364	0.1171	0.0300	0.0630	0.1807
PU-SAT-BI	0.7593	0.3612	0.3981	0.0461	0.3097	0.4894
PU-T-BI	0.7593	0.6419	0.1174	0.0275	0.0654	0.1738
PEOU-PU-SAT	0.4528	0.2156	0.2372	0.0349	0.1714	0.3091
PEOU-PU-T	0.3034	0.1945	0.1089	0.0412	0.0357	0.1951

Similarly, the role of PU as a mediator between PEOU and SAT in the form of hypothesis 14b, $\beta = 0.2372$ (0.1714, 0.3091) and PU as a mediator between PEOU and T in the form of hypothesis 14c, $\beta = 0.1089$ (0.0357, 0.1951) is also significant, supporting hypotheses H14b and H14c, respectively.

The outcomes indicate total, direct, and indirect effects of the perceived ease-of-use (PEOU) and perceived usefulness (PU) on behavioral intention (BI) toward using PCT, with user trust (T) as a mediator. The findings reveal substantial indirect effects of PEOU on BI, with T as a mediator $\beta = 0.1171$ (0.0630, 0.1807), and PU on BI, with T as a mediator $\beta = 0.1174$ (0.0654, 0.1738), which strongly supporting hypotheses H15a and H15b, respectively. Similarly, results also show total, direct, and indirect effects of the perceived usefulness (PU) and perceived ease-of-use (PEOU) on behavioral intention (BI) toward using PCT, with user satisfaction (SAT) as a mediator. As per the results of Table 6, user satisfaction plays a significant role as a mediator between PU and BI $\beta = 0.3981$ (0.3097, 0.4894), and between PEOU and BI $\beta = 0.3246$ (0.2483, 0.4062), supporting hypotheses H16a and H16b, respectively.

4.7. Discussion

This study examines the association between the perceived usefulness and perceived ease-of-use of hypothesized external factors and the behavioral intention toward using PCT. Table 5 demonstrates that most of the hypotheses are supported. The goodness-of-fit (GOF) measurements of the proposed model confirm that it can adequately reflect the obtained data and assist in understanding the behavioral intention of construction practitioners toward adopting PCT. The relevance of every model construct is then addressed, as revealed by hypothesis testing.

The study's findings show that several hypotheses (H1, H2, H4, H5, H6, H7, H8, H9, H10, H11, H12, and H13) are supported and only one (H3) is not supported. The proposed model hypothesized that PR directly affects PCT's perceived usefulness (H1). It means stakeholders are more likely to consider PCT technologies useful if they believe they have adequate project resources. Other researchers have empirically found [56,58] that the project resources of PCT are a determinant of perceived usefulness. This result also validated the claim of M. Li et al. (2017) [88] that prefabricated construction is the way of the future for China's construction sector, as it is a green building type that offers energy savings and environmental protection. Similarly, researchers assumed site management has a direct positive impact on prefabricated construction's perceived usefulness (H2). As a result, stakeholders are more likely to perceive PCT technologies as useful if they perceive competent site management. Other researchers have also empirically found [42,57,89] that the site management of prefabricated construction is a determinant of perceived usefulness. The findings of this study support the hypothesis that PCT in Pakistan has a good influence on site management and perceived usefulness. Construction practitioners who know that employing technological approaches in the construction industry can boost their benefits tend to project resources and management. These findings also suggest that stakeholders tend to go for innovative approaches in the construction industry when they can benefit from those technologies.

Contrary to expectations, project coordination and collaboration did not affect PCT's perceived ease-of-use (H3). This finding contradicts many previous studies that found the same belief system persists. Refs. [14,58,90] hypothesized and found that project coordination and collaboration play a significant role in affecting perceived ease-of-use based on many past research findings. However, this study found no significance for project coordination and collaboration in perceived ease-of-use prediction. This conclusion is surprising since it seems to disagree with the broadly accepted idea that projects with a higher level of simplification and ease of management are more likely to influence stakeholders' preferences.

The findings support hypothesis H4, which states that technological features directly influence PCT's perceived ease-of-use. Among the external variables, the path coefficient signifies the strongest significant relation between technological features and perceived ease-of-use. Thus, it confirms the importance of technological features in PCT usage in Pakistan and shows the importance of technological factors in ease-of-use. The numerous features of this technology, such as flexibility and adaptability, are the reason for the significance of this construct. Hence, they also find ease in their use while using these technologies because of their flexibility and usage irrespective of weather conditions, which in turn increases their motivation to accept further automation in construction. Prefabrication is a new technology in Pakistan's construction sector, so stakeholders need easy-to-use PCT. This finding is consistent with many construction-related empirical studies [42,61]. Technological features have a considerable favorable effect on the perceived ease-of-use of PCT in Pakistan, according to this research.

Trust was assumed to have a positive effect on PCT behavioral intention (H12). The result shows that trust and behavioral intention have the strongest direct relationship. This research is in line with the previous research carried out by Gefen and Straub (2004), which found that consumers were primarily motivated by the technology's ability to give trust [91]. On the other hand, user satisfaction was hypothesized to influence PCT behavioral intention positively (H13). The results show that user satisfaction improves behavioral intention. Thus, the SAT–BI relationship is one of the strongest direct relationships. Many studies imply that the SAT–BI relationship is strongest [22,92]. These findings suggest that user trust and satisfaction are strong predictors of user satisfaction toward using PCT due to the fact that they find PCT useful and easy to use. Thus, construction industry professionals who have trust and satisfaction when using PCT will be more likely to accept and recommend it to others.

Perceived usefulness is defined as the extent to which stakeholders believe that utilizing PCT is useful (Davis, 1986) [28]. Prefabricated construction is more useful for promoting sustainable development regarding the economy, society, and the environment when compared to conventional construction methods [93], which also aligns with the intention to attain sustainable development on a global scale. These benefits align with the principles of sustainable and green building, which prioritize minimizing the environmental impact of building projects while creating healthy and functional spaces for people to live and work. The study hypothesized that perceived usefulness directly affects PCT behavioral intention (H6). The results reveal that perceived usefulness positively affects behavioral intention, which is consistent with previous literature [26,50], where the technology's usefulness and functionalities were the primary motivators for users. Thus, this study confirms the favorable influence of perceived usefulness on behavioral intention to use PCT in Pakistan. Correspondingly, this study concludes a significant positive influence of perceived usefulness on the satisfaction of PC (H8), which is also consistent with previous literature [61], where the satisfaction and features provided by the technology were the primary motivators for users. Thus, our study confirms the strong and favorable influence of perceived usefulness on PCT customer satisfaction in Pakistan. Similarly, the positive impact of perceived usefulness on the trust of PCT was also hypothesized (H7), which was followed previous literature [69,94] where the technology's usefulness and trustworthiness were the primary motivators for individuals. As a result, the findings of this study show the existence of a strong and favorable influence of perceived usefulness regarding trust to use PCT in Pakistan.

For the current study, perceived ease-of-use is defined as the extent to which construction industry stakeholders perceive that using PCT does not involve substantial effort [28]. Perceived ease-of-use was hypothesized to have a direct impact on PCT's perceived usefulness in this study (H5). The results show that perceived ease-of-use positively affects perceived usefulness, confirming that people perceive PCT as useful when it is easy to use. With less effort required to use these technologies, stakeholders perceive them as more useful. Prefabrication reduces the amount of time and effort required for construction workers to learn new skills. In line with other technology models, such as the TAM [25]. On the other hand, the perceived ease-of-use of PCT is assumed to directly impact behavioral intention (H11). This research shows that perceived ease-of-use has a favorable effect on behavioral intention. This result may be explained by the fact that prefabrication in construction is new in Pakistan, and thus user-friendliness is critical. This study's findings support previous research on models such as TAM, TAM2, the PEOU determinants model, and TAM 3. This study confirms the positive effect of perceived ease-of-use on PCT intention in Pakistan. This study hypothesized that perceived ease-of-use has a direct positive influence on user satisfaction with PCT (H10). According to this study, Perceived ease-of-use has a direct impact on PCT user satisfaction (H10). This indicates that perceived ease-of-use has the strongest direct influence on user satisfaction. These findings support prior research [64,65] that found that users were largely motivated by the level of satisfaction they had with the technology and its features. Thus, this study supports a strong and favorable influence of perceived ease-of-use on PCT user satisfaction in Pakistan. Similarly, perceived ease-of-use was assumed to have a positive influence on the trust of PCT (H9). The results demonstrate that perceived ease-of-use had a favorable influence on trust, as anticipated [60,69,95], because individuals were mostly driven by the ease-of-use and trust provided by the technology. So, this study shows that perceived ease-of-use has a favorable effect on trust when using PCT.

An important contribution of this research is the addition of user trust and satisfaction to the technology acceptance model to explain the behavioral intention toward adopting PCT. This study's findings demonstrated that the intention to embrace PCT was strongly influenced by both trust and satisfaction (H12, H13). The findings provide insight to the role of perceived usefulness and perceived ease-of-use in influencing PCT adoption by identifying the mediator variables. It is observed that trust and satisfaction (H15a,b, and H16a,b) mediated the relations between perceived usefulness, perceived ease-of-use, and the intention to utilize PCT. Prior researchers who used these constructs did not study how trust and satisfaction function as mediating factors in influencing the intention to use [22]. As a result, this article proved a model by developing mediation between perceived usefulness, perceived ease-of-use, and intention to use. A further benefit of the model is that it accounts for 73% of the variance in the dependent variable (BI). Further study is required to update the technology acceptance model in light of the constant evolution of construction technology and the emergence of newer technologies to automate the construction industry. As a first step, the proposed model is likely to inspire subsequent research in a variety of domains. Figure 3 illustrates the final model.

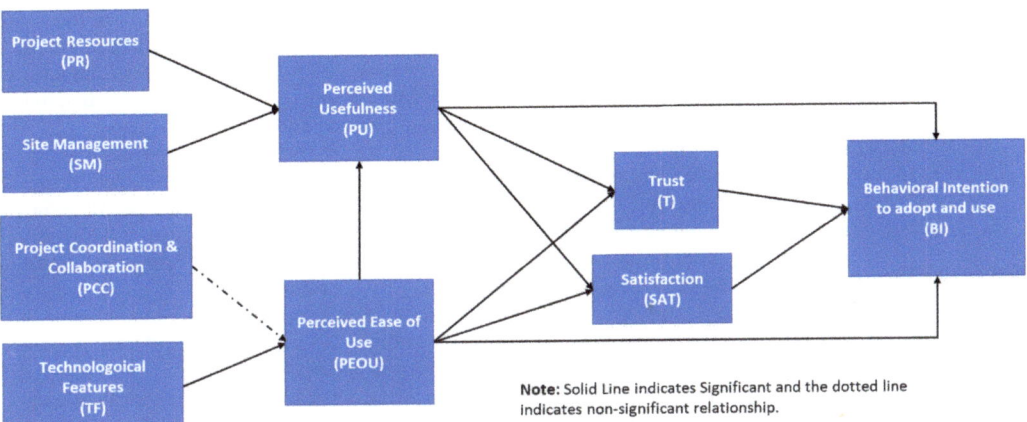

Figure 3. Final model.

5. Conclusions and Recommendations

The first objective of the study was met by proposing an extended version of the TAM-based hypothesized model duly supported by previous researchers. The usage of technology in Pakistani construction is rising, and stakeholders are expected to take advantage of it. To account for factors affecting PCT uptake in developing countries, an extended version of the technology acceptance model is proposed. Project resources, site management, and technological factors have a significantly favorable effect on using PCT technologies. Project coordination and collaboration do not seem to influence the use of these technologies by stakeholders. Similarly, user trust and satisfaction play a significant role as mediators in the proposed model.

The second objective of the study was met by empirically validating the extended version of the TAM-based hypothesized model, which was duly supported by the goodness-of-fit and other statistical parameters. The projected result of this study is expected to function as a guide for forthcoming research on construction technology in Pakistan. The provision of assistance will aid professionals in the construction industry, consultants, and contractors with international operations in Pakistan to efficiently strategize their project planning. The recently developed 36-item PCT scale, based on the established hypothesis, can serve as a valuable tool for construction stakeholders in Pakistan to help prioritize their attempts toward boosting prefabrication. The outcomes of the research are limited to the construction sector of Pakistan; however, they have the potential to be generalized to other civil engineering ventures and emerging markets that share comparable working conditions. Moreover, it is anticipated that the PCT scale and the technology acceptance model that have been formulated will possess validity in developing countries that exhibit comparable work settings, customary practices, and geographical principles. Similarly, the methodologies adopted for this study, i.e., identification of the PCT variables, SEM analysis, and investigation of the mediation relationship using PROCESS can be used in every sector and any part of the world.

The findings of this research emphasize the potential of accepting sustainable PCT via the technology acceptance model. According to Jin et al. (2018) [96], the endorsement of prefabricated construction is deemed a significant advancement toward attaining sustainable development within the construction sector. Prefabricated construction has been deemed a significant advancement toward attaining sustainable development in the construction sector, as it facilitates more efficient use of materials and resources, reduces waste, shortens construction times, and lowers costs while encouraging sustainable construction practices. The close interconnection between sustainable PCT and sustainable and green building practices points out the promising pathway that prefabricated construction offers for accomplishing sustainability goals in the building sector. In addition, the controlled environment of a factory setting enables better quality control, which leads to buildings that are more energy efficient and environmentally friendly. As such, the adoption of sustainable PCT can play a key part in advancing sustainability in the construction industry and contributing to a more sustainable built environment.

A new model based on the technology acceptance model (TAM) is proposed in this research, which includes trust and satisfaction as mediating factors to better explain the factors that contribute to the acceptance and utilization of PCT in the construction industry in Pakistan. As the paper concludes, "the findings can be used to help construction companies plan for Prefabricated Construction Technology adoption by identifying relevant influential factors and providing a theoretical framework for future research." Despite this, the study does have a few limitations, such as the fact that it relies solely on quantitative data, employs a cross-sectional study approach, and focuses primarily on Pakistan's construction industry. To better understand the factors influencing PCT acceptance and usage in developing countries, the study suggests conducting additional research employing longitudinal studies, qualitative approaches, and other external variables and mediating factors.

Author Contributions: Conceptualization, M.H. and R.W.A.; formal analysis, M.H. and R.W.A.; methodology, K.M.M.; software, Z.A.M. and S.H.K.; investigation, M.H.; writing—original draft preparation, M.H.; writing—review and editing, R.W.A., K.M.M. and A.M.; data curation, A.M. and Z.A.M.; validation, S.H.K.; supervision, R.W.A. and K.M.M.; project administration, B.S., A.M. and S.H.K.; funding acquisition, B.S. and S.H.K. All authors have read and agreed to the published version of the manuscript.

Funding: This research received no external funding.

Institutional Review Board Statement: Not applicable.

Informed Consent Statement: Informed consent was obtained from all respondents involved in this study.

Data Availability Statement: These data are available from the corresponding author and can be shared upon reasonable request.

Acknowledgments: The authors extend their appreciation to the researchers from Prince Sultan University, Saudi Arabia and paying the article processing charges for this publication.

Conflicts of Interest: The authors declare no conflict of interest.

Appendix A

Table A1. Model constructs with measurement items.

Construct	Statements	Adopted from
Project Resources (PR)	PR1 "I find prefabricated construction technologies affordable/ low costly." PR2 "I think using prefabricated construction yields savings due to on-site labor reduction." PR3 "I find reduction in construction waste while adopting and using prefabricated construction." PR4 "Using prefabricated construction reduces energy consumption."	[89,97]
Site Management (SM)	SM1 "I find more control of construction quality while using prefabricated construction." SM2 "I find it feasible to adopt and use prefabricated construction as it enhances easy transport of heavy components (size restrictions) and increases site accessibility." SM3 "Using prefabricated construction enhances speed of construction." SM4 "I find reduction in elevated work and dangerous activities by adopting and using prefabricated construction."	[89,97,98]
Project Coordination and Collaboration (PCC)	PCC1 "Using prefabricated construction enhances logistics processes coordination/Supply chain coordination." PCC2 "It is desirable to use prefabricated construction involving information transparency." PCC3 "I find prefabricated construction to enhance simplification of the construction process." PCC4 "Given the ease of handling, I intend to use prefabricated construction."	[14,57,90,99]
Technological Features (TF)	TF1 "Given the industry knowledge, experience, awareness, availability modular producers and suppliers, I expect that I would use it." TF2 "Using prefabricated construction enhances reduction of material use." TF3 "I think adopting and using prefabricated construction to be feasible due to improved flexibility and adaptability." TF4 "It is desirable to use prefabricated construction due to work on-site continues irrespective of whether the product is being made elsewhere."	[57,61,97]
Satisfaction (SAT)	SAT1 "I am satisfied with the performance of prefabricated construction technologies." SAT2 "I am pleased with the experience of using the prefabricated construction technologies." SAT3 "I am satisfied with the prefabricated construction efficiency and effectiveness." SAT4 "Overall, I am satisfied with the Prefabricated construction."	[100]
Trust (T)	T1 "Based on my experience with the prefabricated construction technologies will have integrity." T2 "Centered on my experience with the prefabricated construction will be reliable." T3 "Overall, Prefabricated construction techniques will be trustworthy." T4 "Based on my experience with Prefabricated construction, I believe that this technology will provide good services."	[69]
Perceived Usefulness (PU)	PU1 "Using prefabricated construction would improve Construction Industry performance" PU2 "Using prefabricated construction would increase construction productivity." PU3 "Using Prefabricated Construction would enhance the effectiveness of the construction industry." PU4 "I would find Prefabricated Construction useful in the Construction industry."	[25]
Perceived Ease-of-Use (PEOU)	PEOU1 "Learning to operate Prefabrication techniques would be easy for me." PEOU2 "I would find it easy to get Prefabricated construction to do what I want it to do." PEOU3 "It would be easy for me to become skillful at using Prefabricated Construction." PEOU4 "I would find Prefabricated Construction easy to use."	[25]
Behavioral Intention (BI)	BI1 "I support the adoption and use of Prefabricated Construction in construction industry projects." BI2 "I intend to increase my use of Prefabricated Construction to perform construction activities in the future." BI3 "Given that I had access to Prefabricated Construction (technology, materials and equipment's), I predict that I would use it in construction industry projects." BI4 "I will recommend others to use Prefabricated Construction in performing construction activities."	[101]

References

1. Imran, M.; Memon, N.A.; Hameed, A. *To Explore the Level of Adoption of Prefabrication in Construction Industry of Pakistan*; ResearchGate: Berlin, Germany, 2019; pp. 5–10.
2. Da Trindade, E.L.G.; Lima, L.R.; Alencar, L.H.; Alencar, M.H. Identification of Obstacles to Implementing Sustainability in the Civil Construction Industry Using Bow-Tie Tool. *Buildings* **2020**, *10*, 165. [CrossRef]
3. Durdyev, S. Review of Construction Journals on Causes of Project Cost Overruns. *Eng. Constr. Archit. Manag.* **2020**, *28*, 1241–1260. [CrossRef]
4. Lu, W.; Chen, K.; Xue, F.; Pan, W. Searching for an Optimal Level of Prefabrication in Construction: An Analytical Framework. *J. Clean. Prod.* **2018**, *201*, 236–245. [CrossRef]
5. Bertram, N.; Fuchs, S.; Mischke, J.; Palter, R.; Strube, G.; Woetzel, J. Modular Construction: From Projects to Products. In *Capital Projects & Infrastructure*; McKinsey & Company: New York, NY, USA, 2019; pp. 1–30.
6. Riggs, W.; Sethi, M.; Meares, W.L.; Batstone, D. Prefab Micro-Units as a Strategy for Affordable Housing. *Hous. Stud.* **2022**, *37*, 742–768. [CrossRef]
7. Lee, N.; Kim, S.J. Factors Influencing the Construction Industry's Shift to Modular Construction. In Proceedings of the 54th ASC Annual International Conference, Minneapolis, MN, USA, 18–21 April 2018; pp. 529–536.
8. Karthik, S.; Sharareh, K.; Behzad, R. Modular Construction vs. Traditional Construction: Advantages and Limitations: A Comparative Study. In Proceedings of the Creative Construction e-Conference, Virtual, 28 June–1 July 2020; pp. 11–19. [CrossRef]
9. Seo, M.; Lin, K.-Y. Environmental Impacts of Prefabricated Construction: CO_2 Emissions Comparison of Precast and Cast-in-Place Concrete Case Study. Ph.D. Thesis, University of Washington, Seattle, WA, USA, 2020.
10. Khahro, S.H.; Memon, N.A.; Ali, T.H.; Memon, Z.A. Adoption of Prefabrication in Small Scale Construction Projects. *Civ. Eng. J.* **2019**, *5*, 1099–1104. [CrossRef]
11. Adindu, C.C.; Yisa, S.N.; Yusuf, S.O.; Makinde, J.K.; Kamilu, A.M. Knowledge, Adoption, Prospects and Challenges of Prefabricated Construction Method in Nigeria: An Empirical Study of North Central Geo-Political Zone. *J. Art Archit. Built Environ.* **2020**, *3*, 1–24. [CrossRef]
12. El-Abidi, K.M.A.; Ofori, G.; Zakaria, S.A.S.; Aziz, A.R.A. Using Prefabricated Building to Address Housing Needs in Libya: A Study Based on Local Expert Perspectives. *Arab. J. Sci. Eng.* **2019**, *44*, 8289–8304. [CrossRef]
13. Wu, G.; Yang, R.; Li, L.; Bi, X.; Liu, B.; Li, S.; Zhou, S. Factors Influencing the Application of Prefabricated Construction in China: From Perspectives of Technology Promotion and Cleaner Production. *J. Clean. Prod.* **2019**, *219*, 753–762. [CrossRef]
14. Steinhardt, D.A.; Manley, K. Exploring the Beliefs of Australian Prefabricated House Builders. *Constr. Econ. Build.* **2016**, *16*, 27–41. [CrossRef]
15. AL-DWAIRI, R.; AL-ALI, O. The role of trust and satisfaction as mediators on users' continuous intention to use mobile payments: Empirical study. *J. Theor. Appl. Inf. Technol.* **2022**, *100*, 3035–3047.
16. Elshafey, A.; Saar, C.C.; Aminudin, E.B.; Gheisari, M.; Usmani, A. Technology Acceptance Model for Augmented Reality and Building Information Modeling Integration in the Construction Industry. *J. Inf. Technol. Constr.* **2020**, *25*, 161–172. [CrossRef]
17. Man, S.S.; Alabdulkarim, S.; Chan, A.H.S.; Zhang, T. The Acceptance of Personal Protective Equipment among Hong Kong Construction Workers: An Integration of Technology Acceptance Model and Theory of Planned Behavior with Risk Perception and Safety Climate. *J. Saf. Res.* **2021**, *79*, 329–340. [CrossRef] [PubMed]
18. Al-Khasawneh, M.; Sharabati, A.-A.; AL-Haddad, S.; Tbakhi, R.; Abusaimeh, H. The Adoption of TikTok Application Using TAM Model. *Int. J. Data Netw. Sci.* **2022**, *6*, 1389–1402. [CrossRef]
19. Son, H.; Park, Y.; Kim, C.; Chou, J.-S. Toward an Understanding of Construction Professionals' Acceptance of Mobile Computing Devices in South Korea: An Extension of the Technology Acceptance Model. *Autom. Constr.* **2012**, *28*, 82–90. [CrossRef]
20. Kaushik, A.K.; Agrawal, A.K.; Rahman, Z. Tourist Behaviour towards Self-Service Hotel Technology Adoption: Trust and Subjective Norm as Key Antecedents. *Tour. Manag. Perspect.* **2015**, *16*, 278–289. [CrossRef]
21. Rezvani, A.; Chang, A.; Wiewiora, A.; Ashkanasy, N.; Jordan, P.; Zolin, R. Manager Emotional Intelligence and Project Success: The Mediating Role of Job Satisfaction and Trust. *Int. J. Proj. Manag.* **2016**, *34*, 1112–1122. [CrossRef]
22. Durdyev, S.; Ihtiyar, A.; Banaitis, A.; Thurnell, D. The Construction Client Satisfaction Model: A PLS-SEM Approach. *J. Civ. Eng. Manag.* **2018**, *24*, 31–42. [CrossRef]
23. Anderson, J.C.; Gerbing, D.W. Structural Equation Modeling in Practice: A Review and Recommended Two-Step Approach. *Psychol. Bull.* **1988**, *103*, 411–423. [CrossRef]
24. Fishbein, M.; Ajzen, I. *Belief, Attitude, Intention and Behaviour: An Introduction to Theory and Research*; Addison-Wesley: Boston, MA, USA, 1975; Volume 27.
25. Davis, F.; Bagozzi, R.; Warshaw, P. User Acceptance of Computer Technology: A Comparison of Two Theoretical Models. *Manag. Sci.* **1989**, *35*, 982–1003. [CrossRef]
26. Venkatesh, V.; Davis, F. A Theoretical Extension of the Technology Acceptance Model: Four Longitudinal Field Studies. *Manag. Sci.* **2000**, *46*, 186–204. [CrossRef]
27. Venkatesh, V.; Morris, M.; Davis, G.; Davis, F. User Acceptance of Information Technology: A Unified Model. *MIS Q.* **2003**, *27*, 425–478. [CrossRef]
28. Davis, F. Technology Acceptance Model for Empirically Testing New End-User Information Systems: Theory and Results. Ph.D. Thesis, Massachusetts Institute of Technology, Cambridge, MA, USA, 1986.

29. Jong, D.; Wang, T. Student Acceptance of Web-Based Learning System. *Inf. Syst. J.* **2009**, *8*, 533–536.
30. Venkatesh, V. Determinants of Perceived Ease of Use: Integrating Control, Intrinsic Motivation, and Emotion into the Technology Acceptance Model. *Inf. Syst. Res.* **2000**, *11*, 342–365. [CrossRef]
31. Venkatesh, V.; Bala, H. Technology Acceptance Model 3 and a Research Agenda on Interventions. *Decis. Sci.* **2008**, *39*, 273–315. [CrossRef]
32. Song, Y.; Wang, J.; Guo, F.; Lu, J.; Liu, S. Research on Supplier Selection of Prefabricated Building Elements from the Perspective of Sustainable Development. *Sustainability* **2021**, *13*, 6080. [CrossRef]
33. Qi, X.; Wang, Y.; Sun, C. Analysis of Factors Influencing the Application of Prefabricated Concrete Structure Based on Structure Equation Modeling. In *ICCREM 2018: Sustainable Construction and Prefabrication*; American Society of Civil Engineers: Reston, VA, USA, 2018; pp. 164–176.
34. Shin, J.; Choi, B. Design and Implementation of Quality Information Management System for Modular Construction Factory. *Buildings* **2022**, *12*, 654. [CrossRef]
35. Tang, X. Research on Comprehensive Application of BIM in Green Construction of Prefabricated Buildings. *IOP Conf. Ser. Earth Environ. Sci.* **2021**, *760*, 012006. [CrossRef]
36. Ai, X.; Meng, Q.; Li, Z.; Liu, W. How to Effectively Prevent Alienation Behavior of Prefabricated Construction Developers: An Optimization Analysis of Regulatory Strategies. *Environ. Sci. Pollut. Res.* **2023**, *30*, 59282–59300. [CrossRef]
37. Zhou, J.; Ren, D. A Hybrid Model of External Environmental Benefits Compensation to Practitioners for the Application of Prefabricated Construction. *Environ. Impact Assess. Rev.* **2020**, *81*, 106358. [CrossRef]
38. Darko, A.; Zhang, C.; Chan, A.P.C. Drivers for Green Building: A Review of Empirical Studies. *Habitat Int.* **2017**, *60*, 34–49. [CrossRef]
39. Li, Z.; Zhang, S.; Meng, Q.; Hu, X. Barriers to the Development of Prefabricated Buildings in China: A News Coverage Analysis. *Eng. Constr. Archit. Manag.* **2021**, *28*, 2884–2903. [CrossRef]
40. Rahardjo, H.A.; Dinariana, D. Towards Green Building with Prefabricated Systems on Flat Development in Indonesia. *Int. J. Eng. Technol.* **2016**, *8*, 1. [CrossRef]
41. Razkenari, M.; Fenner, A.; Shojaei, A.; Hakim, H.; Kibert, C. Perceptions of Offsite Construction in the United States: An Investigation of Current Practices. *J. Build. Eng.* **2020**, *29*, 101138. [CrossRef]
42. Hu, X.; Chong, H.; Wang, X.; London, K. Understanding Stakeholders in Off-Site Manufacturing: A Literature Review. *J. Constr. Eng. Manag.* **2019**, *145*, 03119003. [CrossRef]
43. Gan, X.; Chang, R.-D.; Zuo, J.; Wen, T.; Zillante, G. Barriers to the Transition towards Off-Site Construction in China: An Interpretive Structural Modeling Approach. *J. Clean. Prod.* **2018**, *197*, 8–18. [CrossRef]
44. Wang, Y.; Li, H.; Wu, Z. Attitude of the Chinese Public toward Off-Site Construction: A Text Mining Study. *J. Clean. Prod.* **2019**, *238*, 117926. [CrossRef]
45. Wuni, I.Y.; Shen, G.Q. Fuzzy Modelling of the Critical Failure Factors for Modular Integrated Construction Projects. *J. Clean. Prod.* **2020**, *264*, 121595. [CrossRef]
46. Gumusburun Ayalp, G.; Ay, I. Model Validation of Factors Limiting the Use of Prefabricated Construction Systems in Turkey. *Eng. Constr. Archit. Manag.* **2021**, *28*, 2610–2636. [CrossRef]
47. Nanyam, V.P.S.N.; Sawhney, A.; Gupta, P.A. Evaluating Offsite Technologies for Affordable Housing. *Procedia Eng.* **2017**, *196*, 135–143. [CrossRef]
48. Jiang, W.; Huang, Z.; Peng, Y.; Fang, Y.; Cao, Y. Factors Affecting Prefabricated Construction Promotion in China: A Structural Equation Modeling Approach. *PLoS ONE* **2020**, *15*, e0227787. [CrossRef]
49. Zhou, J.; He, P.; Qin, Y.; Ren, D. A Selection Model Based on SWOT Analysis for Determining a Suitable Strategy of Prefabrication Implementation in Rural Areas. *Sustain. Cities Soc.* **2019**, *50*, 101715. [CrossRef]
50. Davis, F.D. Perceived Usefulness, Perceived Ease of Use, and User Acceptance of Information Technology. *MIS Q.* **1989**, *13*, 319–340. [CrossRef]
51. Ajzen, I. *From Intentions to Actions: A Theory of Planned Behavior*; Springer: Berlin/Heidelberg, Germany, 1985.
52. Sue, V.; Ritter, L. Conducting Online Surveys 2012. *Qual. Quant.* **2006**, *40*, 435–456.
53. Bell, E.; Bryman, A.; Harley, B. *Business Research Methods*; Oxford University Press: Oxford, UK, 2018; ISBN 0198809875.
54. Sekaran, U.; Bougie, R. *Research Methods for Business: A Skill Building Approach*; John Wiley & Sons: Hoboken, NJ, USA, 2016; ISBN 1119165555.
55. Dillman, D.A. *Mail and Internet Surveys: The Tailored Design Method*, 2nd ed.; John Wiley & Sons: Hoboken, NJ, USA, 2007.
56. Liu, Y.; Dong, J.; Shen, L. A Conceptual Development Framework for Prefabricated Construction Supply Chain Management: An Integrated Overview. *Sustainability* **2020**, *12*, 1878. [CrossRef]
57. Wuni, I.Y.; Shen, G.Q.P. Holistic Review and Conceptual Framework for the Drivers of Offsite Construction: A Total Interpretive Structural Modelling Approach. *Buildings* **2019**, *9*, 117. [CrossRef]
58. Bendi, D. Developing an Offsite Readiness Framework for Indian Construction Organanisations. Ph.D. Thesis, University of Salford, Stanford, UK, 2017; pp. 5–9.
59. Jaillon, L.; Poon, C.S. Life Cycle Design and Prefabrication in Buildings: A Review and Case Studies in Hong Kong. *Autom. Constr.* **2014**, *39*, 195–202. [CrossRef]

60. Zhang, T.; Tao, D.; Qu, X.; Zhang, X.; Lin, R.; Zhang, W. The Roles of Initial Trust and Perceived Risk in Public's Acceptance of Automated Vehicles. *Transp. Res. Part C Emerg. Technol.* **2019**, *98*, 207–220. [CrossRef]
61. Kamal, A.; Azfar, R.W.; Salah, B.; Saleem, W.; Abas, M.; Khan, R.; Pruncu, C.I. Quantitative Analysis of Sustainable Use of Construction Materials for Supply Chain Integration and Construction Industry Performance through Structural Equation Modeling (SEM). *Sustainability* **2021**, *13*, 522. [CrossRef]
62. Nadim, W.; Goulding, J.S. Offsite Production: A Model for Building down Barriers: A European Construction Industry Perspective. *Eng. Constr. Archit. Manag.* **2011**, *18*, 82–101. [CrossRef]
63. Kim, Y.; Peterson, R.A. A Meta-Analysis of Online Trust Relationships in E-Commerce. *J. Interact. Mark.* **2017**, *38*, 44–54. [CrossRef]
64. Liao, C.; Chen, J.-L.; Yen, D.C.-C. Theory of Planning Behavior (TPB) and Customer Satisfaction in the Continued Use of e-Service: An Integrated Model. *Comput. Hum. Behav.* **2007**, *23*, 2804–2822. [CrossRef]
65. Ling, C.-P.; Ding, C.G. Evaluating Group Difference in Gender During the Formation of Relationship Quality and Loyalty in ISP Service. *J. Organ. End User Comput.* **2006**, *18*, 38–62. [CrossRef]
66. Liew, Y.S.; Falahat, M. Factors Influencing Consumers' Purchase Intention towards Online Group Buying in Malaysia. *Int. J. Electron. Mark. Retail.* **2019**, *10*, 60–77. [CrossRef]
67. Ihtiyar, A.; Ahmad, F.S. Intercultural Communication Competence as a Key Activator of Purchase Intention. *Procedia—Soc. Behav. Sci.* **2014**, *150*, 590–599. [CrossRef]
68. Sadiq, M.; Adil, M. Ecotourism Related Search for Information over the Internet: A Technology Acceptance Model Perspective. *J. Ecotourism* **2021**, *20*, 70–88. [CrossRef]
69. Akbari, M.; Rezvani, A.; Shahriari, E.; Zúñiga, M.A.; Pouladian, H. Acceptance of 5 G Technology: Mediation Role of Trust and Concentration. *J. Eng. Technol. Manag.* **2020**, *57*, 101585. [CrossRef]
70. Phu, N.H.; Van Nhan, H.; Yen, H.T.P.; Tam, N.Q. Technology Acceptance and Future of Internet Banking in Vietnam. *Foresight STI Gov.* **2018**, *12*, 36–48. [CrossRef]
71. Abd Ghani, M.; Mohd Yasin, N.; Alnaser, F. Adoption of Internet Banking: Extending the Role of Technology Acceptance Model (TAM) with E-Customer Service and Customer Satisfaction Technology Acceptance Model View Project The Influence of Services Marketing Mix (7 Ps.) and Subjective Norms on Custom. *World Appl. Sci. J.* **2017**, *35*, 1918–1929. [CrossRef]
72. Fink, A. *How to Conduct Surveys: A Step-by-Step Guide*; Sage Publications: Thousand Oaks, CA, USA, 2015; ISBN 1506347134.
73. Hair, J.F.; Celsi, M.; Money, A.; Samouel, P.; Page, M. *The Essentials of Business Research Methods*; Routledge: Abington-on-Thames, UK, 2016.
74. Hill, R. What Sample Size Is Enough Pilot Study. *Electron. J. 21st Century* **1998**, *6*, 1–10.
75. Hair, J.F.; Anderson, R.E.; Babin, B.J.; Black, W.C. *Multivariate Data Analysis: A Global Perspective*; Pearson Education: London, UK, 2010; Volume 7.
76. Oke, A.E.; Ogunsami, D.R.; Ogunlana, S. Establishing a Common Ground for the Use of Structural Equation Modelling for Construction Related Research Studies. *Australas. J. Constr. Econ. Build.* **2012**, *12*, 89–94. [CrossRef]
77. Podsakoff, P.M.; MacKenzie, S.B.; Lee, J.-Y.; Podsakoff, N.P. Common Method Biases in Behavioral Research: A Critical Review of the Literature and Recommended Remedies. *J. Appl. Psychol.* **2003**, *88*, 879. [CrossRef] [PubMed]
78. Hui, Y.J.B.; Hamzah, A.-R.; Chen, W. Preventive Mitigation of Overruns with Project Communication Management and Continuous Learning: PLS-SEM Approach. *J. Constr. Eng. Manag.* **2018**, *144*, 4018025. [CrossRef]
79. Gaskin, J. Sem Series Part 3: Exploratory Factor Analysis (EFA). 2016. Available online: https://www.youtube.com/watch?v=VBsuEBsO3U8 (accessed on 12 January 2023).
80. Williams, B.; Onsman, A.; Brown, T. Exploratory Factor Analysis: A Five-Step Guide for Novices. *Australas. J. Paramed.* **2010**, *8*, 1–13. [CrossRef]
81. Kaiser, H.F. An Index of Factorial Simplicity. *Psychometrika* **1974**, *39*, 31–36. [CrossRef]
82. Hair, J.F.; Black, W.C.; Babin, B.J.; Anderson, R.E.; Tatham, R.L. Pearson New International Edition. In *Multivar. Data Anal*, 7th ed.; Pearson Educ. Ltd.: Harlow, UK, 2014.
83. Hu, L.; Bentler, P.M. Cutoff Criteria for Fit Indexes in Covariance Structure Analysis: Conventional Criteria versus New Alternatives. *Struct. Equ. Model. Multidiscip. J.* **1999**, *6*, 1–55. [CrossRef]
84. Hanif, A.; Jamal, F.Q.; Imran, M. Extending the Technology Acceptance Model for Use of E-Learning Systems by Digital Learners. *IEEE Access* **2018**, *6*, 73395–73404. [CrossRef]
85. Cheng, E.W.L. SEM Being More Effective than Multiple Regression in Parsimonious Model Testing for Management Development Research. *J. Manag. Dev.* **2001**, *20*, 650–667. [CrossRef]
86. Preacher, K.J.; Rucker, D.D.; Hayes, A.F. Addressing Moderated Mediation Hypotheses: Theory, Methods, and Prescriptions. *Multivar. Behav. Res.* **2007**, *42*, 185–227. [CrossRef]
87. Hayes, A.F. *Introduction to Mediation, Moderation, and Conditional Process Analysis: A Regression-Based Approach*; Guilford Publications: New York, NY, USA, 2017; ISBN 1462534651.
88. Li, M.; Li, G.; Huang, Y.; Deng, L. Research on Investment Risk Management of Chinese Prefabricated Construction Projects Based on a System Dynamics Model. *Buildings* **2017**, *7*, 83. [CrossRef]
89. Kamali, M.; Hewage, K. Life Cycle Performance of Modular Buildings: A Critical Review. *Renew. Sustain. Energy Rev.* **2016**, *62*, 1171–1183. [CrossRef]

90. Zhai, Y.; Zhong, R.Y.; Li, Z.; Huang, G. Production Lead-Time Hedging and Coordination in Prefabricated Construction Supply Chain Management. *Int. J. Prod. Res.* **2017**, *55*, 3984–4002. [CrossRef]
91. Gefen, D.; Straub, D.W. Consumer Trust in B2C E-Commerce and the Importance of Social Presence: Experiments in e-Products and e-Services. *Omega* **2004**, *32*, 407–424. [CrossRef]
92. Askarany, D.; Smith, M. Diffusion of Innovation and Business Size: A Longitudinal Study of PACIA. *Manag. Audit. J.* **2008**, *23*, 900–916. [CrossRef]
93. Zhou, J.; Li, Y.; Ren, D. Quantitative Study on External Benefits of Prefabricated Buildings: From Perspectives of Economy, Environment, and Society. *Sustain. Cities Soc.* **2022**, *86*, 104132. [CrossRef]
94. Barnes, S.J.; Mattsson, J. Understanding Collaborative Consumption: Test of a Theoretical Model. *Technol. Forecast. Soc. Chang.* **2017**, *118*, 281–292. [CrossRef]
95. Gefen, D.; Karahanna, E.; Straub, D.W. Trust and TAM in Online Shopping: An Integrated Model. *MIS Q.* **2003**, *27*, 51–90. [CrossRef]
96. Jin, R.; Gao, S.; Cheshmehzangi, A.; Aboagye-Nimo, E. A Holistic Review of Off-Site Construction Literature Published between 2008 and 2018. *J. Clean. Prod.* **2018**, *202*, 1202–1219. [CrossRef]
97. Gibb, A.G.F.; Isack, F. Re-Engineering through Pre-Assembly: Client Expectations and Drivers. *Build. Res. Inf.* **2003**, *31*, 146–160. [CrossRef]
98. Brissi, S.G.; Debs, L.; Elwakil, E. A Review on the Factors Affecting the Use of Offsite Construction in Multifamily Housing in the United States. *Buildings* **2021**, *11*, 5. [CrossRef]
99. Tee, R.; Davies, A.; Whyte, J. Modular Designs and Integrating Practices: Managing Collaboration through Coordination and Cooperation. *Res. Policy* **2019**, *48*, 51–61. [CrossRef]
100. Estriegana, R.; Medina-Merodio, J.-A.; Barchino, R. Student Acceptance of Virtual Laboratory and Practical Work: An Extension of the Technology Acceptance Model. *Comput. Educ.* **2019**, *135*, 1–14. [CrossRef]
101. Sheglabo, J.; McGill, T.; Dixon, M. An Investigation of the Factors That Impact the Intention to Adopt and Use MICT in the Libyan Construction Industry. *J. Constr. Dev. Ctries.* **2017**, *22*, 55–74. [CrossRef]

Disclaimer/Publisher's Note: The statements, opinions and data contained in all publications are solely those of the individual author(s) and contributor(s) and not of MDPI and/or the editor(s). MDPI and/or the editor(s) disclaim responsibility for any injury to people or property resulting from any ideas, methods, instructions or products referred to in the content.

Article

Principal Component Analysis (PCA)–Geographic Information System (GIS) Modeling for Groundwater and Associated Health Risks in Abbottabad, Pakistan

Tahir Ali Akbar [1,*], Azka Javed [2], Siddique Ullah [1,*], Waheed Ullah [2], Arshid Pervez [2], Raza Ali Akbar [3,4], Muhammad Faisal Javed [1], Abdullah Mohamed [5] and Abdeliazim Mustafa Mohamed [6,7]

1. Department of Civil Engineering, COMSATS University Islamabad, Abbottabad Campus, Abbottabad 22060, Khyber Pakhtunkhwa, Pakistan
2. Department of Environmental Sciences, COMSATS University Islamabad, Abbottabad Campus, Abbottabad 22060, Khyber Pakhtunkhwa, Pakistan
3. Department of Medicine, Hamad General Hospital, Doha P.O. Box 3050, Qatar
4. Clinical Medicine, Department of Medical Education, Weill Cornell Medicine, Doha P.O. Box 24144, Qatar
5. Research Centre, Future University, New Cairo 11745, Egypt
6. Department of Civil Engineering, College of Engineering, Prince Sattam bin Abdulaziz University, Alkharj 11942, Saudi Arabia
7. Building & Construction Technology Department, Bayan College for Science and Technology, Khartoum 210, Sudan
* Correspondence: drtahir@cuiatd.edu.pk (T.A.A.); siddiqullah142@gmail.com (S.U.)

Citation: Akbar, T.A.; Javed, A.; Ullah, S.; Ullah, W.; Pervez, A.; Akbar, R.A.; Javed, M.F.; Mohamed, A.; Mohamed, A.M. Principal Component Analysis (PCA)–Geographic Information System (GIS) Modeling for Groundwater and Associated Health Risks in Abbottabad, Pakistan. *Sustainability* 2022, 14, 14572. https://doi.org/10.3390/su142114572

Academic Editor: Fernando António Leal Pacheco

Received: 16 August 2022
Accepted: 25 October 2022
Published: 5 November 2022

Publisher's Note: MDPI stays neutral with regard to jurisdictional claims in published maps and institutional affiliations.

Copyright: © 2022 by the authors. Licensee MDPI, Basel, Switzerland. This article is an open access article distributed under the terms and conditions of the Creative Commons Attribution (CC BY) license (https://creativecommons.org/licenses/by/4.0/).

Abstract: Drinking water quality is a major problem in Pakistan, especially in the Abbottabad region of Pakistan. The main objective of this study was to use a Principal Component Analysis (PCA) and integrated Geographic Information System (GIS)-based statistical model to estimate the spatial distribution of exceedance levels of groundwater quality parameters and related health risks for two union councils (Mirpur and Jhangi) located in Abbottabad, Pakistan. A field survey was conducted, and samples were collected from 41 sites to analyze the groundwater quality parameters. The data collection includes the data for 15 water quality parameters. The Global Positioning System (GPS) Essentials application was used to obtain the geographical coordinates of sampling locations in the study area. The GPS Essentials is an android-based GPS application commonly used for collection of geographic coordinates. After sampling, the laboratory analyses were performed to evaluate groundwater quality parameters. PCA was applied to the results, and the exceedance values were calculated by subtracting them from the World Health Organization (WHO) standard parameter values. The nine groundwater quality parameters such as Arsenic (As), Lead (Pb), Mercury (Hg), Cadmium (Cd), Iron (Fe), Dissolved Oxygen (DO), Electrical Conductivity (EC), Total Dissolved Solids (TDS), and Colony Forming Unit (CFU) exceeded the WHO threshold. The highly exceeded parameters, i.e., As, Pb, Hg, Cd, and CFU, were selected for GIS-based modeling. The Inverse Distance Weighting (IDW) technique was used to model the exceedance values. The PCA produced five Principal Components (PCs) with a cumulative variance of 76%. PC-1 might be the indicator of health risks related to CFU, Hg, and Cd. PC-2 could be the sign of natural pollution. PC-3 might be the indicator of health risks due to As. PC-4 and PC-5 might be indicators of natural processes. GIS modeling revealed that As, Pb, Cd, CFU, and Hg exceeded levels 3, 4, and 5 in both union councils. Therefore, there could be greater risk for exposure to diseases such as cholera, typhoid, dysentery, hepatitis, giardiasis, cryptosporidiosis, and guinea worm infection. The combination of laboratory analysis with GIS and statistical techniques provided new dimensions of modeling research for analyzing groundwater and health risks.

Keywords: GIS; PCA; groundwater quality; health risk; solid waste

1. Introduction

Drinking contaminated water is recognized as a major problem in undeveloped countries, although rarely highlighted in developing countries [1,2]. There are two types of water contamination sources in terms of their origins: point and diffuse. Industrial sites, municipalities, agricultural installations, manure storage, and dumping sites are all examples of significant point sources. They are easier to identify and regulate than diffuse (non-point) sources, such as nitrates and pesticides leaching into surface and groundwater due to rainfall, soil infiltration, and surface runoff from agricultural land. The diffuse sources generate significant changes in the pollutant load of water over time [3]. Pakistan is a water-stressed country, where access to fresh drinking water is around 1200 m^3 per capita, and around 70% of Pakistan's population relies on groundwater for domestic purposes [4]. Groundwater in Pakistan is contaminated by toxic chemicals and disease-causing microbiological organisms found in household and industrial wastes and effluent. This increases the local population's diseases such as cholera, typhoid, dysentery, hepatitis, giardiasis, cryptosporidiosis, and guinea worm infection. For example, poor water quality is responsible for 30% of all diseases and 40% of all fatalities in Pakistan [5]. Groundwater is contaminated by both anthropogenic (industrial and household waste) and geological sources (underlying surface composition and topography) [6,7].

The most common source of groundwater pollution is open dumping sites. Most dumping sites are used as final disposal locations for various types of garbage, including solid and liquid municipal trash and industrial wastes. These operate without proper technical controls around the world. These sites frequently operate illegally and without environmental permits, and authorities fail to take preventative measures [8]. The dumping sites that are uncontrolled and poorly designed are regarded as serious potential polluters of groundwater contamination [8,9]. However, the pollution level could be affected by several factors, including the quantity and composition of the leachate, the length of time for which the site has been operational, the soil type, groundwater level, and distance from agricultural land or water sources [10]. These factors contribute to groundwater contamination, which could have many health and environmental consequences, especially in developing countries such as Pakistan. Toxic chemicals and disease-causing microbiological organisms could be present in open dumping sites. The poor water quality, for example, is said to be responsible for many diseases and fatalities in Pakistan [11]. It directly influences the ecosystem, community health, and the economy [12]. In Pakistan, the open disposal of municipal solid waste is a common practice [13]. An open dumping site is a land disposal site that does not protect or shield the territory or domain, and it is subject to open burning and visible to community vectors and scavengers [14]. Open dumping deteriorates the natural environment. Due to open dumping water, land, air, and health are the most affected areas [15]. Pakistan is the world's sixth most populous country, and as a result, it generates a large amount of wastes [16]. According to the [17] Pakistan Environmental Protection Agency, Pakistan produces about 48.5 million tons of solid waste annually, and this figure has been increasing by more than 2% each year. Pakistan estimates that 87,000 tons of solid waste are produced per day, mostly from urban areas [18]. The increase in the civil population is causing an increase in the amount of garbage produced, which must be handled.

Many studies focused on groundwater contamination from geological and anthropogenic sources and the ramifications for public health [19,20]. They concluded that anthropogenic sources constituted a greater risk to groundwater quality compared to geological sources. A study on groundwater quality indicated that unconsolidated deposits could reduce pathogen numbers to acceptable levels as the contaminated water flows through them [21]. Therefore, bacterial contamination occurs when the water table is shallow, or the contaminants directly contact the groundwater through open wells [22]. This means that groundwater's bacteriological contamination in deep wells could be due to direct and localized factors such as poor sanitation around the well and discharge of industrial and domestic wastes. It was reported by [23] that the chemical composition of groundwater could be controlled by many factors, which include the timing of precipitation, groundwater recharging, depth to the groundwater, soil type, presence of organic matter, and moisture content. They concluded that the effect of the combination of these factors could create diverse water types that could change in composition spatially and temporally. Water samples collected from major cities of Pakistan, such as Karachi, Faisalabad, Kasur, Gujrat, and Rawalpindi, revealed that the analyzed samples from these cities were unfit to drink. The major source of water contamination was the anthropogenic source from household and industrial wastes [24]. Similar findings were reported by [25], who found that around four million acre feet (MAF) of industrial and household waste and effluent every year in Pakistan is discharged directly into water bodies, aside from a small proportion of 3% that is brought under treatment.

Several methods, including statistical and Geographic Information System (GIS)-based techniques, were used for groundwater studies [26,27]. Statistical analysis is a powerful and commonly used technique for analyzing water quality data [28]. Various researchers employed statistical techniques such as Principal Component Analysis (PCA), Hierarchical Cluster Analysis (HCA), and Multivariate Analysis of Variance (MANOVA) for water studies [29,30]. PCA is a commonly used technique in water quality studies for reducing the dimensionality of datasets and increasing interpretability, while minimizing information loss. For example, [31] studied the hydrochemistry of groundwater in Syria's Upper Jezireh Basin using PCA and found that PC analysis reduced 20 variables into four PCs (F1, F2, F3, and F4) that explained 81.9% of the total variance. The F1 (47.1%) explained the groundwater mineralization, whereas F2 (17%) showed isotopic enrichment and nitrate pollution. A similar study was conducted in Semarang, Central Java, (Indonesia), using 19 water quality parameters and a PC-based cluster analysis. It was found that anthropogenic factors mainly affected water quality, and PC analysis was able to explain all the significant factors [32]. A PC-based water quality parameters study was also carried out in Abu Dhabi, United Arab Emirates (UAE) [33]. The study concluded that employing multivariate statistical approaches in conjunction with GIS would produce better results compared to a single method for assessing water quality parameters. The main disadvantage of statistical methods is their inability to include the spatial dimension of water quality parameters and nonlinear relationships [34]. GIS-based techniques are effective tools for spatial groundwater modeling. Application of GIS with Global Positioning System (GPS) is helpful for the identification of sample sites and covering spatial dimensions of the water quality parameters [35]. The data obtained from the GPS survey might be transferred into GIS software for additional modeling and analysis. For example, groundwater modeling in Central Antalya, Turkey, was done using GIS and the Analytic Hierarchy Process (AHP) [36]. Some researchers used hybrid models for groundwater mapping, such as Ground Water Potential (GWP) with Random Subspace (RS), Multilayer Perception (MLP), Naïve Bayes Tree (NBTree), and Classification and Regression tree (CART) algorithms [37–40].

Groundwater quality maps were created using novel ensemble Weights-of-Evidence (WoE) with Logistic Regression (LR) and Functional Tree (FT) models in the Ningtiaota region of Shaanxi Province, China [41]. Overall, all three models performed well for groundwater spring potential evaluation. However, the FT model's prediction capability was better compared to those of other models [41]. A similar study was conducted by [42], who developed a ground potential map of river effluent across rocky terrain using a statistical–GIS-based technique. The river effluent was divided into three potential zones. The ground potential map was found to be closer to the field observations. The current study employed a novel integrated statistical–GIS-based technique to estimate groundwater quality parameter exceedance levels and related health risks in Abbottabad.

Abbottabad is a major city located along the China–Pakistan Economic Corridor (CPEC) route. The city experienced significant anthropogenic and developmental activities in the past three decades. Furthermore, there was a large influx of people from surrounding districts and cross-border migration from Afghanistan, resulting in the degradation of local water infrastructure and water quality. Abbottabad's groundwater is regarded as the third worst for drinking purposes in Khyber Pakhtunkhwa (KPK) province, Pakistan. Therefore, it is critical to identify groundwater contamination and its effects on public health to improve the local community's health and well-being. For this purpose, two union councils (Mirpur and Jhangi) were selected based on population and level of discharge from homes and industrial sources. Based on population, the research aimed to use PCA and GIS to model spatial exceedance levels of groundwater quality parameters and related health risks. The specific objectives of this study were three-fold: (i) assess the quality of groundwater parameters using laboratory analysis, (ii) model the groundwater data using PCA to find major PCs, and, finally, (iii) the geospatial modeling of the exceedance parameters and related health risks.

2. Materials and Methods

2.1. Study Area

The study area includes Jhangi and Mirpur union councils of Abbottabad district located in the KP province of Pakistan, as presented in Figure 1a,b. The study sites are located at the base of the Himalayan range, in the active monsoon zone. It is surrounded by the Sarban hills and has a cool temperate climate for most of the year. The average annual maximum and minimum temperatures in the Abbottabad district are 22.76 °C and 11.41 °C, respectively, with an annual precipitation of 1366 mm. The temperature drops below 0 °C during the winter, with significant snowfall on the surrounding forest-covered hills. The district of Abbottabad's average relative humidity was recorded at 56% [43]. The elevation range of the whole area is from 1191 m to 2626 m and its total area is 1967 km^2 [44].

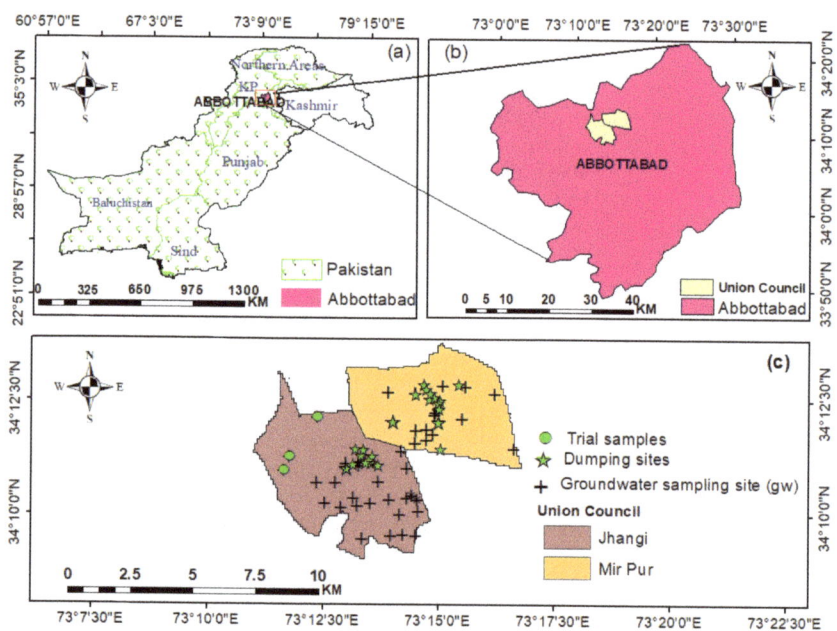

Figure 1. Location of: (**a**) Abbottabad in Pakistan; (**b**) union councils in Abbottabad; (**c**) groundwater-sampling sites (gw) and dumping sites in union councils.

The groundwater data were collected from 41 sampling sites. Boring wells in Mirpur and Jhangi union councils were used as sampling locations (Figure 1). The groundwater samples were collected from the entire study area, near and far from the dumping sites, shown as groundwater sampling sites (gw) in Figure 1. The depth of the groundwater wells varied from 80 ft to 100 ft on average. Groundwater in the study area is contaminated by anthropogenic (industrial and household waste) and geological sources (underlying surface composition and topography). The common sources of contamination are industrial sites, domestic effluent, and open dumping sites. The collection of solid waste was accomplished in a few selected locations under the jurisdiction of Tehsil Municipal Administration (TMA) of Abbottabad. The solid waste is mostly dumped in open dumping sites by most communities living in Abbottabad city. These scattered open dumping sites affect the beauty of the city and they create: (i) bad smell; (ii) air pollution; (iii) soil contamination; (iv) water pollution; (v) poor aesthetic, and (vi) health risks. Open dumping sites could cause great damage to the ecosystem by releasing toxic compounds such as dioxins and furans into the air [45].

Until now, no study was done to model the impact of solid waste and natural factors on groundwater wells in Pakistan and in Abbottabad, specifically. Therefore, considering the importance of the research and study area, the Jhangi and Mirpur union councils of Abbottabad were selected. According to the 2017 census, the population of Abbottabad is 1.33 million. The population of Jhangi is 18,037, and for Mirpur, the total population is 46,206 [44]. Both union councils have residential neighborhoods with narrow streets, open drains, and small shops. The residents normally select open plots, open spaces, grounds, and the vicinity of shopping areas for dumping solid waste. These solid wastes include plastic bags, papers, clothes, cans, plastic bottles, kitchen garbage, etc.

2.2. Data

The groundwater data were collected from 41 sampling sites. Boring wells in Mirpur and Jhangi union councils were used as sampling locations (Figure 1). The data include 15 water quality parameters, which are pH, Total Dissolved Solids (TDSs), Electrical Conductivity (EC), Dissolved Oxygen (DO), Colony Forming Unit (CFU), Turbidity (Tur), Arsenic (As), Mercury (Hg), Lead (Pb), Cadmium (Cd), Calcium (Ca), Zinc (Zn), Potassium (K), Sodium (Na), and Iron (Fe). The samples were also collected near the three dumping sites in both union councils. Two dump sites were in Mirpur and one in the Jhangi union council area. The data for all these parameters were tested in the laboratory by dividing them into three groups, which are: (i) physicochemical analysis; (ii) microbiological analysis; and (iii) heavy metal analysis. The data obtained based on three types of analyses for groundwater, were used for PCA and exceedance. For PCA, this dataset was normalized by using their maximum values. The normalized dataset was used to obtain PCA [46,47]. The Landsat 8 Operational Land Imager (OLI) satellite image (30 m spatial resolutions) extracted the built-up area, prepared by [48], on 25 May 2017, and overlaid on classified parameter exceedance maps. The OLI is a sensor mounted on the Landsat 8 satellite that collects non-thermal data for LULC classification. The entire scene's (34.1688° N, 73.2215° E) cloud cover for the Landsat 8 data was 13% but it was near zero over our studied region. The data were downloaded from the United State Geological Survey website (https://earthexplorer.usgs.gov). This provided support for analyzing the potential population of built-up areas that could have exposure to health risks due to parameter exceedance.

2.3. Methods

The schematic diagram of the methods adopted in this study is presented in Figure 2. The main components of the methodology are: (i) field survey and sampling; (ii) laboratory analysis; (iii) PCA; and (iv) GIS-based modeling. From Figure 2, it is obvious that the field survey was accomplished using GPS Essentials application. This application was used for collecting geographic coordinates (GCS_WGS_1984) of open dumping and groundwater sampling sites. The groundwater samples, collected in the field, were used for laboratory analysis. The laboratory analysis was accomplished for seven physicochemical parameters, one microbiological parameter, and seven parameters related to heavy metals. Figure 2 provides further information on the data obtained from laboratory analysis, which was used for: (i) preparation of graphs against WHO standards; (ii) normalization of data and its application for PCA to obtain five PCs; (iii) exceedance and GIS modeling; (iv) production of maps with exceedance levels; and (v) interpretation of maps for health risks in the population of the study area. The details of all methods are provided in subsequent sections.

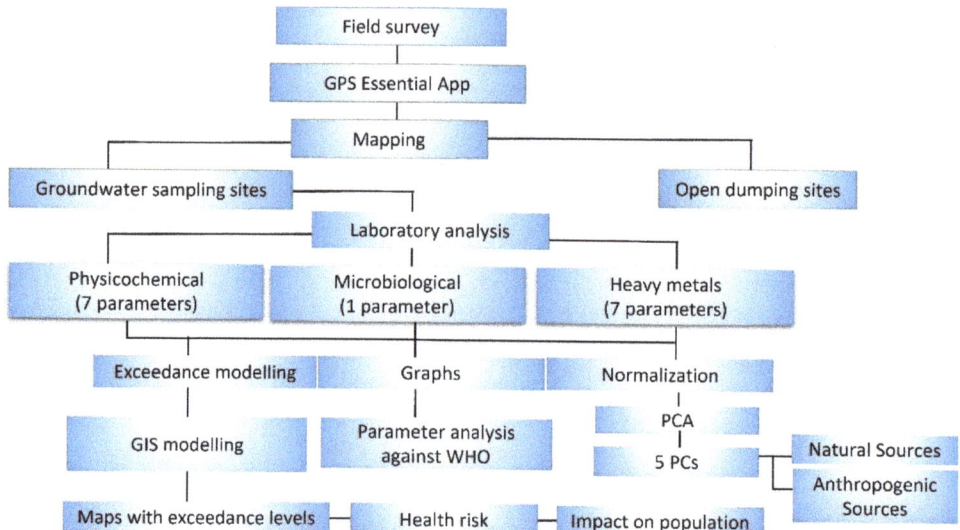

Figure 2. Schematic diagram of the methods.

2.3.1. Field Survey and Sampling

The groundwater samples were collected from 41 sampling sites, as discussed in Section 2.1. All these sampling sites, i.e., "gw", are shown in Figure 1c. The gw represents groundwater well samples collected during the field survey. The groundwater was collected from bore wells in disinfected bottles after siphoning for 5–15 min [49]. The samples were collected once in a month from January to March 2021. The winter season was chosen for sampling because temperature plays an important role in sample preservation, and microbial growth is reduced at low temperatures. Three samples (gw1, gw2, and gw3) from a depth of 90 feet were used as control samples. The average distance between the two sample points was approximately 0.5 km. The distance between dumping sites was approximately 1.6 km. The dump sites were close to residential areas. These samples were brought to the laboratory, stored in plastic polythene containers, and cleaned with nitric acid. These samples were used to analyze physico-chemical, microbiological, and heavy metals for 15 parameters. In physicochemical analysis, seven parameters were tested: pH, TDS, EC, DO, Tur, Ca, and Na. CFU was considered for microbiological analysis. In heavy metals, seven parameters were analyzed: Cd, As, Hg, Zn, Pb, K, and Fe.

2.3.2. Laboratory Analysis

For physicochemical analysis, pH, TDS, and EC of the groundwater samples were tested using a multi-parameter probe called HANNA HI9828, while DO and Tur were obtained using the DO meter and turbidity meter, respectively. The physicochemical parameters such as Na and Ca were analyzed through Atomic Absorption Spectroscopy. The serial dilution technique was used for microbiological analysis. Heavy metals were detected and measured by Atomic Absorption Spectroscopy (AAS) [50]. The details about the instruments and Quality Assurance and Quality Control (QA/QC) used in the current study's laboratory analysis are given in Table 1.

Table 1. The instruments and QA/QC used in the current study's laboratory analysis.

Parameters	Instrument Detection Limit
As	1 (mg/L)
Hg	4.2 (mg/L)
Pb	0.45 (mg/L)
Cd	0.028 (mg/L)
Fe	0.1 (mg/L)
K	0.043 (mg/L)
Zn	0.018 (mg/L)
Ca	0.092 (mg/L)
Na	0.012 (mg/L)
pH	0–14
EC	99.9 µs/cm
TDS	50 ppm
Turbidity	0.05 NTU/FNU

2.3.3. Principal Component Analysis (PCA)

PCA was accomplished for groundwater by using SPSS 23 (Statistical Package for the Social Sciences) is a software program used by researchers in various disciplines for quantitative analysis of complex data. It was developed by International Business Machines Corporation (IBM) located in Armonk, New York, USA [47]. PCA was used to the extract main PCs using an eigenvalue of 1 as a cut-off [47]. The varimax normalized rotation derived the loading values for all the parameters under the major PCs [15,16]. Each PC comprises a set of associated parameters with positive and negative loading values, which are used to interpret the primary processes involved in analyzing and characterizing water quality. The variance was obtained for all PCs, and the cumulative variance (%) was also achieved. In statistics, variance is used to measure dispersion, which indicates the distances of numbers from their average values [51]. The loading values of all PCs were categorized into three classes, i.e., strong > 0.75, 0.75 > moderate > 0.5, and 0.5 > weak > 0.4, with parameter loading values less than 0.40 not being considered because of their lesser significance. These PCs were used to interpret the natural and anthropogenic processes involved in water. The results for all the parameters, obtained through laboratory analysis, were normalized by dividing their values by their respective maximum values [47,52]. These sets of normalized values were used in PCA.

2.3.4. GIS-Based Modeling

An exceedance equation was developed based on original values obtained from the laboratory analysis and WHO standard parameter values for groundwater wells, as given in Table 2. The parameter exceedance values obtained from the exceedance equation were imported into GIS software (ArcMap 10.5) and these were used for modeling exceedance in GIS. For modeling, the Inverse Distance Weighting (IDW) interpolation technique was used to develop parameter exceedance maps. IDW is an interpolation technique in which interpolation is estimated based on the values at neighboring locations primarily weighted by distance from the interpolation location. The IDW technique was used in various studies [53–55]. With this modeling, the exceedance values for all the study area locations could be obtained in case of no data. The IDW is better than kriging because it is easy to understand, and it does not suffer outliers like Kriging. Ref. [56] found the accuracy of the measured and estimated arsenic level using IDW and Kriging techniques. They found that the correlation coefficient between estimated and measured arsenic was greater with IDW compared to Kriging. Ref. [57] showed that IDW is superior to Kriging in estimating whole landfill methane flux. Ref. [55] believed that ordinary Kriging is not accurate, since it requires uniform distribution, which can rarely be met. Five classes were generated in the maps of all parameters. These classes were interpreted as: (i) Level 1 exceedance; (ii) Level 2 exceedance; (iii) Level 3 exceedance; (iv) Level 4 exceedance; and (v) Level 5 exceedance. The exceedance pattern for all parameters was displayed in the maps based on these classes. The built-up area was extracted from the LULC classified image mentioned in Section 2.2. The Land Use/Land Cover (LULC) map was prepared using the artificial intelligence technique of Support Vector Machine (SVM) in Envi 5.3 software by [48]. Finally, these GIS maps were used to study the health risks and impacts of exceeded parameters on the population residing in the study area.

Table 2. WHO standards for drinking water [58].

Parameters	WHO Drinking Water Standards
Physicochemical Parameters	
pH	6.5–8.5
TDS (ppm)	1000
Tur (NTU)	<5
EC (μS cm^{-1})	400
DO (ppm)	6.5–8
Na (mg/L)	200
Ca (mg/L)	100
Microbiological Parameter	
Total coliforms	0 (CFU/mL)
Heavy Metals	
As (mg/L)	0.05
Hg (mg/L)	0.001
Pb (mg/L)	0.01
Cd (mg/L)	0.003
Zn (mg/L)	3
K (mg/L)	12
Fe (mg/L)	0.3

3. Results and Discussion

3.1. Results of Laboratory Analysis for Parameters

The laboratory analysis indicated that six parameters (i.e., pH, Tur, Na, Ca, Zn, and K) were within the permissible limit of WHO guidelines. It was found that nine parameters (i.e., DO, TDS, CFU, As, Hg, Pb, Cd, Fe, and EC) exceeded WHO guidelines. DO and TDS were according to WHO guidelines for 37 wells, and these were exceeded only for four wells, gw12, gw22, gw31, and gw34. The major exceeded parameters were CFU, As, Hg, Pb, Cd, Fe, and EC. The CFU range was from 1000 CFU/mL to 126,000 CFU/mL for 35 wells. It was $\geq$30,000 CFU/mL for gw10, gw14, gw22, gw38, and gw40. Tables 3 and 4 indicate the values of parameters for 41 groundwater wells.

Table 3. Physiochemical parameters for groundwater wells.

Locations	pH (0–14)	Turbidity NTU	TDS (ppm)	EC (μS cm^{-1})	Sodium (mg/L)	Calcium (mg/L)
gw1	7.03	0.48	120	226	11.12	145.8
gw2	7.26	0.71	112	213	9.62	134.5
gw3	6.73	0.51	119	226	11.73	136.2
gw4	8.14	1.03	103	192	9.63	100.3
gw5	7.61	0.69	52	98	8.99	23.82
gw6	7.94	1.69	48.3	97	5.71	76.28
gw7	7.51	1.84	227	455	12.44	71.39
gw8	7.01	1.51	46.7	93	3.03	60.31
gw9	7.78	1.79	272	546	17.73	54.49
gw10	7.34	2.8	284	568	9.83	78.97
gw11	7.32	2.88	316	633	0.99	73.49
gw12	6.93	2.65	572	1144	15.34	114
gw13	7.56	2.15	349	697	11.74	97.94
gw14	7.42	1.67	312	624	10.77	85.13
gw15	7.11	2.04	450	901	11.96	95.09
gw16	7.02	2.39	364	730	7.88	80.47
gw17	7.84	1.87	385	770	9.57	82.72
gw18	7.14	1.64	272	544	6.35	70.23
gw19	7.4	1.91	365	730	9.17	80.22
gw20	7.17	2.58	325	651	8.45	74.22
gw21	7.47	1.34	265	531	7.28	49.79
gw22	7.07	4.4	633	1266	14.28	76.85
gw23	7.97	3.01	393	786	24.95	45.03
gw24	7.17	1.64	278	556	5.82	71.22
gw25	6.99	0.92	334	662	13.77	111.21
gw26	7.05	1.24	377	744	5.15	134.32
gw27	6.6	0.8	256	514	22.35	98.17
gw28	6.96	3.45	424	849	10.84	76.92
gw29	6.95	3.26	320	641	18.78	82.73
gw30	6.91	2.76	72	146	7.84	45.29
gw31	6.88	1.73	511	1019	17.52	23.56
gw32	6.39	0.8	476	938	11.23	39.48
gw33	6.76	3.78	391	770	8	44.76
gw34	6.87	2.74	512	1009	4.89	62.87
gw35	6.84	0.65	475	936	4.52	77.21
gw36	8.49	2.34	85	160	6.159	39.81
gw37	8.58	3.02	91	161	5.829	48.58
gw38	8.45	2.15	114	216	12.21	76.47
gw39	8.5	2.18	90	173	4.006	46.32
gw40	8.16	2.4	98	188	9.776	69.41
gw41	8.36	2.08	101	198	11.43	74.21

Table 4. Heavy metals and microbial concentrations for groundwater wells.

Locations	K (mg/L)	Pb (mg/L)	Fe (mg/L)	Zn (mg/L)	Cd (mg/L)	Hg (mg/L)	As (mg/L)	CFU
gw1	1.73	0.223	2.7	0.421	0.066	1.851	17.7	0
gw2	0.89	0.25	1.2	0.194	0.017	1.821	13	0
gw3	1.41	0.418	3.04	0.058	0.021	1.058	13.5	5000
gw4	1.2	0.535	1.59	0.061	0.121	1.653	15.83	7000
gw5	0.64	0.274	2.68	0.226	0.085	2.558	13.59	3000
gw6	1.74	0.545	1.96	0.24	0.063	1.488	9.97	2000
gw7	3.55	0.506	2.52	0.247	0.104	2.16	12.63	1000
gw8	0.62	0.326	2.76	0.202	0.082	2.358	12.04	6000
gw9	2.3	0.481	3.93	0.246	0.007	4.012	8.63	3000
gw10	3.34	0.603	3.67	0.188	0.088	2.178	10.93	30,000
gw11	1.88	0.564	4.76	0.19	0.031	1.78	9.62	7000
gw12	3.65	0.624	3.76	0.185	0.023	2.045	9.06	3000
gw13	2.33	0.828	4.86	0.129	0.068	2.469	7.28	1000
gw14	1.68	0.643	4.12	0.154	0.019	2.753	9.38	30,000
gw15	3.7	0.706	2.24	0.215	0.036	2.113	13.43	21,000
gw16	2.38	0.656	1.95	0.229	0.093	2.442	11.74	3000
gw17	3.29	0.754	2.67	0.099	0.135	2.482	10.71	10,000
gw18	2.21	0.775	4.37	0.057	0.041	2.339	11.89	6000
gw19	2.25	0.678	0.79	0.214	0.101	3.556	6.72	7000
gw20	2.18	0.814	1.59	0.084	0.081	2.999	12.7	1000
gw21	2.34	0.738	3.47	0.197	0.001	3.363	14.38	4000
gw22	3.02	0.84	3.32	0.09	0.027	2.209	16.02	32,000
gw23	2.34	0.958	2.6	0.149	0.115	3.089	13.46	21,000
gw24	1.74	0.85	4.98	0.136	0.121	3.113	10.99	0
gw25	1.06	0.739	1.03	0.821	0.133	1.789	3.28	1000
gw26	1.88	0.278	3.67	1.021	0.05	1.014	15.14	6000
gw27	1.92	0.442	0.6	0.527	0.071	1.212	3.87	0
gw28	3.7	0.466	3.66	1.188	0.001	0.778	16.85	14,000
gw29	1.85	0.241	2.54	0.565	0.138	1.719	12.39	5000
gw30	1.35	0.332	3	1.133	0.067	0.791	6.58	2000
gw31	0.79	0.675	0.98	1.068	0.04	1.938	14.12	5000
gw32	1.81	0.71	3.39	0.555	0.038	1.561	17.12	1000
gw33	1.86	0.507	4.46	0.646	0.016	0.96	0	5000
gw34	3.44	0.343	4.26	0.252	0.084	1.995	5.59	3000
gw35	2.22	0.517	2.31	0.963	0.044	0.564	5.58	0
gw36	2.305	0.571	0.016	0.222	0.129	1.242	2.746	3000
gw37	2.179	0.568	0.084	1.642	0.143	1.756	3.585	0
gw38	3.169	0.245	0.602	0.366	0.366	13.85	2.981	126,000
gw39	1.907	0.504	0.112	0.838	0.145	3.219	3.337	4000
gw40	2.751	0.323	0.217	0.293	0.142	4.977	0.387	50,000
gw41	1.585	0.378	0.122	1.365	0.129	3.455	0.361	8000

The value of Hg ranges from 0.558 mg/L to 13.849 mg/L for 41 wells. The highest value of Hg (i.e., 13.849 mg/L) was for gw38. gw40 also showed a higher value of 4.976 mg/L. The minimum limit of Pb was 0.213 mg/L, its maximum limit was 0.948 mg/L for 40 wells, and its value was gw23. The range of Cd was from 0.004 mg/L to 0.363 mg/L for 39 wells, and its highest value was for gw38. The range was from 0.3 mg/L to 4.68 mg/L for Fe, and its highest value was for gw24. EC was from 155 $\mu S\ cm^{-1}$ to 966 $\mu S\ cm^{-1}$, and its highest value was for gw22. The exceeded values of exceeded parameters were utilized for GIS modeling. The study findings reveal that there are two types of groundwater well pollution sources: natural and anthropogenic. Open dumping sites are the most obvious anthropogenic source, as evidenced by high levels of groundwater parameter values found in the samples collected from groundwater wells near the open dumping sites. In comparison to non-dumping areas, samples collected from groundwater wells near open dumping sites contained higher levels of As, Hg, Pb, CFU, and Cd. Similar results were reported by [56], who assessed groundwater parameters such as As, Hg, and CFU, and found that groundwater sites near dumping sites had higher levels of pollution compared to non-dumping sites. Cd contamination might come from plastics and electronic waste, whereas Pb contamination could be from batteries and old lead-based paints [59]. Groundwater contamination was also discovered in non-dumping sites in both union councils, which might be caused by natural sources such as geological formation and rock weathering.

3.2. PCA

Five PCs were obtained based on PCA. These PCs were PC-1, PC-2, PC-3, PC-4, and PC-5, which showed variance of 27.663%, 18.592%, 12.992%, 9.439%, and 7.292%, respectively, as given in Table 5. The cumulative variance for all these PCs was 75.978%. PC-1 revealed that four parameters (i.e., CFU, Hg, Cd, and pH) were correlated with each other. All parameters of PC-1 were strongly positively loaded, except pH, which was moderate with a positive value. PC-1 could be related to health risks due to toxic elements such as CFU, Hg, and Cd. Anthropogenic activities might include pollution related to heavy metals and microbial contamination in groundwater. From PC-1, it is obvious that heavy metals such as CFU, Hg, and Cd could contribute to pollution. Their values were found well above WHO guidelines according to laboratory analysis. This could be related to residential and commercial wastes from all dumping sites. Other factors that could contribute to the presence of toxic elements in groundwater wells include agricultural runoff and household effluents, which might have contributed to high CFU levels in groundwater wells. Fecal pollution in open dumping sites might also be linked to human and bird feces [60]. The high levels of Hg and Cd in the study area could be attributed to geological and soil formation. The exceeded Hg might cause diseases such as kidney damage, insomnia, and memory loss [61]. The higher exposure to Hg could also lead to death [62]. Cd is carcinogenic, and it might cause death due to multiorgan failure [60]. pH might be affected by chemicals in the water, determining the solubility and biological availability of chemical components such as nutrients (i.e., phosphorus, nitrogen, and carbon) and heavy metals [63]. PC-1 also exhibited a strong positive value of CFU, which might be an indicator of fecal pollution related to fecal coliforms [6]. Fecal coliform is a sub-group of coliform bacteria, which live and reproduce in the gut of humans and other warm-blooded animals. The CFU showed very high values according to laboratory analysis, except for a few sampling sites. The study area residents are using this water for drinking and consumptive purposes. The consumption of such water could be extremely harmful, and it could lead to diseases such as diarrhea, cholera, typhoid, paratyphoid, hepatitis A, dermatitis, and enteric fever [61]. Overall, PC-1 could be an indicator of health risks due to CFU, Hg, and Cd.

Table 5. PCs with loading values for 15 groundwater quality parameters.

Var.	PC-1	PC-2	PC-3	PC-4	PC-5
CFU	0.898				
Hg	0.913				
Cd	0.825				
pH	0.590			0.488	
EC		0.880			
TDS		0.879			
K		0.711			
Tur		0.599		0.463	
Fe		0.537	0.415		
Zn			−0.809		
DO			0.791		
As			0.554		
Ca				−0.808	
Pb		0.420		0.528	
Na					0.859
Variance (%)	27.663	18.592	12.992	9.439	7.292
Cumul. (%)	27.663	46.255	59.246	68.686	75.978

According to Table 5, PC-2 revealed that three parameters (i.e., EC, TDS, K) positively correlated with each other. High TDS levels might be related to calcium and sodium found in natural sources, sewage, and urban runoff [64]. EC could be related to the dissolution of minerals in groundwater [65]. Fe could be associated to weathering of rocks and disposal of toxic effluents from open dumping sites [43]. The correlated Tur in PC-2 could be related to urban runoff from the road [66]. PC-2 could indicate natural processes due to EC, TDS, K, and Tur. PC-3 in Table 5 indicates that DO, As, and Fe are positively correlated, while Zn is found to be strong with a negative value. Long-term exposure to As through drinking water could cause arsenic poisoning, leading to skin cancer [67]. According to WHO, the biggest threat to public health is groundwater contamination due to As which is available in organic and inorganic forms. Organic arsenic is found in less toxic seafood, whereas inorganic As is found in groundwater and is highly toxic and carcinogenic. Acute arsenic poisoning could cause diarrhea, vomiting, abdominal pain, and muscle cramping. In some cases, it might cause death [68]. Based on all these facts, PC-3 could be an indicator of health risks due to As pollution. The other strongly positively correlated parameter is DO. DO is a crucial metric in determining water quality as it represents the physical and biological processes in the water [69]. Mine wastes and leachate water from overburdened dumps of coal mines in the vicinity might cause the elevated Fe levels found in the groundwater. From PC-3, it could be interpreted that DO could determine the solubility and toxicity of As, as it was very high for all gws, which could be a major health threat for the population living in the study area. PC-3 might be the indicator of natural processes due to the presence of Zn and could be found in trace amounts in almost all igneous rocks. The most prevalent zinc ores are sulfides such as sphalerite and wurtzite [70].

According to Table 5, PC-4 revealed that pH, Tur, Ca, and Pb are correlated. pH and Tur are found to be loaded as weak (i.e., less than 0.5) with positive values. While Ca is loaded as strong (i.e., >0.75) with a negative value. Clay minerals, including montmorillonite, illite, and chlorite, could be responsible for the high Ca, Na, and Mg concentrations in groundwater. Most calcium ions in groundwater come from limestone, dolomite, gypsum, and anhydrite leaching, although they could also come through the cation exchange process [71]. Therefore, PC-4 indicates natural processes due to pH, Tur, and Ca.

PC-5 shows that Na is loaded as strong (i.e., >0.75) with a positive value. Clay minerals, including montmorillonite, illite, and chlorite, are responsible for the high Ca, Na, and Mg concentrations in groundwater [72]. Therefore, PC-5 could be an indicator of natural processes due to Na.

3.3. GIS Modeling for Exceeded Parameters

The exceedance equation (Equation (1)) for groundwater (GW) was developed based on measured values and WHO standards for groundwater parameters.

$$\text{Exceedance (GW)} = \text{Measured values (GW)} - \text{WHO Standards (GW)} \quad (1)$$

where exceedance (GW) represents the groundwater quality parameters that were exceeded, measured values (GW) are the values obtained in the laboratory, and WHO Standards (GW) are the groundwater parameter values recommended by WHO, as provided in Table 1.

Furthermore, the parameters of groundwater wells that contributed to exceedance levels (Level 1 to Level 5) are presented in the form of spatial distribution maps (Figures 3–7). Figure 3 depicts the results for As in the union councils of Jhangi and Mirpur. The wells gw1, gw2, gw3, gw4, gw5, gw20, gw21, gw22, gw23, and gw32 contributed to Level 5 As in the Jhangi union council area (Figure 3). The wells that contributed to Level 4 in Jhnagi are gw10, gw11, gw14, and gw17, while gw12 contributed to exceedance Level 3. The results for Jhangi indicate that Level 2 exceedances are attributed to groundwater wells gw9, gw13, and gw19, while Level 1 exceedances are attributed to gw33, gw34, gw35, gw36, gw40, and gw41, as shown in Figure 3. The highest As (Level 5) was found in both residential and non-residential areas, which showed the contribution of both anthropogenic and natural factors. The anthropogenic factors could be As-containing wastewater from residential and agriculture sources. In contrast, natural factors such as geological formations (e.g., sedimentary deposits/rocks, volcanic rocks, and soils) might have contributed to the high As level in Jhangi.

In Mirpur, gw7, gw8, gw15, gw16, gw22, gw25, gw26, gw28, and gw29 contributed to Level 5 exceedance of As. As shown in Figure 3, the groundwater wells in Mirpur that contributed to Level 4 are gw10, gw11, and gw14, while gw17 contributed to Level 3 (Mirpur). Level 2 is attributed to gw9, gw13, and gw19, while Level 1 is attributed to gw33, gw34, gw35, gw36, gw37, gw40, and gw41 in the Mirpur area.

The analysis revealed that some groundwater wells in Jhangi union council, such as gw5, gw40, and gw41, were located near the open dumping sites, whereas gw18, gw29, gw30, gw38, and gw39 were located near the open dumping sites in Mirpur union council areas. The level of exceedance caused by open dumping sites in Mirpur was higher than in Jhangi union council, as shown in Figure 3. For example, exceedance levels in Mirpur near dumping sites were at Levels 4 and 5, whereas Level 1 was observed in Jhangi union council areas. The high level of As in Mirpur in the groundwater wells such as gw7, gw8, gw15, gw18, gw29, and gw30 could be due to the potential leachate from the nearby dumping sites. The study findings reveal that, as we moved away from the dumping sites, the level of groundwater exceedance decreased. Similar findings were reported by [73], who found that those concentrations of pollutants decreased significantly within 324 m of the dumping sites. The total distance travelled by pollutants is proportional to the material of the aquifer. This groundwater contamination could endanger local groundwater resource users, as well as the natural environment and human health [9].

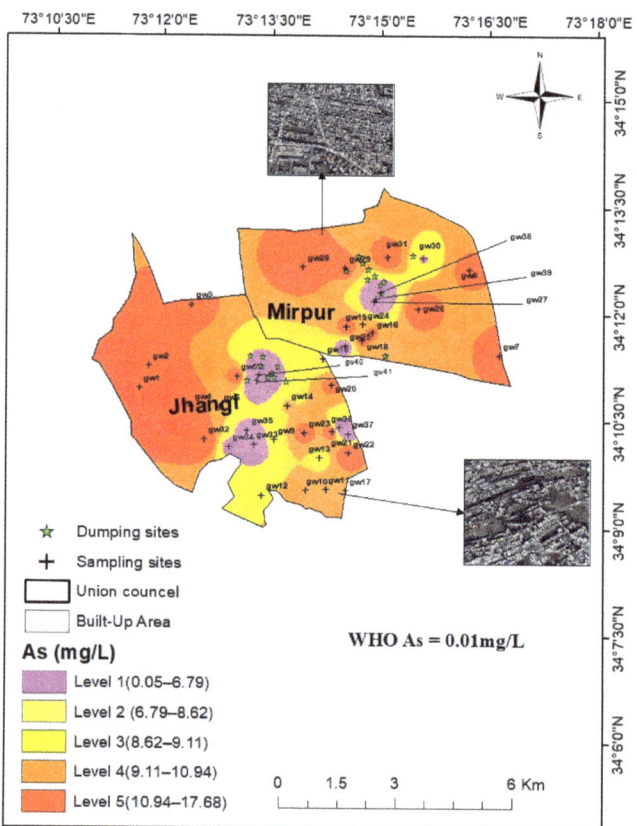

Figure 3. Exceedance modeling for As in groundwater.

Figure 4 depicts the Pb results in the Jhangi union council. The findings for Jhangi revealed that gw1, gw2, gw3, gw5, gw35, gw42, and gw41 contributed to Pb Level 1. As shown in Figure 4, the wells in Jhangi that contributed to Level 4 were gw12, gw14, gw20, and gw33, while Level is contributed by gw10, gw11, gw27, and gw38, and the wells in Jhangi that contributed to Level 2 were gw9, gw34, and gw36. The results indicate that wells that contributed to Level 5 are gw13, gw17, gw21, gw22, and gw24 in the areas of Jhangi. The Pb services line pipes for water distribution near the residential area could be the reason for a high Pb concentration in groundwater. In Mirpur, gw15, gw25, and gw26 contributed to Level 5 exceedance of Pb. The wells in Jhangi that contributed to Level 4 were gw15, gw16, and gw31, while Level 2 was attributed to gw7, gw29, and gw39, as shown in Figure 4. Level 1 exceedance of Pb was attributed to gw8, gw27, gw30, and gw38 in the Mirpur area.

The study results indicate that some groundwater wells in Jhangi union council, such as gw5, gw40, and gw41, were located near open dumping sites, whereas gw31 and gw18 were located near dumping sites in Mirpur union council. The level of exceedance caused by open dumping sites in Mirpur was higher compared to Jhangi union council, as shown in Figure 4. Pb exceedance in groundwater near a dumping site in Mirpur, for example, was at Level 4, while Level 1 exceedance was discovered in Jhangi union council. The high levels of Pb in Mirpur groundwater wells such as gw31 and gw18 could be caused by potential leachate from a nearby dumping site. The Pb present in hospital and household waste dumping sites might contaminate groundwater through precipitation and surface runoff in the study area. Several studies reported similar findings concerning groundwater contamination from open waste dumping sources [74–76], leading to the general conclusion that heavy metals such as Pb pose a significant risk to local groundwater resource users, as well as the natural environment and human health [9]. The health consequences include constipation, abdominal pain, tiredness, headache, loss of appetite, memory loss, and pain or tingling in the hands and/or feet, which are all symptoms.

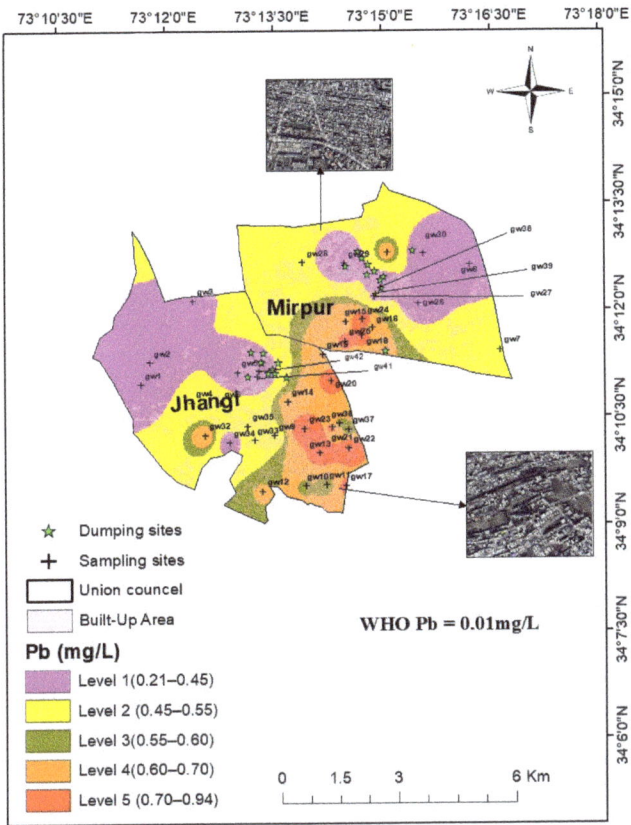

Figure 4. Exceedance modeling for Pb in groundwater.

Figure 5 shows the exceedance level for Hg in the Jhangi union council. According to the findings for Jhangi, no well contributed to Hg Level 5. The groundwater wells in Jhangi that contributed to Level 4 are gw9 and gw40, as shown in Figure 5. Level 3 is caused by gw5, gw14, gw19, gw20, gw21, gw23, and gw41 in the Jhangi area, while Level 2 is attributed to gw10, gw12, gw13, gw17, gw31, and gw34, while Level 1 is due to gw1, gw2, gw3, gw4, gw11, gw32, gw33, gw35, gw36, and gw37. A similar pattern is seen in low-income areas, which could be attributed to natural Hg deposits in the local rock formation. Excess Hg in groundwater could cause various problems, including kidney damage, insomnia, and memory loss.

In Mirpur, gw38 contributed to Level 5 exceedance of Hg. The wells in Mirpur that contributed towards Level 3 were gw24 and gw39, as shown in Figure 5. Level 2 is attributed to gw7, gw8, gw15, gw16, gw18, gw27, and gw31, while Level 5 is caused by gw25, gw26, gw28, gw29, and gw30 in the Mirpur area. According to the study findings, there were no wells that contributed to Level 4 in the Mirpur union council areas.

The analysis reveals that some groundwater wells in Jhangi union council, such as gw5, gw40, and gw41, were located near an open waste dumping site, whereas in Mirpur, gw18, gw29, gw38, gw39, and gw27 were located near the dumping sites. As shown in Figure 5, the level of exceedance caused by dumping sites in Mirpur was higher than in Jhangi union council. The exceedance level in Mirpur near the dumping site was at Level 5 and Level 4 near gw27, gw38, and gw39, while in Jhangi, gw5, gw40, and gw41 contributed to Level 4. The remaining groundwater wells in both union councils contributed to Level 3 exceedance. The high level of Hg in Mirpur in groundwater wells such as gw38, gw39, and gw27 is due to the potential leachate from the nearby dumping site. The Hg present in the wastes is from sources such as electrical switches, fluorescent light bulbs, batteries, thermometers, and some medical waste. After use, mercury-containing products form part of the municipal solid waste (MSW) that is collected and disposed of into open dumping sites, where mercury and other pollutants could pollute groundwater sources through leaching and percolation. Consequently, mercury might become an important constituent of the resulting leachate which is often characterized by high loads of dissolved organic matter, inorganic macrocomponents, metals, and other xenobiotic organic compounds [77,78]. Usually, leachate accumulates percolate through the soil to contaminate the groundwater [76]. Exposure to mercury might cause brain, liver, kidney, and developmental disorders, particularly in young children and developing fetuses [79]. It poses a significant risk to local groundwater resource users, as well as the natural environment and health [9].

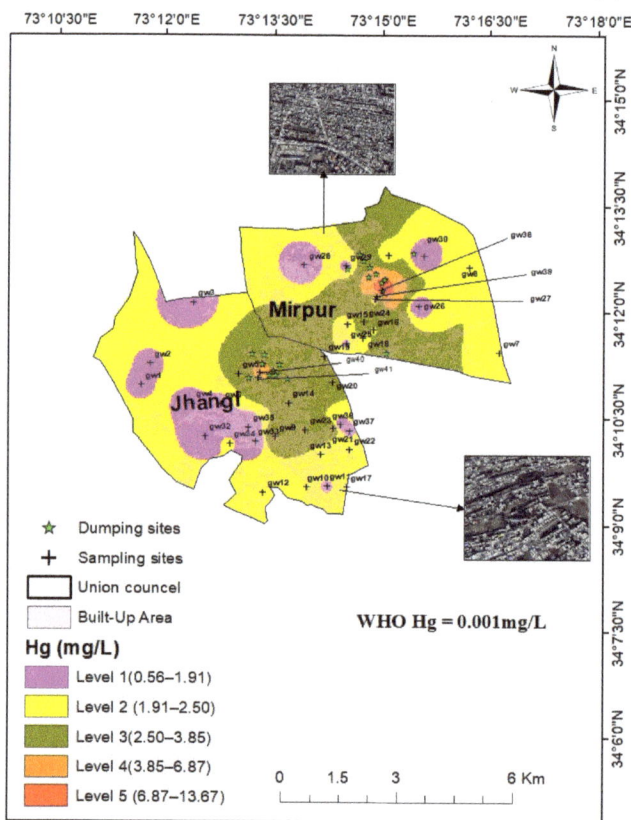

Figure 5. Exceedance modeling for Hg in groundwater.

Figure 6 shows the results for Cd in the Jhangi union council. In Jhangi, gw37 and gw40 contributed to Cd Level 4. As shown in Figure 6, wells in Jhangi that contributed to Level 3 were gw4, gw5, gw10, gw17, gw19, gw23, gw36, and gw41, while gw1, gw13, and gw20 contributed to Level 2. According to the findings, Level 1 is caused by the wells gw2, gw3, gw9, gw11, gw12, gw14, gw21, gw22, gw32, gw33, and gw35 in the Jhangi area. The high values are recorded near agricultural fields and vegetation in the Jhangi union council; therefore, they could be related to Cd-containing fertilizer which might contaminate groundwater through percolation processes. The sources of groundwater contamination in Jhangi might be industrial waste or fertilizer contamination.

In Mirpur, gw38 contributed to Level 5 exceedance of Cd. As shown in Figure 6, the wells in Mirpur that contributed to Level 4 were gw29 and gw39, while gw7, gw8, gw16, gw24, and gw25 contributed to Level 3 (Mirpur). Level 2 is attributed to gw30, while Level 1 is attributed to gw15, gw18, gw26, gw28, and gw31 in Mirpur areas.

The study findings indicate that high levels of Cd (Level 4 and Level 5) were found near gw27, gw38, gw39, and gw29, which were located near open dumping sites in Mirpur union council areas. Level 4 is only found near gw5, gw40, and gw41 in Jhangi union council areas. Level 3 exceedance values were found in the remaining wells in both union councils (Figure 6). The high Cd levels in the waste could be attributed to sources such as batteries, plastics, and cards. Exposure to Cd might result in several negative health effects, including cancer, acute inhalation exposure to cadmium (high levels in a short period of time), and flu-like symptoms (chills, fever, and muscle pain), as well as lung damage. It also endangers local groundwater resource users, as well as the natural environment [74]. After use, Cd-containing products such as plastics and pigments form part of the MSW that is collected and disposed of into open dumping sites where Cd could be leached into landfill leachates [80].

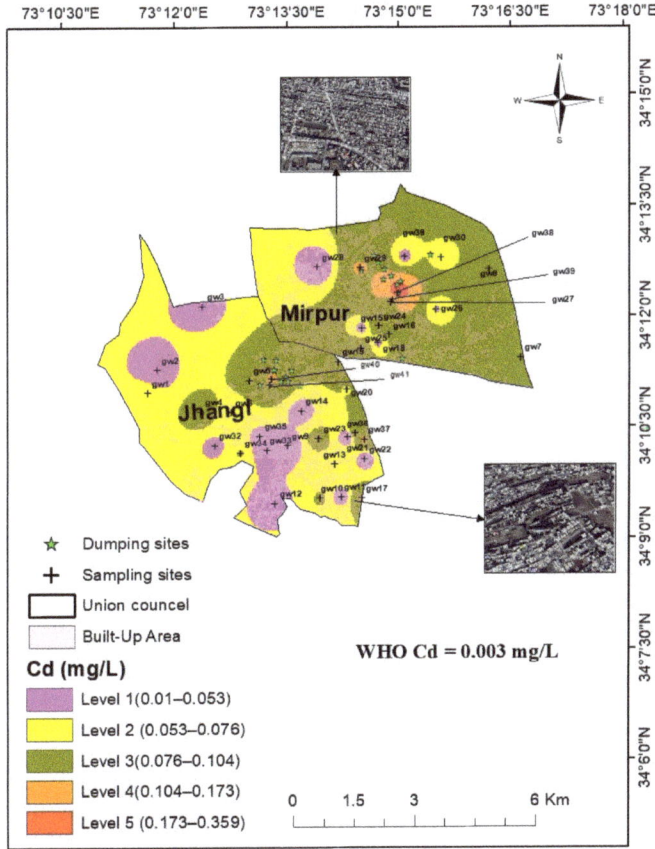

Figure 6. Exceedance modeling for Cd in groundwater.

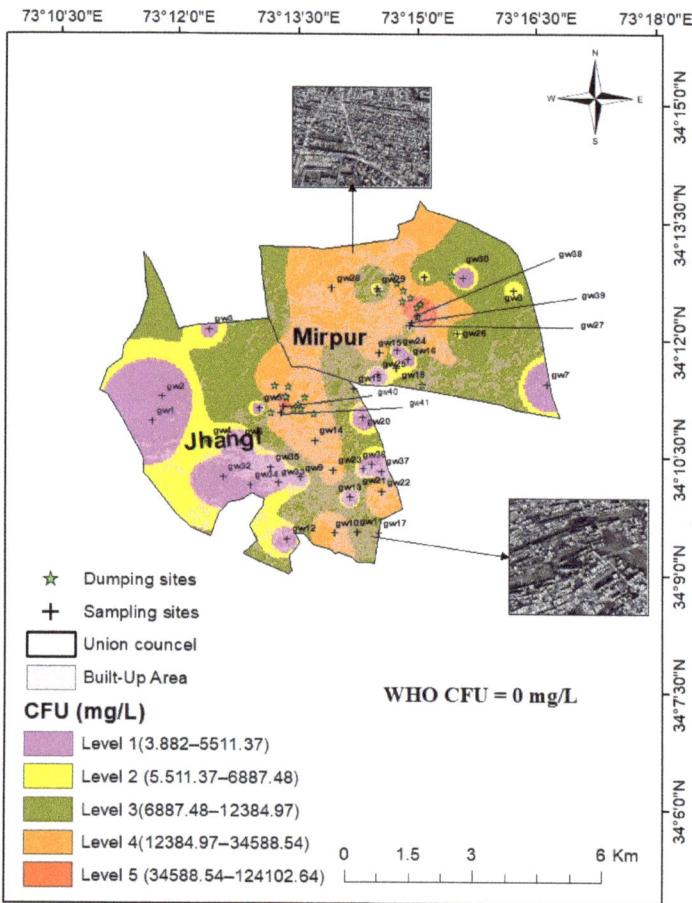

Figure 7. Exceedance modeling for CFU in groundwater.

Figure 7 shows the results for CFU in the Jhangi union council. According to the findings for Jhangi, gw40 contributed to CFU Level 5. As shown in Figure 7, groundwater wells in Jhangi that contributed to Level 4 were gw10, gw14, gw22, gw23, and gw41, while Level 3 is attributed to gw4, gw11, and gw17 in the Jhangi area. Level 1 is attributed to the remaining wells in the Jhangi areas. There were no wells near Level 2 in the Jhangi union council. Coliform bacteria washed into the ground by rain are usually filtered out as the water goes through the soil and into groundwater systems. However, poorly constructed, cracked, or unsealed wells could allow coliform bacteria to enter groundwater and contaminate drinking water.

In Mirpur, gw38 contributed to Level 5 exceedance of CFU. The wells in Mirpur that contributed to Level 4 were gw15 and gw28, while gw8, gw18, and gw26 contributed to Level 2 (Mirpur), as shown in Figure 7. Level 1 is attributed to gw7, gw16, gw18, gw24, and gw25 in the Mirpur areas; no wells were contributed to Level 3 in the Mirpur union council areas.

The study findings showed that the level of pollution caused by dumping sites in Mirpur was higher than in Jhangi union council, as shown in Figure 7. The CFU levels were found to be at Level 5 in Mirpur, close to the dumping site, and Level 4 in Jhangi union council areas. Both union councils' remaining groundwater wells contributed to Level 3 exceedance. The high level of CFU in Mirpur groundwater wells such as gw38, gw39, and gw27 is attributed to nearby dumping sites. Household, hospital, and food waste are the sources of high CFU in dumping sites, which has several health implications. This water is used for drinking and consumption by the population of the study area. Consuming such water is exceedingly dangerous and could result in illnesses such as diarrhoea, cholera, typhoid, paratyphoid, hepatitis A, dermatitis, and enteric fever [81].

The human health risk maps were produced by stimulating groundwater vulnerability assessment using a GIS-based system called the ArcPRZM-3 tool [81]. The health risk maps with three categories of health risk, i.e., low, moderate, and high, were generated for Independence and Randolph counties of Arkansas. It was found that the percentage of areas under high health risk was 5high compared to other health risk classes. The health risk area under Levels 4 and 5 was very high in Mirpur for As, CFU, and Hg. The health risk at Level 3 and Level 4 was high for both Jhangi and Mirpur. The human health risk assessment was accomplished using fuzzy WQI related to groundwater contamination. According to this study, the carcinogenic health risk was high due to Cd, As, and Hg.

4. Conclusions

The combination of field sampling, laboratory analysis, GIS, and advanced statistical modeling in water quality is a new contribution to research and science. PCA modeling provided five PCs that represented the natural processes and anthropogenic pollution. Groundwater was contaminated due to microbial and heavy metal pollution related to natural and anthropogenic sources, especially open dumping sites. In both union councils, As, CFU and Pb were found to be Level 5.

The field-collected groundwater contamination data were utilized for studying the impact on a larger area with the help of remote sensing satellite data. Overall, the communities living in the study area are exposed to hazardous metals such as As and Pb, which are carcinogenic. GIS-based modeling provided effective maps for exceedance levels linked to health risks in the study area. The study area population is at high health risk due to the consumption of unfit water. Both open dumping sites and natural sources could affect the groundwater. It is recommended to analyze water wells near and away from dumping sites for the whole of Abbottabad and obtain information on groundwater contamination and health risks. The number of open dumping sites in Abbottabad city provides information on the negligence of relevant agencies for cleaning the city.

Author Contributions: Methodology, T.A.A., A.J., S.U., A.P., R.A.A., M.F.J., A.M. and A.M.M.; writing—original draft, T.A.A., A.J. and S.U.; supervision T.A.A. and S.U.; software and formal analysis, S.U.; validation, and visualization, W.U.; resources, A.P.; writing—review & editing, R.A.A., M.F.J., A.M. and A.M.M.; funding acquisition, A.M. and A.M.M. All authors have read and agreed to the published version of the manuscript.

Funding: This research received no external funding.

Institutional Review Board Statement: Not applicable.

Informed Consent Statement: Not applicable.

Data Availability Statement: Available on request.

Acknowledgments: Overall excellent support was provided by the chairman of the civil engineering department, COMSATS University, Islamabad. His tremendous support is highly appreciated. The support provided by departments of civil engineering and environmental sciences in this study is acknowledged. LULC map of study area provided by Siddique Ullah is also recognized. The support provided by Abdullah Mohamed and Abdeliazim Mustafa Mohamed is also acknowledged.

Conflicts of Interest: The authors declare no conflict of interest.

References

1. Ashbolt, N.J. Microbial contamination of drinking water and disease outcomes in developing regions. *Toxicology* **2004**, *198*, 229–238. [CrossRef] [PubMed]
2. Bain, R.; Cronk, R.; Hossain, R.; Bonjour, S.; Onda, K.; Wright, J.; Bartram, J. Global assessment of exposure to faecal contamination through drinking water based on a systematic review. *Trop. Med. Int. Health* **2014**, *19*, 917–927. [CrossRef] [PubMed]
3. Wu, L.; Long, T.Y.; Liu, X.; Guo, J.S. Impacts of climate and land-use changes on the migration of non-point source nitrogen and phosphorus during rainfall-runoff in the Jialing River Watershed, China. *J. Hydrol.* **2012**, *475*, 26–41. [CrossRef]
4. Malik, M.A.; Azam, M.; Saboor, A. *Water Quality Status of Upper KPK and Northern Areas of Pakistan*; Pakistan Council of Research in Water Resources (PCRWR), Water Resources Research Centre: Peshawar, Pakistan, 2010.
5. Ahmed, A.; Iftikhar, H.; Chaudhry, G.M. Water resources and conservation strategy of Pakistan. *Pak. Dev. Rev.* **2007**, *2007*, 997–1009. [CrossRef]
6. Akhtar, N.; Syakir Ishak, M.I.; Bhawani, S.A.; Umar, K. Various natural and anthropogenic factors responsible for water quality degradation: A review. *Water* **2021**, *13*, 2660. [CrossRef]
7. Ejaz, N.; Akhtar, N.; Hashmi, H.; Naeem, U.A. Environmental impacts of improper solid waste management in developing countries: A case study of Rawalpindi city. *Sustain. World* **2010**, *142*, 379–387.
8. Abdel-Shafy, H.I.; Mansour, M.S. Solid waste issue: Sources, composition, disposal, recycling, and valorization. *Egypt. J. Pet.* **2018**, *27*, 1275–1290. [CrossRef]
9. Abiriga, D.; Vestgarden, L.S.; Klempe, H. Groundwater contamination from a municipal landfill: Effect of age, landfill closure, and season on groundwater chemistry. *Sci. Total Environ.* **2020**, *737*, 140307. [CrossRef]
10. Chhatwal, G.R. *Dictionary of Environmental Chemistry*; Anmol Publications: New Delhi, India, 1990.
11. Daud, M.K.; Nafees, M.; Ali, S.; Rizwan, M.; Bajwa, R.A.; Shakoor, M.B.; Zhu, S.J. Drinking water quality status and contamination in Pakistan. *Biomed. Res. Int.* **2017**, *2017*, 7908183. [CrossRef]
12. Gunarathne, V.; Ashiq, A.; Ramanayaka, S.; Wijekoon, P.; Vithanage, M. Biochar from municipal solid waste for resource recovery and pollution remediation. *Environ. Chem Lett.* **2019**, *17*, 1225–1235. [CrossRef]
13. Rehman, A.; Saeed, S.; Aslam, M.S.; Khan, M.W. Municipal solid waste management crises in the developing countries: A case study of Peshawar city. *Int. J. Basic Appl. Sci.* **2018**, *5*, 23–32.
14. Ebistu, T.A.; Minale, A.S. Solid waste dumping site suitability analysis using geographic information system (GIS) and remote sensing for Bahir Dar Town, Northwestern Ethiopia. *Afr. J. Environ. Sci. Technol.* **2013**, *7*, 976–989. [CrossRef]
15. Ali, S.M.; Pervaiz, A.; Afzal, B.; Hamid, N.; Yasmin, A. Open dumping of municipal solid waste and its hazardous impacts on soil and vegetation diversity at waste dumping sites of Islamabad city. *J. King. Saud. Univ. Sci.* **2014**, *26*, 59–65. [CrossRef]
16. Yasin, H.; Usman, M. Site investigation of open dumping site of municipal solid waste in Faisalabad. *Earth Sci. Pak.* **2017**, *1*, 23–25. [CrossRef]
17. Pakistan Environmental Protection Agency. Pakistan Environmental Protection Act (PEPA). 1997. Available online: www.environment.gov.pk (accessed on 5 March 2021).
18. Usman, M.; Yasin, H.; Nasir, D.A.; Mehmood, W. A case study of groundwater contamination due to open dumping of municipal solid waste in Faisalabad, Pakistan. *Earth Sci. Pak.* **2017**, *1*, 15–16. [CrossRef]
19. Kundi, B. *Pakistan's Water Crisis: Why a National Water Policy Is Needed*; The Asia Foundation: San Francisco, CA, USA, 2017.
20. ADB. Solid Waste Management Sector in Pakistan. 2022. Available online: https://www.adb.org/sites/default/files/publication/784421/solid-waste-management-pakistan-road-map.pdf (accessed on 7 July 2021).
21. Talib, M.A.; Tang, Z.; Shahab, A.; Siddique, J.; Faheem, M.; Fatima, M. Hydrogeochemical characterization and suitability assessment of groundwater: A case study in Central Sindh, Pakistan. *Int. J. Environ. Health Res.* **2019**, *16*, 886. [CrossRef]
22. Khatri, N.; Tyagi, S. Influences of natural and anthropogenic factors on surface and groundwater quality in rural and urban areas. *Front. Life Sci.* **2015**, *8*, 23–39. [CrossRef]
23. Srinkanth, R.; Rao, A.M.M.; Khanum, A.; Reddy, S.R.P. Mercury contamination of groundwater around Hussain Sagar Lake, India. *Bull. Environ. Contam. Toxicol.* **1993**, *51*, 96–98.
24. Dey, N.C.; Parvez, M.; Dey, D.; Saha, R.; Ghose, L.; Barua, M.K.; Chowdhury, M.R. Microbial contamination of drinking water from risky tubewells situated in different hydrological regions of Bangladesh. *Int. J. Hyg. Environ. Health* **2017**, *220*, 621–636. [CrossRef]
25. Bhutta, M.; Ramzan, M.; Hafeez, C.A. *Pakistan Council for Research in Water Resources*; Pakistan Council: Islamabad, Pakistan, 2008.
26. Dumenci, N.A.; Yolcu, O.C.; Temel, F.A.; Turan, N.G. Identifying the maturity of co-compost of olive mill waste and natural mineral materials: Modelling via ANN and multi-objective optimization. *Bioresour. Technol.* **2021**, *338*, 125516. [CrossRef]
27. Haghizadeh, A.; Moghaddam, D.D.; Pourghasemi, H.R. GIS-based bivariate statistical techniques for groundwater potential analysis (an example of Iran). *J. Earth Syst. Sci.* **2017**, *126*, 1–17. [CrossRef]
28. Zeinivand, H.; Ghorbani Nejad, S. Application of GIS-based data-driven models for groundwater potential mapping in Kuhdasht region of Iran. *Geocarto Int.* **2018**, *33*, 651–666. [CrossRef]
29. Khodaparast, M.; Rajabi, A.M.; Edalat, A. Municipal solid waste landfill siting by using GIS and analytical hierarchy process (AHP): A case study in Qom city, Iran. *Environ. Earth Sci.* **2018**, *77*, 1–12. [CrossRef]
30. Akbar, T.A.; Hassan, Q.K.; Achari, G. A methodology for clustering lakes in Alberta on the basis of water quality parameters. *Clean Soil Air Water* **2011**, *39*, 916–924. [CrossRef]

31. Abou Zakhem, B.; Al-Charideh, A.; Kattaa, B. Using principal component analysis in the investigation of groundwater hydrochemistry of Upper Jezireh Basin, Syria. *Hydrol. Sci. J.* **2017**, *62*, 2266–2279. [CrossRef]
32. Amanah, T.R.N.; Putranto, T.T.; Helmi, M. Application of cluster analysis and principal component analysis for assessment of groundwater quality—A study in Semarang, Central Java, Indonesia. In *IOP Conference Series: Earth and Environmental Science*; IOP Publishing: Bristol, UK, 2019; Volume 2019248, p. 012063.
33. Ali, M.; Chandu, V.; Nandini, V.; Mostafa, M.; Alkendi, R. Evaluation of water quality in the households of Baniyas Region, Abu Dhabi using multivariate statistical approach. *Sustain. Water Resour. Manag.* **2019**, *5*, 1579–1592. [CrossRef]
34. Yang, G.; Moyer, D.L. Estimation of nonlinear water-quality trends in high-frequency monitoring data. *Sci. Total Environ.* **2020**, *715*, 136686. [CrossRef]
35. Ramadas, M.; Samantaray, A.K. Applications of remote sensing and GIS in water quality monitoring and remediation: A state-of-the-art review. *Water Remediat.* **2018**, *2018*, 225–246.
36. Ahmadi, H.; Kaya, O.A.; Babadagi, E.; Savas, T.; Pekkan, E. GIS-based groundwater potentiality mapping using AHP and FR models in central antalya, Turkey. *Environ. Sci. Proc.* **2020**, *5*, 11.
37. Ozdemir, A. Using a binary logistic regression method and GIS for evaluating and mapping the groundwater spring potential in the Sultan Mountains (Aksehir, Turkey). *J. Hydrol.* **2011**, *405*, 123–136. [CrossRef]
38. Naghibi, S.A.; Pourghasemi, H.R.; Pourtaghi, Z.S.; Rezaei, A. Groundwater qanat potential mapping using frequency ratio and Shannon's entropy models in the Moghan watershed, Iran. *Earth Sci. Inform.* **2015**, *8*, 171–186. [CrossRef]
39. Aguilera, P.A.; Fernández, A.; Ropero, R.F.; Molina, L. Groundwater quality assessment using data clustering based on hybrid Bayesian networks. *Stoch. Environ. Res. Risk Assess.* **2013**, *27*, 435–447. [CrossRef]
40. Duan, H.; Deng, Z.; Deng, F.; Wang, D. Assessment of Groundwater Potential Based on Multicriteria Decision Making Model and Decision Tree Algorithms. *Math. Probl. Eng.* **2016**, *16*, 1–11. [CrossRef]
41. Chen, W.; Li, H.; Hou, E.; Wang, S.; Wang, G.; Panahi, M.; Ahmad, B.B. GIS-based groundwater potential analysis using novel ensemble weights-of-evidence with logistic regression and functional tree models. *Sci. Total Environ.* **2018**, *634*, 853–867. [CrossRef] [PubMed]
42. Adeyeye, O.A.; Ikpokonte, E.A.; Arabi, S.A. GIS-based groundwater potential mapping within Dengi area, North Central Nigeria. *Egypt. J. Remote Sens. Space Sci.* **2019**, *22*, 175–181. [CrossRef]
43. Ahmed, T.; Pervez, A.; Mehtab, M.; Sherwani, S.K. Assessment of drinking water quality and its potential health impacts in academic institutions of Abbottabad (Pakistan). *Desalination Water Treat.* **2015**, *54*, 1819–1828. [CrossRef]
44. Local Government. Elections and Rural Development Department. District Abbottabad. 2022. Available online: https://www.lgkp.gov.pk/districts/district-abbottabad/ (accessed on 31 January 2022).
45. Talang, R.P.N.; Sirivithayapakorn, S. Environmental and financial assessments of open burning, open dumping and integrated municipal solid waste disposal schemes among different income groups. *J. Clean. Prod.* **2021**, *312*, 127761. [CrossRef]
46. Boateng, T.K.; Opoku, F.; Akoto, O. Heavy metal contamination assessment of groundwater quality: A case study of Oti landfill site, Kumasi. *Appl. Water Sci.* **2019**, *9*, 1–15. [CrossRef]
47. Wilk-Woźniak, E.; Ligęza, S.; Shubert, E. Effect of Water Quality on Phytoplankton Structure in Oxbow Lakes under Anthropogenic and Non-Anthropogenic Impacts. *Soil Air Water* **2014**, *42*, 421–427. [CrossRef]
48. Ullah, S.; Ahmad, K.; Sajjad, R.U.; Abbasi, A.M.; Nazeer, A.; Tahir, A.A. Analysis and simulation of land cover changes and their impacts on land surface temperature in a lower Himalayan region. *J. Environ. Manag.* **2019**, *245*, 348–357. [CrossRef]
49. Goswami, R.; Kumar, M.; Biyani, N.; Shea, P.J. Arsenic exposure and perception of health risk due to groundwater contamination in Majuli (river island), Assam, India. *Environ. Geochem. Health* **2020**, *42*, 443–460. [CrossRef] [PubMed]
50. Nisbet, R.; Elder, J.; Miner, G. *Handbook of Statistical Analysis and Data Mining Applications*; Academic Press: Cambridge, MA, USA, 2009.
51. Wu, Z.G.; Jiang, W.; Nitin, M.; Bao, X.Q.; Chen, S.L.; Tao, Z.M. Characterizing diversity based on nutritional and bioactive compositions of yam germplasm (*Dioscorea* spp.) commonly cultivated in China. *J. Food Drug Anal.* **2016**, *24*, 367–375. [CrossRef] [PubMed]
52. Vongdala, N.; Tran, H.D.; Xuan, T.D.; Teschke, R.; Khanh, T.D. Heavy metal accumulation in water, soil, and plants of municipal solid waste landfill in Vientiane, Laos. *Int. J. Environ. Res. Public Health* **2019**, *16*, 22. [CrossRef] [PubMed]
53. Gu, K.; Zhou, Y.; Sun, H.; Dong, F.; Zhao, L. Spatial distribution, and determinants of PM2.5 in China's cities: Fresh evidence from IDW and GWR. *Environ. Monit. Assess.* **2021**, *193*, 1–22. [CrossRef] [PubMed]
54. Sayadi Shahraki, A.; Boroomand Nasab, S.; Naseri, A.A.; Soltani Mohammadi, A. Estimation of groundwater depth using ANN-PSO, kriging, and IDW models (case study: Salman Farsi Sugarcane Plantation). *Cent. Asian J. Environ. Sci. Technol. Innov.* **2021**, *2*, 91–101. [CrossRef]
55. Gong, G.; Mattevada, S.; O'Bryant, S.E. Comparison of the accuracy of kriging and IDW interpolations in estimating groundwater arsenic concentrations in Texas. *Environ. Res.* **2014**, *130*, 59–69. [CrossRef]
56. Spokas, K.; Graff, C.; Morcet, M.; Aran, C. Implications of the spatial variability of landfill emission rates on geospatial analyses. *J. Waste Manag.* **2003**, *23*, 599–607. [CrossRef]
57. Zhou, Y.; Michalak, A.M. Characterizing attribute distributions in water sediments by geostatistical downscaling. *Environ. Sci. Technol.* **2009**, *43*, 9267–9273. [CrossRef]
58. World Health Organization. *Guidelines for Drinking-Water Quality*; WHO: Geneva, Switzerland, 1993.

59. Hassan, A.; Kura, N.U.; Amoo, A.O.; Adeleye, A.O.; Ijanu, E.M.; Bate, G.B.; Okunlola, I.A. Assessment of Landfill Induced Ground Water Pollution of Selected Boreholes and Hand-Dug Wells around Ultra-Modern Market Dutse North-West, Nigeria. *Environ. Stud.* **2018**, *1*, 1–10.
60. Idrees, N.; Tabassum, B.; Abd_Allah, E.F.; Hashem, A.; Sarah, R.; Hashim, M. Groundwater contamination with cadmium concentrations in some West UP Regions, India. *Saudi J. Biol. Sci.* **2018**, *25*, 1365–1368. [CrossRef]
61. Nabeela, F.; Azizullah, A.; Bibi, R.; Uzma, S.; Murad, W.; Shakir, S.K.; Ullah, W.; Qasim, M.; Häder, D.P. Microbial contamination of drinking water in Pakistan—A review. *Environ. Sci. Pollut. Res.* **2014**, *21*, 13929–13942. [CrossRef] [PubMed]
62. Obasi, P.N.; Akudinobi, B.B. Potential health risk and levels of heavy metals in water resources of lead–zinc mining communities of Abakaliki, southeast Nigeria. *Appl. Water Sci.* **2020**, *10*, 1–23. [CrossRef]
63. Gu, S.; Dai, J.; Qu, T.; He, Z. Emerging roles of microRNAs and long noncoding RNAs in cadmium toxicity. *Biol. Trace Elem. Res.* **2020**, *195*, 481–490. [CrossRef] [PubMed]
64. Sarlinova, M.; Majerova, L.; Matakova, T.; Musak, L.; Slovakova, P.; Skerenova, M.; Kavcova, E.; Halasova, E. Polymorphisms of DNA repair genes and lung cancer in chromium exposure. *Lung Cancer Autoimmune Disord.* **2014**, *833*, 1–8. [CrossRef]
65. Batayneh, A.; Ghrefat, H.; Zaman, H.; Mogren, S.; Zumlot, T.; Elawadi, E.; Laboun, A.; Qaisy, S. Assessment of the physicochemical parameters and heavy metals toxicity: Application to groundwater quality in unconsolidated shallow aquifer system. *Res. J. Environ. Toxicol.* **2012**, *6*, 169. [CrossRef]
66. Abdelwaheb, M.; Jebali, K.; Dhaouadi, H.; Dridi-Dhaouadi, S. Adsorption of nitrate, phosphate, nickel and lead on soils: Risk of groundwater contamination. *Ecotoxicol. Environ. Saf.* **2019**, *179*, 182–187. [CrossRef]
67. WHO. Arsenic. 2022. Available online: https://www.who.int/news-room/fact-sheets/detail/arsenic (accessed on 31 January 2022).
68. Wang, X.; Liu, L.; Zhao, L.; Xu, H.; Zhang, X. Assessment of dissolved heavy metals in the Laoshan Bay, China. *Mar. Pollut. Bull.* **2019**, *149*, 110608. [CrossRef]
69. Batayneh, A.T.; Al-Taani, A.A. Integrated resistivity and water chemistry for evaluation of groundwater quality of the Gulf of Aqaba coastal area in Saudi Arabia. *Geosci. J.* **2016**, *20*, 403–413. [CrossRef]
70. Salam, M.; Alam, F.; Hossain, M.; Saeed, M.A.; Khan, T.; Zarin, K.; Rwan, B.; Ullah, W.; Khan, W.; Khan, O. Assessing the drinking water quality of educational institutions at selected locations of district Swat, Pakistan. *Environ. Earth Sci.* **2021**, *80*, 1–11. [CrossRef]
71. Kopittke, P.M.; Menzies, N.W. Effect of Cu toxicity on growth of cowpea (*Vigna unguiculata*). *Plant Soil* **2006**, *279*, 287–296. [CrossRef]
72. Zanonib, A.E. *Ground-Water Pollution and Sanitary Landfills*; US Government Printing Office: Washington, DC, USA, 1972; p. 87.
73. Reinhart, D.R.; Grosh, C.J. *Analysis of Florida MSW Landfill Leachate Quality*; Report 97-3; Florida Center for Solid and Hazardous Waste Management: Tallahassee, FL, USA, 1998; pp. 31–53.
74. Kjeldsen, P.; Christophersen, M. Composition of leachate from old landfills in Denmark. *Waste Manag. Res.* **2001**, *19*, 249–256. [CrossRef]
75. Han, H. The norm activation model and theory-broadening: Individuals' decision-making on environmentally responsible convention attendance. *J. Environ. Psychol.* **2014**, *40*, 462–471. [CrossRef]
76. Christensen, T.H.; Kjeldsen, P.; Albrechtsen, H.J.R.; Heron, G.; Nielsen, P.H.; Bjerg, P.L.; Holm, P.E. Attenuation of landfill leachate pollutants in aquifers. *Crit. Rev. Environ. Sci. Technol.* **1994**, *24*, 119–202. [CrossRef]
77. Christensen, T.H.; Kjeldsen, P.; Bjerg, P.L.; Jensen, D.L.; Christensen, J.B.; Baun, A.; Heron, G. Biogeochemistry of landfill leachate plumes. *Appl. Geochem.* **2001**, *16*, 659–718. [CrossRef]
78. Rashid, S.; Shah, I.A.; Tulcan, R.X.S.; Rashid, W.; Sillanpaa, M. Contamination, exposure, and health risk assessment of Hg in Pakistan: A review. *Environ. Pollut.* **2022**, *2022*, 118995. [CrossRef] [PubMed]
79. USA EPA. Health Effects of Exposures to Mercury. 2022. Available online: https://www.epa.gov/mercury/health-effects-exposures-mercury (accessed on 14 April 2022).
80. Clark, L. Disease risks posed by wild birds associated with agricultural landscapes. *Prod. Contam. Probl.* **2014**, *2014*, 139–165.
81. Akbar, T.A.; Lin, H. GIS based ArcPRZM-3 model for bentazon leaching towards groundwater. *J. Environ. Sci.* **2010**, *22*, 1854–1859. [CrossRef]

Article

Optimized Thin-Film Organic Solar Cell with Enhanced Efficiency

Waqas Farooq [1], Muhammad Ali Musarat [2,*], Javed Iqbal [1], Syed Asfandyar Ali Kazmi [1], Adnan Daud Khan [3], Wesam Salah Alaloul [2], Abdullah O. Baarimah [2], Ashraf Y. Elnaggar [4], Sherif S. M. Ghoneim [5] and Ramy N. R. Ghaly [6]

[1] Department of Electrical Engineering, Sarhad University of Science and Information Technology, Peshawar 25000, Pakistan; waqasfarooq.ee@gmail.com (W.F.); javed.ee@suit.edu.pk (J.I.); asfandyaralikazmi@gmail.com (S.A.A.K.)
[2] Department of Civil and Environmental Engineering, Universiti Teknologi PETRONAS, Bandar Seri Iskandar 32610, Perak, Malaysia; wesam.alaloul@utp.edu.my (W.S.A.); abdullah_20000260@utp.edu.my (A.O.B.)
[3] Center for Advanced Studies in Energy, University of Engineering & Technology, Peshawar 25000, Pakistan; adnan.daud@uetpeshawar.edu.pk
[4] Department of Food Nutrition Science (Previously Chemistry), College of Science, Taif University, Taif 21944, Saudi Arabia; aynaggar@Tu.edu.sa
[5] Department of Electrical Engineering, College of Engineering, Taif University, Taif 21944, Saudi Arabia; s.ghoneim@tu.edu.sa
[6] Mininstry of Higher Education, Mataria Technical College, Cairo 11718, Egypt; ramyelectric@yahoo.com
* Correspondence: muhammad_19000316@utp.edu.my

Abstract: Modification of a cell's architecture can enhance the performance parameters. This paper reports on the numerical modeling of a thin-film organic solar cell (OSC) featuring distributed Bragg reflector (DBR) pairs. The utilization of DBR pairs via the proposed method was found to be beneficial in terms of increasing the performance parameters. The extracted results showed that using DBR pairs helps capture the reflected light back into the active region by improving the photovoltaic parameters as compared to the structure without DBR pairs. Moreover, implementing three DBR pairs resulted in the best enhancement gain of 1.076% in power conversion efficiency. The measured results under a global AM of 1.5G were as follows: open circuit voltage (V_{oc}) = 0.839 V; short circuit current density (J_{sc}) = 10.98 mA/cm^2; fill factor (FF) = 78.39%; efficiency (η) = 11.02%. In addition, a thermal stability analysis of the proposed design was performed and we observed that high temperature resulted in a decrease in η from 11.02 to 10.70%. Our demonstrated design may provide a pathway for the practical application of OSCs.

Keywords: thin film; organic solar cell; efficiency; DBR; temperature

1. Introduction

Harvesting energy from cleaner sources and protecting the environment from harmful gasses are currently the main targets worldwide in order to protect the ozone layer. In this regard, photovoltaic technology plays a significant role in generating energy from thin-film solar cells (TFSC) by absorbing light from endless sources via the sun. Different materials are available with which to produce TFSC. Researchers from various fields have investigated and promoted photovoltaic technology usage in order to remove unwanted gasses from the environment. Additionally, several materials have been optimized and commercialized. Such materials deliver high performance, although they also release toxic waste into the environment. There is a dire need to investigate, extract, and synthesize the materials from sources with low toxicity to obtain a suitable material for TFSC. In the green energy scenario, the organic photovoltaic (OPV) method is leading the charge for TFSC, especially in solar cells, which have tremendous advantages over other semiconducting materials, such as their low production costs, high mechanical flexibility, ability for roll-to-roll (R2R) production, semi-transparency, and light weight [1,2]. Such materials provide a benchmark for producing energy from green sources. Organic solar cells (OSCs) are also known as polymer solar cells (PSCs) because the base material is composed of two components, i.e., a donor material and acceptor material (D/A).

The donor is responsible for holes and the acceptor is responsible for electrons. The combined (D/A) material in OSCs is known as the bulk heterojunction (BHJ), which is placed between the two electrodes. The cells work on the principle of the photovoltaic effect. The roles of the lowest unoccupied molecular orbital (LUMO) and highest unoccupied molecular orbital (HUMO) are considered in OSCs, as shown in Figure 1. The bonded pair (e–h) in OSCs is known as an exciton, which is disassociated by force, whereby (e–h) moves to the respected functional electrode of the device. The ability to absorb a large amount of light even at low light conditions is observed in OSCs, making them prominent materials in the field of PV. A series of materials have been studied in the literature, which have been synthesized and conjugated with other novel materials to improve certain properties by modifying the morphology of the substrate by introducing wide bandgap materials, which improves the device performance [3].

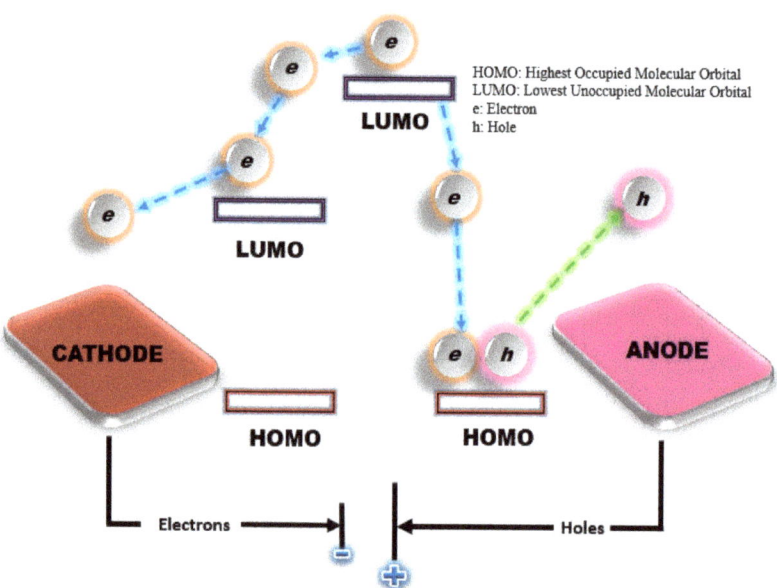

Figure 1. Working principle for OSCs.

OSCs suffer from low efficiency and low stability at high temperatures, making them less effective in terms of commercialization. The transmittance problem in OSCs is also one of the barriers to their commercialization, which needs to be resolved by injecting effective interfacial layers or by sufficiently reducing this transmittance problem and improving the reflection of light and stability in the active region of the cell. The traditional bottom metal electrode also produces heat, which causes thermalization in the cell and results in a decline in the PV parameters [4–6]. DBR pairs are used to replace metal electrodes because metal electrodes in solar cells produce heat, which further generates thermal losses, degrading the performance of the device. Regarding these thermal losses, the DBR plays a significant role not only by reducing the thermalization but also by increasing the electrical photovoltaic performance parameters, for example via high conversion efficiency. However, other approaches utilized to replace metal electrodes only reduce the thermalization but do not significantly impact the device performance parameters. In contrast to this, the DBR pairs increase the cell's thickness, increasing its height and overall cost; as the thickness increases, more material is utilized, meaning the cost increased [7,8].

Thus, low efficiency is always a hindrance to the commercialization of solar cells. Numerous experimental and theoretical modeling studies have been performed to optimize cells so as to increase performance in this context. The theoretical optimized solar cells provide a great opportunity for experimental investigation and could help determine the device's inner performance more precisely. Moreover, numerical modeling is considered a cost-effective technique. From the literature, it is observable that the power conversion efficiency rate of a cell can be remarkably improved by utilizing different approaches, such as inverted structure [9], tandem structure [10], optimization by optical admittance analysis method [11], characterization [12], and DBR [8] techniques.

Ourahmoun computationally examined the impacts of different interfacial layers on organic solar cells and optimized the cell performance by delivering a conversion efficiency rate of 5% [13]. Hamed et al. recorded an η of 6.03% from PSCs [14]. Sartorio et al. fabricated OSCs and reported an efficiency rate of 4.46% [15]. Jiang et al. recorded an efficiency rate of 7.07% from PBDBT-T:PDT solar cells [16]. Park et al. fabricated PTB7:PC$_{71}$BM OSCs and obtained an η of 8.984% [17]. Sun et al. demonstrated an efficiency rate of 9.036% for PTB7:PC$_{71}$BM-based cells [18]. Muhammad et al. reported an η of 1.06% from the organic solar cells [19]. Wu et al. obtained an efficiency rate of 9.09% from non-fluorine organic solar cells [20]. Kazmi et al. investigated CdTe solar cells and used Si/Al$_2$O$_3$ as DBR pairs, observing a conversion efficiency rate of 23.94% under standard testing conditions of 1.5 AM [8]. Ozen incorporated DBR pairs composed of Si/SiO$_2$ on CdTe solar cells and reported an efficiency rate of 10.39% [21]. Rouhbakhshmeghrazi et al. investigated tandem solar cells based on DBR pairs(SiO$_2$/TiO$_2$) and reported a conversion efficiency rate of 28.5% [22]. Yu et al. used ZnSe/LiF as DBR pairs in semitransparent polymer solar cells and reported a conversion efficiency rate of 6.19% [23].

As follows from the above review, there is no effective non-toxic, single-junction organic solar cell option available based on the DBR technique. This article aims to develop a new single-junction organic solar cell based on the DBR technique. The main objectives are as follows:

1. To achieve high conversion efficiency by optimizing the active layer of the cell;
2. To observe the impact of DBR pairing on the performance of the cell;
3. To observe the impact of high temperature on the cell performance parameters.

2. Materials and Methods

Figure 2 shows the proposed schematic of the cell with a combination of different effective layers. Fluorine tin oxide (FTO) was used as the transparent conductive oxide (TCO) at a thickness of 125 nm as the top electrode. The advantages of using FTO are its high transparency, good thermal stability, and high conductivity. The transparent conductive oxide was utilized because it allows maximum transmission of solar light towards the active region. Next, V$_2$O$_5$ was placed as a hole transport layer (HTL) at a thickness of 30 nm. The benefit of using V$_2$O$_5$ as an HTL layer is that it has an excellent hole transporting ability and has outstanding functionality as an electron blocking layer [24].

Moreover, it has excellent thermal stability. Next to the HTL layer, PTB7:PCBM was introduced as an active layer at a thickness of 180 nm. PTB7:PCBM can absorb many photons even in low light conditions because it has a high absorption coefficient [25] and has a narrow bandgap. In addition, PTB7:PCBM is easy to fabricate, involves low manufacturing costs, and has highly flexible mechanical properties. Thus, PTB7:PCBM is an attractive candidate for thin-film technology because of its high molecular weight of 891.4 and broad absorption coefficient value α of 67,237 cm^{-1} [26]. For better quenching of electrons, three electron transporting layers (ETL) (i.e., PCBM (10 nm), zinc oxide (ZnO; 5 nm), and tungsten trioxide, WO$_3$ (70 nm) were utilized to improve the transportation of electrons and achieve high conversion efficiency. ZnO was utilized because of its low cost, large binding energy, and much higher electron mobility as compared to other types of oxides such as TiO_2. The reason for using WO_3 is that it can provide a rapid hopping mechanism for electrons, which further increases the electron mobility, resulting in the collection of enhanced performance parameters [27]. For the back electrode, ZnO-doped silver (Ag) (ZnO:Ag) was used at a thickness of 100 nm. Next, DBR pairs were introduced, namely pairs of WO_3 and lithium fluoride (LiF). The reason for using a DBR is that it can reflect the light from the bottom of the cell, and that the reflected light can be reabsorbed in the active layer, which has a significant effect on the cell's performance [8].

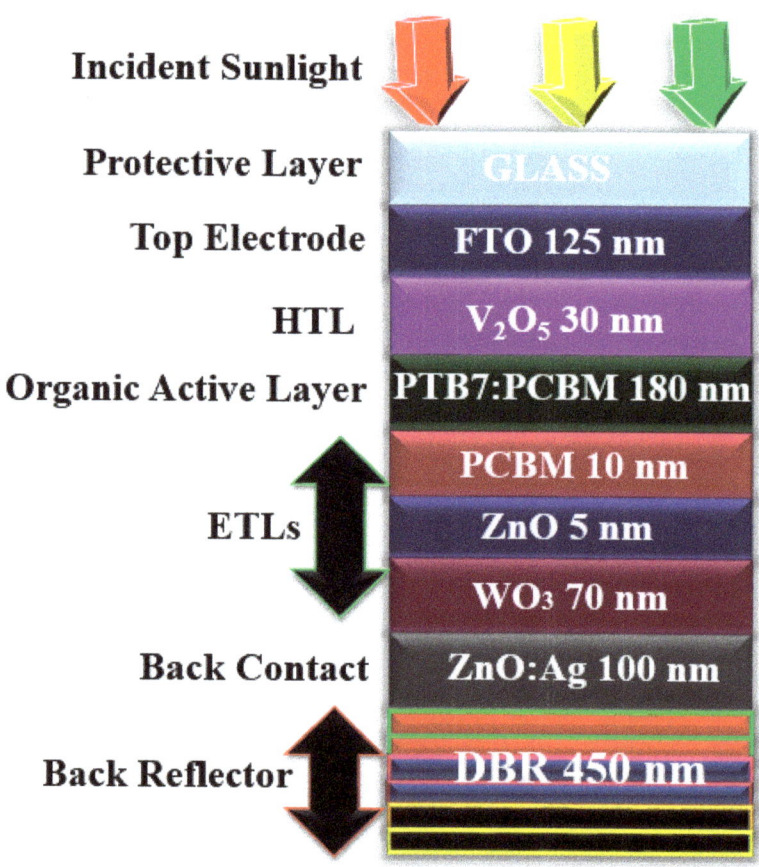

Figure 2. Configuration of the proposed structure with different critical layers and an optimized estimated thickness, in which FTO is used as the TCO; V_2O_5 is used as the HTL; PTB7:PCBM is used as an active layer; PCBM, ZnO, and WO_3 are used as ETLs (for extracting the enhanced amount of electrons); ZnO:Ag is used as a back electrode; and DBR pairs are used as back reflectors.

In this investigation, a general-purpose photovoltaic device model (GPVDM) [28] was used to obtain the PV electrical parameters. The software package utilizes the continuity and Poisson equations for calculations and the drift–diffusion model for transport. Moreover, the Shockley–Read–Hall formalism is used for charge carrier recombination and the Scharfetter–Gummel approach is considered for numerical stability.

The Poisson equation is expressed as (1):

$$\frac{d}{d_x}\varepsilon_o\varepsilon_r.\frac{d_\varphi}{dx} = q(n-p) \tag{1}$$

whereas Equations (2) and (3) represent the drift–diffusion model for holes and electrons, respectively:

$$J_n = q\mu_c n \frac{\partial E_c}{\partial x} + qD_n \frac{\partial n}{\partial x} \quad (2)$$

$$J_p = q\mu_c p \frac{\partial E_v}{\partial x} + qD_p \frac{\partial p}{\partial x} \quad (3)$$

Here, D_n and D_p represent the coefficient of diffusion. The continuity equation can be written as (4) and (5):

$$\frac{\partial J_n}{\partial x} = q\left(R_n - G + \frac{\partial n}{\partial t}\right) \quad (4)$$

$$\frac{\partial J_p}{\partial x} = q\left(R_p - G + \frac{\partial p}{\partial t}\right) \quad (5)$$

For plane waves, the relation between the magnetic and electric fields can be expressed as (6):

$$\nabla \times E = -j\omega\mu H \quad (6)$$

whereas the wave vector is expressed as (7):

$$k = \frac{2\omega}{\lambda} = \frac{\omega n}{c} \quad (7)$$

3. Result and Discussion

3.1. Cell Architecture and Active Layer Optimization

The aim here is to understand the underlying photo physics of the cell architecture, optimize the active layer to achieve high conversion efficiency, and utilize DBR pairs to absorb the reflected light back into the active region. Optimizing the active layer not only helps in improving the performance parameters but also helps to reduce the cost of the material. After optimizing the active layer, DBR pairs help enhance the efficiency by reflecting the light towards the photoactive layer. The composition and role of the effective layers are displayed in Table 1.

Table 1. Materials thickness, composition, and role.

Layers	Thickness (nm)	Composition	Role
FTO	125	Transparent Conductive Oxide	Electrode
V_2O_5	70	Inorganic	HTL
PTB7:PCBM	180	Organic	Active Layer
PCBM	10	Organic	ETL
ZnO	5	Inorganic	ETL
WO_3	70	Inorganic	ETL
Ag	100	Metal	Electrode
DBR	450	Inorganic	Back Reflector

3.2. Optimization of Light-Harvesting Layer (LHL)

The performance of the cell relay on the LHL increased as the incident photons became absorbed in this region. The absorption of photons is the critical aspect of the cell, as the thickness of the LHL is the most crucial aspect of the device architecture. The thickness of the LHL should be very thin [29]. Optimization of the active layer is needed with to reduce the manufacturing costs by utilizing only the optimized amount of layers, which helps in achieving high energy efficiency. The optimization of PTB7:PCBM as the LHL is achieved by modulating its thickness. The thickness of the active region has a direct impact on the performance parameters [30]. Thickness alterations of the active layer are performed between 80–260 nm. As the thickness of the cell increases from 80 nm, the performance parameters, i.e., V_{oc}, J_{sc}, FF, and η, start improving, as shown in Figure 3a–d, respectively. Equation (8) is considered for the calculation of FF:

$$FF = \frac{J_{mp}V_{mp}}{J_{sc}V_{oc}} \tag{8}$$

whereas the η of the cell can be calculated by (9):

$$\eta = \frac{V_{oc} \cdot I_{sc} \cdot FF}{P_{input}} \times 100 \tag{9}$$

where P_{input} is equal to 100 mW/cm^2 or 1000 W/m^2.

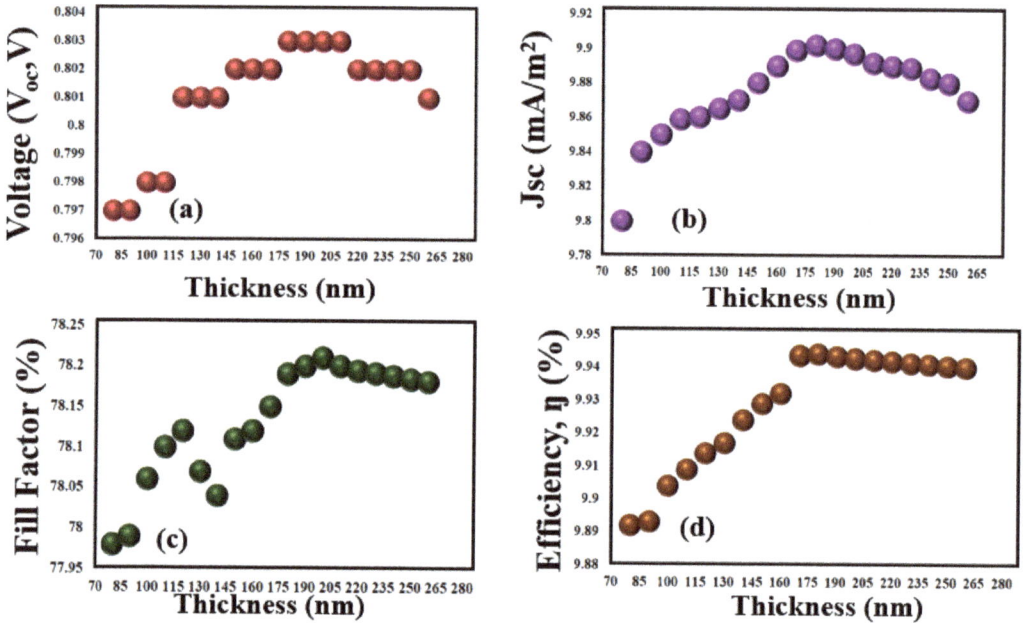

Figure 3. Thickness optimization of the performance parameters: (**a**) optimization of the open circuit voltage (V_{oc}) as a function against the active layer thickness; (**b**) optimization of the short circuit current density (J_{sc}) as a function against the active layer thickness; (**c**) optimization of the fill factor (FF) as a function against the active layer thickness; (**d**) optimization of the efficiency (η) as a function against active layer thickness.

The highest performance parameters are recorded at 180 nm by delivering high values of $V_{oc} = 0.803$ V, $J_{sc} = 9.902$ mA/cm^2, FF = 78.19%, and $\eta = 9.944$%. Further enhancement in the thickness results decreased cell performance parameters. If the cell is very thin the photons will be lost, while at the same time if it is thick then the photons may not travel in the active region for a long distance because of their low lifetime.

The J_{sc} of the cell increases by improving the cell thickness, because many photons are absorbed and excitons are created, which helps achieve high values. The FF started decreasing after attaining the optimized thickness of 180 nm because of series resistance, which appeared due to the thick active layer of the cell [31]. The appearance of the bell curve in Figure 3a–d occurred when the film thickness increased, whereby in addition to the e–h pairs increasing, the recombination and defect in the matrix of the cell also increase, causing the bell curve. Moreover, the decrease in the V_{oc} occurred because of the high recombination of electrons and holes. The high recombination of electrons and holes occurred because of the increased defect centers [32], whereby J_{sc} influences the V_{oc}. Moreover, increased film thickness results in reduced power density. Therefore, a substantial decrease in the PV parameters was obvious.

3.3. Implanting DBR Pairs

After optimizing the cell LHL layer, the DBR pairs composed of WO_3/LiF were introduced, as shown in Figure 4. The DBR pairs help capture the reflected light back into the active region, increasing the cell's performance.

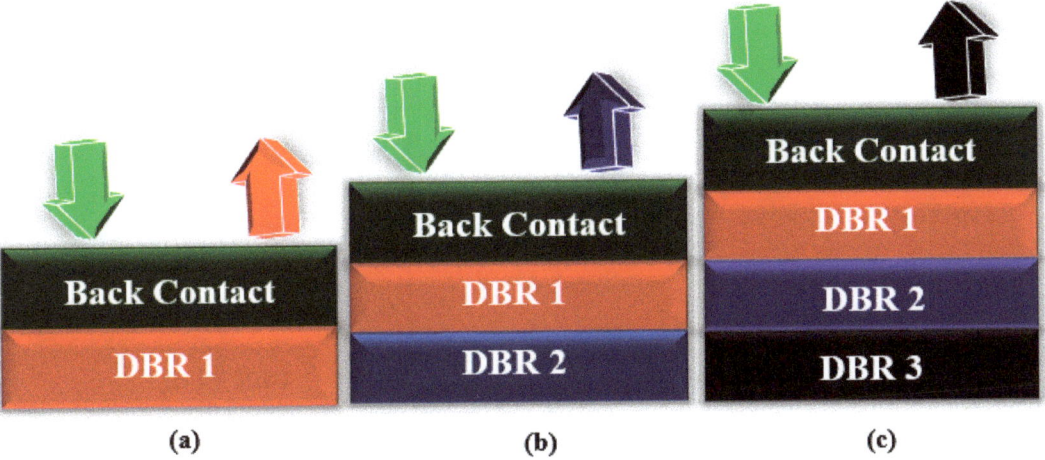

Figure 4. Proposed configuration with DBR pairs: (**a**) single pair; (**b**) double pairs; (**c**) triple pairs.

The improved performance occurred because of the LHL absorption spectrum matching with the DBR reflectance spectrum. By increasing the number of DBR pairs, the performance parameters, including, V_{oc}, J_{sc}, FF, and η, started improving. Due to the insertion of the DBR pairs, the J_{sc} increased from 9.902 to 10.98 mA/cm², V_{oc} from 0.803 to 0.839 V, FF from 78.19 to 78.38%, and η from 9.944 to 11.02%, as shown in Figure 5a–d, respectively, thereby resulting in an enhancement of 1.07% in η. Thus, triple pairs of DBR were found to be more efficient with the proposed architecture, which further suggests that improved performance parameters can be achieved if a properly optimized layer can be used with DBR pairs.

3.4. Thermal Stability Analysis

The device temperature was increased above 300–400 K to test the sensitivity and thermal stability of the cell under high temperatures. As expected, the performance of the cell degraded and the V_{oc} of the cells appeared to blur from 0.839 to 0.802 V. The decreased value of V_{oc} was because of the material E_g, which became unstable, meaning fluctuation occurred in the reverse saturation current due to the increased device temperature. The variation in the reverse saturation current was because of the concentration of the intrinsic carrier n_i, which further depended on the energy of the bandgap E_g, i.e., $n_i^2 = k_1\,e^{-Eg/k}$ [33]. Within the context of temperature, V_{oc} is given as:

$$\frac{d(Voc)}{dT} = \frac{1}{T}\left(Voc - \frac{Eg}{q}\right) \tag{10}$$

E_g/q in Equation (10) will be greater than V_{oc}, which specifies that any adjustment in V_{oc} due to a temperature increase will be negative, i.e., the value of V_{oc} starts fading as the temperature of the device grows. In the present case, if the V_{oc} of the cell decreases, then the η of the cell also decreases from 11.02 to 10.71%, as shown in Figure 6. Thus, the high temperature negatively impacts the cell and makes it less efficient by deteriorating its performance. The small increase in the value of J_{sc} under high temperature is because of the band energy, which decreases, and more electron–hole pairs are created [33]. This suggests that this gain in the value of J_{sc} is so small that it cannot result in any significant improvement in cell performance. It is observable from the results that solar cells appear to be sensitive towards high temperatures. The J_{sc} of the cell under increased temperature can be calculated using Equation (11):

$$J_{sc} \int_{hv=E_g}^{\infty} \frac{dN_{ph}}{dhv} d(hv) \quad (11)$$

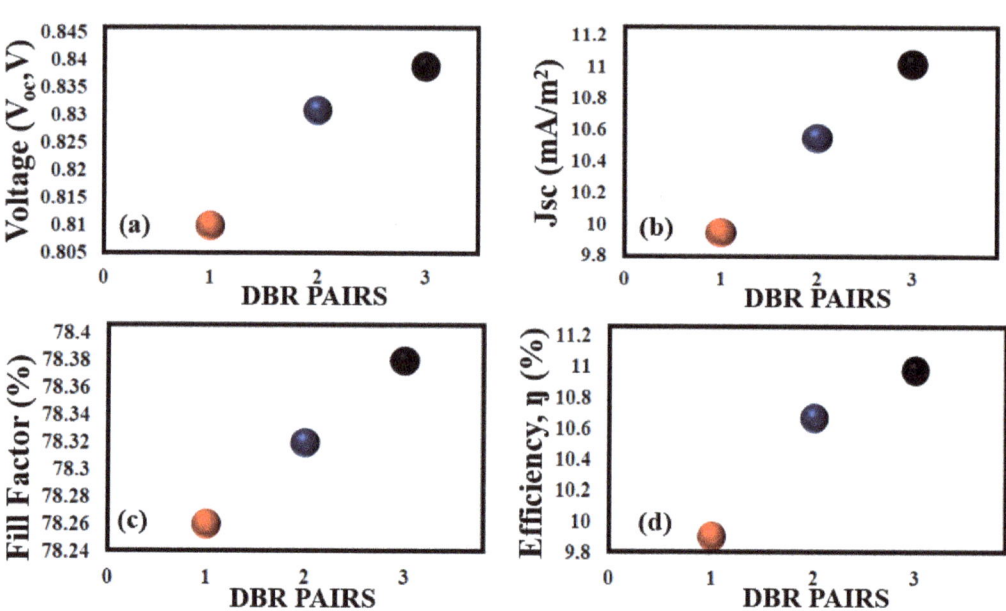

Figure 5. Performance parameters after insertion of DBR pairs composed of (WO$_3$/LiF). (**a**) V_{oc} after insertion of DBR pairs 1, 2, and 3. (**b**) J_{sc} after insertion of DBR pairs 1, 2, and 3. (**c**) FF after insertion of DBR pairs 1, 2, and 3. (**d**) Lastly, η after insertion of DBR pairs 1, 2, and 3.

A comparison with a similar family of thin-film solar cells is presented in Table 2 to highlight the significane of the structure.

Table 2. Comparison of the current study with a similar family of thin-film solar cells with respect to the power conversion efficiency.

Ref	Struture	Method	Junction Mode	η
[34]	Substrate/ITO/PEDOT:PSS/P3HT:PCBM/TiOx/Al	Computational	Single	5.14
[34]	Substrate/ITO/PEDOT:PSS/P3HT:PCBM/PCBM/Al	Computational	Single	4.95
[35]	PET/ITO/PEDOT:PSS/P3HT:PCBM/Al	Computational	Single	4.34
[36]	Glass/FTO/ZnO/doped P3HT:PCBM/Ag	Experimental	Single	4.84
[37]	ITO/ZnO/PTB7:PCBM/PEDOT:PSS/Ag	Computational	Single	5.73
[37]	ITO/ZnO/PTB7:PCBM/MoO$_3$/Ag	Computational	Single	5.92
[38]	ZnO:Al/i-ZnO/CdS/CuInS$_2$/Cu$_2$O/Mo	Computational	Single	22.73

Table 2. Cont.

Ref	Struture	Method	Junction Mode	η
[38]	ZnO:Al/i-ZnO/SnS$_2$/CuInS$_2$/Cu$_2$O/Mo	Computational	Single	21.62
[39]	ITO/GaSe/CIGS-P$^+$/Back Contact/Glass	Computational	Single	33.36
[40]	Glass/ZnO:Al/In$_2$S$_3$/CH$_3$NH$_3$PbI$_3$/Spiro-OMeTAD/Au	Computational	Single	23.05
[41]	AZO/ZnO/CdS/Cu$_2$ZnSnS$_{1.8}$Se$_{2.2}$/Back contact/Glass substrate	Computational	Single	15.3
[42]	SLG/ITO/WS$_2$/CdTe/Au	Computational	Single	20.55
[43]	Al/ITO/Al-ZnO/i-ZnO/CIGS/PbS/Mo	Computational	Single	24.22
Present Work	Glass/FTO/V$_2$O$_5$/PTB7:PCBM/PCBM/ZnO/WO$_3$/Ag/DBR	Computational	Single	11.02

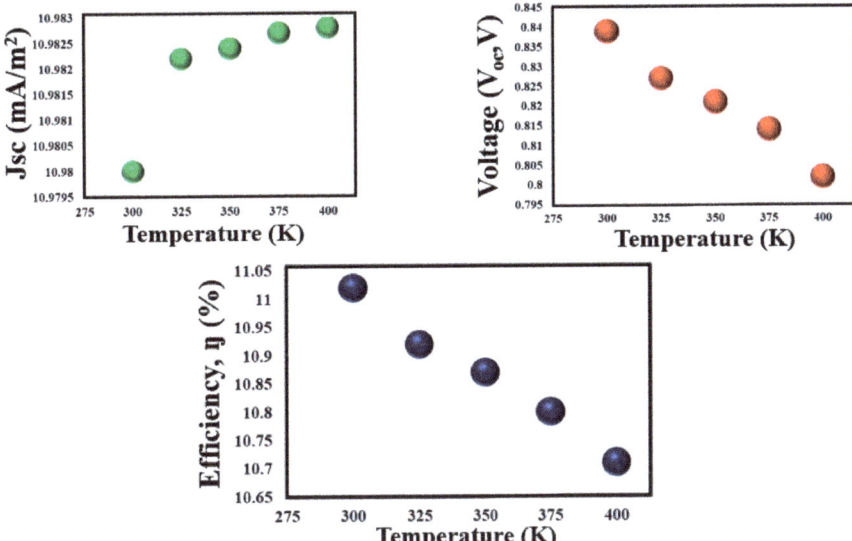

Figure 6. Thermal stability of PV parameters (J_{sc}, V_{occ}, and η) at different operating temperatures.

4. Conclusions

To summarize, we improved the performance of OSCs by computationally injecting DBR pairs in the following structure (glass/FTO/V$_2$O$_5$/PTB7:PCBM/PCBM/ZnO/WO$_3$/Ag/DBR). The DBR pairs composed of WO$_3$/LiF helped in achieving the enhancement of 1.076%. Moreover, the performed thermal stability analysis helped in determining the performance of the cells under high temperatures. We showed that under high temperature, the efficiency rate of a cell deteriorates by 0.32%. In addition, the performed analysis of the thermal stability under different operating temperatures suggested that solar cells are sensitive towards high temperatures and decreased performance parameters. The obtained results provide a smooth pathway for thin-film technology that can be utilized for potential applications in solar-based devices.

Author Contributions: Conceptualization, W.F., M.A.M. and J.I.; methodology, W.F., M.A.M., J.I. and A.D.K.; software, A.Y.E., S.S.M.G. and R.N.R.G.; validation, J.I., S.A.A.K., A.D.K. and W.S.A.; formal analysis, W.F., M.A.M., J.I., S.A.A.K. and A.O.B.; investigation, W.F., M.A.M., J.I. and A.D.K.; resources, A.Y.E., S.S.M.G. and R.N.R.G.; data curation, W.F., S.A.A.K. and A.O.B.; writing—original draft preparation, W.F., J.I., S.A.A.K. and A.D.K.; writing—review and editing, M.A.M. and W.S.A.; supervision, M.A.M., J.I. and A.D.K.; project administration, A.Y.E., S.S.M.G. and R.N.R.G.; funding acquisition, A.Y.E., S.S.M.G. and R.N.R.G. All authors have read and agreed to the published version of the manuscript.

Funding: This study was funded by Taif University Researchers Supporting Project, TURSP-2020/32, Taif University, Taif, Saudi Arabia.

Institutional Review Board Statement: Not applicable.

Informed Consent Statement: Not applicable.

Data Availability Statement: All the data are available within this manuscript.

Acknowledgments: This study was carried out using the facilities and materials of Taif University Research Supporting, TURSP-2020/32, Taif University, Taif, Saudi Arabia.

Conflicts of Interest: The authors declare no conflict of interest.

References

1. Benghanem, M.; Almohammedi, A. Organic Solar Cells: A Review. In *A Practical Guide for Advanced Methods in Solar Photovoltaic Systems*; Springer: Berlin/Heidelberg, Germany, 2020; pp. 81–106.
2. Gu, Y.; Liu, Y.; Russell, T.P. Fullerene-Based Interlayers for Breaking Energy Barriers in Organic Solar Cells. *ChemPlusChem* **2020**, *85*, 751–759. [CrossRef]
3. Firdaus, Y.; He, Q.; Lin, Y.; Nugroho, F.A.A.; Le Corre, V.M.; Yengel, E.; Balawi, A.H.; Seitkhan, A.; Laquai, F.; Langhammer, C.; et al. Novel wide-bandgap non-fullerene acceptors for efficient tandem organic solar cells. *J. Mater. Chem. A* **2020**, *8*, 1164–1175. [CrossRef]
4. Xue, R.; Zhang, J.; Li, Y.; Li, Y. Organic solar cell materials toward commercialization. *Small* **2018**, *14*, 1801793. [CrossRef]
5. Cha, H.; Wu, J. Understanding what determines the organic solar cell stability. *Joule* **2021**, *5*, 1322–1325. [CrossRef]
6. Zhu, C.; Huang, H.; Jia, Z.; Cai, F.; Li, J.; Yuan, J.; Meng, L.; Peng, H.; Zhang, Z.; Zou, Y.; et al. Spin-coated 10.46% and blade-coated 9.52% of ternary semitransparent organic solar cells with 26.56% average visible transmittance. *Sol. Energy* **2020**, *204*, 660–666. [CrossRef]
7. Farooq, W.; Alshahrani, T.; Kazmi, S.A.A.; Iqbal, J.; Khan, H.A.; Khan, M.; Raja, A.A.; Rehman, A.U. Materials optimization for thin-film copper indium gallium selenide (CIGS) solar cell based on distributed braggs reflector. *Optik* **2021**, *227*, 165987. [CrossRef]
8. Kazmi, S.A.A.; Khan, A.D.; Khan, A.D.; Rauf, A.; Farooq, W.; Noman, M.; Ali, H. Efficient materials for thin-film CdTe solar cell based on back surface field and distributed Bragg reflector. *Appl. Phys. A* **2020**, *126*, 46. [CrossRef]
9. He, Z.; Zhong, C.; Su, S.; Xu, M.; Wu, H.; Cao, Y. Enhanced power-conversion efficiency in polymer solar cells using an inverted device structure. *Nat. Photonics* **2012**, *6*, 591–595. [CrossRef]
10. Farooq, W.; Khan, A.D.; Khan, A.D.; Rauf, A.; Khan, S.D.; Ali, H.; Iqbal, J.; Khan, R.U.; Noman, M. Thin-film tandem organic solar cells with improved efficiency. *IEEE Access* **2020**, *8*, 74093–74100. [CrossRef]
11. Ompong, D.; Narayan, M.; Singh, J. Optimization of photocurrent in bulk heterojunction organic solar cells using optical admittance analysis method. *J. Mater. Sci. Mater. Electron.* **2017**, *28*, 7100–7106. [CrossRef]
12. Murugesan, V.S.; Ono, S.; Tsuda, N.; Yamada, J.; Shin, P.-K.; Ochiai, S. Characterization of organic thin film solar cells of PCDTBT: PC71BM prepared by different mixing ratio and effect of hole transport layer. *Int. J. Photoenergy* **2015**, *2015*, 687678. [CrossRef]
13. Ourahmoun, O. Effect of the interfacial materials on the performance of organic photovoltaic cells. *Mater. Today Proc.* **2021**, in press. [CrossRef]
14. Hamed, M.; Oseni, S.O.; Kumar, A.; Sharma, G.; Mola, G.T. Nickel sulphide nano-composite assisted hole transport in thin film polymer solar cells. *Solar Energy* **2020**, *195*, 310–317. [CrossRef]
15. Sartorio, C.; Campisciano, V.; Chiappara, C.; Cataldo, S.; Scopelliti, M.; Gruttadauria, M.; Giacalone, F.; Pignataro, B. Enhanced power-conversion efficiency in organic solar cells incorporating copolymeric phase-separation modulators. *J. Mater. Chem. A* **2018**, *6*, 3884–3894. [CrossRef]
16. Jiang, P.; Lu, H.; Jia, Q.Q.; Feng, S.; Li, C.; Li, H.B.; Bo, Z. Dihydropyreno [1, 2-b: 6, 7-b′] dithiophene based electron acceptors for high efficiency as-cast organic solar cells. *J. Mater. Chem. A* **2019**, *7*, 5943–5948. [CrossRef]
17. Park, S.; Kang, R.; Cho, S. Effect of an Al-doped ZnO electron transport layer on the efficiency of inverted bulk heterojunction solar cells. *Curr. Appl. Phys.* **2020**, *20*, 172–177. [CrossRef]
18. Sun, Y.; Wang, M.; Liu, C.; Li, Z.; Fu, D.; Guo, W. Realizing efficiency improvement of polymer solar cells by using multi-functional cascade electron transport layers. *Org. Electron.* **2020**, *76*, 105482. [CrossRef]
19. Mohammad, T.; Kumar, V.; Dutta, V. Spray deposited indium doped tin oxide thin films for organic solar cell application. *Phys. E Low-Dimens. Syst. Nanostruct.* **2020**, *117*, 113793. [CrossRef]
20. Wu, M.; Shi, L.; Hu, Y.; Chen, L.; Hu, T.; Zhang, Y.; Yuan, Z.; Chen, Y. Additive-free non-fullerene organic solar cells with random copolymers as donors over 9% power conversion efficiency. *Chin. Chem. Lett.* **2019**, *30*, 1161–1167. [CrossRef]
21. Özen, Y. The enhancement in cell performance of CdTe-based solar cell with Si/SiO$_2$ distributed Bragg reflectors. *Appl. Phys. A* **2020**, *126*, 632. [CrossRef]
22. Rouhbakhshmeghrazi, A.; Madadi, M. Novel Design of polycrystalline CdTe/Si Tandem Solar Cells Using SiO$_2$/TiO$_2$ Distributed Bragg Reflector. *Tecciencia* **2020**, *15*, 67–75. [CrossRef]

23. Yu, W.; Fu, X.; Dong, K. Study of semitransparent polymer solar cells with ZnSe/LiF distributed Bragg Reflector. *J. Mater. Sci. Mater. Electron.* **2021**, *32*, 13409–13417. [CrossRef]
24. Wang, H.-Q.; Li, N.; Guldal, N.S.; Brabec, C.J. Nanocrystal V2O5 thin film as hole-extraction layer in normal architecture organic solar cells. *Org. Electron.* **2012**, *13*, 3014–3021. [CrossRef]
25. Brédas, J.-L.; Norton, J.E.; Cornil, J.; Coropceanu, V. Molecular understanding of organic solar cells: The challenges. *Acc. Chem. Res.* **2009**, *42*, 1691–1699. [CrossRef]
26. Stelling, C.; Singh, C.R.; Karg, M.; König, T.A.F.; Thelakkat, M.; Retsch, M. Plasmonic nanomeshes: Their ambivalent role as transparent electrodes in organic solar cells. *Sci. Rep.* **2017**, *7*, 42530. [CrossRef] [PubMed]
27. Roy, A.; Bhandari, S.; Ghosh, A.; Sundaram, S.; Mallick, T.K. Incorporating Solution-Processed Mesoporous WO$_3$ as an Interfacial Cathode Buffer Layer for Photovoltaic Applications. *J. Phys. Chem. A* **2020**, *124*, 5709–5719. [CrossRef] [PubMed]
28. MacKenzie, R.C.I.; Kirchartz, T.; Dibb, G.F.A.; Nelson, J. Modeling nongeminate recombination in P3HT: PCBM solar cells. *J. Phys. Chem. C* **2011**, *115*, 9806–9813. [CrossRef]
29. Khan, A.D.; Iqbal, J.; Rehman, S.U. Polarization-sensitive perfect plasmonic absorber for thin-film solar cell application. *Appl. Phys. A* **2018**, *124*, 610. [CrossRef]
30. Kosyachenko, L.; Savchuk, A.; Grushko, E. Dependence of efficiency of thin-film CdS/CdTe solar cell on parameters of absorber layer and barrier structure. *Thin Solid Film.* **2009**, *517*, 2386–2391. [CrossRef]
31. Kim, M.-S.; Kim, B.-G.; Kim, J. Effective variables to control the fill factor of organic photovoltaic cells. *ACS Appl. Mater. Interfaces* **2009**, *1*, 1264–1269. [CrossRef]
32. Chen, S.; Small, C.E.; Amb, C.M.; Subbiah, J.; Lai, T.-H.; Tsang, S.-W.; Manders, J.R.; Reynolds, J.R.; So, F. Inverted polymer solar cells with reduced interface recombination. *Adv. Energy Mater.* **2012**, *2*, 1333–1337. [CrossRef]
33. Khan, A.D.; Khan, A.D. Optimization of highly efficient GaAs–silicon hybrid solar cell. *Appl. Phys. A* **2018**, *124*, 851. [CrossRef]
34. Sen, S.; Islam, R. Effect of Different Layers on the Performance of P3HT: PCBM-Based Organic Solar Cell. *Braz. J. Phys.* **2021**, *51*, 1661–1669. [CrossRef]
35. Zidan, M.N.; Ismail, T.; Fahim, I.S. Effect of thickness and temperature on flexible organic P3HT: PCBM solar cell performance. *Mater. Res. Express* **2021**, *8*, 095508. [CrossRef]
36. Khairulaman, F.L.; Yap, C.C.; Jumali, M.H.H. Improved performance of inverted type organic solar cell using copper iodide-doped P3HT: PCBM as active layer for low light application. *Mater. Lett.* **2021**, *283*, 128827. [CrossRef]
37. Bendenia, C.; Merad-Dib, H.; Bendenia, S.; Bessaha, G.; Hadri, B. Theoretical study of the impact of the D/A system polymer and anodic interfacial layer on inverted organic solar cells (BHJ) performance. *Opt. Mater.* **2021**, *121*, 111588. [CrossRef]
38. Moujoud, S.; Hartiti, B.; Touhtouh, S.; Rachidy, C.; Belhora, F.; Thevenin, P.; Hajjaji, A. Numerical modeling of copper indium disulfide thin film based solar cells. *Opt. Mater.* **2021**, *122*, 111749. [CrossRef]
39. Al-Hattab, M.; Moudou, L.; Khenfouch, M.; Bajjou, O.; Chrafih, Y.; Rahmani, K. Numerical simulation of a new heterostructure CIGS/GaSe solar cell system using SCAPS-1D software. *Sol. Energy* **2021**, *227*, 13–22. [CrossRef]
40. Jamal, S.; Khan, A.D.; Khan, A.D. High performance perovskite solar cell based on efficient materials for electron and hole transport layers. *Optik* **2020**, *218*, 164787. [CrossRef]
41. Kumar, A. Impact of selenium composition variation in CZTS solar cell. *Optik* **2021**, *234*, 166421. [CrossRef]
42. Islam, A.; Islam, S.; Sobayel, K.; Emon, E.; Jhuma, F.; Shahiduzzaman, M.; Akhtaruzzaman, M.; Amin, N.; Rashid, M. Performance analysis of tungsten disulfide (WS2) as an alternative buffer layer for CdTe solar cell through numerical modeling. *Opt. Mater.* **2021**, *120*, 111296. [CrossRef]
43. Barman, B.; Kalita, P. Influence of back surface field layer on enhancing the efficiency of CIGS solar cell. *Sol. Energy* **2021**, *216*, 329–337. [CrossRef]

MDPI
St. Alban-Anlage 66
4052 Basel
Switzerland
Tel. +41 61 683 77 34
Fax +41 61 302 89 18
www.mdpi.com

Sustainability Editorial Office
E-mail: sustainability@mdpi.com
www.mdpi.com/journal/sustainability

www.ingramcontent.com/pod-product-compliance
Lightning Source LLC
LaVergne TN
LVHW070230100526
838202LV00015B/2113